Impact of Invasive Grasses and Plants on Biodiversity

Sana Basharat
Hainan University, China

IGI Global
Scientific Publishing
Publishing Tomorrow's Research Today

Vice President of Editorial Melissa Wagner
Director of Acquisitions Mikaela Felty
Director of Book Development Jocelynn Hessler
Production Manager Mike Brehm
Cover Design Jose Rosado

Published in the United States of America by
IGI Global Scientific Publishing
701 East Chocolate Avenue
Hershey, PA, 17033, USA
Tel: 717-533-8845
Fax: 717-533-7115
Website: https://www.igi-global.com E-mail: cust@igi-global.com

Library of Congress Cataloging-in-Publication Data

Names: Basharat, Sana, 1992- editor.
Title: Impact of invasive grasses and plants on biodiversity / edited by Sana Basharat.
Description: Hershey, PA : IGI Global Scientific Publishing, [2026] | Includes bibliographical references and index. | Summary: "Addressing the impact of invasive grasses and plants on biodiversity requires a comprehensive approach that includes early detection and rapid response, public awareness and education, and the development of sustainable management practices. Collaborative efforts among scientists, policymakers, and communities are essential to mitigate the adverse effects of invasive species and protect biodiversity for future generations"-- Provided by publisher.
Identifiers: LCCN 2025000140 (print) | LCCN 2025000141 (ebook) | ISBN 9798369387245 (hardcover) | ISBN 9798369387252 (paperback) | ISBN 9798369387269 (ebook)
Subjects: LCSH: Invasive plants. | Plant ecology. | Plant diversity.
Classification: LCC SB613.5 .I46 2025 (print) | LCC SB613.5 (ebook) | DDC 577--dc23/eng/20250218
LC record available at https://lccn.loc.gov/2025000140
LC ebook record available at https://lccn.loc.gov/2025000141

British Cataloguing in Publication Data
A Cataloguing in Publication record for this book is available from the British Library.

Table of Contents

Detailed Table of Contents

Rajeshree P. Patel, Department of Biosciences, Veer Narmad South Gujarat University, Surat, India
Kailash P. Patel, Department of Biosciences, Veer Narmad South Gujarat University, Surat, India
Jagruti K. Barot, Department of Biosciences, Veer Narmad South Gujarat University, Surat, India
Pravinsang P. Dodia, Zoology Department, Sir P.P. Institute of Science, M.K. Bhavnagar University, Bhavnagar, India
Vishal M. Makwana, Department of Biosciences, Veer Narmad South Gujarat University, Surat, India

Invasive grasses exert profound ecological and evolutionary impacts on native ecosystems, altering species composition, disrupting trophic interactions, and modifying fire regimes. Their rapid proliferation outcompetes native flora, leading to habitat homogenization and biodiversity loss. These grasses drive evolutionary shifts in native species through selection pressures such as competition, hybridization, and altered soil microbiomes. Additionally, their influence extends to ecosystem processes, including nutrient cycling, hydrology, and carbon sequestration. This chapter synthesizes current research on the mechanisms by which invasive grasses reshape ecological dynamics and evolutionary trajectories, emphasizing case studies across diverse biomes. Understanding these impacts is critical for developing effective management strategies and mitigating biodiversity threats in a rapidly changing global landscape.

Plants that grow beyond their native geographical range are known as exotic plants. They are introduced into forests either intentionally or unintentionally.These plants have impact on the natural or native plants. Their impact is either positive or negative. The exotic plant species affect the plant diversity as well as the features of the soil in a forest as it changed the microbial activity in a forest. The plant diversity, evenness and richness are also affected by the exotic species. Forest disturbances, provoke the invasion of exotic species.Most of the exotic plant species become invasive and spread very quickly in that particular area. These plant species disturb the whole native vegetation of a forest. To control the prevalence of exotic species is not an easy task as these become invasive with the passage of time.This study reviews the ecological dynamics of exotic plant species in response to forest disturbances, emphasizing mechanisms of invasion and establishment. Understanding these responses is crucial for biodiversity and control and management of these species.

This chapter explores the different strategies for restoring natural environments, highlighting the importance of management practices in considering future climate conditions. The main approaches involve excluding invasive species and restoring native and climate-resilient species to boost ecosystem stability. In addition to this, the chapter also focuses on the role of community and stakeholders' collaboration for restoration projects. This chapter discusses the need for constant monitoring and adaptable strategies to address climate change issues. It demands extensive research on the interactions between native and invasive flora in changing climate conditions. Generally, this chapter's objective is to train scientists and policymakers with the facts and tools necessary to restore natural ecosystems from climate change and invasive species, leading to the conservation of biodiversity and ecosystem resilience.

Plant invasions are a significant ecological challenge, threatening biodiversity, ecosystem functions, and human well-being. This chapter explores ecological restoration as a strategy to mitigate the impacts of invasive plant species. It discusses various restoration techniques, including mechanical, chemical, biological, cultural, and assisted regeneration methods, emphasizing their role in rebuilding native plant communities and restoring ecosystem functions. The chapter highlights the importance of integrating these methods into adaptive management frameworks, incorporating innovative approaches such as geospatial technologies and genetic engineering. Additionally, the role of stakeholder engagement and community-based initiatives in successful restoration is examined. By providing a comprehensive overview, this chapter aims to offer practical solutions for sustainable invasive species management and ecosystem recovery.

In recent decades, the convergence of plant invasions and climate change has emerged as a critical concern for ecosystems and human societies worldwide. Plant invasions occur when non-native species spread into new areas, often outcompeting native flora and disrupting local ecological balances. Conversely, climate change can weaken native plants, making them more susceptible to invasion. Additionally, climate changes can disrupt plant-pollinator relationships, further complicating ecosystem dynamics and potentially accelerating the spread of invasive species. Furthermore, climate change can facilitate the movement of invasive species across regions by altering habitat connectivity and allowing species to colonize new areas that were previously climatically unsuitable. This dynamic poses significant challenges for conservation efforts, as traditional management strategies may become less effective in the face of rapidly changing conditions. Addressing the dual challenges of plant invasions and climate change requires a multifaceted approach.

Amina Farooq, University of Agriculture, Faisalabad, Pakistan
Farooq Ahmad, University of Agriculture, Faisalabad, Pakistan
Ansa Asghar, University of Agriculture, Faisalabad, Pakistan
Sajid Aqeel, University of Agriculture, Faisalabad, Pakistan
Anum Javaid, University of Agriculture, Faisalabad, Pakistan
Zunaira Zulifqar, University of Agriculture, Faisalabad, Pakistan
Usman Tufail, University of Agriculture, Faisalabad, Pakistan
Jazab Shafqat, University of Agriculture, Faisalabad, Pakistan
Mishal Iqbal, University of Agriculture, Faisalabad, Pakistan
Nida Hussain, University of Agriculture, Faisalabad, Pakistan
Muhammad Sufyan, Agriculture University, Beijing, China
Sana Maqsood, University of Agriculture, Faisalabad, Pakistan

Exotic species turns to invasive have negative impacts on native biota and ecosystem functioning. Globalization, development of active transport accelerates plants trade, create threat to global biodiversity. Developing countries like Pakistan invasive species pose economic loss like agricultural productivity, Eucalyptus camaldulensis intentionally introduced for green cover and soil reclamation, Parthenium hysterophorus L. for trade, Lantana camara for horticulture purposes, Broussonetia papyrifera (L.) for landscaping purpose. Invasive plants have adaptability against wide range of environmental conditions, reduced yield by competing with native biodiversity and have econo-medicinal properties fodder, oil, some other allelopathic effects that could pay their damage cost. Pakistan requires specific management program to control their introduction, spread and their impacts on native species. The control and management of invasive species can be implemented by assessment of their risk factors on biodiversity degradation and loss.

Native plants, when to new environments, establish themselves rapidly and spread aggressively, often to the detriment of local ecosystems. Their ability to outcompete native flora and alter habitat structure can have profound effects on biodiversity, ecosystem functions, and ecological processes. Understanding the impacts on preserving ecological balance. Definition and Characteristics of Invasive Plants: characterized by their high growth rates, prolific reproduction, and adaptability to various environmental conditions. Unlike native species that have evolved specific relationships with local fauna and flora, invasive plants often lack natural predators or controls in their new environments, allowing them to proliferate unchecked. Key characteristics of invasive plants include vigorous growth, high seed production, and the ability to outcompete native species for resources. Ecological Impacts: Biodiversity Loss: plants water, significant reductions in plant diversity.

The invasive plant species are non-native and rapidly grow in heterogeneous environments may affect native species. The invasive plant, Justicia adhatoda L,shows complex growth patterns controlled by various above-ground and below-ground factors. The aerial factors include; environmental constraints and biotic relationships that determine the growth patterns, physiological processes and invasion of Justicia adhatoda L. Whereas, the edaphic factors also play a crucial role in root growth and plant health. An interaction of these two factors will help to understand the basic mechanism of survival of Justicia adhatoda L. under different environmental

Uttarakhand, such as climatic conditions, distribution of species, and ecosystem factors. After preprocessing, Particle Swarm Optimization (PSO) was applied to determine the most important features influencing invasion dynamics. An Artificial Neural Network (ANN) model was then built to predict the effect of invasive species on ecosystem diversity. The ANN showed good predictive performance, recovering intricate nonlinear interactions and offering valuable insights into patterns of invasion under varying environmental conditions.

This study explores the Geographic Variation in Growth Patterns of Invasive Plants by integrating ecological field data, advanced feature selection techniques, and deep learning methods. Invasive plant species pose serious threats to biodiversity and agriculture, especially when their growth patterns vary across regions due to environmental and human-influenced factors. Focusing on major spice-growing districts such as Guntur and Krishna, the dataset includes records of species like cinnamon, coriander, cumin, black pepper, turmeric, and chili, along with environmental attributes like soil composition, temperature, and rainfall. After rigorous preprocessing, Particle Swarm Optimization (PSO) was employed to extract the most relevant features influencing invasive species richness. A Gated Recurrent Unit (GRU) model was developed to predict the temporal and spatial growth patterns, capturing complex dependencies across multiple time steps.

Mycorrhizae are symbiotic fungi that live on plant roots and essential for plant communities and ecosystems stability. Mycorrhizal networks are often used by invasive species, which drastically disrupt ecosystems by surpassing native flora, reducing biodiversity and altering ecological processes, to facilitate their establishment. Mycorrhizal associations, enhance nutrient uptake, improve resilience to drought and disease. The complex interaction between mycorrhizae with native, environments and invasive plants are examined in this chapter. The impacts of invasive plants cover biodiversity loss and influence ecosystem functions. Understanding the molecular, ecological, and environmental factors driving invasions is essential to

Preface

Overview of the Subject Matter

Invasive plant species have emerged as one of the most pressing ecological concerns of the 21st century, exerting profound and often irreversible impacts on ecosystems worldwide. Their introduction—whether accidental or intentional—into non-native habitats has triggered cascading effects on biodiversity, ecosystem services, and human livelihoods. These species outcompete native flora, disrupt food webs, alter soil and hydrological properties, and ultimately compromise the ecological integrity of natural and managed systems.

Invasive grasses and broadleaf plants, in particular, are notorious for their adaptability, rapid reproduction, and competitive dominance. They exploit ecological niches by leveraging their physiological flexibility and allelopathic properties, which grant them an edge over native plant communities. Consequently, the spread of invasive species undermines global biodiversity targets, threatens food security, and imposes significant economic costs on agriculture and forestry. Compounding these issues, climate change acts as a catalyst, facilitating range expansions and creating conditions favorable to invasive species proliferation.

This book, "Impact of Invasive Grasses and Plants on Biodiversity," provides an integrated, interdisciplinary perspective on the ecological, evolutionary, economic, and technological dimensions of plant invasions. By drawing upon diverse case studies and research methodologies, the volume examines both fundamental mechanisms and practical solutions, positioning itself as an essential resource for understanding and managing the global challenge of biological invasions.

Relevance in the Contemporary World

In today's era of globalization and environmental change, the problem of invasive species is intricately linked to international trade, agricultural intensification, urbanization, and climate variability. The Anthropocene—a period defined by human-driven alterations of Earth systems—has witnessed unprecedented biotic homogenization, where invasive species spread rapidly across geographical barriers that once maintained ecological distinctiveness.

The socio-economic implications of invasions are staggering. It is estimated that invasive species cost the global economy hundreds of billions of dollars annually through crop yield losses, pasture degradation, and ecosystem service disruption. Moreover, their impacts on biodiversity resonate deeply with global conservation goals, such as the Convention on Biological Diversity (CBD) and the Sustainable Development Goals (SDGs), particularly those aimed at life on land (SDG 15).

Technological innovations in artificial intelligence, remote sensing, and molecular biology have introduced new tools for predicting, monitoring, and mitigating invasions. Nevertheless, integrated strategies that blend ecological insights, economic analysis, and community engagement remain paramount. This book bridges these domains, offering actionable knowledge that addresses both theoretical and applied dimensions of invasive species research.

Target Audience

The intended readership for this book is broad and multidisciplinary, reflecting the complexity of the subject. Primary audiences include:

Researchers and Academics in ecology, conservation biology, forestry, agronomy, and environmental sciences seeking cutting-edge insights into plant invasions and ecosystem dynamics.

Policy Makers and Environmental Managers tasked with developing strategies for invasive species control, habitat restoration, and biodiversity conservation.

Graduate and Postgraduate Students who require a comprehensive resource on plant invasion ecology, evolutionary processes, and management interventions.

Technology Developers and Data Scientists interested in leveraging artificial intelligence and machine learning to predict invasion patterns.

Non-Governmental Organizations (NGOs) and community-based conservation initiatives engaged in ecosystem restoration and advocacy for sustainable land-use practices.

IMPORTANCE OF EACH CHAPTER SUBMISSION

Each chapter in this volume has been meticulously curated to cover a distinct dimension of invasive species ecology, ensuring a coherent yet comprehensive treatment of the topic. Below is an overview of the contributions:

Chapter 1: Ecological and Evolutionary Impacts of Invasive Grasses: Disrupting Ecosystems & Shaping Species Adaptation

This foundational chapter examines how invasive grasses influence ecological networks and evolutionary processes. By outcompeting native flora, altering fire regimes, and modifying soil microbiomes, invasive grasses drive habitat homogenization and biodiversity decline. The authors synthesize global case studies to demonstrate how these species exert selection pressures on native species, ultimately shaping adaptive responses and ecosystem trajectories.

Chapter 2: Response of Exotic Plant Species to Forest Disturbances

This chapter highlights the intricate relationship between forest disturbances—whether natural or anthropogenic—and exotic plant invasions. Disturbance regimes create ecological windows that favor the establishment and spread of non-native species. The authors explore how changes in soil microbial communities, light availability, and competitive dynamics facilitate invasions, emphasizing the challenges of controlling exotic species in disturbed landscapes.

Chapter 3: Restoration of Native Habitats Invaded by Invasive Plants under Changing Climate

Focusing on restoration ecology, this chapter underscores the complexity of rehabilitating ecosystems degraded by invasive species, particularly under the added stressor of climate change. It advocates for climate-resilient restoration strategies that integrate adaptive management, stakeholder participation, and long-term monitoring to enhance ecosystem stability and biodiversity conservation.

Chapter 4: Restoration Techniques to Mitigate Plant Invasions

This chapter provides a comprehensive overview of ecological restoration methods aimed at mitigating plant invasions. It evaluates mechanical, chemical, and biological control techniques alongside emerging approaches such as genetic engineering and geospatial monitoring. The role of community engagement and policy frameworks in ensuring sustainable restoration outcomes is also emphasized.

Chapter 5: Restoration Techniques to Mitigate Plant Invasions under Climate Change

Complementing Chapter 4, this chapter situates restoration efforts within the broader context of climate change. It elucidates how shifts in temperature and precipitation patterns interact with plant invasions, calling for integrated approaches that account for future climate projections in restoration planning.

Chapter 6: Economic Perspectives on Managing Invasive Plant Species in Pakistan

The economic dimension of plant invasions is often underexplored. This chapter fills that gap by quantifying the costs associated with invasive species in Pakistan, including yield losses and management expenditures. It discusses the paradox of species initially introduced for economic benefits—such as Eucalyptus camaldulensis and Lantana camara—that later became ecological liabilities.

Chapter 7: Effects of Invasive Plants on Soil Health and Nutrient Cycling

This chapter investigates the subterranean impacts of invasions, detailing how invasive plants disrupt nutrient cycling, alter soil structure, and modify microbial communities. By linking soil health to ecosystem function, the chapter underscores the need for integrated management strategies that address below-ground ecological processes.

Chapter 8: Role of Aerial and Edaphic Factors in the Invasive Success of Justicia adhatoda L. in Arid Environments

Through a species-specific case study, this chapter examines how aerial (climatic) and edaphic (soil-related) factors jointly influence the invasiveness of Justicia adhatoda L. in arid ecosystems. It provides insights into physiological adaptations and ecological interactions that enable survival under water-limited conditions, informing targeted control strategies.

Chapter 9: Mechanism of Invasion: How Non-Native Species Establish and Spread

This chapter unpacks the biological and ecological mechanisms that underpin successful invasions. It delineates the stages of invasion—introduction, establishment, and spread—and highlights the traits that confer competitive advantages to invasive species. The discussion extends to management frameworks emphasizing early detection and rapid response.

Chapter 10: Predicting Ecological Consequences of Invasive Alien Plants in Kashmir Himalayas with Artificial Intelligence

Here, the authors leverage Artificial Neural Networks (ANNs) and Particle Swarm Optimization (PSO) to predict the ecological consequences of invasive species in the fragile ecosystems of the Kashmir Himalayas. This chapter exemplifies how AI-driven models can capture complex, nonlinear interactions between environmental variables and invasion dynamics.

Chapter 11: Deep Learning-Based Forecasting of Invasive Plant Spread Across Regions

Building on Chapter 10, this contribution applies advanced deep learning models, specifically Gated Recurrent Units (GRUs), to forecast spatial and temporal patterns of invasive species proliferation. The case study on spice-growing districts in India demonstrates the transformative potential of machine learning in ecological forecasting and resource planning.

Chapter 12: The Role of Mycorrhizae in Invasive Plant Ecology

This concluding chapter explores the often-overlooked symbiotic relationships that facilitate invasions. Mycorrhizal associations enhance invasive species' nutrient acquisition and resilience, granting them a competitive edge over native flora. The authors argue for restoration strategies that rebuild native mycorrhizal networks to curb invasions and restore ecosystem functionality.

CONTRIBUTION OF THIS VOLUME

The chapters compiled in this book collectively underscore the multifaceted nature of invasive species ecology, spanning ecological theory, evolutionary biology, soil science, socio-economics, and cutting-edge computational modeling. By integrating diverse disciplinary perspectives, the book offers readers not only a profound understanding of the mechanisms and consequences of plant invasions but also actionable strategies for their management and mitigation.

This volume aspires to advance scientific discourse, inform policy-making, and catalyze practical interventions aimed at conserving biodiversity and sustaining ecosystem resilience in a rapidly changing world. In doing so, it bridges the gap between knowledge generation and real-world application, reaffirming the urgency of addressing invasive species as a cornerstone of global environmental stewardship.

Sana Basharat

Chapter 1
Ecological and Evolutionary Impacts of Invasive Grasses: Disrupting Ecosystems and Shaping Species Adaptation

Rajeshree P. Patel
Department of Biosciences, Veer Narmad South Gujarat University, Surat, India

Kailash P. Patel
Department of Biosciences, Veer Narmad South Gujarat University, Surat, India

Jagruti K. Barot
Department of Biosciences, Veer Narmad South Gujarat University, Surat, India

Pravinsang P. Dodia
Zoology Department, Sir P.P. Institute of Science, M.K. Bhavnagar University, Bhavnagar, India

Vishal M. Makwana
https://orcid.org/0000-0001-6343-9449
Department of Biosciences, Veer Narmad South Gujarat University, Surat, India

ABSTRACT

Invasive grasses exert profound ecological and evolutionary impacts on native ecosystems, altering species composition, disrupting trophic interactions, and modifying fire regimes. Their rapid proliferation outcompetes native flora, leading to habitat homogenization and biodiversity loss. These grasses drive evolutionary shifts in native species through selection pressures such as competition, hybridization, and

DOI: 10.4018/979-8-3693-8724-5.ch001

altered soil microbiomes. Additionally, their influence extends to ecosystem processes, including nutrient cycling, hydrology, and carbon sequestration. This chapter synthesizes current research on the mechanisms by which invasive grasses reshape ecological dynamics and evolutionary trajectories, emphasizing case studies across diverse biomes. Understanding these impacts is critical for developing effective management strategies and mitigating biodiversity threats in a rapidly changing global landscape.

1. INTRODUCTION

Biological invasions by non-native species are among the leading causes of biodiversity loss and ecosystem degradation in the Anthropocene. While much attention has been devoted to invasive woody plants and aquatic weeds, the role of invasive grasses remains under-explored despite their significant and often cascading effects on terrestrial systems (D'Antonio & Vitousek, 1992; Pyšek et al., 2020). Grasses have been intentionally introduced across continents for forage improvement, erosion control, bioenergy crops, and land stabilization, often without fully considering their ecological consequences (Strömberg, 2011). Once established, these grasses can outcompete native vegetation, restructure trophic dynamics, and transform disturbance regimes, especially fire. Their impacts are not only ecological but also evolutionary, as they impose strong selection pressures on native species, drive hybridization, and even affect the evolution of the invasive species themselves (Maron et al., 2004). Despite these broad effects, invasive grasses have received disproportionately less attention in both the literature and policy frameworks, particularly in tropical and subtropical regions such as Sub-Saharan Africa, Southeast Asia, and South America. These areas, often rich in biodiversity and ecological complexity, are increasingly vulnerable to grass-driven habitat homogenization, fire amplification, and loss of ecosystem services (Hardy et al., 2025; Gaertner et al., 2009).

By situating invasive grasses within both ecological and evolutionary dimensions, we argue that managing grass invasions demands not only removal or containment strategies but also an understanding of long-term, system-wide change.

2. CHARACTERISTICS OF INVASIVE GRASSES

Invasive grasses spread rapidly due to their fast growth, prolific seed production, and efficient dispersal through wind, animals, and human activity. Their adaptability allows them to thrive in diverse environments, withstand extreme conditions, and recover from disturbances. Competitive traits such as allelopathy, dense mat formation, and efficient resource use give them an advantage over native species. The absence of natural predators and diseases further facilitates their spread, leading to significant ecological and economic impacts (Gioria & Osborne, 2014). These species alter fire regimes, reduce biodiversity, and change water dynamics, with high genetic diversity enhancing their adaptability. Examples like *Imperata cylindrica* (Cogon grass), *Pennisetum setaceum* (fountain grass), and *Bromus tectorum* (cheatgrass) have transformed ecosystems globally. In temperate grasslands, one of the world's most threatened ecosystems, invasive species pose a major challenge to conservation and restoration efforts. Understanding landscape variables influencing invasions is critical for effective management strategies (Bowen & Stevens, 2020).

2.1 Defining Traits of Invasive Grass Species

Functional traits are widely used in invasion studies to understand mechanisms driving invasiveness across species, aiding both border screening and management (Palma et al., 2021). However, defining universal traits for invasiveness remains challenging due to varying criteria for classification. While many studies compare invasive and non-invasive species, continuous metrics offer a nuanced approach, linking invasiveness to demographic processes like growth, reproduction, and spread. Invasive grasses, such as Cogon grass, Cheatgrass, and Buffelgrass, thrive through adaptability, rapid colonization, and resource efficiency. Their spread disrupts ecosystems, reduces biodiversity, and necessitates early detection and management to mitigate ecological and economic impacts. These defining traits underscore the importance of early detection and targeted management to mitigate the ecological and economic impacts of invasive grass species.

2.2 Mechanisms of Invasion: Dispersal, Reproduction and Competition

Invasive plant species, such as *Imperata cylindrica* (Cogon grass) and *Eichhornia crassipes* (water hyacinth), spread through abiotic dispersal via wind and water. Biotic dispersal occurs through animals, as seen in *Lantana camara*, or human activities, like *Parthenium hysterophorus*. They reproduce sexually through high seed production and extended viability and asexually via rhizomes, stolons, or root fragments, as in *Cynodon dactylon* (Bermuda grass). Species like *Bromus tectorum* (cheatgrass) benefit from early flowering. Invasive plants compete by monopolizing resources, releasing allelopathic chemicals (*Ailanthus altissima*), and altering ecosystems, making early management crucial for controlling their spread.

North American and European grasslands, shaped by human influence, are highly sensitive to ecological changes. Global factors and invasive species alter plant composition, exploiting resource opportunities from disturbances. Successful invasions depend on species traits that outcompete natives, restructure soil microbial communities, and modify trophic interactions, further facilitating their spread and persistence. (*Ailanthus altissima*, *Bromus tectorum*, *Cynodon dactylon*, *Eichhornia crassipes*, *Imperata cylindrica*, *Lantana camara*, *Parthenium hysterophorus*) (Seastedt & Pyšek, 2011).

3. ECOLOGICAL IMPACTS OF INVASIVE SPECIES

Invasive grasses exert a range of ecological impacts that span from local species interactions to landscape-level transformations. These impacts are context-dependent but often follow predictable patterns related to fire dynamics, nutrient cycling, water availability, and biotic competition. Their influence is especially profound in fire-prone, arid, or nutrient-poor ecosystems, where changes in structure and function can occur rapidly following invasion (Vilà et al., 2011; Pyšek et al., 2020).

3.1 Alteration of Fire Regimes

One of the most well-documented ecological impacts of invasive grasses is the modification of fire regimes. Many grasses produce large amounts of fine, continuous fuels that dry quickly and increase the frequency and intensity of fires. This creates a positive feedback loop in which fire-adapted invasive grasses proliferate while native woody or herbaceous species decline.

- *Bromus tectorum* in western North America has shortened fire return intervals from 60–100 years to 3–5 years, resulting in the conversion of sagebrush steppe to near-monocultures of cheatgrass (Balch et al., 2013).
- *Andropogon gayanus* in Australia's savanna increases fire temperatures and intensity, leading to the collapse of native tree seedlings and long-term vegetation shifts (Rossiter et al., 2003).
- In Sub-Saharan Africa, *Hyparrhenia rufa* and *Urochloa spp.* facilitate frequent fires that eliminate native herbs and promote their own dominance.

3.1.1 Case study: Bromus tectorum and North American Rangelands

Bromus tectorum, commonly known as cheatgrass, is a highly invasive grass species native to Europe and southwestern Asia. It was accidentally introduced to North America in the late 19th century, primarily through contaminated grain and ballast soil. Cheatgrass thrives in disturbed environments, particularly arid and semi-arid rangelands, where it has become a dominant species. Its invasiveness is driven by characteristics such as rapid growth, prolific seed production, adaptability to diverse soil types and climates, and its ability to recover quickly after fire events. These traits have allowed *Bromus tectorum* to outcompete native vegetation, monopolize water and nutrients, and reduce habitat quality for wildlife, leading to significant biodiversity loss. Additionally, *Bromus tectorum* contributes to soil degradation by depleting nutrients, reducing soil moisture availability, and promoting erosion. These changes disrupt the ecological balance of rangelands and compromise their ability to support native species and ecosystems.

Addressing *Bromus tectorum* requires integrated management strategies, including prevention through monitoring seed contamination, mechanical removal, targeted herbicide application, and exploration of biological controls. Ecological restoration, such as reintroducing native plants and improving soil health, is essential for long-term recovery. The case of *Bromus tectorum* demonstrates how invasive grasses can fundamentally alter ecosystems, highlighting the importance of proactive and coordinated efforts to prevent and mitigate invasions. Its impact on North American rangelands underscores the need for sustained management and restoration to preserve these critical landscapes and their ecological, economic, and cultural significance.

3.2 Changes in Nutrient Cycling and Soil Chemistry

Invasive grasses significantly alter nutrient cycling and soil chemistry, often leading to long-term degradation of ecosystems. These species can monopolize nutrients such as nitrogen and phosphorus, reducing their availability for native plants. Some invasive grasses, such as *Bromus tectorum* (cheatgrass), contribute to shifts in nitrogen cycling by rapidly uptaking and recycling nitrogen through their fast growth and decomposition rates. In the Brazilian Cerrado, *Melinis minutiflora* generates low-lignin, nutrient-poor litter that disrupts nitrogen cycling by increasing soil nitrate and altering the structure of nitrifying microbial communities (Ribeiro et al., 2017). *Imperata cylindrica*, common in Southeast Asia and parts of Africa, reduces soil fertility by promoting frequent burning and nutrient volatilization (Gibson et al., 2019). Some species also accelerate nitrogen cycling (e.g., *Pennisetum setaceum* in Hawaii), favouring early successional species over late-successional natives (D'Antonio & Vitousek, 1992).

This altered nutrient dynamic often favours the growth of invasive species while hindering native plants that are adapted to slower nutrient cycling systems. These alterations in soil chemistry and nutrient cycling create feedback loops that reinforce the dominance of invasive grasses, making it challenging to restore native ecosystems. Addressing these impacts requires active management, including soil restoration techniques such as nutrient amendments, the introduction of native plant species, and practices that enhance soil microbial diversity. Understanding and mitigating the effects of invasive grasses on soil systems is essential to preserve ecosystem health and resilience.

3.2.1: Effects on Nitrogen and Carbon Cycling

Invasive grasses significantly influence nitrogen and carbon cycling within ecosystems, altering their ecological balance. These grasses often establish dominance through rapid growth and high biomass production, which can suppress native vegetation and modify soil processes. Their litter decomposition rates tend to differ from those of native plants, impacting the release and storage of carbon. (Wang et al., 2015). Moreover, invasive grasses frequently alter nitrogen dynamics by either enhancing nitrogen availability through symbiotic relationships or depleting soil nitrogen through high nutrient uptake. These changes can lead to shifts in microbial communities and nutrient cycling pathways, potentially reducing ecosystem resilience and biodiversity. Understanding these effects is crucial for managing invasive species and mitigating their long-term ecological consequences. (Ehrenfeld, 2003)

3.2.2 Case study: Phragmites australis in Wetland Ecosystems

The ecological impact of invasive grasses is well-illustrated by the case of *Phragmites australis*, a widely invasive grass species in wetland ecosystems. Native to many regions but invasive in North America, *Phragmites australis* has aggressively expanded into wetlands, often forming dense monocultures that outcompete native plant species. This invasion disrupts habitat structure, reduces biodiversity, and alters ecosystem functions. In terms of nutrient cycling, *Phragmites australis* significantly modifies nitrogen and carbon dynamics. (Wails et. al, 2021) It often increases soil organic matter through its high biomass production, influencing carbon storage and altering greenhouse gas emissions. Additionally, its litter decomposes more slowly than native species, leading to changes in soil nutrient availability and microbial activity. Case studies have documented reduced habitat suitability for native fauna, including fish and birds, due to the physical and chemical changes caused by *Phrag-*

mites australis. Effective management of its spread is vital for conserving wetland ecosystems and maintaining their ecological functions. (Mamolos et. al, 2011)

3.3 Hydrological Modifications

Invasive grasses exert profound hydrological modifications within ecosystems, fundamentally altering water dynamics and ecosystem functions. Their dense root systems and high transpiration rates often disrupt natural water flow patterns, leading to changes in groundwater recharge, surface runoff, and soil moisture retention. (Mircea, et al. 2023) For example, species such as *Phragmites australis* in wetlands and *Arundo donax* in riparian zones are notorious for their capacity to deplete water resources through excessive water uptake. The accumulation of their biomass can obstruct water channels, reduce flow efficiency and alter flood regimes. Additionally, the decomposition of invasive grass litter contributes to sedimentation in aquatic systems, which can decrease water quality and habitat availability for native aquatic species. These grasses also affect evapotranspiration rates, sometimes exacerbating water scarcity in arid and semi-arid regions. Such hydrological modifications not only threaten the ecological integrity of affected ecosystems but also pose challenges for water management, agriculture, and human livelihoods. (Enright, 2000).

3.3.1 Impact on water availability and soil moisture

Invasive species significantly impact water availability and soil moisture, disrupting hydrological dynamics and threatening ecosystem balance. Many invasive plants, particularly grasses, exhibit rapid growth and high transpiration rates, leading to excessive water uptake and depletion of soil moisture. (Funk, 2013). For instance, *Tamarix spp.* (salt cedar) in riparian zones extracts large volumes of water, reducing streamflow and groundwater recharge. These changes are often exacerbated by the hydrophobic soil conditions created by the decomposition of invasive plant biomass, further limiting water infiltration and compounding the problem. *Miscanthus sinensis*, a perennial grass native to East Asia, has become invasive in disturbed and water-limited regions of the United States. It frequently outcompetes native vegetation, forming dense stands that reduce biodiversity and disrupt regenerative processes (Dougherty et al., 2014). The species demonstrates high adaptability, surviving in compacted or nutrient-poor soils (Quinn et al., 2012). Similar congeners (e.g., *Miscanthus sacchariflorus*) have been shown to diminish plant species richness and alter soil seed banks (Hager et al., 2015), indicating a broader ecological risk associated with the genus. Dense monocultures of invasive grasses can significantly alter hydrological dynamics. In wetland and riparian zones, *Phragmites australis* reduces surface water through transpiration, displaces emergent vegetation, and alters hydroperiods (Wails et al., 2021). In dryland systems, invasive grasses often reduce infiltration and increase runoff because their dense root mats and uniform canopy decrease soil permeability and surface retention (Le Maitre et al., 2011; Le Maitre et al., 2020). These hydrological changes can cascade through ecosystems, affecting aquatic invertebrates, amphibians, and birds that depend on seasonal water availability. Effective management must therefore include hydrological restoration, for example by removing water-thirsty or soil-compacting invasive species and re-planting deep-rooted native flora to stabilize soils, increase infiltration, and support recharge (Le Maitre, 2004).

3.4 Effects on Native Biodiversity

Invasive grasses threaten native biodiversity by outcompeting native plants for light, water, and nutrients, reducing plant diversity and altering ecosystems (Rout et al., 2013) (Table 1). Urbanization exacerbates biodiversity decline by fragmenting habitats and creating ecological conditions that favour invasive species establishment (Makwana et al, 2024a). Species like *Bromus tectorum* form dense monocultures, disrupting plant communities. This loss of native vegetation affects wildlife, particularly ground-nesting birds and pollinators, while also altering soil chemistry and hydrology (Barney et al., 2015). Additionally, invasive grasses like *Pennisetum setaceum* increase fire frequency, harming fire-sensitive species (Van Wilgen et al., 2022). To mitigate these impacts, targeted management, habitat restoration, and policies preventing their spread are essential for preserving native ecosystems.

3.4.1 Competition with Native Plants:

Invasive grasses outcompete native plants through rapid growth, high reproductive output, and adaptability, leading to ecological disruptions (Iacarella et al., 2015). In arid environments, they aggressively absorb soil moisture, disadvantaging native species with less competitive root systems (Corbin et al., 2004). They also modify soil properties, creating conditions that further suppress native recovery (Vila & Weiner, 2004). This displacement reduces habitat quality, disrupts food webs, and weakens ecosystem resilience. Over time, biodiversity declines, making invasive grass control essential for restoring native vegetation and maintaining ecosystem functionality (Flory & Clay, 2010).

3.4.2 Impact on Trophic Interactions and Food Web

Invasive grasses disrupt trophic interactions and food webs by outcompeting native plants and altering habitats, reducing food and shelter for herbivores and pollinators (McCary, 2016; Makwana et al., 2024b). Species like *Bromus tectorum* and *Imperata cylindrica* form dense monocultures with lower nutritional value (David et al., 2017). They also alter soil properties, affecting decomposer communities and nutrient cycling. Increased fire frequency further harms native species while favouring invasive grasses (Kao, et al., 2014). This biodiversity loss weakens ecosystem resilience, reducing its ability to recover from disturbances (Khan, & Panikkar, 2009).

3.5 Competitive Exclusion and Community Simplification

Through rapid growth, allelopathy, shading, and pre-emption of space and nutrients, invasive grasses frequently outcompete native plants and disrupt plant-pollinator and plant-herbivore networks.

- *Andropogon gayanus* excludes native herbaceous flora in northern Australia and reduces the diversity of understory fauna (Setterfield et al., 2013).
- In South America, *Melinis minutiflora* converts biodiverse savannas into species-poor grasslands, negatively affecting endemic plant and insect species (Gonçalves et al., 2021).
- Sharma and Verma (2016) documented how *Cenchrus ciliaris* achieved high productivity in Western India's arid environments, yielding up to ~29 q/ha across several cultivated strains. The results underscore the species' aggressive colonization and competitive dominance, which are widely understood to suppress native flora and degrade forage resources over time.

Table 1. Ecosystem impacts by invasive grass species and region

Invasive Grass Species	Region	Primary Ecosystem Impact	Affected Ecosystem Type
Bromus tectorum	North America	Increased fire frequency & native shrub loss	Sagebrush steppe
Andropogon gayanus	Australia	High fire intensity, tree seedling mortality	Tropical savanna
Imperata cylindrica	Southeast Asia, Africa	Soil nutrient depletion, fire cycles	Grassland, disturbed forest

continued on following page

Table 1. Continued

Invasive Grass Species	Region	Primary Ecosystem Impact	Affected Ecosystem Type
Melinis minutiflora	Brazil (Cerrado)	Loss of endemic species, altered N cycling	Savanna
Phragmites australis	Global (wetlands)	Hydrological alteration, native exclusion	Wetlands, riparian zones
Cenchrus ciliaris	India, Africa	Competitive exclusion, fodder reduction	Arid grasslands

4. EVOLUTIONARY IMPACTS OF INVASIVE GRASSES

Invasive grasses can drive significant evolutionary changes in both native and non-native species, altering ecological and genetic dynamics. Their introduction creates new selection pressures on native plants and animals, forcing adaptations that can have long-term consequences for ecosystem stability. (Lavergne & Molofsky, 2007). However, these adaptations may come at the cost of reduced genetic diversity and fitness in native species, leaving them more vulnerable to other environmental stressors. (Nguyen et al., 2016). In animal populations, invasive grasses can alter evolutionary trajectories by reshaping habitat structure and the availability of food resources. For example, changes in grassland composition may drive shifts in dietary preferences, foraging behaviour, or nesting habits in herbivores and pollinators. (Mooney & Cleland 2001) Such shifts can lead to rapid phenotypic changes or even evolutionary divergence. (Prentis et al., 2008). In some cases, invasive grasses themselves undergo rapid evolution in their introduced range, adapting to local environmental conditions or acquiring traits that enhance their invasiveness. This phenomenon, known as "evolution of increased competitive ability," can further entrench their dominance and ecological impact. (Overholt, & Franck, 2017). The evolutionary impacts of invasive grasses highlight the dynamic and reciprocal nature of biological invasions (Lambertini, 2019). Monitoring genetic and phenotypic changes in affected species is crucial for understanding these impacts and designing effective conservation strategies.

4.1 Selective Pressures on Native Species

Invasive grasses exert strong selective pressures on native species, driving ecological and evolutionary changes. Competition for light, water, and nutrients forces native plants to develop traits like faster growth and early seed production to survive (Prentis et al., 2008). *Bromus tectorum* increases fire frequency, favouring fire-adapted species and reducing biodiversity (Lavergne & Molofsky, 2007). Native animals also adapt by altering diets and foraging behaviors (Callaway et al., 2005). These pressures weaken ecological resilience, pushing some species toward extinction while others evolve (Oduor, 2013). Understanding these dynamics is essential for effective invasive species management.

4.1.1 Evolutionary Adaptations in Native Plants

One common adaptation is the acceleration of growth rates or phenological shifts, such as earlier flowering and seed production, to complete life cycles before invasive grasses dominate resources (Mooney & Cleland, 2001). For instance, native plants in ecosystems invaded by *Bromus tectorum* (cheatgrass) may adapt to germinate earlier to capitalize on limited water and nutrient availability. Similarly, some native species may evolve increased seed dispersal mechanisms or higher seed output to compete with the prolific seed production of invasive grasses (Mealor & Hild, 2006). Native plants may also develop physiological adaptations to tolerate changes in soil chemistry and water availability caused by invasive grasses. For example, they might evolve to thrive in nutrient-depleted soils or develop deeper or more efficient root systems to access limited water resources. Additionally, native plants in fire-prone areas invaded by grasses like *Imperata cylindrica* (Cogon grass) may evolve fire-resistant traits, such as thicker bark or rapid post-fire regeneration. (Carroll, 2007). These adaptations, however, often come with trade-offs. Rapid changes in traits like phenology or reproduction can reduce genetic diversity, making populations more vulnerable to other environmental challenges. Furthermore, the speed at which these adaptations occur may not keep pace with the rapid spread and dominance of invasive grasses, leading to the decline or local extinction of less adaptable native species.

4.1.2 Case Study: Phenological Shifts in Native Species

Invasive grasses disrupt native plant communities by altering resource availability and habitat conditions, often leading to phenological shifts in native species. Native plants may respond by germinating or flowering earlier to avoid competition, as seen in areas invaded by *Bromus tectorum* (cheatgrass), where species adapt to utilize early spring moisture (Howell et al., 2020). Some plants extend or delay flowering periods to minimize overlap with invasive grasses during peak growth phases (Dyer et al., 2012). Additionally, native species may adjust to changing temperature and water conditions driven by invasions (Carol Adair & Burke, 2010). However, these shifts have limitations, as invasive species like cheatgrass alter ecosystems so profoundly that adaptation alone may not ensure native biodiversity survival. Long-term resilience requires proactive strategies, including invasive species management and ecosystem restoration, to mitigate their ecological impact (Bradford & Lauenroth, 2006).

4.2 Evolutionary Responses in Native Animals

The evolutionary responses of native animals to the spread of invasive grasses are complex and multifaceted, involving changes in behaviour, diet, habitat use, and reproductive strategies. While some species may adapt successfully to the altered environment, others may struggle to keep pace with the rapid spread of invasive species, leading to population declines or even local extinctions. These evolutionary dynamics underscore the importance of managing invasive species to maintain ecosystem balance and preserve native biodiversity. Adaptive management strategies, including habitat restoration and control of invasive grasses, can help mitigate the negative impacts on native wildlife and promote ecological resilience (D'Antonio, & Vitousek, 1992).

4.2.1 Adaptation of Herbivores and Pollinators to Invasive Grasses

Invasive grasses significantly impact ecosystems, forcing herbivores and pollinators to adapt to changes in food availability and habitat structure.

4.2.1.1 Herbivores Adapting to Invasive Grasses: Species like *Bromus tectorum* (cheatgrass) and *Pennisetum setaceum* (fountain grass) replace native vegetation, challenging specialized herbivores. Some adapt through dietary flexibility, incorporating invasive grasses despite their lower nutritional value, while others evolve digestive modifications to process fibrous material (Wolkovich & Cleland, 2014). Increased foraging efforts become necessary as herbivores consume larger quantities to meet energy needs (Bremm et al., 2016). Additionally, physical adaptations such as stronger jaw structures may evolve to handle tougher grasses (Funk, 2013).

4.2.1.2 Pollinators Adapting to Invasive Grasses: The dominance of invasive grasses reduces the availability of native flowering plants, forcing pollinators to shift foraging behavior (Stout & Tiedeken, 2017). Pollinators may expand their search range, switch to invasive plant species, or modify their pollination routes due to habitat changes (Linder et al., 2018). These shifts can lower pollination efficiency, reducing native plant reproduction (Jogesh et al., 2014). Some pollinators may co-evolve with invasive grasses, utilizing their floral resources, but at the cost of native plant-pollinator relationships (Zhang et al., 2015). Although adaptation occurs, these changes often come with trade-offs. Herbivores may experience reduced fitness due to poor nutrition, while pollinators face a decline in floral diversity, impacting their populations. The rapid spread of invasive grasses can outpace the ability of these organisms to adapt effectively, leading to long-term ecological consequences (Harvey & Fortuna, 2012).

4.2.2 Shifts in Herbivore Feeding Strategies

Invasive grasses, such as *Bromus tectorum* (cheatgrass), *Pennisetum setaceum* (fountain grass), *Miscanthus sinensis* (Chinese silver grass), and others, are widespread across ecosystems globally, particularly in grasslands, savannas, and arid regions. As a result, herbivores that rely on native plants for nutrition face substantial challenges. To cope with these changes, many herbivores have had to adjust their feeding strategies, exhibiting shifts in diet, foraging behaviors, and habitat use (Zhang et. al., 2018)

4.2.2.1 Herbivore Feeding Strategies Before the Invasion of Invasive Grasses: Before the spread of invasive grasses, herbivores typically followed established foraging patterns, feeding on a diverse mix of native grasses, shrubs, and forbs. These ecosystems provided a wide variety of food sources throughout the growing season. Herbivores often exhibited selective grazing behaviors, targeting specific plants based on their nutritional content, seasonal availability, and abundance. For example, many herbivores would feed on nutrient-rich young plants during the spring and summer while avoiding more fibrous and lower-quality plant material in fall and winter (Gonzales, 2008).

However, these adaptations may not fully compensate for the reduced nutritional quality of invasive species, and they can have cascading effects on plant communities, herbivore populations, and broader ecosystem dynamics. Effective management of invasive grasses is crucial to mitigate these impacts and restore the balance of native plant and animal communities in affected ecosystems.

4.3 Hybridization and Genetic Introgression

This process can have profound ecological, evolutionary, and genetic consequences for native plant communities. Hybridization with invasive grasses can introduce new traits, such as increased tolerance to stressors like drought, disease, or herbivory, which may allow hybrids to thrive in disturbed environments (Pujolar et al., 2023). However, the genetic mixing between invasive and native species can threaten the integrity of native species, leading to the loss of unique genetic traits and reducing biodiversity (Quilodrán et al., 2020).

4.3.1 Genetic mixing between invasive and native species

Hybridization between invasive and native grasses threatens genetic integrity by leading to genetic assimilation, potentially causing native species extinction (Quilodrán et al., 2020). Some hybrids exhibit hybrid vigour, making them more adaptable to harsh conditions and more competitive than native species, accelerating their spread and ecological impact (Pujolar et al., 2023). Lakoba & Barney, (2020) demonstrated that invasive grasses can rapidly differentiate into distinct ecotypes depending on local habitat conditions. Their study shows that plants from drought-prone and nutrient-poor environments exhibit morphological and physiological adaptations, such as altered root-to-shoot ratios and chlorophyll profiles, that enhance survival in these degraded settings. This supports the claim that invasion processes can produce novel ecotypes better suited to disturbed environments, thereby intensifying displacement of native flora. Additionally, invasive grass hybrids disrupt ecosystem services by altering plant community structures, impacting nutrient cycling, soil stabilization, and water infiltration. They may also be less beneficial to native herbivores and pollinators, weakening ecological interactions and biodiversity (De Santis et al., 2021). These consequences highlight the urgent need for monitoring and managing invasive grass hybridization to protect native species and ecosystem stability.

4.3.2 Hybridization and Genetic Introgression in Invasive Grasses

4.3.2.1 *Bromus tectorum* and Native Brome Species: *Bromus tectorum* (cheatgrass), an invasive species in North America, has hybridized with native brome species like *Bromus marginatus* and *Bromus japonicus*. These hybrids display increased resilience to drought and fire, giving them a competitive advantage in disturbed rangelands. Their higher seed production and germination rates further contribute to their dominance (Quilodrán, et al., 2020).

4.3.2.2 *Pennisetum setaceum* and Native Grass Species in Hawaii: In Hawaii, *Pennisetum setaceum* (fountain grass) has hybridized with native grasses such as *Eragrostis variabilis* and *Cenchrus setaceus*. These hybrids grow faster, tolerate nutrient imbalances, and are more fire-resistant, making them highly competitive against native vegetation (Pujolar et al., 2023). Their spread threatens Hawaii's ecosystems by reducing the survival chances of endemic species. Early detection and monitoring of hybrid zones help track hybrid spread and mitigate their ecological impact. Reintroducing native species and controlling invasive grasses through techniques like controlled burns and reseeding can help prevent hybridization (De Santis et al., 2021; Hessenauer et al., 2020). Programs focused on selective breeding and seed banking can preserve native genetic diversity and aid in restoring pure native strains in affected areas (Jeffries et al., 2024).

4.3.3 Implications for Native Species' Genetic Integrity

The genetic integrity of native species is a fundamental aspect of their evolutionary potential and long-term survival. Hybridization between invasive and native grasses often results in genetic introgression, which can reduce the genetic distinctiveness of native species. Over time, this process may cause native populations to become more similar to invasive species, potentially leading to the loss of key adaptive traits (Quilodrán et al., 2020). Managing invasive grasses is vital to preserving the genetic integrity of native species. For example, *Phragmites australis* management in North America relies on genetic monitoring to identify and control invasive hybrids while protecting native lineages (Hauber et al., 2011). Similarly, studies in Australia show that fire-based control of invasive *Cenchrus ciliaris* affects its population genetic structure—emphasizing the importance of integrating genetic assessment into management programs (Di et al., 2022). Genetic monitoring detects early hybridization signs, while seed banks preserve native genetic material for future restoration (Hamilton & Miller, 2016; Jeffries et al., 2024). These strategies collectively safeguard biodiversity and the unique traits of native plant populations.

5. CASE STUDIES ON INVASIVE GRASSES

5.1 *Imperata cylindrica* (Cogon grass) in Tropical Ecosystems

Imperata cylindrica, commonly known as Cogon grass, is a highly invasive species native to Southeast Asia and parts of Africa. It has spread across tropical ecosystems, particularly in the southern United States, Australia, and parts of Southeast Asia (MacDonald, 2004). Cogon grass is notorious for its ability to outcompete native vegetation and its dramatic impact on fire regimes and biodiversity. Its aggressive growth, high seed production, and rapid spread make it a formidable invader in tropical and subtropical environments, where it often displaces native plants and alters ecosystem functions (Kerns et al., 2020). It thrives in degraded lands, particularly after logging, shifting cultivation, or fire events. In Indonesia and the Philippines, *Imperata* forms persistent grasslands that inhibit forest regeneration by creating a flammable and allelopathic understory. These invasions have not only ecological costs but also socioeconomic ones, as they reduce agricultural productivity and increase the risk of uncontrolled wildfires. Research from Kalimantan and Mindanao indicates that *Imperata* can persist for decades in the absence of intervention, maintaining low soil fertility and impeding both natural succession and reforestation efforts (Van Noordwijk et al., 2020). Its resilience is partly due to its deep rhizome system and rapid recovery after fire, which outpaces that of native shrubs and tree seedlings.

5.2 *Bromus tectorum* (Cheatgrass) in North American Rangelands

The invasion of *Bromus tectorum* (cheatgrass) has significantly altered fire regimes in North American rangelands. Its high flammability, especially during late spring and summer, increases the frequency and intensity of wildfires, making fire-prone landscapes more vulnerable (Bradley et al., 2018). Fires fuelled by cheatgrass burn more frequently and with greater severity than those in native plant communities (Clark, 2020). This shift in fire regimes has led to a transformation in vegetation structure. Historically, native shrubs and grasses experienced infrequent, lower-intensity fires. However, cheatgrass accelerates fire cycles, rapidly colonizing burned areas and outcompeting native vegetation, resulting in landscapes dominated by the invasive species (Fenesi et al., 2016). Additionally, cheatgrass-driven wildfires alter soil properties, making them less suitable for native plant species. Repeated fires degrade soil nutrients and structure, creating conditions unfavourable for native flora but well-suited for cheatgrass, reinforcing its dominance (Harrison et al., 2024).

5.3 *Phragmites australis* (Common Reed) in Wetlands

Phragmites australis, or common reed, is an invasive grass species that has rapidly spread across wetlands in North America, Europe, and Asia. Originally from Eurasia, it alters wetland ecosystems by modifying soil properties and nutrient dynamics (Uddin & Robinson, 2017). Its dense biomass and root systems increase organic matter accumulation, leading to soil acidification and reduced oxygen, inhibiting native plant growth (Crocker et al., 2017; Bickford et al., 2022). Additionally, *Phragmites* absorbs large amounts of nitrogen and phosphorus, disrupting nutrient cycles and potentially causing eutrophication (Uddin et al., 2020; Wang et al., 2015). It also affects hydrology by altering drainage patterns and soil compaction, further impacting biodiversity (Volesky et al., 2018). Low oxygen levels promote anaerobic microbial activity, affecting nitrogen and sulphur cycles (Stager, 2023). In coastal wetlands, *Phragmites* increases soil salinity, outcompeting native salt-tolerant species (Fuchs et al., 2024). The dominance of *Phragmites* results in significant biodiversity loss as it outcompetes native plants and alters wetland structure. Many animal species, including insects, amphibians, birds, and small mammals, struggle to survive in these monocultures due to the lack of diverse food and shelter options (Hunt et al., 2017). Nutrient cycling changes also impact water quality. While *Phragmites* can sequester nitrogen and phosphorus, its seasonal dieback releases these nutrients, contributing to eutrophication, algal blooms, and oxygen depletion in wetland waters (Wails et al., 2021). Lastly, the transformation of wetland hydrology due to soil compaction and oxygen depletion weakens critical ecosystem functions such as water filtration, flood control, and groundwater recharge. The loss of native plant species further reduces the ability of wetlands to mitigate environmental stresses, leading to long-term degradation of ecosystem services (Tercero et al., 2015).

These regional examples highlight both the commonalities and context-specific nuances of grass invasions. Across all three regions, invasive grasses are associated with fire feedbacks, competitive exclusion, and the disruption of ecological succession. Yet the pathways and consequences vary depending on local land use, climate, biodiversity, and governance frameworks. A key challenge is the relative paucity of long-term data and local research capacity in many of these areas, which hinders both understanding and response. Recognizing these geographic gaps is crucial. Conservation science must actively support research in underrepresented regions, promote the integration of local ecological knowledge, and ensure that management strategies are tailored to socio-ecological contexts. Only then can we build a truly global understanding of invasive grass dynamics and design interventions that are both effective and equitable.

6. MANAGEMENT, RESTORATION, AND EVOLUTIONARY-INFORMED STRATEGIES

Given the persistent and multifaceted impacts of invasive grasses on ecosystems, management strategies must move beyond reactive removal and toward more holistic, adaptive, and evolutionarily-informed approaches. The resilience of these species, coupled with their ability to alter disturbance regimes, suppress regeneration, and evolve in novel environments, requires integrative solutions that combine prevention, suppression, and long-term ecological restoration.

Early Detection and Rapid Response (EDRR) remains one of the most effective tools in the invasive species management toolkit. Preventing the establishment and spread of invasive grasses through monitoring, quarantine, and community awareness can reduce control costs and ecological damage. Advances in remote sensing, including hyperspectral imaging and unmanned aerial vehicles (UAVs), now enable fine-scale monitoring of grass invasions across large landscapes (Huang & Asner, 2009). When coupled with citizen science platforms and spatial modelling, these tools allow for earlier and more targeted interventions.

Where invasions are already well established, mechanical, chemical, and fire-based control methods have all been employed, with varying degrees of success. For instance, the suppression of *Imperata cylindrica* often requires repeated mowing or herbicide application, followed by active reforestation or cover crop planting to outcompete regrowth. However, such methods can be resource-intensive and may have non-target effects. Similarly, prescribed fire, while sometimes effective in reducing invasive grass biomass, can also favour grasses over native woody plants if not carefully timed and integrated into broader restoration goals.

Thus, the most promising outcomes often arise from ecological restoration frameworks that incorporate native species reintroduction, soil amelioration, and adaptive management.

6.1.1 Mechanical, chemical, and biological control methods

Invasive species are a major threat to biodiversity, ecosystem health, and human activities. Controlling these species is crucial for ecological restoration, and several methods are used depending on the species, ecosystem, and goals of management (Guo et al., 2018). The three primary categories of control methods are mechanical, chemical, and biological control, each with its benefits, challenges, and applications.

Mechanical control techniques such as cutting, mowing, hand-pulling, grazing, and prescribed burns are most effective for small infestations or in sensitive areas where chemical or biological methods may be inappropriate. These methods are thoroughly reviewed in management toolkits (Meyerson et al., 2022). Chemical control, including the judicious application of herbicides like glyphosate and triclopyr, as well as soil fumigants and selective pesticides, is often essential when addressing large or inaccessible invasions, with careful implementation needed to avoid unintended ecological consequences (Guo et al., 2018). Biological control tools, such as introducing host-specific predators, parasites, or pathogens, have proven effective across numerous classical biocontrol programs, reducing invasive plant biomass and density while allowing native plant communities to recover (Clewley et al., 2012).

6.1.2 Rehabilitating Ecosystem Processes

Restoring ecosystem processes is essential for maintaining resilience, particularly when disrupted by land-use change, invasions, or degradation. Nutrient cycling, water regulation, pollination, seed dispersal, energy flow, and decomposition underpin ecosystem health and biodiversity. However, deforestation and pollution degrade soil fertility by disrupting microbial and nutrient dynamics. Rehabilitation via organic amendments, enhanced microbial activity, and revegetation with native species fosters nutrient balance and functional recovery (Moreno-Mateos et al., 2017).

Watersheds rely on wetlands, riparian buffers, and reforestation to manage floods, recharge aquifers, and stabilize soils—restoration that enhances hydrological resilience (Zedler & Kercher, 2005). Additionally, restoring native vegetation structures pollination networks and increases seed production, boosting the recovery of plant communities (Kaiser-Bunbury et al., 2017). Primary producers and decomposers are foundational: revegetation and soil restoration facilitate energy flow and biomass recycling, as described by Miller et al. (2017). Fire, floods, and grazing maintain biodiversity, but human interventions often disrupt them (Banerjee et al., 2021). Prescribed burns, floodplain restoration, and managed grazing help sustain natural disturbance cycles, supporting ecosystem stability. Effective rehabilitation integrates multiple strategies, continuous monitoring, and adaptive management to restore ecological integrity and ensure long-term sustainability.

6.2 Integrating Evolutionary Principles into Management

Integrating evolutionary principles into ecosystem management is crucial for promoting the long-term health and resilience of ecosystems. A key aspect of this is ensuring that native species maintain genetic diversity and adaptive potential, which are essential for responding to environmental changes, including climate change, habitat degradation, and the spread of invasive species (Lambert & Donihue, 2020). Genetic diversity enables populations to adapt to new selective pressures, ensuring survival and maintaining ecological functions. Management strategies that prioritize genetic health and adaptive capacity can significantly improve the success of conservation and restoration efforts.

6.2.1 Promoting Genetic Diversity in Restoration and Conservation

Integrating genetic diversity into conservation and restoration efforts is crucial for maintaining the variability of native species in both wild and managed habitats. Key strategies include establishing habitat connectivity to enable gene flow and prevent inbreeding (Di Marco, 2019), preventing genetic bottlenecks by ensuring diverse populations in reintroduction programs, and utilizing outcrossing to enhance genetic diversity while avoiding disruption of local adaptations. Ex-situ conservation methods, including seed banking and cryopreservation, serve as genetic reservoirs, providing material to reinforce or restore in situ populations (Nagel et al., 2024). Additionally, small or isolated populations are vulnerable to genetic drift and inbreeding, but assisted gene flow via translocations has been empirically proven to restore genetic diversity and adaptive potential (Pavlova et al., 2017).

6.2.1.1 Enhancing Adaptive Potential Through Evolutionary Considerations: A species' adaptive potential, its ability to evolve in response to environmental changes, is closely tied to genetic diversity. To enhance this potential, conservation strategies should promote natural selection by maintaining healthy ecosystems and managing invasive species. Encouraging ecological processes such as predation and competition helps preserve evolutionary pressures (Di Marco, 2019). Adaptive management, which involves continuous monitoring and adjustments, ensures long-term resilience. Restoring functional ecosystem processes, such as nutrient cycling and pollination, further supports biodiversity and species adaptation.

6.2.1.2 Climate Change and the Need for Adaptive Management: Climate change is accelerating evolutionary pressures, often surpassing the rate at which many species can adapt naturally (Khatsuriya et al., 2024). Consequently, conservationists are prioritizing climate-resilient populations, selecting genotypes best suited to future conditions (Prober et al., 2015). An increasingly applied strategy is assisted migration, the intentional relocation of species to climatically suitable habitats, though it must be carefully managed to avoid ecological harm (McKone et al., 2021). Additionally, restoration of resilient ecosystems, such as peatlands and temperate wetlands, provides refugia for species and helps stabilize ecosystem processes under changing climates (Zou et al., 2022).

6.2.1.3 The Role of Evolutionary Principles in Invasive Species Management: Invasive species threaten native genetic diversity by competing for resources, hybridizing with native populations, and introducing maladaptive genes. Effective management includes monitoring and controlling invasives to prevent genetic erosion and facilitating the coexistence of native species by strengthening ecological niches that limit invasive spread (Banerjee et al., 2021). Promoting genetic diversity and adaptive potential is fundamental to conservation and ecosystem management. Retaining genetic variation ensures species' resilience against environmental changes, climate shifts, and invasive threats. As global challenges intensify, integrating evolutionary principles into conservation strategies will be essential for maintaining biodiversity and ecosystem functionality (Frankham et al., 2010; Hoffmann & Sgrò, 2011). By implementing these strategies, conservation efforts can support long-term evolutionary success and ecosystem stability.

By integrating adaptive management, ecosystem-based strategies, habitat protection, climate resilience, invasive species control, and community involvement, long-term management ensures ecosystem sustainability. These efforts require collaboration, scientific knowledge, and policy integration to maintain biodiversity, support human well-being, and preserve ecological balance for future generations.

7. CONCLUSION & FUTURE RESEARCH DIRECTION

Invasive grasses significantly alter ecosystems by outcompeting native species, disrupting nutrient cycling, modifying fire regimes, and affecting hydrology. Their rapid spread exerts strong selective pressures, driving evolutionary changes in both native flora and fauna. Despite growing awareness of the ecological and evolutionary impacts of invasive grasses, substantial knowledge gaps remain. These gaps constrain our ability to predict invasion trajectories, assess long-term outcomes, and design adaptive management interventions that are effective under future environmental conditions. Based on the synthesis presented in this chapter and areas identified by recent literature, we propose the following research priorities:

1. **Mechanisms and limits of evolutionary rescue**: More empirical studies are needed to determine the speed and scope of adaptive responses in native species confronted with invasive grass dominance. Understanding genetic constraints and plasticity can inform species-specific conservation strategies.
2. **Rapid evolution in invasive grasses**: Investigations into trait shifts, reproductive changes, and local adaptation in introduced populations can shed light on the mechanisms driving invasiveness and inform risk assessments for other species.
3. **Invasional meltdown and co-facilitation networks**: Better characterization of multi-invader interactions (e.g., grass–fungus, grass–shrub, grass–herbivore) is crucial, especially in tropical systems where these dynamics are understudied.
4. **Eco-evolutionary feedbacks and system thresholds**: Research should explore how altered ecological processes (fire, soil fertility, hydrology) reinforce evolutionary responses that further destabilize ecosystems. Identifying thresholds and feedback loops can help prioritize restoration investments.
5. **Tropical and subtropical underrepresentation**: More long-term, field-based research is urgently needed in biodiversity-rich but under-monitored regions such as Sub-Saharan Africa, Southeast Asia, and the Neotropics.
6. **Integration of genomic tools**: High-throughput sequencing and landscape genomics offer powerful means to detect adaptive changes, trace invasion pathways, and characterize population structure in both native and invasive species.
7. **Restoration and resilience frameworks**: Experimental and participatory trials that test combinations of management strategies—including native species reintroduction, soil rehabilitation, and fire management—are essential for improving restoration success under local conditions.
8. **Policy, livelihoods, and governance**: Interdisciplinary studies should examine the socio-economic and policy dimensions of grass invasions, including trade, land use, and incentives for sustainable management.

Taken together, these research directions support a shift from static, one-size-fits-all interventions toward dynamic, evidence-based, and evolutionarily-informed management frameworks. Grass invasions are not transient disruptions; they are catalysts of long-term ecological change and must be addressed with the same urgency and complexity afforded to other global environmental threats.

In conclusion, invasive grasses represent a globally significant yet often underrecognized driver of ecological and evolutionary transformation. Their ability to modify fire regimes, nutrient dynamics, and species interactions makes them formidable agents of change. As this chapter has shown, their impacts are neither uniform nor isolated—they ripple through ecosystems, shaping not only biodiversity and ecosystem function, but also the evolutionary trajectories of entire communities. Addressing these challenges requires coordinated efforts across research, policy, and practice. By embracing an integrative approach that spans from molecular evolution to landscape management, we can begin to restore ecological resilience in the face of grass-driven invasions.

REFERENCES

Balch, J. K., Bradley, B. A., D'Antonio, C. M., & Gómez-Dans, J. (2013). Introduced annual grass increases regional fire activity across the arid western USA (1980–2009). *Global Change Biology*, *19*(1), 173–183. DOI: 10.1111/gcb.12046 PMID: 23504729

Banerjee, A., Khuroo, A. A., Dehnen-Schmutz, K., Pant, V., Patwardhan, C., Bhowmick, A. R., & Mukherjee, A. (2021). An integrated policy framework and plan of action to prevent and control plant invasions in India. *Environmental Science & Policy*, *124*, 64–72. DOI: 10.1016/j.envsci.2021.06.003

Barney, J. N., Tekiela, D. R., Barrios-Garcia, M. N., Dimarco, R. D., Hufbauer, R. A., Leipzig-Scott, P., Nuñez, M. A., Pauchard, A., Pyšek, P., Vítková, M., & Maxwell, B. D. (2015). Global Invader Impact Network (GIIN): Toward standardized evaluation of the ecological impacts of invasive plants. *Ecology and Evolution*, *5*(14), 2878–2889. DOI: 10.1002/ece3.1551 PMID: 26306173

Bickford, W. A., Goldberg, D., Zak, D., Snow, D. S., & Kowalski, K. (2022). Plant effects on and response to soil microbes in native and non-native *Phragmites australis*. *Ecological Applications*, *32*(4), 2565. DOI: 10.1002/eap.2565 PMID: 35138659

Bowen, A. K., & Stevens, M. H. (2020). Temperature, topography, soil characteristics, and NDVI drive habitat preferences of a shade-tolerant invasive grass. *Ecology and Evolution*, *10*(19), 10785–10797. DOI: 10.1002/ece3.6735 PMID: 33072296

Bradford, J. B., & Lauenroth, W. K. (2006). Controls over invasion of Bromus tectorum: The importance of climate, soil, disturbance and seed availability. *Journal of Vegetation Science*, *17*(6), 693–704. DOI: 10.1111/j.1654-1103.2006.tb02493.x

Bradley, B., Curtis, C. A., Fusco, E., Abatzoglou, J., Balch, J., Dadashi, S., & Tuanmu, M. (2018). Cheatgrass (*Bromus tectorum*) distribution in the Intermountain Western United States and its relationship to fire frequency, seasonality, and ignitions. *Biological Invasions*, *20*(6), 1493–1506. DOI: 10.1007/s10530-017-1641-8

Bremm, C., Carvalho, P. C., Fonseca, L., Amaral, G. A., Mezzalira, J. C., Perez, N. B., Nabinger, C., & Laca, E. A. (2016). Diet switching by mammalian herbivores in response to exotic grass invasion. *PLoS One*, *11*(2), e0150167. DOI: 10.1371/journal.pone.0150167 PMID: 26919613

Callaway, R. M., Ridenour, W. M., Laboski, T., Weir, T., & Vivanco, J. M. (2005). Natural selection for resistance to the allelopathic effects of invasive plants. *Journal of Ecology*, *93*(3), 576–583. DOI: 10.1111/j.1365-2745.2005.00994.x

Carol Adair, E., & Burke, I. C. (2010). Plant phenology and life span influence soil pool dynamics: Bromus tectorum invasion of perennial C 3–C 4 grass communities. *Plant and Soil, 335*(1-2), 255–269. DOI: 10.1007/s11104-010-0413-3

Carroll, S. P. (2007). Natives adapting to invasive species: Ecology, genes, and the sustainability of conservation. *Ecological Research, 22*(6), 892–901. DOI: 10.1007/s11284-007-0352-5

Clark, S. L. (2020). Using herbicides to restore native species and improve habitat on rangelands and wildlands. *Outlooks on Pest Management, 31*(2), 57–60. DOI: 10.1564/v31_apr_02

Clewley, G. D., Eschen, R., Shaw, R. H., & Wright, D. J. (2012). The effectiveness of classical biological control of invasive plants. *Journal of Applied Ecology, 49*(6), 1287–1295. DOI: 10.1111/j.1365-2664.2012.02209.x

Corbin, J. D., & D'Antonio, C. M. (2004). Competition between native perennial and exotic annual grasses: Implications for an historical invasion. *Ecology, 85*(5), 1273–1283. DOI: 10.1890/02-0744

Crocker, E., Nelson, E., & Blossey, B. (2017). Soil conditioning effects of Phragmites australis on native wetland plant seedling survival. *Ecology and Evolution, 7*(15), 5571–5579. DOI: 10.1002/ece3.3024 PMID: 28808539

D'Antonio, C. M., & Vitousek, P. M. (1992). Biological invasions by exotic grasses, the grass/fire cycle, and global change. *Annual Review of Ecology and Systematics, 23*(1), 63–87. DOI: 10.1146/annurev.es.23.110192.000431

David, P., Thebault, E., Anneville, O., Duyck, P. F., Chapuis, E., & Loeuille, N. (2017). Impacts of invasive species on food webs: A review of empirical data. *Advances in Ecological Research, 56*, 1–60. DOI: 10.1016/bs.aecr.2016.10.001

De Santis, V., Quadroni, S., Britton, R. J., Carosi, A., Gutmann Roberts, C., Lorenzoni, M., Crosa, G., & Zaccara, S. (2021). Biological and trophic consequences of genetic introgression between endemic and invasive *Barbus* fishes. *Biological Invasions, 23*(12), 3351–3368. DOI: 10.1007/s10530-021-02577-6 PMID: 34054333

Di, B., Firn, J., Buckley, Y. M., Lomas, K., Pausas, J. G., & Smith, A. L. (2022). Impact of roadside burning on genetic diversity in a high-biomass invasive grass. *Evolutionary Applications, 15*(5), 790–803. DOI: 10.1111/eva.13369 PMID: 35603028

Di Marco, S. (2019). *Integrating ecology and evolutionary theory: A game changer for biodiversity conservation?* Biodiversity Conservation Science.

Dougherty, R. F., Quinn, L. D., Endres, A. B., Voigt, T. B., & Barney, J. N. (2014). Natural-history survey of the ornamental grass *Miscanthus sinensis* in the introduced range. *Invasive Plant Science and Management*, *7*(1), 113–120. DOI: 10.1614/IPSM-D-13-00037.1

Dyer, A. R., Hardison, J. L., & Rice, K. J. (2012). Phenology constrains opportunistic growth response in Bromus tectorum L. *Plant Ecology*, *213*(1), 103–112. DOI: 10.1007/s11258-011-0010-4

Ehrenfeld, J. G. (2003). Effects of exotic plant invasions on soil nutrient cycling processes. *Ecosystems (New York, N.Y.)*, *6*(6), 503–523. DOI: 10.1007/s10021-002-0151-3

Enright, W. D. (2000). The effect of terrestrial invasive alien plants on water scarcity in South Africa. *Physics and Chemistry of the Earth. Part B: Hydrology, Oceans and Atmosphere*, *25*(3), 237–242. DOI: 10.1016/S1464-1909(00)00010-1

Fenesi, A., Saura-Mas, S., Blank, R. R., Kozma, A., Lózer, B. M., & Ruprecht, E. (2016). Enhanced fire-related traits may contribute to the invasiveness of downy brome (Bromus tectorum). *Invasive Plant Science and Management*, *9*(3), 182–194. DOI: 10.1614/IPSM-D-16-00006.1

Flory, S. L., & Clay, K. (2010). Non-native grass invasion alters native plant composition in experimental communities. *Biological Invasions*, *12*(5), 1285–1294. DOI: 10.1007/s10530-009-9546-9

Frankham, R., Ballou, J. D., & Briscoe, D. A. (2010). *Introduction to conservation genetics*. Cambridge University Press. DOI: 10.1017/CBO9780511809002

Fuchs, A., Davidson, I. C., Megonigal, J., Devaney, J. L., Simkanin, C., Noyce, G., Lu, M., & Cott, G. M. (2024). Stronger increase of methane emissions from coastal wetlands by non-native Spartina alterniflora than non-native Phragmites australis. PLANTS, PEOPLE, PLANET.

Funk, J. L. (2013). The physiology of invasive plants in low-resource environments. *Conservation Physiology*, *1*(1), cot026. DOI: 10.1093/conphys/cot026 PMID: 27293610

Gaertner, M., Den Breeyen, A., Hui, C., & Richardson, D. M. (2009). Impacts of alien plant invasions on species richness in Mediterranean-type ecosystems: A meta-analysis. *Progress in Physical Geography*, *33*(3), 319–338. DOI: 10.1177/0309133309341607

Gibson, D. J., & Newman, J. A. (Eds.). (2019). *Grasslands and climate change*. Cambridge University Press. DOI: 10.1017/9781108163941

Gioria, M., & Osborne, B. A. (2014). Resource competition in plant invasions: Emerging patterns and research needs. *Frontiers in Plant Science*, *5*, 501. DOI: 10.3389/fpls.2014.00501 PMID: 25324851

Gonçalves, R. V. S., Cardoso, J. C. F., Oliveira, P. E., & Oliveira, D. C. (2021). Changes in the Cerrado vegetation structure: Insights from more than three decades of ecological succession. *Web Ecology*, *21*(1), 55–64. DOI: 10.5194/we-21-55-2021

Gonzales, E. K. (2008). *The effects of herbivory, competition, and disturbance on island meadows* (Doctoral dissertation, University of British Columbia).

Guo, Q., Brockway, D. G., Larson, D. L., Wang, D., & Ren, H. (2018). Improving ecological restoration to curb biotic invasion—A practical guide. *Invasive Plant Science and Management*, *11*(4), 163–174. DOI: 10.1017/inp.2018.29

Hager, H. A., Rupert, R., Quinn, L. D., & Newman, J. A. (2015). Escaped *Miscanthus sacchariflorus* reduces the richness and diversity of vegetation and the soil seed bank. *Biological Invasions*, *17*(6), 1833–1844. DOI: 10.1007/s10530-014-0839-2

Hamilton, J., & Miller, J. M. (2016). Adaptive introgression as a resource for management and genetic conservation in a changing climate. *Conservation Biology*, *30*(1), 256–268. DOI: 10.1111/cobi.12574 PMID: 26096581

Hardy, N. G., Kuebbing, S. E., Duguid, M. C., Ashton, M. S., Sheban, K. C., Inman, S. E., & Martin, M. P. (2025). Non-native invasive plants in tropical dry forests: A global review of presence, impacts, and management. *Restoration Ecology*, *33*(1), e14288. DOI: 10.1111/rec.14288

Harrison, G. R., Jones, L. C., Ellsworth, L. M., Strand, E. K., & Prather, T. S. (2024). Cheatgrass alters flammability of native perennial grasses in laboratory combustion experiments. *Fire Ecology*, *20*(1), 103. DOI: 10.1186/s42408-024-00338-z

Harvey, J. A., & Fortuna, T. M. (2012). Chemical and structural effects of invasive plants on herbivore–parasitoid/predator interactions in native communities. *Entomologia Experimentalis et Applicata*, *144*(1), 14–26. DOI: 10.1111/j.1570-7458.2012.01252.x

Hauber, D. P., Saltonstall, K., White, D. A., & Hood, C. S. (2011). Genetic variation in the common reed, Phragmites australis, in the Mississippi River Delta marshes: Evidence for multiple introductions. *Estuaries and Coasts*, *34*(4), 851–862. DOI: 10.1007/s12237-011-9391-9

Hessenauer, P., Fijarczyk, A., Martin, H., Prunier, J., Charron, G., Chapuis, J., Bernier, L., Tanguay, P., Hamelin, R., & Landry, C. (2020). Hybridization and introgression drive genome evolution of Dutch elm disease pathogens. *Nature Ecology & Evolution, 4*(4), 626–638. DOI: 10.1038/s41559-020-1133-6 PMID: 32123324

Hoffmann, A. A., & Sgrò, C. M. (2011). Climate change and evolutionary adaptation. *Nature, 470*(7335), 479–485. DOI: 10.1038/nature09670 PMID: 21350480

Howell, A., Winkler, D. E., Phillips, M. L., McNellis, B., & Reed, S. C. (2020). Experimental warming changes phenology and shortens growing season of the dominant invasive plant Bromus tectorum (cheatgrass). *Frontiers in Plant Science, 11*, 570001. DOI: 10.3389/fpls.2020.570001 PMID: 33178240

Huang, C., & Asner, G. P. (2009). Applications of remote sensing to alien invasive plant studies. *Sensors (Basel), 9*(6), 4869–4889. DOI: 10.3390/s90604869 PMID: 22408558

Hunt, V., Fant, J., Steger, L., Hartzog, P. E., Lonsdorf, E., Jacobi, S. K., & Larkin, D. (2017). PhragNet: Crowdsourcing to investigate ecology and management of invasive Phragmites australis in North America. *Wetlands Ecology and Management, 25*(5), 607–618. DOI: 10.1007/s11273-017-9539-x

Iacarella, J. C., Mankiewicz, P. S., & Ricciardi, A. (2015). Negative competitive effects of invasive plants change with time since invasion. *Ecosphere, 6*(7), 1–14. DOI: 10.1890/ES15-00147.1

Jeffries, D. L., Lawson-Handley, L., Lamatsch, D., Olsén, K. H., Sayer, C. D., & Hänfling, B. (2024). Towards the conservation of the crucian carp in Europe: Prolific hybridization but no evidence for introgression between native and non-native taxa. *Molecular Ecology, 33*(19), 17515. DOI: 10.1111/mec.17515 PMID: 39212263

Jogesh, T., Stanley, M. C., & Berenbaum, M. R. (2014). Evolution of tolerance in an invasive weed after reassociation with its specialist herbivore. *Journal of Evolutionary Biology, 27*(11), 2334–2346. DOI: 10.1111/jeb.12469 PMID: 25417738

Kaiser-Bunbury, C. N., Mougal, J., Whittington, A. E., Valentin, T., Gabriel, R., Olesen, J. M., & Blüthgen, N. (2017). Ecosystem restoration strengthens pollination network resilience and function. *Nature, 542*(7640), 223–227. DOI: 10.1038/nature21071 PMID: 28135718

Kao, Y. C., Adlerstein, S., & Rutherford, E. (2014). The relative impacts of nutrient loads and invasive species on a Great Lakes food web: An Ecopath with Ecosim analysis. *Journal of Great Lakes Research, 40*, 35–52. DOI: 10.1016/j.jglr.2014.01.010

Kerns, B. K., Tortorelli, C., Day, M. A., Nietupski, T., Barros, A. M., Kim, J. B., & Krawchuk, M. A. (2020). Invasive grasses: A new perfect storm for forested ecosystems? *Forest Ecology and Management*, *463*, 117985. DOI: 10.1016/j.foreco.2020.117985

Khan, M. F., & Panikkar, P. (2009). Assessment of impacts of invasive fishes on the food web structure and ecosystem properties of a tropical reservoir in India. *Ecological Modelling*, *220*(18), 2281–2290. DOI: 10.1016/j.ecolmodel.2009.05.020

Khatsuriya, P. A., Makwana, V. M., & Dodia, P. P. (2024). Biodiversity Resilience Amidst Climate Shifts: A Global Perspective. In *Global Challenges for the Environment and Climate Change* (pp. 75-99). IGI Global. DOI: 10.4018/979-8-3693-2845-3.ch005

Lakoba, V. T., & Barney, J. N. (2020). Home climate and habitat drive ecotypic stress response differences in an invasive grass. *AoB Plants*, *12*(6), plaa062. Advance online publication. DOI: 10.1093/aobpla/plaa062 PMID: 33408848

Lambert, M., & Donihue, C. (2020). Urban biodiversity management using evolutionary tools. *Nature Ecology & Evolution*, *4*(7), 903–910. DOI: 10.1038/s41559-020-1193-7 PMID: 32393868

Lambertini, C. (2019). Why are tall-statured energy grasses of polyploid species complexes potentially invasive? A review of their genetic variation patterns and evolutionary plasticity. *Biological Invasions*, *21*(10), 3019–3041. DOI: 10.1007/s10530-019-02053-2

Lavergne, S., & Molofsky, J. (2007). Increased genetic variation and evolutionary potential drive the success of an invasive grass. *Proceedings of the National Academy of Sciences of the United States of America*, *104*(10), 3883–3888. DOI: 10.1073/pnas.0607324104 PMID: 17360447

Le Maitre, D. C. (2004). Predicting invasive species impacts on hydrological processes: The consequences of plant physiology for landscape processes. *Weed Technology*, *18*(4), 1408–1410. DOI: 10.1614/0890-037X(2004)018[1408:PISIOH]2.0.CO;2

Le Maitre, D. C., Blignaut, J. N., Clulow, A., Dzikiti, S., Everson, C. S., Görgens, A. H., & Gush, M. B. (2020). Impacts of plant invasions on terrestrial water flows in South Africa. In *Biological Invasions in South Africa* (pp. 431–457). Springer International Publishing. DOI: 10.1007/978-3-030-32394-3_15

Le Maitre, D. C., Gaertner, M., Marchante, E., Ens, E. J., Holmes, P. M., Pauchard, A., O'Farrell, P. J., Rogers, A. M., Blanchard, R., Blignaut, J., & Richardson, D. M. (2011). Impacts of invasive Australian acacias: Implications for management and restoration. *Diversity & Distributions*, *17*(6), 1015–1029. DOI: 10.1111/j.1472-4642.2011.00816.x

Linder, H. P., Lehmann, C. E., Archibald, S., Osborne, C. P., & Richardson, D. M. (2018). Global grass (P oaceae) success underpinned by traits facilitating colonization, persistence and habitat transformation. *Biological Reviews of the Cambridge Philosophical Society*, *93*(2), 1125–1144. DOI: 10.1111/brv.12388 PMID: 29230921

MacDonald, G. E. (2004). Cogongrass (Imperata cylindrica)—Biology, ecology, and management. *Critical Reviews in Plant Sciences*, *23*(5), 367–380. DOI: 10.1080/07352680490505114

Makwana, V. M., Patel, F. P., & Dodia, P. P. (2024a). Urbanization's Ecological Toll: Exploring Its Role in Biodiversity Decline. In *Impact of Societal Development and Infrastructure on Biodiversity Decline* (pp. 245-270). IGI Global. DOI: 10.4018/979-8-3693-6950-0.ch014

Makwana, V. M., Thakkar, K., & Dodia, P. P. (2024b). Exploring the Synergy of Soil-Animal Interactions: A Blueprint for Biodiversity Preservation. In *Ecological Aspects of Soil and Land Preservation* (pp. 30-50). IGI Global. DOI: 10.4018/979-8-3693-3374-7.ch002

Mamolos, A. P., Nikolaidou, A. E., Pavlatou-Ve, A. K., Kostopoulou, S. K., & Kalburtji, K. L. (2011). Ecological threats and agricultural opportunities of the aquatic cane-like grass Phragmites australis in wetlands. *Genetics, Biofuels and Local Farming Systems*, 251-275.

Maron, J. L., Vilà, M., Bommarco, R., Elmendorf, S., & Beardsley, P. (2004). Rapid evolution of an invasive plant. *Ecological Monographs*, *74*(2), 261–280. DOI: 10.1890/03-4027

McCary, M. A., Mores, R., Farfan, M. A., & Wise, D. H. (2016). Invasive plants have different effects on trophic structure of green and brown food webs in terrestrial ecosystems: A meta-analysis. *Ecology Letters*, *19*(3), 328–335. DOI: 10.1111/ele.12562 PMID: 26757702

McKone, M. J., & Hernández, D. L. (2021). Community-level assisted migration for climate-appropriate prairie restoration. *Restoration Ecology*, *29*(7), e13416. DOI: 10.1111/rec.13416

Mealor, B. A., & Hild, A. L. (2006). Potential selection in native grass populations by exotic invasion. *Molecular Ecology*, *15*(8), 2291–2300. DOI: 10.1111/j.1365-294X.2006.02931.x PMID: 16780441

Meyerson, L. A., Pauchard, A., Brundu, G., Carlton, J. T., Hierro, J. L., Kueffer, C., & Packer, J. G. (2022). Moving toward global strategies for managing invasive alien species. In *Global plant invasions* (pp. 331–360). Springer International Publishing. DOI: 10.1007/978-3-030-89684-3_16

Miller, B. P., Sinclair, E. A., Menz, M. H., Elliott, C. P., Bunn, E., Commander, L. E., Dalziell, E., David, E., Davis, B., Erickson, T. E., Golos, P. J., Krauss, S. L., Lewandrowski, W., Mayence, C. E., Merino-Martín, L., Merritt, D. J., Nevill, P. G., Phillips, R. D., Ritchie, A. L., & Stevens, J. C. (2017). A framework for the practical science necessary to restore sustainable, resilient, and biodiverse ecosystems. *Restoration Ecology*, *25*(4), 605–617. DOI: 10.1111/rec.12475

Mircea, D. M., Calone, R., Estrelles, E., Soriano, P., Sestras, R. E., Boscaiu, M., Sestras, A. F., & Vicente, O. (2023). Responses of different invasive and non-invasive ornamental plants to water stress during seed germination and vegetative growth. *Scientific Reports*, *13*(1), 13281. DOI: 10.1038/s41598-023-40517-7 PMID: 37587264

Mooney, H. A., & Cleland, E. E. (2001). The evolutionary impact of invasive species. *Proceedings of the National Academy of Sciences of the United States of America*, *98*(10), 5446–5451. DOI: 10.1073/pnas.091093398 PMID: 11344292

Moreno-Mateos, D., Barbier, E. B., Jones, P. C., Jones, H. P., Aronson, J., López-López, J. A., McCrackin, M. L., Meli, P., Montoya, D., & Rey Benayas, J. M. (2017). Anthropogenic ecosystem disturbance and the recovery debt. *Nature Communications*, *8*(1), 14163. DOI: 10.1038/ncomms14163 PMID: 28106039

Nagel, M., Pence, V., Ballesteros, D., Lambardi, M., Popova, E., & Panis, B. (2024). Plant cryopreservation: Principles, applications, and challenges of banking plant diversity at ultralow temperatures. *Annual Review of Plant Biology*, *75*(1), 75. DOI: 10.1146/annurev-arplant-070623-103551 PMID: 38211950

Nguyen, M. A., Ortega, A. E., Nguyen, K. Q., Kimball, S., Goulden, M. L., & Funk, J. L. (2016). Evolutionary responses of invasive grass species to variation in precipitation and soil nitrogen. *Journal of Ecology*, *104*(4), 979–986. DOI: 10.1111/1365-2745.12582

Oduor, A. M. (2013). Evolutionary responses of native plant species to invasive plants: A review. *The New Phytologist*, *200*(4), 986–992. DOI: 10.1111/nph.12429 PMID: 24712050

Overholt, W. A., & Franck, A. R. (2017). The invasive legacy of forage grass introductions into Florida. *Natural Areas Journal*, *37*(2), 254–264. DOI: 10.3375/043.037.0214

Palma, E., Mabey, A. L., Vesk, P. A., & Catford, J. A. (2021). Characterising invasive species. In *Routledge Handbook of Biosecurity and Invasive Species* (pp. 15–39). Routledge. DOI: 10.4324/9781351131599-3

Pavlova, A., Beheregaray, L. B., Coleman, R., Gilligan, D., Harrisson, K. A., Ingram, B. A., Kearns, J., Lamb, A. M., Lintermans, M., Lyon, J., Nguyen, T. T. T., Sasaki, M., Tonkin, Z., Yen, J. D. L., & Sunnucks, P. (2017). Severe consequences of habitat fragmentation on genetic diversity of an endangered Australian freshwater fish: A call for assisted gene flow. *Evolutionary Applications*, *10*(6), 531–550. DOI: 10.1111/eva.12484 PMID: 28616062

Prentis, P. J., Wilson, J. R., Dormontt, E. E., Richardson, D. M., & Lowe, A. J. (2008). Adaptive evolution in invasive species. *Trends in Plant Science*, *13*(6), 288–294. DOI: 10.1016/j.tplants.2008.03.004 PMID: 18467157

Prober, S. M., Byrne, M., McLean, E. H., Steane, D. A., Potts, B. M., Vaillancourt, R. E., & Stock, W. D. (2015). Climate-adjusted provenancing: A strategy for climate-resilient ecological restoration. *Frontiers in Ecology and Evolution*, *3*, 65. DOI: 10.3389/fevo.2015.00065

Pujolar, J., Breitburg, D., Lee, J., Decker, M. B., & Jaspers, C. (2023). Hybridization and adaptive introgression in a marine invasive species in native habitats. *iScience*, *26*(12), 108430. DOI: 10.1016/j.isci.2023.108430 PMID: 38077133

Pyšek, P., Hulme, P. E., Simberloff, D., Bacher, S., Blackburn, T. M., Carlton, J. T., Dawson, W., Essl, F., Foxcroft, L. C., Genovesi, P., Jeschke, J. M., Kühn, I., Liebhold, A. M., Mandrak, N. E., Meyerson, L. A., Pauchard, A., Pergl, J., Roy, H. E., Seebens, H., & Richardson, D. M. (2020). Scientists' warning on invasive alien species. *Biological Reviews of the Cambridge Philosophical Society*, *95*(6), 1511–1534. DOI: 10.1111/brv.12627 PMID: 32588508

Quilodrán, C., Tsoupas, A., & Currat, M. (2020). The spatial signature of introgression after a biological invasion with hybridization. *Frontiers in Ecology and Evolution*, *8*, 8. DOI: 10.3389/fevo.2020.569620

Quinn, L. D., Stewart, J. R., Yamada, T., Toma, Y., Saito, M., Shimoda, K., & Fernández, F. G. (2012). Environmental tolerances of *Miscanthus sinensis* in invasive and native populations. *BioEnergy Research*, *5*(1), 139–148. DOI: 10.1007/s12155-011-9163-1

Ribeiro, P. C., Menendez, E., da Silva, D. L., Bonieck, D., Ramírez-Bahena, M. H., Resende-Stoianoff, M. A., Peix, A., Velázquez, E., Mateos, P. F., & Scotti, M. R. (2017). Invasion of the Brazilian campo rupestre by the exotic grass Melinis minutiflora is driven by the high soil N availability and changes in the N cycle. *The Science of the Total Environment*, *577*, 202–211. DOI: 10.1016/j.scitotenv.2016.10.162 PMID: 27823824

Rossiter, N. A., Setterfield, S. A., Douglas, M. M., & Hutley, L. B. (2003). Testing the grass-fire cycle: Alien grass invasion in the tropical savannas of northern Australia. *Diversity & Distributions*, *9*(3), 169–176. DOI: 10.1046/j.1472-4642.2003.00020.x

Rout, M. E., Chrzanowski, T. H., Smith, W. K., & Gough, L. (2013). Ecological impacts of the invasive grass Sorghum halepense on native tallgrass prairie. *Biological Invasions*, *15*(2), 327–339. DOI: 10.1007/s10530-012-0289-7

Seastedt, T. R., & Pyšek, P. (2011). Mechanisms of plant invasions of North American and European grasslands. *Annual Review of Ecology, Evolution, and Systematics*, *42*(1), 133–153. DOI: 10.1146/annurev-ecolsys-102710-145057

Setterfield, S. A., Rossiter-Rachor, N. A., Douglas, M. M., Wainger, L., Petty, A. M., Barrow, P., Shepherd, I. J., & Ferdinands, K. B. (2013). Adding fuel to the fire: The impacts of non-native grass invasion on fire management at a regional scale. *PLoS One*, *8*(5), e59144. DOI: 10.1371/journal.pone.0059144 PMID: 23690917

Sharma, S. K., & Verma, C. M. (2016). Performance of *Cenchrus ciliaris* Linn. strains in an arid rangeland of western Rajasthan. *Annals of Arid Zone*, *55*(1), 1–9.

Stager, M. (2023). Predicting the Great Lakes wetlands' resilience to climate change in response to Phragmites australis subsp. australis removal. *Surgery Journal (New York, N.Y.)*.

Stout, J. C., & Tiedeken, E. J. (2017). Direct interactions between invasive plants and native pollinators: Evidence, impacts and approaches. *Functional Ecology*, *31*(1), 38–46. DOI: 10.1111/1365-2435.12751

Strömberg, C. A. (2011). Evolution of grasses and grassland ecosystems. *Annual Review of Earth and Planetary Sciences*, *39*(1), 517–544. DOI: 10.1146/annurev-earth-040809-152402

Tercero, M. C., Álvarez-Rogel, J., Conesa, H., Ferrer, M., Calderón, A., López-Orenes, A., & González-Alcaraz, M. (2015). Response of biogeochemical processes of the water-soil-plant system to experimental flooding-drying conditions in a eutrophic wetland: The role of Phragmites australis. *Plant and Soil*, *396*(1-2), 109–125. DOI: 10.1007/s11104-015-2589-z

Uddin, M., & Robinson, R. (2017). Changes associated with Phragmites australis invasion in plant community and soil properties: A study on three invaded communities in a wetland, Victoria, Australia. *Limnologica, 66*, 24–30. DOI: 10.1016/j.limno.2017.07.006

Uddin, M. N., Robinson, R. W., & Asaeda, T. (2020). Nitrogen immobilization may reduce invasibility of nutrient enriched plant community invaded by Phragmites australis. *Scientific Reports, 10*(1), 1601. DOI: 10.1038/s41598-020-58523-4 PMID: 32005878

Van Noordwijk, M., Ekadinata, A., Leimona, B., Catacutan, D., Martini, E., Tata, H. L., Öborn, I., Hairiah, K., Wangpakapattanawong, P., Mulia, R., Dewi, S., Rahayu, S., & Zulkarnain, T. (2020). Agroforestry options for degraded landscapes in Southeast Asia. *Agroforestry for Degraded Landscapes: Recent Advances and Emerging Challenges-Vol., 1*, 307–347. DOI: 10.1007/978-981-15-4136-0_11

Van Wilgen, N. J., Faulkner, K. T., Robinson, T. B., South, J., Beckett, H., Janion-Scheepers, C., . . . Richardson, D. M. (2022). Climate change and biological invasions in South Africa. In *Invasive species and global climate change* (pp. 158-187). GB: CABI.

Vilà, M., Espinar, J. L., Hejda, M., Hulme, P. E., Jarošík, V., Maron, J. L., Pergl, J., Schaffner, U., Sun, Y., & Pyšek, P. (2011). Ecological impacts of invasive alien plants: A meta-analysis of their effects on species, communities and ecosystems. *Ecology Letters, 14*(7), 702–708. DOI: 10.1111/j.1461-0248.2011.01628.x PMID: 21592274

Vila, M., & Weiner, J. (2004). Are invasive plant species better competitors than native plant species?–evidence from pair-wise experiments. *Oikos, 105*(2), 229–238. DOI: 10.1111/j.0030-1299.2004.12682.x

Volesky, L. A., Iqbal, S., Kelly, J., & Geddes, P. (2018). Relationships of native and exotic strains of Phragmites australis to wetland ecosystem properties. *Wetlands, 38*(3), 577–589. DOI: 10.1007/s13157-018-1001-1

Wails, C. N., Baker, K., Blackburn, R., Del Vallé, A., Heise, J., Herakovich, H., Holthuijzen, W. A., Nissenbaum, M. P., Rankin, L., Savage, K., Vanek, J. P., & Jones, H. P. (2021). Assessing changes to ecosystem structure and function following invasion by Spartina alterniflora and Phragmites australis: A meta-analysis. *Biological Invasions, 23*(9), 2695–2709. DOI: 10.1007/s10530-021-02540-5

Wang, C., Xiao, H., Liu, J., Wang, L., & Du, D. (2015). Insights into ecological effects of invasive plants on soil nitrogen cycles. *American Journal of Plant Sciences, 6*(1), 34–46. DOI: 10.4236/ajps.2015.61005

Wolkovich, E. M., & Cleland, E. E. (2014). Phenological niches and the future of invaded ecosystems with climate change. *AoB Plants*, *6*, plu013. DOI: 10.1093/aobpla/plu013 PMID: 24876295

Zedler, J. B., & Kercher, S. (2005). Wetland resources: Status, trends, ecosystem services, and restorability. *Annual Review of Environment and Resources*, *30*(1), 39–74. DOI: 10.1146/annurev.energy.30.050504.144248

Zhang, Z., Pan, X., Blumenthal, D., Van Kleunen, M., Liu, M., & Li, B. (2018). Contrasting effects of specialist and generalist herbivores on resistance evolution in invasive plants. *Ecology*, *99*(4), 866–875. DOI: 10.1002/ecy.2155 PMID: 29352479

Zhang, Z., Pan, X., Zhang, Z., He, K. S., & Li, B. (2015). Specialist insect herbivore and light availability do not interact in the evolution of an invasive plant. *PLoS One*, *10*(9), e0139234. DOI: 10.1371/journal.pone.0139234 PMID: 26407176

Zou, J., Ziegler, A. D., Chen, D., McNicol, G., Ciais, P., Jiang, X., Zheng, C., Wu, J., Wu, J., Lin, Z., He, X., Brown, L. E., Holden, J., Zhang, Z., Ramchunder, S. J., Chen, A., & Zeng, Z. (2022). Rewetting global wetlands effectively reduces major greenhouse gas emissions. *Nature Geoscience*, *15*(8), 627–632. DOI: 10.1038/s41561-022-00989-0

KEY TERMS AND DEFINITIONS

Assisted Migration (Managed Relocation): The intentional movement of species or populations to areas outside their historical range to mitigate risks from climate change or habitat loss.

Eco-Evolutionary Feedback: Reciprocal interactions where ecological changes (e.g., species invasions, altered fire cycles) drive evolutionary adaptations, which in turn influence ecosystem dynamics.

Evolutionary Rescue: The rapid adaptive evolution of a population that prevents extinction in response to severe environmental change, such as altered habitats from invasive species or climate change.

Habitat Connectivity: The degree to which landscapes allow the movement of organisms and gene flow between populations, critical for maintaining genetic diversity and ecosystem resilience.

Invasional Meltdown: A process in which the establishment of one invasive species facilitates the invasion and spread of others, leading to accelerated ecosystem degradation.

Invasive Species: Non-native organisms that establish, spread, and cause ecological, economic, or social harm in their introduced environments. Invasive grasses alter fire regimes, nutrient cycling, and native biodiversity.

Restoration Ecology: The scientific discipline focused on repairing and managing ecosystems that have been degraded, damaged, or destroyed, often incorporating control of invasive species and re-establishment of native biota.

Chapter 2
Response of Exotic Plant Species to Forests Disturbances

Sadia Gull
https://orcid.org/0009-0003-3343-4016
University of Agriculture, Faisalabad, Pakistan

Ansa Asghar
University of Agriculture, Faisalabad, Pakistan

Farooq Ahmad
University of Agriculture, Faisalabad, Pakistan

Muhammad Sajid Aqeel Ahmad
University of Agriculture, Faisalabad, Pakistan

Amina Ameer
College of Geography and Environmental Sciences, Henan University, Kaifeng, China

Usman Tufail
https://orcid.org/0009-0000-2798-3113
University of Agriculture, Faisalabad, Pakistan

Zahida Parveen
University of Agriculture, Faisalabad, Pakistan

Jazab Shafqat
University of Agriculture, Faisalabad, Pakistan

ABSTRACT

Plants that grow beyond their native geographical range are known as exotic plants. They are introduced into forests either intentionally or unintentionally.These plants have impact on the natural or native plants. Their impact is either positive or negative.The exotic plant species affect the plant diversity as well as the features of the soil in a forest as it changed the microbial activity in a forest. The plant diversity, evenness and richness are also affected by the exotic species. Forest disturbances, provoke the invasion of exotic species.Most of the exotic plant species become inva-

DOI: 10.4018/979-8-3693-8724-5.ch002

sive and spread very quickly in that particular area. These plant species disturb the whole native vegetation of a forest. To control the prevalence of exotic species is not an easy task as these become invasive with the passage of time.This study reviews the ecological dynamics of exotic plant species in response to forest disturbances, emphasizing mechanisms of invasion and establishment. Understanding these responses is crucial for biodiversity and control and management of these species.

INTRODUCTION

Introduction of species either intentionally or unintentionally outside their native range has been burgeoning for last few decades and is expected to increase in the near future (Pyšek et al., 2020). The rapid increase in exotic species is due to international trade, travel, global networking and the alteration of native ecosystems. At present, more than 13,000 species of vascular plants have successfully established themselves in regions outside their indigenous habitats (Wan & Zhang, 2021). A portion of these species has effectively surmounted survival constraints and is currently proliferating into novel territories where they are classified as invasive species. (Sedibana et al., 2025) examined that 2500 plant species are recognized as exotic invasive. Approximately 6,000 species of native flowering plants are found in Pakistan, 400 of which are endemic. In Pakistan, four hundred plant species are non-indigenous, constituting about 6.6% flora of the country. Exotic flora includes invasive (27%), casual (45%) and naturalized (29%) species in Pakistan (Zargar et al., 2025).

Disturbances such as wildfire, herbivory, overgrazing alter the structure and composition of plant communities and modifying the species habitat. Disturbance increases the species diversity by excluding competitors. Influence of disturbance on species abundance results from complicate ecological dynamics (Zhang et al., 2025). Disturbance displaces the colonizing species and nutrient assimilation and helps in the invasion of introduced species. The increase in nutrients promotes the invasion of non-native species. Exotic species show better resilience than local species and dominate endemic species due to better adaptibility these are capable to survive in disturbance affected regions than endemic species, irrespective of their reproductive attributes. Localized species are found in habitats that are less disrupted while exotics are frequent in disturbed areas (Taylor et al., 2025).

Grazing and browsing influence the organization of natural ecosystem. Trampling by hoofed animals significantly modify the community dynamics (Chiariotti et al., 2025). Browsing by livestock was initially believed to have no effect, but with the passage of time it is considered as top-notch ecological driver in communities. Herbivory promotes the invasion of non-indigenous species. It also facilitates formation of seedbeds and effective transport of seeds (Natale & Kiviat, 2025). Trampling displaces the surface soil and favors the growth of alien species. In addition, it has been observed that mycorrhizal association between fungi and exotic plants are positively correlated.

Human mediated disturbances influence the forest ecosystem. Logging, overgrazing pose serious threats on the composition of community (Babu et al., 2025). Anthropogenic activities such as silviculture is linked with reduced species abundance, especially in US grasslands. Human land use in various forms has created many disturbed areas which promotes the establishment of alien species. Non-native species explicitly outcompete with localized flora for resource acquisition such as shelter, nutrients, light, water and pollinating agents (Wafer, 2025). After successful invasion in a novel region, these non-native species threaten the endemic species. Introduced grass species increase wildfire regimes and after fire alien flora become more conspicuous. This fire-driven grass cycle results in the establishment of grassland. The invasion of non-native species is inevitable process and once established they become a part of whole ecosystem (Holland et al., 2025).

Non-native species have substantial effects on biological diversity. This type of biological pollution occurred due to massive exchange of items among different countries, enhancing the scattering of species through vehicles. Non- indigenous species cause extirpations of many species mainly on islands and are responsible for habitat degradation (Bushmann et al., 2025). Introduced species act as a double-edged sword. Conservationists, however, sometimes done deliberate introductions for various agricultural purposes. A multitude of tree species are being introduced in various countries for commercial use. Rapidly growing exotic species are used for food, fuel and timber and ameliorating the needs of burgeoning population (Ighalo et al., 2025). Dominant tree species such as *Paulownia*, *Eucalyptus* and *Poplar* grown on cultivated lands. Non-native species provide clean water, manage pesticides, mitigate natural calamities and sometimes used for pollination (Dibben & Raskin, 2025). They are used as aesthetics, education, scientific purposes, traditional customs, tourism and other cultural values. While, successful introduction of species, cautionary measures should be considered. Species should only be introduced under absolute need and should be accompanied by techniques that can evaluate the extent of harm to native diversity. This chapter aims to address impact of forest disturbances on exotic plants (Couch et al., 2025). How natural and anthropogenic disturbances alter the composition of community and the way plants respond to these disturbances.

2. DISTRIBUTIONAL PATTERN OF EXOTIC PLANTS IN PAKISTAN

Exotic species are distributed across all six geographic regions of Pakistan, with the highest concentration (117) observed within the fertile regions of Punjab (Kaur et al., 2025). Disturbed areas provide a diverse range of non-native flora, accommodating nearly all introduced species regardless of their invasibility. Pakistan comprises of about four hundred non-indigenous species. Out of these four hundred species, some are naturalized, invasive and casual aliens. Family Asteraceae and Poaceae contain maximum number of invasive species (Oancea, 2025). Non-native species introduced from many countries in Pakistan as shown in **figure 1**. These may include North America, Africa and Temperate Asia. Baluchistan contains the minimum number of non-native species, while Punjab contains the maximum diversity of non -natives. The most populated region of Pakistan is Punjab. The high species richness in Punjab could be attributed to low elevation plains which is beneficial for farming and establishment of non-indigenous species (Kumar et al., 2025). Moreover, numerous introduced species are introduced as seed contaminants in the cultivated zones. Exotic species are mainly found in grasslands, meadows, agricultural lands and along roadsides and disturbed habitats.

Table 1. Introduction pathways of Introduced Flora in Pakistan

Introduction Pathways	*Introduction Purpose*	*Invasive*	*Naturalized*	*Casual Alien*	*Total*
	Food	1	1	5	1.72
	Ethnobotanical use	0	2	1	0.74
	Plantation	2	10	28	9.83
	Land Restoration	2	1	4	1.72
	Forage	1	1	2	0.98
	Decorative use	4	3	24	7.37
	Dune Stability	2	2	0	0.98
	Gardening use	0	1	0	0.25
Accidental	Indirectly	19	12	9	9.83
	Seed deterioration	1	0	0	0.25
Unknown pathways	Unknown	73	87	108	67.32

Table 2. Some Exotic Species In Pakistan (https://griis.org) Global Register of Introduced and Invasive Species (GRII)

Sr No.	*Botanical Name*	*Common Name*	*Family*	*Habitat*
1	*Typha domingensis*	Southern cattail	Typhaceae	Terrestrial
2	*Paspalum distichum* L.	Knotgrass	Poaceae	Freshwater
3	*Morus alba*	White mulberry	Moraceae	Brackish
4	*Malvastrum coromandelianum*	False mallow	Malvaceae	Brackish
5	*Amaranthus viridis* L.	Slender amaranth	Amaranthaceae	Terrestrial
6	*Agave americana* L.	American Aloe	Asparagaceae	Terrestrial
7	*Alternanthera pungens*	Khaki joyweed	Amaranthaceae	Terrestrial
8	*Arundo donax*	Giant reed	Poaceae	Terrestrial
9	*Avena fatua* L.	Wild oat	Poaceae	Terrestrial
10	*Bromus unioloides*	Rescuegrass	Poaceae	Terrestrial
11	*Broussonetia papyrifera*	Paper mulberry	Moraceae	Terrestrial
12	*Cannabis sativa* L.	Hemp	Cannabaceae	Terrestrial
13	*Carthamus oxyacanthus*	Wild safflower	Asteraceae	Terrestrial
14	*Chenopodium album* L.	White goosefoot	Amaranthaceae	Terrestrial
15	*Cirsium arvense*	Creeping thistle	Asteraceae	Terrestrial
16	*Corchorus aestuans*	Jute	Malvaceae	Terrestrial
17	*Datura stramonium* L.	Jimson weed	Solanaceae	Terrestrial
18	*Eucalyptus sideroxylon*	Red ironbark	Myrtaceae	Terrestrial
19	*Ipomoea carnea*	Morning glory	Convolvulaceae	Freshwater
20	*Lantana camara* L.	Yellow Sage	Verbenaceae	Terrestrial, Freshwater

Figure 1. Geographic distribution of Invasive, Casual Alien and Naturalized Exotic Species in Pakistan

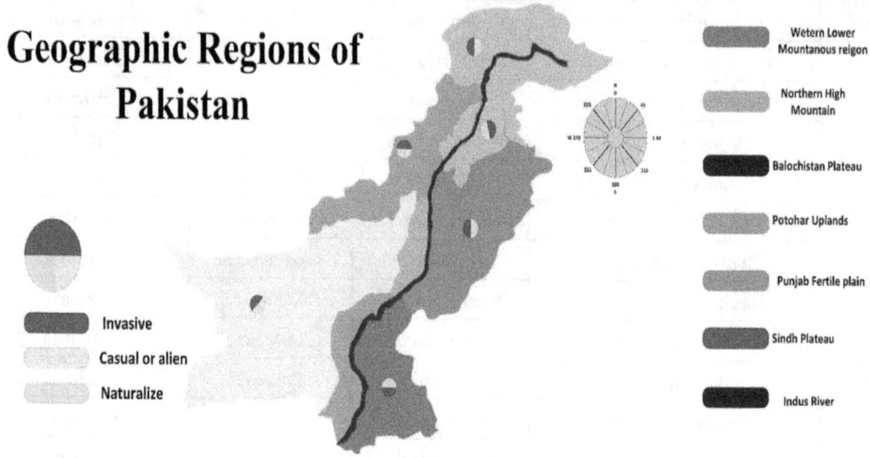

3. ECOLOGY OF NON-NATIVE PLANTS

Non-native plants are greatly benefited from investigations done in forests and pastures from ecological perspective (Hardy et al., 2025). The detrimental effects that invasive species have on the makeup of native communities, the production and palatability of fodder, and the related economic values have motivated this investigation. Forest managers should be aware about the recent advancements related to forest invasions and experimental setup which is practiced over ages (Morales-Rodríguez et al., 2025). Interruption in forest ecosystem results in the production of exotic plants. Exotic plant invasions are successful up to 1% only (Mungi et al., 2025). For healthy and heterogenous forest ecosystems, disturbance is a key element. For instance, moderate fire regimes in open environmental conditions play a vital role to maintain number of west side forests with dynamic belowground herbaceous vegetation. In the past decades, number of forests turn into thick and bare apertures, with understory plantations (Gao et al., 2025). Techniques like pruning and controlled fire applied in that kind of forests help in controlling the build-up of fire catalysts and boost up the local species specially growth of less frequent endemic species also encourage exotic plant invasion. Techniques like burning and thinning produce lot of essential nutrients like nitrogen which brings life for native and exotic plants, but these affects are not available for long span (Wangchu et al.).

4. INTERACTIONS BETWEEN NATIVE AND EXOTIC VEGETATION

Exotic plant species exert influence on grasslands in several methods i.e. by displacing significant endemic species in grasslands (Zheng & Li, 2025), manipulating altered fire patterns driven by fuel dynamics, lowering endemic diversity, and obstructing the process of tree reestablishment. Studies predict the negative relationship between alien and natural species. Furthermore, a competing theory suggests that humid, high yielding fertile areas may decrease biodiversity because the invaded species acquire the nutrients and eliminate neighboring species (Mishra et al., 2025). The 'invasion paradox' suggests different ideas to elucidate how exotic plants dominate novel environments and exist with native species simultaneously. The 'Niche Differentiation Hypothesis' highlights that exotic plants are different from endemic in some characteristics that enable the non-native species to dominate new locations and as a result of contrasting trait structures among communities. The 'Habitat Filtering Hypothesis' states that exotic species possess features that permit them to acclimate novel niches which suggests that exotic species have same features to localized vegetation.

Various investigations suggest novel-weapons hypothesis indicating that allelopathic introduced species hinder the growth of endemic species (Souza et al., 2025). Chemicals secreted by allelopathic species hinder growth and germination of neighboring plants. These native plants are not adapted to allelopathic substances, so exotic species dominate. The allelochemicals may produce through root exudates, foliar metabolites etc. The diversification of introduced species occur with changing climatic patterns, habitat alteration and global transport (Zambelli et al., 2025). Invasion by introduced plants identified as crucial threat to natural local species and these species altered communities by replacing endemic species. Many studies have revealed the disastrous effects (Mendes et al., 2025). For example, *Fallopia japonica* invade the habitat and eliminate other local species. Anthropogenic activities have led to the alterations in species diversity and composition. Introduced exotic plants of the Britain flora have increased relative to resident species due to environmental alterations (Sharma & Kumar, 2025). Exotic plants have been introduced into local communities to improve eroded soil, for fodder or ornamental purposes. Exotic species are also used for foraging.

5. IMPACT OF EXOTIC PLANTS ON FOREST ECOSYSTEM SERVICES

Forests are important constituents of ecosystem because they offer large quantity of resources for usage. These provide food for various communities and also provide clean water to more than one eighty million people in the US. They also offer the mechanisms for holding rainwater, transfer, and vaporization as it moves through many communities' at large scale (Sarker & Leta, 2025). Introduction of non-native species into forests may disrupt forest ecosystem by outcompeting with native species and replacing them. Dominant species in the community may change ecological patterns and distribution of resources. The disturbance is fulfilled when exotic species vary from local flora in particular characteristics like recycling of nutrients, seasonal patterns and chemistry (Walia et al., 2025). It is observed that all species are not able to alter ecosystem at large scale. Existing ecosystem services may change with invasion of exotic species that become dominant species with time. While introducing species into ecosystem, it is important to note that which species pose the greatest threats to forest biota (Iglesias-Carrasco et al., 2025).

6. NAVIGATING THE DUAL ROLE OF EXOTIC PLANT SPECIES IN ECOLOGICAL RESTORATION:

Positive Impacts of Exotic Species in Ecological Restoration

Since non-native species restore the degraded habitats, their introduction must be promoted for ecological restoration. Alien grasses are introduced into places affected with soil erosion where they rapidly establish cover (Weidlich et al., 2020). In places where extensive mining and grazing have resulted in infertile soil, exotic species that are involved in nitrogen fixation have been utilized to overcome such problems. (Mungi et al., 2025) observed that nitrogen-fixing trees from Asia could survive in disturbed lands of Puerto Rico and ultimately increased the number of naturalized rainforests. Exotic plants can improve the degraded sites, resulting in regeneration of endemic flora . (Qureshi & Anwar, 2024)proposed that using non-native species may assist in improving deteriorated lands infested by naturalized grass *Imperata cylindrica* (cogon grass) in the southeast Asian tropics. Agriculturists often choose to practice burn and slash farming in woodlands for restoring the region infested with *I. cylindrica (Bolfrey-Arku, 2004)*. Cogon grass causes soil hardening, alter hydrological patterns, nutrient inadequacy and its vulnerability to landfire hinders successional development by damaging seedlings and symbiotic microbes.

Numerous studies showed that non-indigenous species usually perform the functions of endemic species. Removal of alien species could have detrimental impacts on colonizing species because they provide food and shelter to imperiled endemic species. For instance, *Tamarix* species commonly known as salt cedar, displaced local vegetation that threatened willow flycatchers in the US, the birds then nested on *Tamarix*, although its reproductive potential was quite lower relative to their native habitats (County & District, 2025). *Tamarix* is highly water demanding species that negatively affect the indigenous riparian diversity. Non-native plants usually preserve indigenous plant diversity. (Ishihara et al., 2018) pointed out that *Leucaena leucocephala* was introduced in Saipan to improve soil fertility but later on, it was used as shelter for threatened *Acrocephala luscinia luscinia* because it provided the similar conditions used by the species.

Negative Impacts

For past two decades, invasive exotics have been considered as one of the main reasons in declining native plant diversity. Exotic plants outcompete with natural vegetation, making them susceptible to those diseases for which they have no resistance or modify ecological functions (Cucu et al., 2025). The main aim of environmental specialists is to create a heterogenous habitat involving particular community structure, ecological functions and community structure. Conservation practitioners reported that they spend a lot of money for eradication of these introduced species. Managers in Hawaii National Park, estimate that eighty percent of their budget is spent on removing these invasive exotics (Sanders et al., 2025). Broadly speaking, non-indigenous plants pose a serious threat in 194 of the 368 National Parks in America. Introduction of those species should be promoted which can be easily controlled without much efforts. In 1800s, *Schinus terebinthofolius* and *Melaleuca quinquenervia* were introduced to Florida but they did not become widespread until 1950s (Ewel, 2012). In these conditions, non-native species that surviving in naturalized habitats should overlooked until their number has multiplied and removal has become more difficult. This is possibly occurred during the initial invasion periods in vast natural areas. Therefore, it is crucial that reserves managers should be vigilant about threats within their region and controlled monitoring of species should be done (Utkhamthiang et al., 2025).

7. EFFECT OF DISTURBANCE ON THE MOVEMENT OF NATIVE AND EXOTIC PLANTS UNDER A CHANGING CLIMATE

Corridors

(Utkhamthiang et al., 2025) investigated that corridors are involved in the successful introduction of exotic species. Transportation corridors promote the establishment of exotic species by removing obstructions in many ways. Corridor edges change the disturbance patterns in plant communities by altering the structure of plant communities and by forming strips and gaps between them and indirectly by changing ecological conditions such as soil moisture and availability of light. Also, road traffic act as vectors for transporting seeds of non-native species thus ensure their successful introduction into surrounding areas. Large proportion of exotic species have been found near transportation corridors representing that these corridors serve as microhabitats for various alien species (Manu & Li, 2025). Much of the work has been done on the impact of exotics along these corridors in semi-arid, arid, disturbed and open habitats which provide essential conditions for invasion of exotics. Numerous research has demonstrated roads as corridors for distribution of exotic plant species.

2. Disturbed Habitats

Patchy disturbance promotes the growth of road side plants. Russian Arctic regions, disturbance favor the growth of apophytes and ruderals, a trend observed almost thirty-five years ago. Patchy disturbance is usually ignored in context of global change, yet it is the local scale that is most important in terms of plant acclimation (Su et al., 2025). There will be no migration without establishment of plant species. In the High Arctic, passage of a tracked vehicle in summer can exert disastrous effects on tundra water regimes. Repeated passage of vehicles alters the ground hydrology and redirects moisture away from wet sedge moss meadows. Desiccation in northern Baffin Island results in the regional extinction of sedges and some bryophytes. After many years, this plantation has been replaced by lichens and graminoids. These patchy areas indicated higher soil pH than their controls. Such disturbances are characterized by warmer soils and increases in available nutrients (Georgiou et al., 2025).

3. Browsing

Tyser and Worley (1992) worked on temperate grasslands, provided evidence that low level grazing by animals promoted the growth of alien species in otherwise intact stands of indigenous plantation (Samways et al., 2025). Research have shown that extreme browsing of tundra can cause destruction of non-vascular plants, especially *Sphagnum* mosses and crustose, fruticose lichens and further invasion by apophytic native plants. 4567 observed the recovery of lichens is quite difficult after minimal disturbance. In the Yamalo-Nenetsky district of northwest Siberia, 46% of the tundra is utilized as meadows (Sokolov et al., 2012). Large portions of the Gydan and southern Yamal Peninsulas indicated signs of herbivory.

Other factors

Climate is an important factor that influences the distributional range of endemic and exotic species. With increasing temperature, several other factors will come into play. Snow periods may influence seed production and flowering. Establishment of new communities is strongly dependent on soil development. Even nearby disturbances can increase the soil pH and regulate the nutrients uptake in otherwise oligotrophic soils. (Taylor et al., 2025) observed that European weed communities approaching their northern range limits tend to favor calcareous soils. However, within their optimal climatic regions, they exhibit no particular preference for this soil type. Similar soil preferences can also be observed among certain elements of both arctic-alpine and non-arctic floras.

8. FOREIGN PLANT SPECIES INVADE HOTSPOTS OF NATIVE FLORA

Introduced species are increasing rapidly due to anthropogenic activities. Widespread invasion of exotic species influences native species, disturbance patterns and functioning of ecosystem. These invasions may lead to global change (Biancolini & Rondinini, 2025). Evidence from various studies suggest that native plants, fertile soils and unique habitats are more vulnerable to infestation by introduced flora. In Europe, about 40% species are neophytes and the proportion increases with city area and level of urbanization. This demonstrates that human activities favor the growth of foreign species (Russo et al., 2025). Introduced plants are found to alter ecosystem functions by reducing endemic species, changing fire regimes and water cycles, altering soil nutrients and geomorphological processes.

Exotic plants invade native species that thrive in high diversity areas and regions having good soil fertility. Tallgrass prairie systems possess large number of native species and biodiversity so they face higher risk of land cover change, patchy habitat and wildfire control. Exotic plants invade endemic tallgrass communities that have been severely imperiled by segmentation and change in land-use methods (Endress et al., 2020). Many areas of United States such as California, Florida and Hawaii that contain threatened species are hotspots of invasive species.

Many researchers argue that exotic plants invade endemic species and alter ecological communities. Exotic invasives alter fire regimes, forage quality, wildlife grazing and recycling of nutrients. Invasion of the poisonous weed *Euphorbia esula* decreases the productivity and edibility. It decreases the pastures capacity by 50-75% in some regions (Lemma & Hailu, 2022).

9. THE ROLE OF WILDFIRE IN SHAPING EXOTIC PLANT COMMUNITIES

Fire is a kind of disturbance and ecosystems are characterized on the basis of disturbance regimes of intensity, extent, frequency, type and seasonality. Fire has been categorized into three types. This may include ground fire, surface fire and crown fire. Invasion by exotic species change fire patterns and so are considered as significant revolutionary species globally (Wolsiffer, 2025). Fire determines the structure and composition of species in a community, making an intricate association with alien plant species as shown in **Figure 2**. Altered fire patterns promote the growth of non-native species. Exotic species regenerates more frequently in burned areas. Patterns of wildfire can be altered in any of its parts such as severity, frequency, range, class and seasonal variations (Rafaqat et al., 2025).

Biodiversity is usually increased by disturbances. Disturbance results in the invasion of exotic species. (Maust, 2025) examined that wildfire increases species diversity of non-native plants such as *Lactuca serriola* and *Cirsium arvense* and fire intensity also increases. Vegetation cover increased in fire-prone areas and was greatest in critical burning areas.

Figure 2. Impact of Wildfire on Exotic Plant Species

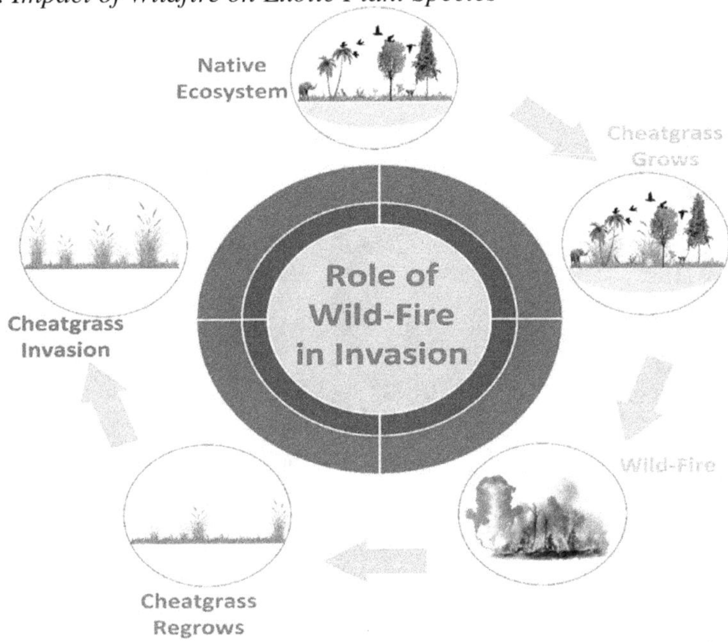

10. ALTERATIONS IN SOIL NUTRIENT CYCLING DUE TO EXOTIC FLORA

The establishment of exotic species in an area is recognized as critical threat to species loss and ecological stability. Each species has been shown to influence a number of factors such as carbon cycle, nitrogen, growth rate per plant, total productivity, litter deposition rate, chemical characteristics and carbon and nutrients cycling rates (Ping et al., 2025). Studies suggest that invasion by non-native flora change the structure and distribution of species, there might be alterations in soil nutrient fluxes and biogeochemical cycles as shown in **Figure 3**. Introduction of non-native flora alters the microbes in the soil related to specific plant species which in turn change the soil nutrient cycling processes. In addition, these exotic species alter the soil physical characteristics (Chang et al., 2025). Types of plants in area such as herbs vs trees, non-nitrogen-fixing versus nitrogen fixing plants, C_3 versus C_4 plants are also correlated with alterations in soil nutrient patterns. Figure 3

Figure 3. Invasive Plants Remove Native Biodiversity

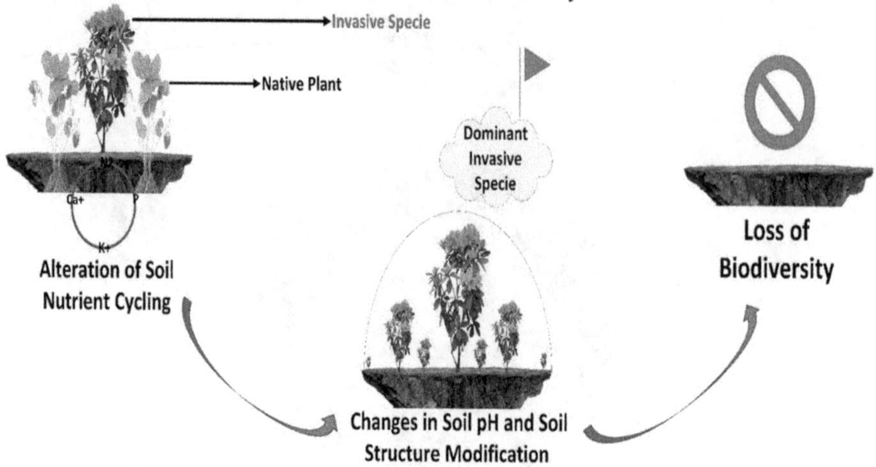

11. ANTHROPOGENIC DISTURBANCES AND NON-NATIVE PLANTS

The effects of anthropogenic activities and exotic plants on biological diversity is of major significance (Liu et al., 2025). It is essential to comprehend the relation between human disturbances and non-indigenous plants for conservation and management practices. Human mediated disturbances increase the proportion of exotic plant species. Extensive logging and farming practices influence non-indigenous plant diversity. Studies have shown that human activities like farming, logging and tourism influence exotic plant species. It is observed that roads and other human access areas may increase the invasion intensity of exotic species in an ecosystem but its susceptibility may vary with surroundings (Barros et al., 2025). While the dispersal potency of alien plants into fragmented communities is better understood, there is limited research quantifying the impact of various disturbances on plant densities.

Shifts in Land Use and Cover Patterns

With increasing globalization, increased demand for resources influences native species negatively and promote the growth of non-indigenous species (Ahmed et al., 2025). The grasslands are increasing but the total area is becoming smaller due to fragmentation. Transformation of grasslands and pastures into urban lands can increase fragmented landscapes and introduce non-native species. About eight percent of the total US population is expected to be urban by 2050. Due to travelling and trade these urbanized forests act as top-notch sites for exotic species. Strategic control of these species should implement as it will help in the management efforts (Chaudhary Tharu & Gyawali, 2025).

12. RESPONSE OF EXOTIC PLANTS TO MYCORRHIZAL FUNGI

The mechanism of symbiosis promotes the growth and invasion of introduced species. Arbuscular mycorrhizal fungi make symbiotic relationships with plant roots and provide them nutrients, protect from pests and maintain soil fertility (Chaudhary Tharu & Gyawali, 2025) Fungus mycorrhizal establishment specifies ecological coexistence of species and spread of reserves among coexisting species. Introduced as well as endemic species can form mycorrhizal relationships for resource acquisition and other benefits. Non-native species can associate with different mycorrhizal fungi than localized species, potentially giving them benefit over colonizing species in introduced communities. For example, arbuscular mycorrhizal fungi *Glomus etunicatum* and *G. geosperum* increases the growth of *Solidago canadensis* which ultimately increases the invasibility of *S. canadensis* and decreases the abundance of localized *Kummerowia striata* species (Dong et al., 2021).

13. RESPONSE OF EXOTIC SPECIES TO HURRICANE

Exotic plants show positive relationships with natural disturbances such as hurricanes and strong winds. Highly disturbed areas harbor more exotic species as compared to low disturbed areas. Disturbance of any kind promotes invisibility. Flood, wildfire and hurricanes disturb forests and provide resources to non-native seedlings, enhancing invasion and growth. This wind whirl usually occurs at the outcome of growing period and it facilitates the dispersion of exotic seeds (Grenz & Clements, 2023). The dissemination of introduced species has been associated with these strong winds. In the mountainous regions of Jamaica, forest disturbance such as typhoon promoted the distribution of alien *Pittosporum undulatum* (Cheesewood) (Lhoumeau et al., 2025). Similarly, in Hawaii, these hurricanes facilitated the growth of non-indigenous *Schefflera actinophylla* (Octopus tree). Thus, disturbances could have serious implications on the growth and development of native vegetation.

14. EXOTIC PLANT SPECIES AND COMMUNITY DYNAMICS

Exotic species may change the community dynamics. Different communities become alike due to the massive dominance of exotic species in that area. Invasive exotics prevalence displaces the native flora in invaded ecosystems Surprisingly, there are relatively few studies that assess the effects of invasive plants at the community level. Establishment of non-native species in terrestrial ecosystems is of quite important for conservationists and environmentalists (Singh et al., 2025). The invasion of terrestrial communities by non-native species is a great concern for ecologists and natural resource managers. Non-indigenous species are becoming a prominent component of local flora with increasing intentional or unintentional introduction of species throughout the world (Hardy et al., 2025). Moreover, alien species encompass many plant communities and sometimes, even regional communities.

There is antagonistic relation between non-native plants and species richness in various ecological communities. This relation has been elaborated by two community-level patterns that may vary in the cause/consequence association between richness and invasion. In the first pattern, species rich communities are more prone to invasion, showing the diversity/invasion association. Numerous ecologists support this pattern through experimental and theoretical evidences (Wang et al., 2025). In the second method, prevalence of exotic species led to the reduction in community diversity by excluding resident species from the region. In this context, changes in diversity result from invasions. Conservationists often use this perspective to monitor and manage non-native vegetation. *Lonicera maackii* showed the response of exotic plants to invasibility (Ranabhat, 2025). This shrub is endemic to Asia but through birds it become transported to eastern United States. This specie poses negative effects on disturbed and deciduous forests and usually thrives in forest understory canopy. It is observed that this invasive shrub causes decline in seedbeds and in the richness of forest herbaceous vegetation.

15. ALIEN PLANT SPECIES, BIODIVERSITY AND HERBIVORY

Human mediated disturbances (logging, herbivory) influence various communities such as Grand fir (*Abies grandis*), Douglas fir (*Pseudotsuga menziesii*) and Ponderosa pine (*Pinus ponderosa*) (Hilger-Tisdale, 2025). Disruption of Pacific Northwest Forest communities, whether natural or management induced, is recognized to promote non-native species and ultimately decline in species diversity. Grazing by hoofed animals effect the diversity of endemic species. Grazing may change the structure of plant communities. (Segrestin et al., 2025) investigated the role of herbivory and species richness. She observed that herbivory influences growth, establishment and death rates of plants and do so in ways associated with vegetation frequency, cover or with other characteristics, or through competitive traits. (Preston et al., 2025) recorded that grazing play a key role in shaping ecosystems by creating spatial variation, influencing ecological diversity and regulating the transition of communities between different states. (Wehr et al., 2025) noticed an increase in deer density with decline in species richness.

Hoofed animals alter the nutrient cycling by influencing soil organic matter and by depositing nitrogen to soil through feces and urine. Response of plant species to different communities may vary (Shen et al., 2025). This is important especially for Pacific Forests where nitrogen is limiting to herbaceous layers of forests. Exotic plants outcompete with native species and absorb more nutrients. Increased availability of nitrogen through urine and faeces of animals led to the invasion of exotic species. Controlled experiments indicated that increased availability of nutrients or water may enhance the capability of non-indigenous plants to compete with endemic species (Hardy et al., 2025).

Herbivory not only stimulates the invasion of exotic species but also involved in the seed establishment and transportation. Trampling displaces the soil, contributing to the formation of seedbed for exotic plants (Natale & Kiviat, 2025). Furthermore, epizoochory and endozoochory involved in the dispersal of seeds from affected to non-effected sites. Grazing is involved in the invasion, settling and dispersal of alien species. Limited knowledge is available on ungulate grazing due to inadequacy of experimental investigations. Herbivory is specie specific and is dependent on size of body, grazer or browser and wild or domestic, so it is quite difficult to generalize (Hoff et al., 2025). However, the main concerns regarding to the distribution of exotic plants are the physiological effects and the soil displacement caused by trafficking.

16. RESPONSE OF EXOTIC PLANTS TO INVASIBILITY IN TROPICAL FORESTS

Invasive nature of non-native plants is a main threat all around the world. Each year, alien species are becoming more frequent, establishing themselves in naturalized areas. Tropical forests possess remarkable biodiversity yet, they are also being degraded rapidly. (Carmes et al., 2025) reported that each tropical forests contain at least 1 exotic species. Tropical forests may include various forest types such as dry forests, rain forests and cloud forests etc that respond to exotic species in different ways. Two attributes of tropical forests distinguish them from temperate forests.

- Firstly, tropical forests have increased biodiversity. They contain much more species than temperate forests.

- Second, some researchers reported that these forests are heavily infested by alien plants.

Non-native diversity is more prevalent in disturbed tropical grasslands. Deciduous tropical grasslands are recognized as the most threatened forests in the world (Hardy et al., 2025). Human activities transform these forests into rangelands and agricultural lands by introducing fire. Fire curtails the germination of endemic species and supports the growth of exotic species particularly grasses as shown in **Figure 4**. Exotic species thus involved in degrading native habitat that promotes their dominance.

Figure 4. The Invasion patterns of Exotic Plants

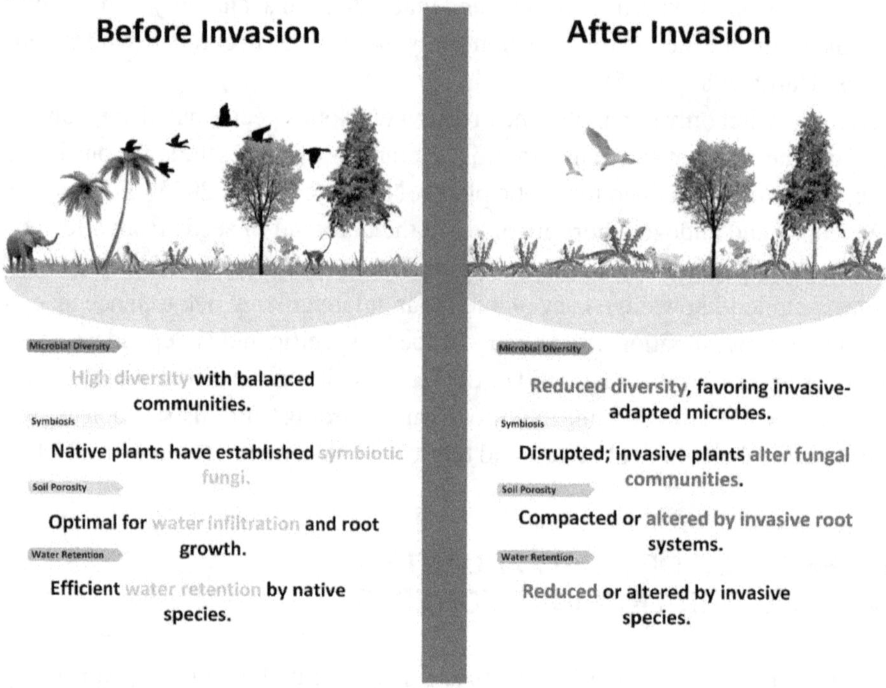

17. ALIEN SPECIES: ECONOMIC AND ENVIRONMENTAL IMPLICATIONS

Exotic species that have become invasive are extremely harmful for the community both from ecological and economical perspective (Santosa, 2025). Introduced plant species pose serious threats to cultivated land, grasslands and meadows and altering various ecological communities. Moreover, they are blocking water channels and modifying the role of aquatic biota. Most of these plants are recognized as toxic. Native species are moving towards extirpation due to massive introduction of these species. Invasive species are harmful for nurseries, animals, marine organisms as well as for other trees (Roychoudhury, 2025). In addition, invasive non-indigenous species are also harmful for humans. About 1.5 million species are introduced by humans all around the world.

Infestation of native species by exotic species is disrupting the cultural, ecological and recreational values of ecosystem (Zilio et al., 2025). Most of the introduction is probably due to travelling and trading on international levels. These species are causing serious harm to economy. For instance, introduced species cause loss of 137.2 billion dollars yearly in the US (Hulme, 2021). A loss of about 248 billion dollars occurred due to agricultural crops. These species are also harmful for industries, aquatic life and navigation. Increasingly, these are displacing endemic species. Despite mitigation strategies, invasion by non-native species is still increasing (Ahmed et al., 2025). The correlation between the introduced species in the US and establishment of alien mollusks, phytopathogens, and insects have been investigated. Later reports revealed that three species of mollusks, five species of phytopathogens and 115 species of alien insects have become established within the United States (Milián-García et al., 2025). Some other models yielded projections that were over tenfold greater as shown in **Figure 5**. The increasing rate of internal commerce and transportation, encompassing the exchange of organisms facilitated via the Internet, further implies that the dissemination of alien species within the United States and globally will be substantial.

Figure 5. Impact of Invasive Plant Species

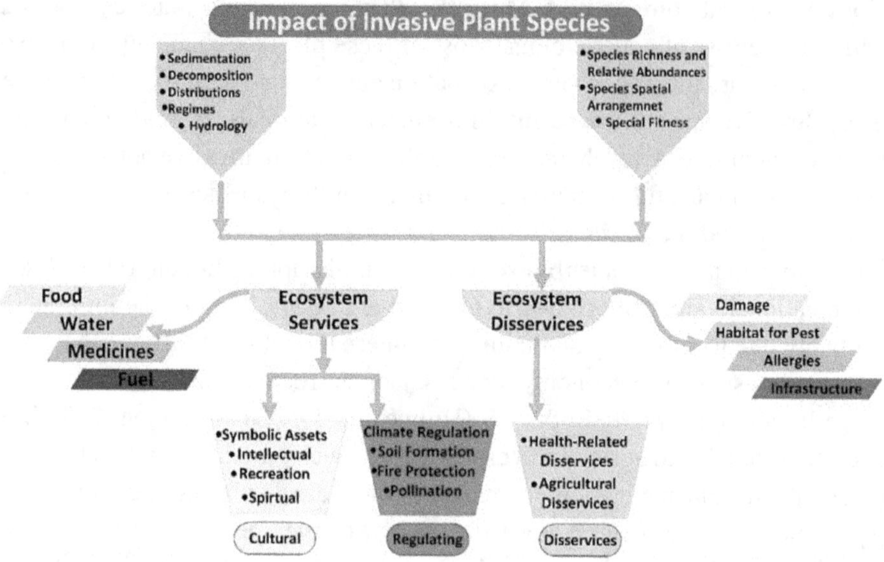

18. MANAGING EXOTIC PLANT SPECIES

Some non-indigenous species cause extreme damage to ecosystem. Management of non-native species is quite difficult task. Park managers must also focus on the effect of removing the species. In some cases, eradication of these species often creates disturbed areas that are then inhabited by exotic species. While controlling exotic plants, their span and vigor must also be considered (Zeng et al., 2025). Managers should implement policies for management of exotic species. Species that cause little damage should be less prioritized. However, those species should be greatly monitored that apparently cause no harm to resources but possess features associated with weedy or colonizing species (Natale & Kiviat, 2025).

Methods of Control

Table 3. shows different control methods of exotic species

Sr no.	Methods of Control	
1.	**Physical Control**	It includes removal of exotic plant species by hand, using ploughing or bulldozing practices. Hand practice is time consuming and quite difficult method. The latter methods cause accidental harm and leave bare areas that become prone to infestation by exotics. Cutting may inhibit growth of flowers and decreases belowground reserves but will hardly establish perennial species .
2.	**Chemical Control**	This method is usually easier than physical method. Specific chemicals are present for each crop. Some herbicides are specially manufactured for weed control. Unintentional damage may occur as a consequence of particle drift or long-term effects of herbicides that have been applied to the soil. Use of same herbicide or from same class induce resistance in exotics.
3.	**Ecological Control**	Mulching and competing plants are effective ecological methods for exotic plants management. Animals trampling, altered fire regimes and hydrological patterns influence ecosystems. Over-grazing may remove exotic plant species before flowering. Controlled use of fertilizers may facilitate the establishment of desired species over non-native species.
4.	**Biological Control**	In this method, biological agents are used such as microorganisms and insects. This method is time consuming due to the effective preparation of agent. The classical biological control method, which involves using exotic organisms, is especially effective because the biological control agent naturally locates the target plant, eliminating the need for human intervention in detection.

Gaps and Future Directions

Although considerable progress has been made in understanding the response of exotic plant species to forest disturbances, several critical gaps remain. One major limitation is the lack of long-term ecological studies. Most existing research is short-term and does not capture the cumulative or delayed impacts of exotic species on native biodiversity, forest regeneration, or ecosystem functioning. Tropical and subtropical forests despite being biodiversity hotspots remain underrepresented in invasion ecology studies. Expanding research in these regions is essential, especially as they are increasingly affected by both anthropogenic and natural disturbances. Another significant gap is the limited mechanistic understanding of how exotic species succeed in disturbed environments. The roles of plant-soil interactions, allelopathy, reproductive strategies, and adaptive traits are still not fully understood. The interaction between forest disturbances and climate change has not been adequately explored. Rising temperatures, altered precipitation patterns, and extreme weather events could further facilitate exotic species invasions, yet this area remains poorly studied. Research focusing on functional traits can offer predictive insights into which exotic species are likely to dominate following specific disturbances, but such approaches have been sparsely applied. There is a need to integrate socioeconomic and policy perspectives, including how land-use practices, forest governance, and public awareness shape invasion dynamics and management outcomes.

In terms of management implications, forest conservation efforts must incorporate species-specific responses to disturbances and prioritize the use of native, resilient species in restoration programs. Minimizing human-induced disturbances such as deforestation, logging, and infrastructure development is also critical to preventing exotic species establishment. Establishing early detection and rapid response (EDRR) frameworks post-disturbance can enhance management efficiency. Future research must be interdisciplinary, context-specific, and policy-relevant to effectively address the growing challenge of exotic plant invasions in disturbed forest ecosystems.

CONCLUSION:

Introduced, non-native and non-indigenous species are similar terms. Introduction of foreign plant species influence the ecological conditions of forests. Any kind of disturbance has severe impact on forest ecosystem. These species are further classified into casual alien, naturalized and invasive species, on the basis of their nature. Wildfire, grazing, human land cover and land use change, mycorrhizal association, hurricanes and strong winds provoke the invasion of exotic species. The introduction of species is both a blessing as well as curse. This may decrease native diversity by outcompeting with them. The introduction of species either intentionally or unintentionally displaces the endemic communities. Contrarily, these species may also provide ecological restoration services, clean water and several other benefits. Sometimes, these species are intentionally introduced for ornamental purposes, decorative use, forage, food, fodder and medicinal purposes. Generally, it is assumed that high diversity habitats are less prone to invasion. Disturbed sites harbor more exotic species. These species also alter soil nutrient dynamics. There should be effective measures for control and maintenance of these non-indigenous species. High priority should be given to those species that may cause severe harm in future. There are several methods for the removal of exotic species such as biological control, physical control, chemical control and ecological control. Park managers and conservationists should notice the distribution of these species, so that their invasion may not cause harm.

REFERENCES

Ahmed, D. A., Sousa, R., Bortolus, A., Aldemir, C., Angeli, N. F., Błońska, D., Briski, E., Britton, J. R., Cano-Barbacil, C., Clark-Ginsberg, A., Culic, I., Cuthbert, R. N., Dick, J., Dimarco, R. D., Essl, F., Everts, T., García-Berthou, E., Hauer, M., Kouba, A., & Haubrock, P. J. (2025). Parallels and discrepancies between non-native species introductions and human migration. *Biological Reviews of the Cambridge Philosophical Society*, *100*(3), 1365–1395. DOI: 10.1111/brv.70004 PMID: 39980263

Babu, S., Niveditha, E., Sooraj Dev, K., & Shamil Haque, K. (2025). Forest Degradation: Loss of Biodiversity and Ecosystem Services. In Forest Degradation and Management: An Indian Perspective (pp. 125-137). Springer.

Barros, A., Fuentes Lillo, E., Aschero, V., Pauchard, A., Alvarez, M. A., Wedegärtner, R., Clavel, J., Müllerová, J., Pergl, J., Zong, S., Vítková, M., Klinerová, T., Cavieres, L. A., Larson, C., Rew, L. J., Seipel, T., Meffre, C., Arellano, T., Essl, F., & Lembrechts, J. J. (2025). Beyond the Trail—Understanding Non-Native Plant Invasions in Mountain Ecosystems. *Global Ecology and Biogeography*, *34*(6), e70060. DOI: 10.1111/geb.70060

Biancolini, D., & Rondinini, C. (2025). Global Enhancers and Constraints of Alien Range Size in Mammals: The Roles of Species Attributes, Invasion History and Ecological Contexts. *Global Ecology and Biogeography*, *34*(7), e70081. DOI: 10.1111/geb.70081

Bolfrey-Arku, G. E.-K. (2004). Management of the Noxious Weed Speargrass (Imperata Cylindrica (L.) Beauv.) in the Forest and Forest-Savanna Transition Agro-Ecological Zones of Ghana University of Cape Coast].

Bushmann, S. L., Drummond, F. A., & Lage, C. (2025). A survey of wild bees (Hymenoptera: Apoidea) on Maine islands: Possible refugia for rare native and non-native species. *Northeastern Naturalist*, *32*(2), 278–302. DOI: 10.1656/045.032.0215

Carmes, A. A., Sühs, R. B., Silva, G. M., Fiaschi, P., & Dechoum, M. S. (2025). Factors explaining species richness of invasive plants in Neotropical forests. *Acta Botanica Brasílica*, *39*, e20240012. DOI: 10.1590/1677-941x-abb-2024-0012

Chang, X., Wang, W., & Zhou, H. (2025). Nitrogen acquisition by invasive plants: Species preferential N uptake matching with soil N dynamics contribute to its fitness and domination. *Plants*, *14*(5), 748. DOI: 10.3390/plants14050748 PMID: 40094724

Chaudhary Tharu, S., & Gyawali, S. (2025). Effect of adventure tourism in Nepal's environment and sustainability.

Chiariotti, A., Borghese, A., Boselli, C., & Barile, V. L. (2025). Water Buffalo's Adaptability to Different Environments and Farming Systems: A Review. *Animals (Basel), 15*(11), 1538. DOI: 10.3390/ani15111538 PMID: 40509004

Couch, C., Simbiano, F. J., Diawara, M., Loua, E., Mamy, L., Cheek, M., & Magassouba, S. (2025). Developing a threat risk register based on the IUCN threat hierarchy for five tropical Important Plant Areas in Guinea. *PeerJ, 13*, e19629. DOI: 10.7717/peerj.19629 PMID: 40677745

County, D. A., & District, N. (2025). Environmental Assessment Border Highway Connector Project.

Cucu, M. A., Choudhary, R., Trkulja, V., Garg, S., & Matić, S. (2025). Utilizing Environmentally Friendly Techniques for the Sustainable Control of Plant Pathogens: A Review. *Agronomy (Basel), 15*(7), 1551. DOI: 10.3390/agronomy15071551

Dibben, A., & Raskin, B. (2025). *Silvohorticulture: A grower's guide to integrating trees into crops*. Chelsea Green Publishing.

Dong, L.-J., Ma, L.-N., & He, W.-M. (2021). Arbuscular mycorrhizal fungi help explain invasion success of Solidago canadensis. *Applied Soil Ecology, 157*, 103763. DOI: 10.1016/j.apsoil.2020.103763

Endress, B. A., Averett, J. P., Naylor, B. J., Morris, L. R., & Taylor, R. V. (2020). Non-native species threaten the biotic integrity of the largest remnant Pacific Northwest Bunchgrass prairie in the United States. *Applied Vegetation Science, 23*(1), 53–68. DOI: 10.1111/avsc.12464

Ewel, J. (2012). 13. Invasibility: Lessons from South Florida. *Ecology of Biological Invasions of North America and Hawaii, 58*, 214–230. DOI: 10.1007/978-1-4612-4988-7_13

Gao, Y., Wang, X., Liu, Q., Zhang, J., Han, X., Yang, S.-I., & Lu, D. (2025). Converting larch monocultures to mixed-species forests: Ten-year impacts of canopy manipulation on understory regeneration. *Forest Ecology and Management, 585*, 122611. DOI: 10.1016/j.foreco.2025.122611

Georgiou, K., Angers, D., Champiny, R. E., Cotrufo, M. F., Craig, M. E., Doetterl, S., Grandy, A. S., Lavallee, J. M., Lin, Y., Lugato, E., Poeplau, C., Rocci, K. S., Schweizer, S. A., Six, J., & Wieder, W. R. (2025). Soil carbon saturation: What do we really know? *Global Change Biology, 31*(5), e70197. DOI: 10.1111/gcb.70197 PMID: 40345163

Grenz, J., & Clements, D. R. (2023). Water, wind, and fire: extreme climate events enhance the spread of invasive plants in sensitive North American ecosystems. In *Plant invasions and global climate change* (pp. 113–137). Springer. DOI: 10.1007/978-981-99-5910-5_6

Hardy, N. G., Kuebbing, S. E., Duguid, M. C., Ashton, M. S., Sheban, K. C., Inman, S. E., & Martin, M. P. (2025). Non-native invasive plants in tropical dry forests: A global review of presence, impacts, and management. *Restoration Ecology*, *33*(1), e14288. DOI: 10.1111/rec.14288

Hilger-Tisdale, A. N. (2025). *Effects of Plant Community Composition*. Nutrition, and Human Disturbance on the Behavior and Distribution of Large Herbivores University of Idaho.

Hoff, H. K., Littleford-Colquhoun, B. L., Kartzinel, R. Y., Anderson, H. M., Geremia, C., McGarvey, L. M., Segal, C., & Kartzinel, T. R. (2025). The apportionment of dietary diversity in wildlife. *Proceedings of the National Academy of Sciences of the United States of America*, *122*(29), e2502691122. DOI: 10.1073/pnas.2502691122 PMID: 40658848

Holland, O. C., Stark, J. S., Wilson, K. A., Baker, C. M., Shaw, J. D., & Helmstedt, K. J. (2025). Invasive species pose a threat to native species' abundances in an East Antarctic coastal marine ecosystems. *Journal of Applied Ecology*, 1365-2664.70062. DOI: 10.1111/1365-2664.70062

Hulme, P. E. (2021). Unwelcome exchange: International trade as a direct and indirect driver of biological invasions worldwide. *One Earth*, *4*(5), 666–679. DOI: 10.1016/j.oneear.2021.04.015

Ighalo, J. O., Ohoro, C. R., Ojukwu, V. E., Oniye, M., Shaikh, W. A., Biswas, J. K., Seth, C. S., Mohan, G. B. M., Chandran, S. A., & Rangabhashiyam, S. (2025). Biochar for ameliorating soil fertility and microbial diversity: From production to action of the black gold. *iScience*, *28*(1), 111524. DOI: 10.1016/j.isci.2024.111524 PMID: 39807171

Iglesias-Carrasco, M., Torres, J., Cruz-Dubon, A., Candolin, U., Wong, B. B., & Velo-Antón, G. (2025). Global impacts of exotic eucalypt plantations on wildlife. *Biological Reviews of the Cambridge Philosophical Society*, *100*(4), 1734–1753. DOI: 10.1111/brv.70022 PMID: 40159998

Ishihara, K. L., Honda, M. D., Bageel, A., & Borthakur, D. (2018). *Leucaena leucocephala: a leguminous tree suitable for eroded habitats of Hawaiian Islands. In ravine lands: greening for livelihood and environmental security*. Springer.

Kaur, M., Kaur, J., Sood, A., Katnoria, J. K., Nagpal, A. K., Bhatia, A., Verma, P., Kumar, V., Singh, I., & Bhatti, S. S. (2025). Potentially toxic metal contamination in semi-arid agricultural soils: Sources, risk analysis, and spatial distribution. *Environmental Monitoring and Assessment, 197*(7), 783. DOI: 10.1007/s10661-025-14237-6 PMID: 40542220

Kumar, R., Salaria, N., Kumari, I., & Hajam, Y. A. (2025). *Plant Tissue Culture Technology: An Innovative Tool for the Propagation, Management, and Conservation of Himalayan Biodiversity*. CRC Press. DOI: 10.1201/9781779643445

Lemma, F., & Hailu, H. (2022). *IMPACTS OF Lantana camaraON ELEPHANT HABITAT USE AND FEED AVAILABILITY IN BABILE ELEPHANT SANCTUARY*. EASTERN ETHIOPIA Haramaya University.

Lhoumeau, S., Elias, R. B., Seidel, D., Gabriel, R., & Borges, P. A. (2025). Landscape to microhabitat: Uncovering the multiscale complexity of native and exotic forests on Terceira Island (Azores, Portugal). *PLoS One, 20*(6), e0326304. DOI: 10.1371/journal.pone.0326304 PMID: 40522958

Liu, M., Chen, X., Zhang, Y., Zhang, X., & Chen, Y. (2025). Nonlinear relationship between diversity of rare and common species and ecosystem multifunctionality in alpine meadows along an altitude gradient. *Journal of Environmental Management, 376*, 124374. DOI: 10.1016/j.jenvman.2025.124374 PMID: 39914212

Manu, M., & Li, J. (2025). Urban pavement crevices as microhabitats: Soil properties and spontaneous vegetation in roadside environments. *International Journal of Sustainable Development and World Ecology*, •••, 1–13. DOI: 10.1080/13504509.2025.2531435

Maust, A. (2025). Quantifying the Effects of Wildfire Reintroduction on Native Pollinators in Ponderosa Pine Forests of the Pacific Northwest University of Washington].

Mendes, S. B., Olesen, J. M., Nogales, M., Marrero, P., Romero, J., Nieves, C., & Heleno, R. (2025). Urbanization of seed dispersal networks. *Conservation Biology*, •••, 70097. DOI: 10.1111/cobi.70097 PMID: 40605764

Milián-García, Y., Pyne, C., Chen, A., Lindsay, K., & Hanner, R. H. (2025). Uncovering the hidden within shipping containers: Molecular biosurveillance confirms a pathway for introducing multiple regulated and invasive species. *Biological Invasions, 27*(3), 91. DOI: 10.1007/s10530-025-03549-w

Mishra, A., Rajak, P., Raghubanshi, A. S., & Singh, H. (2025). Understanding competition between two invasive woody plants of India under an altered rainfall regime. *Plant Ecology, 226*(1), 27–45. DOI: 10.1007/s11258-024-01470-1

Morales-Rodríguez, C., Vannini, A., Scanu, B., González-Moreno, P., Turco, S., Drais, M. I., Brandano, A., Varo Martínez, M. Á., Mazzaglia, A., Deidda, A., Cidre, A., Gomes-Laranjo, J., Liberati, D., & Ruiz-Gómez, F. J. (2025). Challenges to Mediterranean Fagaceae ecosystems affected by Phytophthora cinnamomi and climate change: Integrated pest management perspectives. *Current Forestry Reports*, *11*(1), 9. DOI: 10.1007/s40725-024-00237-1

Mungi, N. A., Galetti, M., Ratnam, J., Chala, D., Sanín, M. J., & Svenning, J.-C. (2025). Alien plants and novel ecosystems in the Greater Tropics. Nature Reviews Biodiversity, 1-17.

Natale, K., & Kiviat, E. (2025). BEST MANAGEMENT PRACTICES FOR PRIORITY INVASIVE PLANTS IN THE LOWER HUDSON VALLEY.

Oancea, S. (2025). Occurrence, Pharmacological Properties, Toxic Effects, and Possibilities of Using Berries from Selected Invasive Plants. *Antioxidants*, *14*(4), 399. DOI: 10.3390/antiox14040399 PMID: 40298647

Ping, Q., Xu, S., He, X., & Sun, S. (2025). Nitrogen addition alters litter chemical traits to regulate decomposition: A meta-analysis. *The Science of the Total Environment*, *983*, 179705. DOI: 10.1016/j.scitotenv.2025.179705 PMID: 40403540

Preston, J., Debney, A., Gamble, C., Hardy, M., Underwood, G., Garbutt, A., Harley, J., Baker, R., Dunk, R., & Grigg, M. (2025). Seascape connectivity: evidence, knowledge gaps and implications for temperate coastal ecosystem restoration practice and policy. npj Ocean Sustainability, 4(1), 33.

Pyšek, P., Hulme, P. E., Simberloff, D., Bacher, S., Blackburn, T. M., Carlton, J. T., Dawson, W., Essl, F., Foxcroft, L. C., Genovesi, P., Jeschke, J. M., Kühn, I., Liebhold, A. M., Mandrak, N. E., Meyerson, L. A., Pauchard, A., Pergl, J., Roy, H. E., Seebens, H., & Richardson, D. M. (2020). Scientists' warning on invasive alien species. *Biological Reviews of the Cambridge Philosophical Society*, *95*(6), 1511–1534. DOI: 10.1111/brv.12627 PMID: 32588508

Qureshi, H., & Anwar, T. (2024). Role of allelopathy in invasion of invasive species. *Allelopathy Journal*, *62*(2), 115–138. DOI: 10.26651/allelo.j/2024-62-2-1490

Rafaqat, W., Sanchez, P., Botnen, D., & Fernandez-Anez, N. (2025). Analysing historical events and current management strategies of wildfires in Norway. *Scientific Reports*, *15*(1), 24905. DOI: 10.1038/s41598-025-08760-2 PMID: 40640312

Ranabhat, G. (2025). Vegetation Diversity in Southwestern Illinois Sand Prairies Southern Illinois University at Edwardsville].

Roychoudhury, N. (2025). Forest Invasive Species and Their Management. In Textbook of Forest Science (pp. 315-331). Springer. DOI: 10.1007/978-981-97-8289-5_14

Russo, A., Esperon-Rodriguez, M., St-Denis, A., & Tjoelker, M. G. (2025). Native vs. Non-Native Plants: Public Preferences, Ecosystem Services, and Conservation Strategies for Climate-Resilient Urban Green Spaces. *Land (Basel), 14*(5), 954. DOI: 10.3390/land14050954

Samways, M. J., Gaigher, R., Deacon, C., & Pryke, J. S. (2025). Heterogeneous conservation corridors of remnant vegetation protect biodiversity in South African timber mosaics. *Biodiversity and Conservation, 34*(2), 357–379. DOI: 10.1007/s10531-024-02993-9

Sanders, S. K., van Kleunen, M., Allan, E., & Thakur, M. P. (2025). Effects of extreme drought on the invasion dynamics of non-native plants. *Trends in Plant Science, 30*(3), 291–300. DOI: 10.1016/j.tplants.2024.10.009 PMID: 39523141

Santosa, E. B. (2025). Blessing in disguise of the Invasive Alien Species (IAS): Environmental problem into local-based advantages potentials. Media Konservasi, 30(2).

Sarker, S., & Leta, O. T. (2025). Review of Watershed Hydrology and Mathematical Models. *Eng, 6*(6), 129. DOI: 10.3390/eng6060129

Sedibana, L., Kganyago, M., Jaca, T., & Yessoufou, K. (2025). First Spectral Reflectance Dataset of a Highly Invasive and Aggressive Alien Plant Species, Equisetum hyemale (snakegrass), in South Africa. Towards Early Detection and Effective Eradication. *Data in Brief, 60*, 111582. DOI: 10.1016/j.dib.2025.111582 PMID: 40534718

Segrestin, J., Lisner, A., Götzenberger, L., Hájek, T., Janíková, E., Jílková, V., Konečná, M., Švancárová, T., & Lepš, J. (2025). Biodiversity loss disrupts seasonal carbon dynamics in a species-rich temperate grassland. *Ecology, 106*(5), e70091. DOI: 10.1002/ecy.70091 PMID: 40342156

Sharma, S. K., & Kumar, K. (2025). Forest Fire Effects on Structure, Composition, and Functioning of Ecosystem. In Forest Fire and Climate Change: Insights into Science (pp. 79-99). Springer.

Shen, Y., Fang, Y., Oelmann, Y., Vancov, T., Chen, H., Du, Z., Chang, S. X., & Cai, Y. (2025). Livestock excretion alleviates soil microbial carbon and phosphorus limitations relative to trampling and defoliation in an alpine meadow. *Agriculture, Ecosystems & Environment, 389*, 109687. DOI: 10.1016/j.agee.2025.109687

Singh, V., Mina, U., Meena, R., & Koul, M. (2025). Understanding the Nuances and Mitigation Measures of Plant Invasion: A Step Towards Sustainable Development. Green Equilibrium: Deciphering Earth's Ecosystems for Sustainable Tomorrow, 197-224.

Sokolov, V., Ehrich, D., Yoccoz, N. G., Sokolov, A., & Lecomte, N. (2012). Bird communities of the Arctic shrub tundra of Yamal: Habitat specialists and generalists. *PLoS One*, *7*(12), e50335. DOI: 10.1371/journal.pone.0050335 PMID: 23239978

Souza, C. S., Marins, G., Camargo, I. F. L., Lima, L. B., Gomes, A. S., & Borghetti, F. (2025). Allelopathic hypotheses revisited: The interactions between native and exotic species in the Brazilian savanna. *Anais da Academia Brasileira de Ciências*, *97*(3), e20240252. DOI: 10.1590/0001-3765202520240252 PMID: 40561269

Su, Y., Zhang, C., Cescatti, A., Yu, K., Ciais, P., Smith, T., Shang, J., Carnicer, J., Liu, J., Chen, J. M., Green, J. K., Wu, J., Ponce-Campos, G. E., Zhang, Y., Zuo, Z., Liao, J., Wu, J., Lafortezza, R., Yan, K., & Zhou, W. (2025). Pervasive but biome-dependent relationship between fragmentation and resilience in forests. *Nature Ecology & Evolution*, •••, 1–15. DOI: 10.1038/s41559-025-02776-7 PMID: 40629088

Taylor, R. M., Stavert, J. R., Pou, D. L., & Wilmshurst, J. M. (2025). Ecology, history, threats, and management of gumland ecosystems in Aotearoal New Zealand. New Zealand Journal of Ecology, 49(1), 3589.

Utkhamthiang, K., Chaimanee, V., Hermhuk, S., & Kamyo, T. (2025). Assessment of Invasive Species Severity along the Nature Trail at the Doi Chiang Dao Biosphere Reserve, Chiang Mai Province. *Thai Forest Ecological Research Journal*, *9*(1), 55–72. DOI: 10.34044/tferj.2025.9.1.6271

Wafer, A. (2025). Into the weeds: integrating evolutionary ecology into weed science The Ohio State University].

Walia, K., & Kumari, Y. Garima, & Mehta, A. (2025). Ecosystem Recovery and Resilience After Forest Fires. In Forest Fire and Climate Change: Insights into Science (pp. 119-145). Springer.

Wan, X., & Zhang, L. (2021). Global new species of vascular plants published in 2020. *Shengwu Duoyangxing*, *29*(8), 1003–1010. DOI: 10.17520/biods.2021156

Wang, J., Raes, B., Debrabandere, C., van Aken, V., Jaramillo-Toro, S., Waldherr, S., Horemans, B., & Springael, D. (2025). Niche creation improves bioaugmentation of an organic micropollutant degrader in oligotrophic waters. *The ISME Journal*, •••, wraf140. DOI: 10.1093/ismejo/wraf140 PMID: 40613637

Wangchu, L., Tasung, A., Kumar, A., Angami, T., Suryawanshi, M. A., Jini, D., Singh, R., Alone, R. A., Bhagawati, S. K., & Khan, R. TRAINING MANUAL ON.

Wehr, N. H., Boone, H. M., van den Bosch, M., Zubiria Perez, A., Wehr, K. L., Wehr, S. R., & Belant, J. L. (2025). Interspecific carnivore competition and ungulate predation correlate with predator species richness. *Ecology*, *106*(6), e70136. DOI: 10.1002/ecy.70136 PMID: 40528493

Weidlich, E. W., Flórido, F. G., Sorrini, T. B., & Brancalion, P. H. (2020). Controlling invasive plant species in ecological restoration: A global review. *Journal of Applied Ecology*, *57*(9), 1806–1817. DOI: 10.1111/1365-2664.13656

Wolsiffer, S. K. (2025). *Wildland fuel loading estimates along a gradient of forest cover types and landscape factors in Harpers Ferry National Historic Park, West Virginia and Rock Creek Park*. USA Virginia Tech.

Zambelli, A., Nocito, F. F., & Araniti, F. (2025). Unveiling the multifaceted roles of root exudates: Chemical interactions, allelopathy, and agricultural applications. *Agronomy (Basel)*, *15*(4), 1–30. DOI: 10.3390/agronomy15040845

Zargar, S. A., Malik, R. A., Khuroo, A. A., Ganie, A. H., & Reshi, Z. A. (2025). Alien flora of the Himalayan highlands: Naturalised and invasive plants in the Trans-Himalayan region of Ladakh, India. *Biological Invasions*, *27*(4), 1–19. DOI: 10.1007/s10530-025-03562-z

Zeng, Z., Liang, Z., Chen, Y., Shu, Q., Li, J., Tso, N., Chen, M., Zhang, S., Tan, X., Qiong, L., & Wang, J. (2025). Adaptation analysis of two Asteraceae invasive plants in Lhasa, Tibet. *PeerJ*, *13*, e19667. DOI: 10.7717/peerj.19667 PMID: 40642327

Zhang, X., Fang, F., Sha, L., & Nizamani, M. M. (2025). Ecological dynamics and microbial community composition of Heliotropium arboreum in the coastal ecosystems of Hainan province. *Frontiers in Microbiology*, *16*, 1611262. DOI: 10.3389/fmicb.2025.1611262 PMID: 40678047

Zheng, J., & Li, C. (2025). Plant functional diversity as an important mediator of invasive effects on productivity in riparian grasslands. *Biological Invasions*, *27*(5), 128. DOI: 10.1007/s10530-025-03586-5

Zilio, M. I., Restrepo, A. F. S., Gervazoni, P., Minuti, G., Muzon, J., Djeddour, D., Mc Kay, F., & Sosa, A. (2025). Economic loss of ecosystem services caused by yellow flag iris and white ginger lily, two invasive alien plants in natural ecosystems of Argentina. *Journal for Nature Conservation*, *84*, 126789. DOI: 10.1016/j.jnc.2024.126789

Chapter 3
Restoration of Native Habitats Invaded by Invasive Plants Under Changing Climate

safura Bibi
https://orcid.org/0000-0002-1715-1871
University of Agriculture, Faisalabad, Pakistan

Muhammad Sajid Aqeel Ahmad
University of Agriculture, Faisalabad, Pakistan

Mansoor Hameed
University of Agriculture, Faisalabad, Pakistan

Zunaira Naeem
https://orcid.org/0009-0005-8092-088X
University of Agriculture, Faisalabad, Pakistan

Muaz Ameen
https://orcid.org/0000-0003-1388-2113
University of Agriculture, Faisalabad, Pakistan

ABSTRACT

This chapter explores the different strategies for restoring natural environments, highlighting the importance of management practices in considering future climate conditions. The main approaches involve excluding invasive species and restoring native and climate-resilient species to boost ecosystem stability. In addition to this, the chapter also focuses on the role of community and stakeholders' collaboration

DOI: 10.4018/979-8-3693-8724-5.ch003

for restoration projects. This chapter discusses the need for constant monitoring and adaptable strategies to address climate change issues. It demands extensive research on the interactions between native and invasive flora in changing climate conditions. Generally, this chapter's objective is to train scientists and policymakers with the facts and tools necessary to restore natural ecosystems from climate change and invasive species, leading to the conservation of biodiversity and ecosystem resilience.

INTRODUCTION

Since 1900, global average surface temperatures have increased by 0.7 °C per century, with a more rapid rise of 0.16 °C per decade since 1970. This warming is primarily attributed to higher atmospheric concentrations of greenhouse gases due to anthropogenic activities (Bilgili, Tumse, & Nar, 2024). Temperature increases have been more pronounced during winter than summer and are mainly reflected in rising minimum temperatures. Changes in precipitation patterns have also been noted, though these shifts are moreover variable and less consistent than temperature changes (Gao et al., 2018). Even with conservative emission estimates, future climate projections indicate further warming, substantial drying in regions, and a rise in the intensity and frequency of extreme weather events. For instance, climate models suggest annual temperatures in the United States could increase by 3-9 °C over the next century, alongside dormant reductions in summer rainfall in certain areas (Jones et al., 2023).

Native to a particular area, invasive species are plants or animals that overrun a new ecosystem. Most often, humans plant them in new climates, either purposefully or inadvertently. An invasive alien species has been introduced to a new location and has started to cause problems in its new environment. These species threaten native biodiversity because they outcompete local species for resources, and they are spreading moreover quickly due to an increase in international travel and trade. A recent study found numerous biodiversity hotspots, and one-sixth of the Earth's land surface is particularly vulnerable to invasion (Smith, 2024). Taxa that endanger biological variety, food security, human health, and well-being when they are introduced and spread by human activity outside of their natural distribution. The term invasive denotes a species present outside of its native habitat. Invasive refers to a tendency to spread into and alter newly introduced habitats. As a result, a species can be invasive without being alien, or it can spread rapidly within a region and become foreign without actually being alien. The threat of invasive species (IS) to many ecosystems has spread worldwide. The escalating danger is causing great global anxiety and severe global activities and coverage through agreements, treaties, governments, and organizations (Dindaroglu, 2024).

In 1992, the United Nations approved the Convention on Biological Diversity. Invasive alien species have intensified as a severe issue in recent years, and the world community has committed to focusing on this issue as part of an international effort to lessen its adverse effects. Invasive species pose the second most significant danger to biodiversity. Biological pollution, known as biological invasion, is undoubtedly moreover harmful than chemical pollution. As a result, after habitat degradation, it is ranked as the second most significant danger to biodiversity worldwide (Brodie et al., 2021). Invasive species are widely dispersed throughout ecosystems and coexist with all living things.

Nevertheless, the most prevalent invasive species in terrestrial ecosystems are plants, animals, and insects. Due to their enormous biomass, plants are among the worst invaders in the world. These species may have various negative environmental consequences, such as increased soil erosion, decreased soil value, altered natural flora and fauna, choked water bodies that influence water quality and fish populations, economic losses, and human health risks (Meyer et al., 2021).

Ecological restoration is the method of supporting the recovery of ecosystems that have been damaged. This recovery involves reestablishing the ecosystem's original structure, composition, and function as much as possible, reflecting its state before degradation. It also includes restoring the ecosystem's ability to adapt to changing conditions. In contrast, restoration differs from rehabilitation, which focuses on maintaining functionality and delivering specific services rather than fully reinstating the pre-disturbance ecosystem with all its biological complexity (Gerwing et al., 2022). Rehabilitation is often the only viable option when ecosystems have degraded beyond recovery, for example, when species have gone extinct or seed and soil biota have been lost. Choosing between restoration and rehabilitation involves trade-offs, such as deciding which species to prioritize, which ecosystem functions to target, and whether to rely on passive or active interventions to generate services (Puspitaloka et al., 2020). Today's economic systems tend to prioritize short-term gains over long-term sustainability. A clear example is the exploitation and destruction of vast areas of tropical, temperate, and boreal forests despite their critical role in supporting local and global ecosystems. This short-term perspective is fundamentally at odds with ecological restoration, which seeks to build a sustainable recovery path for biodiversity and ecosystem services that benefit as many people as possible for as long as possible, ideally forever. From both ecological and economic perspectives, rehabilitation is often the most practical approach when all stakeholders agree that the condition of an ecosystem is degraded. However, long-term rehabilitation can also be an initial step toward full-scale restoration (Raihan & Bari, 2024).

1. INVASIVE SPECIES

Invasion refers to the process by which a species is introduced, intentionally or inadvertently, into a new topographical area where it can thrive and survive. Species that are outside the historical range they originally evolved are termed invasive. Most introduced non-native organisms do not manage to establish themselves successfully, and those that do often spread slowly and have little noticeable impact on their surroundings. However, a small proportion of non-native species become invasive, meaning they can spread rapidly or have significant environmental effects. Invasive species pose a global challenge, threatening native biodiversity, disrupting ecosystems, affecting natural resources, harming regional economies, and impacting human health (Reaser, Burgiel, et al., 2020).

2. ECOLOGICAL IMPACTS OF INVASIVE SPECIES

With no indication of saturation across most taxonomic groupings, the introduction and spread of invading species are growing along with world trade. Invasive species affect local and national economies, natural processes, and ecosystems. They can significantly change natural terrestrial and aquatic systems' content, organization, or purpose. Affecting genetic material, individual organisms, populations, communities, and ecosystems, their impacts could be direct or indirect (**Table 1**). Below is a synopsis of the ecological effects of invading species in terrestrial and aquatic habitats, highlighting the spectrum of problems covered in this evaluation. The evaluation also covers management techniques and the possible financial consequences of spreading invasive species (Borden & Flory, 2021).

Table 1. List of Invasive Plant Species, Primary Habitat Affected and Impacts from around the globe

Sr No.	Scientific Name	Common Name	Primary Habitat Affected	Impacts	Citations
1	*Ailanthus altissima*	Tree of Heaven	Urban areas, forests	Produces chemicals inhibiting native plant growth	(Lafaie et al., 2024)
2	*Bromus tectorum*	Cheatgrass	Grasslands, deserts	Increases wildfire frequency, suppresses native grasses	(Molvar et al., 2024)
3	*Cenchrus ciliaris*	Buffelgrass	Deserts, drylands	Fuels wildfires, displaces native species	(Siller-Clavel et al., 2022)

continued on following page

Table 1. Continued

Sr No.	Scientific Name	Common Name	Primary Habitat Affected	Impacts	Citations
4	*Centaurea solstitialis*	Yellow Star Thistle	Grasslands, rangelands	Reduces forage quality, outcompetes natives	(Tsynkevych, 2018)
5	*Eichhornia crassipes*	Water Hyacinth	Freshwater bodies	Depletes oxygen in water, affects aquatic life	(Macedo et al., 2024)
6	*Fallopia japonica*	Japanese Knotweed	Riparian zones, urban areas	Erodes riverbanks, inhibits native plant growth	(Li et al., 2024)
7	*Heracleum mantegazzianum*	Giant Hogweed	Riparian zones, roadsides	Outcompetes natives, toxic to humans	(Murphy & Hall, 2024)
8	*Impatiens glandulifera*	Himalayan Balsam	Riverbanks, moist soils	Outcompetes native plants, promotes erosion	(Coakley & Petti, 2021)
9	*Lantana camara*	Lantana	Forests, grasslands	Toxic to livestock, suppresses native plants	(Kumar et al., 2024)
10	*Lythrum salicaria*	Purple Loosestrife	Wetlands, marshes	Alters hydrology, reduces biodiversity	(Kotta, 2023)
11	*Parthenium hysterophorus*	Parthenium Weed	Agricultural lands, grasslands	Toxic to animals, reduces crop yields	(Tiawoun et al., 2024)
12	*Pennisetum setaceum*	Fountain Grass	Urban areas, arid regions	Alters fire regimes, invades natural areas	(Areces-Berazain, 2023)
13	*Phragmites australis*	Common Reed	Wetlands, marshes	Forms dense monocultures, degrades habitat quality	(Robichaud, Basso, & Rooney, 2022)
14	*Pueraria montana*	Kudzu	Forest edges, open areas	Smothers native vegetation, alters soil composition	(Khattak et al., 2024)
15	*Tamarix spp.*	Saltcedar (Tamarisk)	Riparian zones, arid regions	Lowers water tables, salinizes soil	(Gaffke et al., 2022)

Invasive terrestrial plants can disrupt key ecological functions, including nitrogen cycling and production; therefore, they impact native biota at higher levels via the food chain. Many times, accelerating the carbon cycle, these invading plants might raise the energy availability for other organisms. Still, the impacts rely on the features of the invader, the native species in the invaded habitat, and the surroundings. Significant carbon storage losses may result from the invading species' upsetting disturbance patterns. Usually, plants that invade grasslands and forests raise the nitrogen levels in the plant tissues above ground and the availability of nitrogen in the soils they invade, encouraging microbial activity (Bertelsmeier, 2021). Invasion can change the water quality by changing the shape of stream channels and, by extension, hydrologic cycles, such as the number of floods.

Invasive plants can significantly affect freshwater habitats moreover than terrestrial ones; they can decrease sunlight penetration, drop oxygen levels, and alter water temperature, pH, and nutrient content. The variety of invading aquatic species significantly affects biological organization, from genes to individuals to populations to ecosystems. Invasive aquatic plants can make it hard to navigate, affecting the flow of water needed for flood control, irrigation, or hydropower. They can also get in the way of recreational activities and make water bodies look bad. They can also lower property values and discourage tourists (Costello et al., 2022) (**Figure 1**).

Figure 1. Diagram showing the invasion process of invasive species under various climate conditions

Non-native plant species have a long history in Pakistan. The primary reason for the presence of most foreign invasive species in the country is to address the demand for lumber and fuelwood. Fortunately, the situation regarding invasive species in Pakistan is not as severe as in some other countries; however, the influence of foreign species on local biodiversity has not received enough attention (N. Khan et al., 2023). Limited studies have identified 700 invasive vascular plant species, categorized as high-impact land invasives, including *Broussonetia papyrifera, Lantana camara,*

Prosopis juliflora, and *Parthenium hysterophorus. Eucalyptus camaldulensis,* introduced to the Malakand Hills, has been observed to affect the growth of grasses and shrubs negatively. Due to allelopathic effects, *Dodonea viscosa,* a related species in scrub forests, has wholly disappeared (Jehangir et al., 2024) **(Table 2)**.

Table 2. List of Invasive Plant Species, Primary Habitat Affected and Impacts in Pakistan

Sr No.	Scientific Name	Common Name	Primary Habitat Affected	Impacts	Citations
1	*Arundo donax*	Giant Reed	Riverbanks, wetlands	Forms dense thickets, disrupts water flow	(Yang et al., 2024)
2	*Asphodelus tenuifolius*	Wild Onion Weed	Agricultural fields	Reduces crop yield, competes with cultivated species	(I. Khan et al., 2023)
3	*Broussonetia papyrifera*	Paper Mulberry	Urban areas, road verges	Causes severe pollen allergies, displaces native flora	(Lin et al., 2023)
4	*Cuscuta spp.*	Dodder	Agricultural fields, gardens	Parasitic, affects crops and ornamental plants	(Meena & Sharma, 2024)
5	*Eichhornia crassipes*	Water Hyacinth	Lakes, rivers, wetlands	Blocks waterways, depletes oxygen in water	(Izhar et al., 2023)
6	*Lantana camara*	Lantana	Forests, urban areas	Reduces biodiversity, toxic to grazing animals	(Sharma et al., 2022)
7	*Parthenium hysterophorus*	Parthenium Weed	Roadsides, agricultural lands	Toxic to livestock, causes allergies in humans	(Kumar & Aggarwal, 2024)
8	*Prosopis juliflora*	Mesquite	Arid and semi-arid regions	Depletes groundwater, displaces native vegetation	(Balcha, 2022)
9	*Sesbania punicea*	Red Sesbania	Wetlands, riversides	Alters wetland hydrology, displaces native plants	(Bologo, 2023)
10	*Sorghum halepense*	Johnson Grass	Agricultural fields, grasslands	Competes with crops, toxic to livestock	(Velez-Gavilan, 2024)
11	*Tamarix spp.*	Saltcedar (Tamarisk)	Riparian areas, arid zones	Increases soil salinity, depletes water resources	(Khalid, Naveed, & Khan, 2023)

3. ECOLOGICAL RESTORATION

Ecological restoration involves returning an ecosystem to a previous condition or trajectory. However, because ecosystems are dynamic and influenced by external factors like climate change, achieving complete fidelity or a full recovery is seldom possible. Along with a longer and moreover detailed list of the nine traits of a restored ecosystem discussed in the SER primer, they give us a way to judge the effectiveness and classification of various projects. One could expand this framework to handle particular ecological services (Zhu et al., 2021).

The Millennium Ecosystem Assessment found four partially overlapping types of natural capital: (1) renewable, comprising well-functioning ecosystems and their innate biodiversity; (2) cultivated, referred to as native crop varieties and livestock breeds; (3) replenishable, such as clean air, water, and fertile soils; and (4) nonrenewable, including petroleum, copper, coal, and diamonds. Environmental restoration primarily addresses the first two kinds (Genes & Dirzo, 2022).

4. ECOLOGICAL RESTORATION OF TERRESTRIAL ECOSYSTEMS

Some systems degrade progressively and predictably in a direct manner as environmental conditions change, leading to proportional shifts in plant species composition and function. In other cases, systems may appear resilient to ecological changes until a critical threshold is reached, causing sudden and rapid changes. Identifying the trigger or driver behind such abrupt shifts can be crucial, as removing the driver might allow the system to recover. On the other hand, some regions may undergo the same ecological changes by establishing several constant conditions, a phenomenon known as hysteresis (Weidlich et al., 2020).

Restoring these systems can be particularly challenging since the path to recovery may differ significantly from the path leading to the degraded state. Recent reviews of weed management methods, including herbicide use, mechanical removal, and biocontrol, suggest that controlling weeds alone often does not lead to full ecosystem recovery. This failure may stem from not addressing the underlying type of system degradation (Kerns et al., 2020). Similarly, efforts to remove invasive aquatic plants in riparian systems are temporary fixes unless underlying issues, such as nutrient loading, are corrected. In these cases, active restoration may be necessary after invasive species control. It could involve reintroducing desired species or utilizing a predictable successional trajectory to achieve the desired plant community gradually. The more the site is degraded, the more likely the latter approach will be required. Before beginning the active restoration, assess whether passive recovery, spontaneous regeneration from the native seed bank, may occur once the invader is removed (Raghurama & Sankaran, 2021).

In grasslands dominated by annual grasses in the USA, restoration efforts focus on building resilience to wildfire and invasive species. Resistance to invasion and resilience after disturbance in these semi-arid ecosystems are closely tied to soil moisture and temperature regimes. For example, areas with low resistance and resilience to cheatgrass (Bromus tectorum) may benefit from wildfire suppression to prevent cheatgrass dominance following fires. A significant challenge in restoring innate plant communities is secondary invasion by other invasive species once the primary target invader has been controlled (Arasumani, Bunyan, & Robin, 2021).

Recovery is moreover likely if secondary invasions are anticipated and actively managed alongside the primary invasive species. In some cases, restoration may also require addressing changes to soil conditions caused by invaders. These changes can include shifts in soil pH, nutrient levels, microbial composition, or the addition of allelopathic compounds-alterations that may inhibit native plant recovery while promoting further invasion by non-native species. Amending degraded soils is especially critical when soil changes trigger threshold responses (Gaskin et al., 2021). Overgrazing by herbivores and unsustainable forestry practices have depleted native seed banks in some areas, reducing resilience to disturbances and increasing the risk of secondary invasions. In these cases, gaps left by invasive species must be quickly filled with native plants or non-invasive external species. Establishing a diverse native plant community can enhance resilience to reinvasion, but it is worth noting that species-rich communities in resource-rich environments are often moreover vulnerable to invasion. Therefore, increasing native species richness may not prevent further invasions (Miniat et al., 2021).

Moreover, an effective approach may involve planting aggressive, early-successional native species capable of outcompeting invasive plants, especially in large numbers. Early, mid, and late successional species can be introduced to guide the system along a predictable successional trajectory. Succession ecology offers valuable strategies for quickly filling gaps left by invaders and steering the system toward a stable, native-dominated state. However, achieving this goal requires addressing the underlying drivers of degradation. In highly degraded sites, it may be necessary to accept “arrested novel communities” as new, stable ecosystems dominated by non-native species. These communities can still increase innate species richness and provide specific environmental services (Török et al., 2021).

A common obstacle to restoration is the failure to reintroduce mid- and late-successional native species. Using plants derived from local, diverse seed sources generally raises the probability of successful establishment. The distance of the seed source from the restoration site should be carefully considered, and seeds should be chosen based on environmental gradients to enhance diversity and improve adaptability to degraded conditions or future climate changes. One method to support the survival of native species across varying environmental conditions is the development of provisional seed zones. These zones incorporate species-specific fitness data and can guide species selection for restoration efforts. Provisional seed zones have been developed for some species, but further testing is needed to evaluate their effectiveness, even in areas where they appear locally suitable (Buisson et al., 2022)

An accidental consequence of using commercially produced native seed mixes from broad environmental gradients is the introduction of non-site-specific genotypes. While this may increase genetic diversity at the restoration site, it can compromise the local genetic integrity of species and potentially reduce fitness. Even when restoration efforts appear to successfully reestablish the desired plant species composition, significant gaps remain in understanding how ecosystem processes, such as nutrient cycling and trophic interactions, are restored (Inácio et al., 2022).

In many cases, current restoration practices do not fully recover the original level of ecosystem function. Globally, there is increasing use of payment for ecosystem services (PES) programs to fund restoration projects, including efforts in the Florida Everglades, Costa Rica, Madagascar, and China's initiative to convert sloping cropland into forests and pasture (Ghasemi, Charrahy, & González-García, 2023).

However, long-term sustainability remains a concern, as these programs often depend on continuous funding. Additionally, PES may prioritize certain ecosystem services over others, potentially leading to restoration projects that fail to address the full range of ecosystem functions previously provided at a site (Hardy et al., 2024).

5. RESTORATION STRATEGIES OF INVASIVE SPECIES

Invasive plant species are managed and controlled using various techniques, such as mechanical removal, chemical control, and biological control, using the invasive species' natural enemies. Prevention efforts include stricter regulations on importing and trading potentially invasive plants, early detection and rapid response programs, and manual and mechanical removal of existing invasive species (Jamil et al., 2022). It is efficiently controlled by mechanical tree cutting in combination with the administration of a non-selective herbicide like glyphosate in the case of small plants and through injectors in the case of big trees (Waheed et al., 2023) (**Figure 2**).

Figure 2. Flow chart showing different restoration approaches and challenges under changing climate

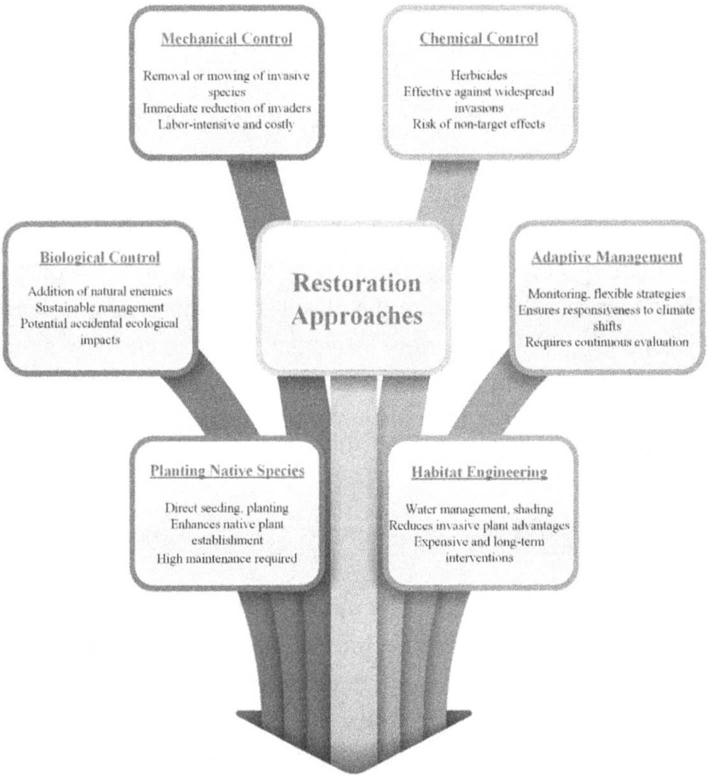

5.1 Mechanical Control Approaches

Mechanical techniques are usually the first choice when organizing an intrusive plant removal project. Since these methods don't require a specific license, it is best to avoid using chemicals in the surrounding areas. In some cases, permits, such as those involving wetland development, may be necessary. Mechanical clearance works well, but if it's not managed correctly, it can cause much harm to the site and be very labor-intensive, which could lead to fast reinvasion. Common (Beaury et al., 2020) techniques are pulling and excavating. If found in tiny amounts, several herbaceous plants and some woody species can grow up to around one inch in diameter. Eliminating the root system as much as possible is crucial since even a small amount left behind might start the infection. Tools can trim off portions of the root system, allowing for regeneration, but you can pull plants by hand or using a digging fork. Larger woody stems (roughly three inches in diameter) call for instruments like a weed wrench, root jack, or root talon. Different companies manufacture these instruments to remove the aboveground sections of the plant and its entire root system. The optimal season for this management is spring or early summer since the soil is usually moist, and removing plants is straightforward (Kaufman & Kaufman, 2023).

5.1.1 Suffocation:

Try drowning tiny seedlings and herbaceous plants. Cover the area with two or three layers of strong UV-stabilized plastic sheeting, either black or transparent, then fasten it with masses. Make sure the plastic spans both sides at least five feet beyond the boundary of the infestation. Leave the plastic in place for a minimum of two years. This approach will kill all (invasive and non-invasive) vegetation under the plastic. Remove the plastic, then plant a cover crop like annual rye to stop fresh incursions (Zolper et al., 2022).

5.1.2 Cutting or mowing

This approach is most appropriate for sites you can routinely visit and treat. For up to five years, you must mow or trim infested areas three or four times a year to be effective. Removing as much leafy material as possible will help prevent the plant from photosynthesizing (Reis et al., 2021). Ground-level plant-cutting will help eliminate all the resultant waste from the site. This treatment may cause the infestation to worsen first; hence, you will have to be as relentless as the invading plants themselves. The root system gets somewhat bigger every time you cut the plants back, but it depends on its energy stores to drive fresh development. You will run out of these stores, and the plants will perish. It could take many years; thus, once you start, you must be dedicated to this procedure; otherwise, the treatment may backfire and aggravate the situation (Knapp et al., 2023).

5.2 Chemical Control Approaches

Among the most successful and economical tools for controlling invading species are herbicides. Usually, only two herbicides—glyphosate (the active ingredient in RoundupTM and RodeoTM) and triclopyr (the active ingredient in Brush-BGoneTM and GarlonTM) can be used to treat the most often known invading plants. Glyphosate, being non-selective, kills any plant material it comes into contact with. Conversely, Triclopyr is selective and does not damage lilies, grasses, orchids, or monocots. For environmental and human safety, closely observe instructions and pay particular attention to labeling (Kenis et al., 2022). While these herbicides are relatively benign, their improper application can lead to short and long-term environmental problems and health concerns. Applications in wetland conditions call for unique aquatic formulations. Applying these chemicals to land you do not own also requires a state-issued pesticide applicator license. Typically, this license is issued by the Department of Agriculture. You can contact your state's pesticide control division to obtain further information about pesticide regulations in your area. Generally speaking, the Wetlands Protection Act calls for extra permits in wetland zones (Kaufman & Kaufman, 2023).

Various other herbicides are being proposed, such as Chwastox, Buctril Super, Atrazil 38% SC, Ametryn + Atrazine 80WP, Bromoxynil + MCPA 40EC, Butachlor 60EC, and metribuzin, specifically targeting Parthenium weed. Organic pesticides that could control parthenium grass include allelopathic aqueous extracts from *Desmostachya bipinnata, Dicanthium annulatum, Cenchrus pennisetorum*, and *Sorghum halepense*. Aqueous leaf extracts from allelopathic trees like *Azadirachta indica, Ficus bengalensis, Melia azedarach, Mangifera indica*, and *Syzygium cumini* may also help prevent *Parthenium* weed seedling germination and growth (Shabbir et al., 2023). The aqueous extract of *Tagetes erecta's* root, stalk, and flower has anti-inflammatory properties. In contrast, the leaf and bark extract from *Alstonia scholaris*, along with aqueous extracts from the roots and shoots of *Datura metel*, shoots of *Withania somnifera*, and leaves of *Azadirachta indica, Ficus bengalensis, Melia azedarach*, and *Syzygium cumini*, along with metabolites from *Trichoderma spp*., can be exploited for managing *Parthenium* weed (Peng et al., 2024).

5.2.1 Foliar applications

This approach typically utilizes a tiny handheld spray bottle or a backpack sprayer when dealing with minor difficulties. It either permits spot treatment of specific plants challenging to remove manually, such as goutweed, swallow-wort, or purple loosestrife, or targets vast monocultures of herbaceous plants. Growing in dense masses across large regions, some woody species—including Japanese barberry, multiflora rose, Japanese honeysuckle, and Oriental bittersweet—can also be effectively targeted (Ilyas et al., 2022; Manzoor et al., 2023).

The herbicide combination should have no moreover than five percent active ingredient; hence, following product label directions is essential. This treatment is most effective while the plants are actively growing-ideally during flowering or when they begin to develop fruit. Studies reveal that, notably for Japanese knotweed, plants frequently respond better if current stems are removed before treating the regrowth. On a day when rain is not expected during the next 24 to 48 hours, make sure the target plants are wholly soaked with herbicide (de Castro et al., 2024).

5.2.2 Cut stem treatments

This kind of treatment calls for moreover active component concentration than foliar treatments. Cut stem treatments should use a solution containing 25 to 35 percent active ingredient; however, reading and adhering to all label instructions remains crucial. Except for Oriental bittersweet, which requires the use of triclopyr, glyphosate is typically the most effective herbicide (Huebner & Jones, 2022). This approach works with all woody stems, phragmites, and Japanese knotweed. Treatments for woody stems work best in late summer and autumn, especially between late August and November. You should chop the stems close to the ground, but not so low that it becomes difficult to locate them. Apply the herbicide immediately after cutting the stems. Delaying the application will less effectively treat the condition. One can apply the herbicide via a sponge, paintbrush, or spray bottle (Kaufman & Kaufman, 2023).

5.3 Biological Control Methods

A traditional biological control strategy has often been used, which involves releasing native natural enemies as biological control agents after assessing their host range. However, some agents have occasionally been found to be present due to unintended or accidental introductions. In Pakistan, few biological control methods are available, and no agents have been intentionally introduced to manage Parthenium. *Zygogramma bicolorata* Pallister, the Mexican or Parthenium leaf beetle, was discovered in Pakistan in 2006 and has since spread widely there. This beetle was deliberately introduced from Mexico to India in 1983 (Pratt et al., 2021). Although biological controls are still emerging, they are becoming increasingly important as a control method. Currently, the only widely applied biocontrol method relates to purple loosestrife (Newsome et al., 2024).

6. POLICIES FOR MANAGING INVASIVE SPECIES

Policy issues differ from environmental issues in that invasive species have at least six obstacles.

1) First, as populations increase and develop, the effects of invading species usually grow with time. On the other hand, discontinuing the activity often reduces the severity of different environmental problems, such as the discharge of chemical pollutants (Reaser, Simpson, et al., 2020). Thus, fast-removing new populations are the best way to control invading species. But often, until a species has expand-

ed and become invasive, the biological data required to justify eradication and the tools and political will to act are not readily available. By then, population control becomes considerably more challenging. By that time, controlling the population becomes much moreover difficult. The resources required and the risk of unintended effects on other parts of the ecosystem are usually too great (Thompson, Olden, & Converse, 2021).

2) Second, many environmental issues, such as forestry or mineral exploration, can be managed without much concern for the policies of a neighboring region. However, invasive species can easily cross borders if a suitable habitat exists nearby. Preventing invasions, therefore, requires coordination with neighboring jurisdictions. For example, as European economies have become increasingly integrated, border checks that once slowed the spread of invasive species are now less common (Venette et al., 2021).
3) Third, invasive species are often transported unintentionally through valuable trade activities. For instance, large ships carry ballast water when not fully loaded with cargo. This ballast water, taken on at one port and any organisms present, is discharged at a different port as cargo is loaded. The unintentional movement of organisms through ballast tanks is now the leading cause of marine invasions and a significant factor in freshwater invasions (García-Díaz et al., 2021). Addressing this issue would require restricting ship movements or installing expensive ballast water treatment systems on all ships. Both options would increase shipping costs, making managing invasive species even more complicated due to trade-offs with other priorities (Ziska, 2022).
4) Fourth, forecasting the precise circumstances causing an invasion is often impossible. Risky events like gathering ballast water in one port and discharging it into another could usually happen without causing an invasion until one happens. For example, because the Laurentian Great Lakes are connected to the sea by land and sea, zebra mussels (*Dreissena polymorpha*) may have been present in those areas for over a hundred years before they came to the lakes (Olden et al., 2022). Furthermore, the mussels' fit for the Great Lakes' circumstances was well-known. Still unclear, though, is why the invasion took so long and why it occurred where it did. This uncertainty makes it harder for managers to make choices because they have to weigh the risks of unpredictable invasions against economic activities that pay off right away (García-Díaz et al., 2022).
5) Fifth, controlling the spread of invading species usually calls for global cooperation, which might prove difficult. The International Maritime Organization (IMO) enacted the International Convention for the Control and Management of Ships Ballast Water and Sediments for the threat of ballast water invasions in February 2004 (Chung et al., 2022). This agreement calls for ships to adopt ballast water management technologies to stop the spread of live entities. Ship

size and age determined the adoption timelines for these systems; deadlines were established between 2009 and 2016. Although the convention is almost finished, and a consensus exists on its ability to lower invasion rates, it has not been implemented because too few nations have signed it. Though such strategies are hardly discussed or carried out, other possible invasion routes could benefit from international accords (Meinard et al., 2022; Ziska, 2022).

6) At last, there are often notable time gaps spanning centuries between the introduction of non-native species and their dissemination. For terrestrial plants, these lags are shockingly long. According to a recent study, "invasion debt," where human activity from decades or even centuries before has already set off invasions not yet seen is caused by Though many species were imported during the latter half of the 20th century, the study revealed that for many plant and animal groups, socioeconomic conditions in Europe around 1900 were moreover closely linked to present levels of invasion than conditions in 2000. This implies that today's operations could prepare the ground for subsequent invasions (Wehi et al., 2023).

7. BIOECONOMIC APPROACHES TO POLICY AND MANAGEMENT OF INVASIVE SPECIES

Over the past few decades, ecological studies on species invasions have become increasingly plentiful and sophisticated. This has helped determine the number of species that have evolved outside their natural habitats and their effects. Acknowledging these effects has increased demands for more robust policies and management (Faria et al., 2023). Many countries have rules to lessen the impact of invasive species. However, the fact that invaders are still coming in faster shows that policy and management changes based on ecological studies are not working well enough. Apart from developments in ecology, economists have also focused on the problem of invasive species and the trade-offs required for their control. Economists might investigate, for instance, the advantages of a commerce route and the invasions it helps to enable. They can then evaluate potential changes or limitations to the commerce route, aiming to minimize invasion risks and enhance societal welfare (Lieurance et al., 2023). Though they regularly handle related issues, such as reducing the effects of invasive species, ecologists and economists have hardly worked together. This results in part from the fields' misunderstandings. Ecologists could object to giving environmental elements monetary value. Simultaneously, even ecologists may struggle to comprehend intricate ecological systems, which may not align with conventional economic theories. As said, bringing non-native species is intimately related to trade and travel globalization (Kearney et al., 2023).

Globalization has many advantages, as do many non-native species exploited in agriculture, horticulture, forestry, fisheries, landscaping, and the pet trade. Policies meant to avoid and control invasions must thus strike a balance between these advantages and the expenses of restrictions, including possible trade restrictions. To reach this balance, ecologists and economists need to work together to make sensible rules, and people need to be very aware of how invasions affect both the environment and the economy (Finley, Dovciak, & Dean, 2023).

General principles for policy and management suggest that preventing introductions is the most effective way to reduce the overall impacts of invasive species. Studies demonstrate that stopping invaders' spread between lakes or across countries via trade is often moreover cost-effective than managing established species after their arrival. When complete prevention is not feasible, the goal should shift to reducing propagule pressure, which can help prevent some invasions. Prevention is moreover effective than post-establishment management for several reasons (Bradley et al., 2023).

First, it is the only sure approach to prevent the effects of invasive species; once established, they are hardly eliminated. Second, management options are available before an invasion occurs, such as laws, regulations, quarantine measures, and punitive actions for non-compliance. Finally, prevention allows the costs of reducing impacts to be borne by those who benefit most from the species trade. For instance, while the pet trade has introduced many harmful invaders, governments and society typically shouldered the costs of controlling these invaders rather than the businesses that profit from the trade (Yadav et al., 2024). However, regulations can be implemented before introductions to ensure that trades do not transport potential invaders, thereby internalizing the external costs of trade. Once a species is established, the fantastic range of invading species, the ecosystems they influence, and the various ways humans value those ecosystems make it difficult to provide uniform solutions for management and policy (Zenni et al., 2024).

Reasonable control of damaging invading species depends much on the surroundings. Among the options are hand removal, that is, removing plants, trapping or killing, shooting, and releasing other species for biological control, such as poisons, herbicides, pesticides, and piscicides. The particular use of these techniques depends on the situation, so they are not investigated much here since different European nations have different criteria for applying them. Still, preventive initiatives to halt or stop population growth are the most affordable ways to reduce overall effects (de Carvalho-Souza et al., 2024).

CONCLUSIONS

In ecosystems, invasive species are becoming a significant threat. Intentionally or inadvertently brought into the nation, these non-native species endanger the natural biodiversity and ecological equilibrium. This analysis clarified the effects of invasive species in Pakistan by highlighting a few high-impact invaders that disrupt local ecosystems and displace native flora, like *Broussonetia papyrifera* and *Parthenium hysterophorus*. Chemical control, mechanical removal, and applying organic pesticides are strategies employed in the fight against invasive species. Introducing biological control agents has demonstrated potential in managing these invaders, such as the Mexican parthenium leaf beetle. Invading species need to be dealt with right away. Ongoing research and monitoring are necessary to comprehend their impact and create management plans that work fully. Doing this can lessen these invaders' ecological risks and protect Pakistan's distinctive ecosystems and biodiversity for generations.

Future Perspectives

1. Looking ahead, the successful restoration and management of native habitats under the pressure of invasive plants and climate change will rely on proactive, integrated, and collaborative approaches. Prioritizing early detection and rapid response (EDRR) systems is essential to prevent the establishment and spread of new invasive species. Continued research into plant-plant, plant-microbe, and plant-pathogen interactions will provide deeper insights for tailoring management and restoration strategies to changing ecological contexts. The adoption of climate-smart restoration techniques should be expanded, emphasizing the use of locally adapted, genetically diverse, and climate-resilient native species to enhance ecosystem stability and resilience.
2. Policy development must keep pace with emerging threats, enforcing strict regulations on the introduction and movement of potentially invasive species through coordinated international and regional efforts. Innovative, sustainable biotechnological solutions—including biological control agents and eco-friendly agro-technologies—should be further explored to minimize ecological disruption and support long-term habitat recovery. Additionally, fostering strong stakeholder collaboration—engaging local communities, government agencies, scientists, and policymakers—will be vital for the success of restoration projects.
3. Ultimately, adaptive management supported by ongoing monitoring, knowledge sharing, and capacity building will strengthen our ability to conserve biodiversity, safeguard ecosystem services, and enhance ecosystem resilience in the face of invasions and a rapidly changing climate.

REFERENCES

Arasumani, M., Bunyan, M., & Robin, V. (2021). Opportunities and challenges in using remote sensing for invasive tree species management, and in the identification of restoration sites in tropical montane grasslands. *Journal of Environmental Management, 280*, 111759. DOI: 10.1016/j.jenvman.2020.111759 PMID: 33298397

Areces-Berazain, F. (2023). Pennisetum setaceum (fountain grass).

Balcha, L. D. (2022). Prosopis juliflora Distribution, Impacts, and Control Methods Available in Ethiopia. *Int. J. Nat. Resour. Ecol. Manag, 7*, 132–144.

Beaury, E. M., Fusco, E. J., Jackson, M. R., Laginhas, B. B., Moreoverlli, T. L., Allen, J. M., Pasquarella, V. J., & Bradley, B. A. (2020). Incorporating climate change into invasive species management: Insights from managers. *Biological Invasions, 22*(2), 233–252. DOI: 10.1007/s10530-019-02087-6

Bertelsmeier, C. (2021). Globalization and the anthropogenic spread of invasive social insects. *Current Opinion in Insect Science, 46*, 16–23. DOI: 10.1016/j.cois.2021.01.006 PMID: 33545436

Bilgili, M., Tumse, S., & Nar, S. (2024). Comprehensive Overview on the Present State and Evolution of Global Warming, Climate Change, Greenhouse Gasses and Renewable Energy. *Arabian Journal for Science and Engineering, 49*(11), 14503–14531. DOI: 10.1007/s13369-024-09390-y

Bologo, N. (2023). *Impact of the invasive alien plant species clearing programme on socio-economic benefits and plant biodiversity along the northern Drakensberg*. Mpumalanga Province.

Borden, J. B., & Flory, S. L. (2021). Urban evolution of invasive species. *Frontiers in Ecology and the Environment, 19*(3), 184–191. DOI: 10.1002/fee.2295

Bradley, B. A., Beaury, E. M., Fusco, E. J., & Lopez, B. E. (2023). Invasive species policy must embrace a changing climate. *Bioscience, 73*(2), 124–133. DOI: 10.1093/biosci/biac097

Brodie, J. F., Lieberman, S., Moehrenschlager, A., Redford, K. H., Rodríguez, J. P., Schwartz, M., Seddon, P. J., & Watson, J. E. (2021). Global policy for assisted colonization of species. *Science, 372*(6541), 456–458. DOI: 10.1126/science.abg0532 PMID: 33926936

Buisson, E., Archibald, S., Fidelis, A., & Suding, K. N. (2022). Ancient grasslands guide ambitious goals in grassland restoration. *Science, 377*(6606), 594–598. DOI: 10.1126/science.abo4605 PMID: 35926035

Chung, H. I., Choi, Y., Yoo, Y., Engler, R., Lee, K., & Jeon, S. W. (2022). Integrated spatial model based evaluation methodology for optimal invasive species management: Common ragweed in the Republic of Korea. *Environmental Research Letters, 17*(3), 034047. DOI: 10.1088/1748-9326/ac4dc7

Coakley, S., & Petti, C. (2021). Impacts of the invasive Impatiens glandulifera: Lessons learned from one of Europe's top invasive species. *Biology (Basel), 10*(7), 619. DOI: 10.3390/biology10070619 PMID: 34356476

Costello, K. E., Lynch, S. A., McAllen, R., O'Riordan, R. M., & Culloty, S. C. (2022). Assessing the potential for invasive species introductions and secondary spread using vessel movements in maritime ports. *Marine Pollution Bulletin, 177*, 113496. DOI: 10.1016/j.marpolbul.2022.113496 PMID: 35272109

de Carvalho-Souza, G. F., Kourantidou, M., Laiz, I., Nuñez, M. A., & González-Ortegón, E. (2024). How to deal with invasive species that have high economic value? *Biological Conservation, 292*, 110548. DOI: 10.1016/j.biocon.2024.110548

de Castro, A. F., Boanares, D., de Medeiros Sarmento, P. S., Caldeira, C. F., Michelan, T. S., Ramos, S. J., & Gastauer, M. (2024). Fertilization during mineland rehabilitation may shift competitive outcomes toward invasive species. *Restoration Ecology, 32*(4), e14108. DOI: 10.1111/rec.14108

Dindaroglu, T. (2024). Ecosystemic Alienation from the Perspective of Paraecology. *SilvaWorld, 3*(1), 28–43. DOI: 10.61326/silvaworld.v3i1.250

Faria, L., de Carvalho, B. M., Carneiro, L., Miiller, N. O. R., Pedroso, C. R., Occhi, T. V. T., Tonella, L. H., & Vitule, J. R. S. (2023). Invasive species policy in Brazil: A review and critical analysis. *Environmental Conservation, 50*(1), 67–72. DOI: 10.1017/S0376892922000406

Finley, D., Dovciak, M., & Dean, J. (2023). A data driven method for prioritizing invasive species to aid policy and management. *Biological Invasions, 25*(7), 2293–2307. DOI: 10.1007/s10530-023-03041-3

Gaffke, A. M., Dudley, T. L., Bean, D. W., Drus, G. M., Johnson, M. J., Knutson, A. E., Weaver, D. K., Sing, S. E., Orr, B. K., & Thompson10, D. C. (2022). Tamarix Biological Control in North America. *CLASSICAL BIOLOGICAL CONTROL*, 329.

Gao, L., Huang, J., Chen, X., Chen, Y., & Liu, M. (2018). Contributions of natural climate changes and human activities to the trend of extreme precipitation. *Atmospheric Research, 205*, 60–69. DOI: 10.1016/j.atmosres.2018.02.006

García-Díaz, P., Cassey, P., Norbury, G., Lambin, X., Montti, L., Pizarro, J. C., Powell, P. A., Burslem, D. F., Cava, M., Damasceno, G., Fasola, L., Fidelis, A., Huerta, M. F., Langdon, B., Linardaki, E., Moyano, J., Núñez, M. A., Pauchard, A., Phimister, E., & Tomasevic, J. A. (2021). Management policies for invasive alien species: Addressing the impacts rather than the species. *Bioscience, 71*(2), 174–185. DOI: 10.1093/biosci/biaa139

García-Díaz, P., Montti, L., Powell, P. A., Phimister, E., Pizarro, J. C., Fasola, L., Langdon, B., Pauchard, A., Raffo, E., Bastías, J., Damasceno, G., Fidelis, A., Huerta, M. F., Linardaki, E., Moyano, J., Núñez, M. A., Ortiz, M. I., Rodríguez-Jorquera, I., Roesler, I., & Lambin, X. (2022). Identifying priorities, targets, and actions for the long-term social and ecological management of invasive non-native species. *Environmental Management, 69*(1), 140–153. DOI: 10.1007/s00267-021-01541-3 PMID: 34586487

Gaskin, J. F., Espeland, E., Johnson, C. D., Larson, D. L., Mangold, J. M., McGee, R. A., Milner, C., Paudel, S., Pearson, D. E., Perkins, L. B., Prosser, C. W., Runyon, J. B., Sing, S. E., Sylvain, Z. A., Symstad, A. J., & Tekiela, D. R. (2021). Managing invasive plants on Great Plains grasslands: A discussion of current challenges. *Rangeland Ecology and Management, 78*, 235–249. DOI: 10.1016/j.rama.2020.04.003

Genes, L., & Dirzo, R. (2022). Restoration of plant-animal interactions in terrestrial ecosystems. *Biological Conservation, 265*, 109393. DOI: 10.1016/j.biocon.2021.109393

Gerwing, T. G., Hawkes, V. C., Gann, G. D., & Murphy, S. D. (2022). Restoration, reclamation, and rehabilitation: On the need for, and positing a definition of, ecological reclamation. *Restoration Ecology, 30*(7), e13461. DOI: 10.1111/rec.13461

Ghasemi, M., Charrahy, Z., & González-García, A. (2023). Mapping cultural ecosystem services provision: An integrated model of recreation and ecotourism opportunities. *Land Use Policy, 131*, 106732. DOI: 10.1016/j.landusepol.2023.106732

Hardy, N. G., Kuebbing, S. E., Duguid, M. C., Ashton, M. S., Sheban, K. C., Inman, S. E., & Martin, M. P. (2024). Non-native invasive plants in tropical dry forests: A global review of presence, impacts, and management. *Restoration Ecology*, •••, 14288.

Huebner, C. D., & Jones, T. (2022). Invasive plants field & reference guide: an ecological perspective of plant invaders of forests & woodlands. *US Department of Agriculture, Forest Service, Northern Research Station and Eastern Region State & Private Forestry. R9-TP-001-22.*

Ilyas, M., Shah, S., Lai, Y.-W., Sher, J., Bai, T., Zaman, F., Bibi, F., Koul, M., Wani, S. H., Majrashi, A., Alharby, H. F., Hakeem, K. R., Wang, Y.-J., & Rather, S. A. (2022). Leaf functional traits of invasive grasses conferring high-cadmium adaptation over natives. *Frontiers in Plant Science, 13*, 869072. DOI: 10.3389/fpls.2022.869072 PMID: 35720536

Inácio, M., Gomes, E., Bogdzevič, K., Kalinauskas, M., Zhao, W., & Pereira, P. (2022). Mapping and assessing coastal recreation cultural ecosystem services supply, flow, and demand in Lithuania. *Journal of Environmental Management, 323*, 116175. DOI: 10.1016/j.jenvman.2022.116175 PMID: 36088764

Izhar, S. K., Rizvi, S. F., Siddiqui, S., & Afaq, U. (2023). Invasive weeds and their management in India. *Journal of Survey in Fisheries Sciences, 10*(2S), 2571–2596.

Jamil, M. D., Waheed, M., Akhtar, S., Bangash, N., Chaudhari, S. K., Majeed, M., Hussain, M., Ali, K., & Jones, D. A. (2022). Invasive plants diversity, ecological status, and distribution pattern in relation to edaphic factors in different habitat types of district Mandi Bahauddin, Punjab, Pakistan. *Sustainability (Basel), 14*(20), 13312. DOI: 10.3390/su142013312

Jehangir, S., Khan, S. M., Ejaz, U., Zahid, N., Rashid, N., Noshad, Q., Din, Z. U., & Shoukat, A. (2024). Alien flora of Pakistan: Taxonomic composition, invasion status, geographic origin, introduction pathways, and ecological patterns. *Biological Invasions, 26*(8), 1–17. DOI: 10.1007/s10530-024-03311-8

Jones, A. D., Rastogi, D., Vahmani, P., Stansfield, A. M., Reed, K. A., Thurber, T., Ullrich, P. A., & Rice, J. S. (2023). Continental United States climate projections based on thermodynamic modification of historical weather. *Scientific Data, 10*(1), 664. DOI: 10.1038/s41597-023-02485-5 PMID: 37770463

Kaufman, S. R., & Kaufman, W. (2023). *Invasive plants: guide to identification and the impacts and control of common North American species*. Rowman & Littlefield.

Kearney, S. G., Watson, J. E., Reside, A. E., Fisher, D. O., Maron, M., Doherty, T. S., Legge, S. M., Woinarski, J. C., Garnett, S. T., Wintle, B. A., Ritchie, E. G., Driscoll, D. A., Lindenmayer, D., Adams, V. M., Ward, M. S., & Carwardine, J. (2023). Threat-abatement framework confirms habitat retention and invasive species management are critical to conserve Australia's threatened species. *Biological Conservation, 277*, 109833. DOI: 10.1016/j.biocon.2022.109833

Kenis, M., Benelli, G., Biondi, A., Calatayud, P.-A., Day, R., Desneux, N., Harrison, R. D., Kriticos, D., Rwomushana, I., & Van den Berg, J. (2022). Invasiveness, biology, ecology, and management of the fall armyworm, Spodoptera frugiperda.

Kerns, B. K., Tortorelli, C., Day, M. A., Nietupski, T., Barros, A. M., Kim, J. B., & Krawchuk, M. A. (2020). Invasive grasses: A new perfect storm for forested ecosystems? *Forest Ecology and Management*, *463*, 117985. DOI: 10.1016/j.foreco.2020.117985

Khalid, N., Naveed, S., & Khan, I. U. (2023). 47. Allelopathic effect of Tamarix aphylla on the emergence and growth parameters of Triticum aestivum and Zea mays. [PAB]. *Pure and Applied Biology*, *12*(1), 540–554. DOI: 10.19045/bspab.2023.120056

Khan, I., Ibrahim, M., Khan, M. I., Hashim, S., Afzal, S., & Nawab, K. (2023). Management of Wild Onion (Asphodelus tenuifolius Cav.) in Chickpea Crop at District Karak. *Pakistan Journal of Weed Science Research*, *29*(3).

Khan, N., Ullah, R., Okla, M. K., Abdel-Maksoud, M. A., Saleh, I. A., Abu-Harirah, H. A., AlRamadneh, T. N., & AbdElgawad, H. (2023). Spatial distribution of the four invasive plants and their impact on natural communities' dynamics across the arid and semi-arid environments in northwest Pakistan. *Frontiers in Plant Science*, *14*, 1207222. DOI: 10.3389/fpls.2023.1207222 PMID: 37692447

Khattak, W. A., Sun, J., Hameed, R., Zaman, F., Abbas, A., Khan, K. A., Elboughdiri, N., Akbar, R., He, F., Ullah, M. W., Al-Andal, A., & Du, D. (2024). Unveiling the resistance of native weed communities: Insights for managing invasive weed species in disturbed environments. *Biological Reviews of the Cambridge Philosophical Society*, *99*(3), 753–777. DOI: 10.1111/brv.13043 PMID: 38174626

Knapp, L. S. P., Coyle, D. R., Dey, D. C., Fraser, J. S., Hutchinson, T., Jenkins, M. A., Kern, C. C., Knapp, B. O., Maddox, D., & Pinchot, C. (2023). Invasive plant management in eastern North American Forests: A systematic review. *Forest Ecology and Management*, *550*, 121517. DOI: 10.1016/j.foreco.2023.121517

Kotta, J. (2023). *Biological Invasions in a Changing World: Introduction to the Special Issue* (Vol. 15). MDPI.

Kumar, N., & Aggarwal, N. K. (2024). A review on valorization, management, and applications of the hazardous weed Parthenium hysterophorus. *Systems Microbiology and Biomanufacturing*, *4*(2), 607–619. DOI: 10.1007/s43393-023-00226-8

Kumar, R., Guleria, N., Deeksha, M. G., Kumari, N., Kumar, R., Jha, A. K., Parmar, N., Ganguly, P., de Aguiar Andrade, E. H., Ferreira, O. O., de Oliveira, M. S., & Chandini, . (2024). From an Invasive Weed to an Insecticidal Agent: Exploring the Potential of Lantana camara in Insect Management Strategies—A Review. *International Journal of Molecular Sciences*, *25*(23), 12788. DOI: 10.3390/ijms252312788 PMID: 39684506

Lafaie, L., Poenou, G., Hanon, O., López-Jiménez, L., Nieto, J. A., Lorenzo, A., Porras, J. A., Lumbierres, M., Bertoletti, L., & Monreal, M. (2024). Anticoagulation and venous thromboembolism in patients aged 90 years and older: Data from the RIETE registry. *Journal of the American Geriatrics Society*, *72*(1), 113–125. DOI: 10.1111/jgs.18626 PMID: 37814983

Li, S., Jiang, Y., Xu, J., Wang, D., Ding, Z., Zhu, T., Chen, B., Yang, Y., Wei, M., Guo, R., Hou, Y., Chen, Y., Sun, C., Wei, K., Qaid, S. M. H., Lu, H., Tan, H., Di, D., Chen, J., & Yuan, M. (2024). High-efficiency and thermally stable FACsPbI3 perovskite photovoltaics. *Nature*, *635*(8037), 1–3. DOI: 10.1038/s41586-024-08103-7 PMID: 39348872

Lieurance, D., Canavan, S., Behringer, D. C., Kendig, A. E., Minteer, C. R., Reisinger, L. S., Romagosa, C. M., Flory, S. L., Lockwood, J. L., Anderson, P. J., Baker, S. M., Bojko, J., Bowers, K. E., Canavan, K., Carruthers, K., Daniel, W. M., Gordon, D. R., Hill, J. E., Howeth, J. G., & Wanamaker, C. (2023). Identifying invasive species threats, pathways, and impacts to improve biosecurity. *Ecosphere*, *14*(12), e4711. DOI: 10.1002/ecs2.4711

Lin, J., Zhang, B., Zou, J., Luo, Z., Yang, H., Zhou, P., Chen, X., & Zhou, W. (2023). Induction of tetraploids in Paper Mulberry (Broussonetia papyrifera (L.) L'Hér. ex Vent.) by colchicine. *BMC Plant Biology*, *23*(1), 574. DOI: 10.1186/s12870-023-04487-2 PMID: 37978431

Macedo, M. I., Ferreira, F. A., Dabić, M., & Ferreira, N. C. (2024). Structuring and analyzing initiatives that facilitate organizational transformation processes: A sociotechnical approach. *Technological Forecasting and Social Change*, *209*, 123739. DOI: 10.1016/j.techfore.2024.123739

Manzoor, R., Zafar, M., Yaqoob, T., Ahmad, M., Ramadan, M. F., Althobaiti, A. T., Demirpolat, A., Çobanoğlu, D. N., Sultana, S., Makhkamov, T., Mamarakhimov, O., Yuldashev, A., Khakimova, D., Nizomova, M., Ochilov, U., & Majeed, S. (2023). Micromorphological sculptural diversity in foliar epidermis and trichomes features among invasive species. *Microscopy and Microanalysis*, *29*(4), 1531–1555. DOI: 10.1093/micmic/ozad063 PMID: 37488823

Meena, B., & Sharma, K. (2024). Biological Control of Weeds. *International Journal of Environment, Agriculture and Biotechnology*, *9*(4), 091–100. DOI: 10.22161/ijeab.94.13

Meinard, Y., Dereniowska, M., Philippot, V., Glatron, S., Maris, V., & Georges, J.-Y. (2022). A heuristic for innovative invasive species management actions and strategies. *Ecology and Society*, *27*(4), art24. DOI: 10.5751/ES-13615-270424

Meyer, S. E., Callaham, M. A., Stewart, J. E., & Warren, S. D. (2021). Invasive species response to natural and anthropogenic disturbance. *Invasive species in forests and rangelands of the United States: A comprehensive science synthesis for the United States forest sector*, 85-110.

Miniat, C. F., Fraterrigo, J. M., Brantley, S. T., Callaham, M. A., Cordell, S., Dukes, J. S., Giardina, C. P., Jose, S., & Lovett, G. (2021). Impacts of invasive species on forest and grassland ecosystem processes in the United States. *Invasive species in forests and rangelands of the United States: A comprehensive science synthesis for the United States forest sector*, 41-55.

Molvar, E. M., Rosentreter, R., Mansfield, D., & Anderson, G. M. (2024). Cheatgrass invasions: History, causes, consequences, and solutions.

Murphy, B. A., & Hall, J. A. (2024). How a strong measurement validity review can go astray: A look at and recommendations for future measurement-focused reviews. *Clinical Psychology Review*, *114*, 102506. DOI: 10.1016/j.cpr.2024.102506 PMID: 39615954

Newsome, T., Cairncross, R., Cunningham, C. X., Spencer, E. E., Barton, P. S., Ripple, W. J., & Wirsing, A. J. (2024). Scavenging with invasive species. *Biological Reviews of the Cambridge Philosophical Society*, *99*(2), 562–581. DOI: 10.1111/brv.13035 PMID: 38148253

Olden, J. D., Chen, K., García-Berthou, E., King, A. J., South, J., & Vitule, J. R. (2022). Invasive species in streams and rivers. In *Encyclopedia of Inland waters* (Vol. 2). DOI: 10.1016/B978-0-12-819166-8.00083-9

Peng, W., Ochilov, A., Mirzaev, K., Sharafetdinovich, T. J., Nizamov, A., Ergashev, E., Ergashev, T. K., Artikova, S., Babakulov, B., & Ergashev, T. (2024). Operational management science of sustainable brick manufacturing using fluoride-enriched marble slurry and parthenium weeds. *Fluoride*, *57*(8), 1–9.

Pratt, P. D., Moran, P. M., Pitcairn, M. J., Reddy, A., & O'Brien, J. (2021). Biological control of invasive plants in California's Delta: Past, present, and future. *Journal of Aquatic Plant Management*, *59*(s), 55–66. DOI: 10.57257/japm-20215955

Puspitaloka, D., Kim, Y. S., Purnomo, H., & Fulé, P. Z. (2020). Defining ecological restoration of peatlands in Central Kalimantan, Indonesia. *Restoration Ecology*, *28*(2), 435–446. DOI: 10.1111/rec.13097

Raghurama, M., & Sankaran, M. (2021). Restoring tropical forest–grassland mosaics invaded by woody exotics. *Restoration Ecology*, *29*(8), e13491. DOI: 10.1111/rec.13491

Raihan, A., & Bari, A. M. (2024). Energy-economy-environment nexus in China: The role of renewable energies toward carbon neutrality. *Innovation and Green Development*, *3*(3), 100139. DOI: 10.1016/j.igd.2024.100139

Reaser, J. K., Burgiel, S. W., Kirkey, J., Brantley, K. A., Veatch, S. D., & Burgos-Rodríguez, J. (2020). The early detection of and rapid response (EDRR) to invasive species: A conceptual framework and federal capacities assessment. *Biological Invasions*, *22*(1), 1–19. DOI: 10.1007/s10530-019-02156-w

Reaser, J. K., Simpson, A., Guala, G. F., Morisette, J. T., & Fuller, P. (2020). Envisioning a national invasive species information framework. *Biological Invasions*, *22*(1), 21–36. DOI: 10.1007/s10530-019-02141-3

Reis, B. P., Kövendi-Jakó, A., Szitár, K., Török, K., & Halassy, M. (2021). Long-term effect of mowing on the restoration of Pannonian sand grassland to replace invasive black locust plantation. *Restoration Ecology*, *29*(S1), e13152. DOI: 10.1111/rec.13152

Robichaud, C. D., Basso, J. V., & Rooney, R. C. (2022). Control of invasive Phragmites australis (European common reed) alters macroinvertebrate communities. *Restoration Ecology*, *30*(4), e13548. DOI: 10.1111/rec.13548

Shabbir, A., Zalucki, M. P., Dhileepan, K., Khan, N., & Adkins, S. W. (2023). The current and potential distribution of parthenium weed and its biological control agent in Pakistan. *Plants*, *12*(6), 1381. DOI: 10.3390/plants12061381 PMID: 36987069

Sharma, J., Singh, R., Garai, S., Rahaman, S. M., Khatun, M., Ranjan, A., Mishra, S. N., & Tiwari, S. (2022). Climate change and dispersion dynamics of the invasive plant species Chromolaena odorata and Lantana camara in parts of the central and eastern India. *Ecological Informatics*, *72*, 101824. DOI: 10.1016/j.ecoinf.2022.101824

Siller-Clavel, P., Badano, E. I., Villarreal-Guerrero, F., Prieto-Amparán, J. A., Pinedo-Alvarez, A., Corrales-Lerma, R., Álvarez-Holguín, A., & Hernández-Quiroz, N. S. (2022). Distribution patterns of invasive buffelgrass (Cenchrus ciliaris) in Mexico estimated with climate niche models under the current and future climate. *Plants*, *11*(9), 1160. DOI: 10.3390/plants11091160 PMID: 35567161

Smith, L. A. (2024). Wizards of Climate Science: Ancient Magicians in the Court of Big Science. *Zygon, 59*(1). Advance online publication. DOI: 10.16995/zygon.11699

Thompson, B. K., Olden, J. D., & Converse, S. J. (2021). Mechanistic invasive species management models and their application in conservation. *Conservation Science and Practice, 3*(11), e533. DOI: 10.1111/csp2.533

Tiawoun, M. A., Malan, P. W., Moshobane, M. C., Ramarumo, L. J., Comole, A. A., & Naidoo, K. K. (2024). Ecological traits and socio-economic impacts of the alien invader weed Parthenium hysterophorus L. in South Africa's rangeland ecosystems: A review. *Diversity, 16*(4), 205. DOI: 10.3390/d16040205

Török, P., Brudvig, L. A., Kollmann, J. N., Price, J., & Tóthmérész, B. (2021). The present and future of grassland restoration. *Restoration Ecology, 29*, e13378. DOI: 10.1111/rec.13378

Tsynkevych, S. (2018). Modeling the risks and damages from a "potential" invasive plant species: yellow starthistle in British Columbia.

Velez-Gavilan, J. (2024). *Sorghum Halepense (Johnson Grass)*. CABI Compendium. DOI: 10.1079/cabicompendium.50624

Venette, R. C., Gordon, D. R., Juzwik, J., Koch, F. H., Liebhold, A. M., Peterson, R. K., Sing, S. E., & Yemshanov, D. (2021). Early intervention strategies for invasive species management: connections between risk assessment, prevention efforts, eradication, and other rapid responses. *Invasive species in forests and rangelands of the United States: A comprehensive science synthesis for the United States forest sector*, 111-131.

Waheed, M., Haq, S. M., Arshad, F., Bussmann, R. W., Ali, H. M., & Siddiqui, M. H. (2023). Phyto-ecological distribution patterns and identification of alien invasive indicator species in relation to edaphic factors from semi-arid region. *Ecological Indicators, 148*, 110053. DOI: 10.1016/j.ecolind.2023.110053

Wehi, P. M., Kamelamela, K. L., Whyte, K., Watene, K., & Reo, N. (2023). Contribution of Indigenous Peoples' understandings and relational frameworks to invasive alien species management. *People and Nature, 5*(5), 1403–1414. DOI: 10.1002/pan3.10508

Weidlich, E. W., Flórido, F. G., Sorrini, T. B., & Brancalion, P. H. (2020). Controlling invasive plant species in ecological restoration: A global review. *Journal of Applied Ecology, 57*(9), 1806–1817. DOI: 10.1111/1365-2664.13656

Yadav, S. P. S., Mehata, D. K., Pokhrel, S., Ghimire, N. P., Gyawali, P., Katel, S., & Timilsina, U. (2024). Invasive alien plant species (Banmara): Investigating its invasive potential, ecological consequences on biodiversity, and management strategies. *Journal of Agriculture and Food Research*, *15*, 101031. DOI: 10.1016/j.jafr.2024.101031

Yang, F., Liu, M., Wang, X., Hong, Y., Yao, Q., Chang, X., Shi, G., Chen, W., Tian, B., & Hegazy, A. (2024). Differences in the Microbial Composition and Function of the Arundo donax Rhizosphere Under Different Cultivation Conditions. *Microorganisms*, *12*(12), 2642. DOI: 10.3390/microorganisms12122642 PMID: 39770844

Zenni, R. D., Ziller, S. R., da Rosa, C. A., Sühs, R. B., Puechagut, P. B., Marterer, B. T., Heringer, G., de Matos, A. C. L., Bueno, M. L., Almeida, J. W., Silva, C. H. T., Legracie-Júnior, J. R., & Chapla, T. E. (2024). Invasive non-native species in Brazil: An updated overview. *Biological Invasions*, *26*(8), 1–9. DOI: 10.1007/s10530-024-03302-9

Zhu, T. R., Litton, C. M., Giardina, C. P., & Trauernicht, C. (2021). Moisture availability and ecological restoration limit fine fuels and modelled wildfire intensity following non-native ungulate removal in Hawaii. *Journal of Applied Ecology*, *58*(10), 2207–2219. DOI: 10.1111/1365-2664.13952

Ziska, L. (2022). *Invasive species and global climate change*. CABI. DOI: 10.1079/9781800621459.0000

Zolper, T. J., Smith, D. L., Jackson, P. R., & Cupp, A. R. (2022). Performance of a carbon dioxide injection system at a navigation lock to control the spread of aquatic invasive species. *Journal of Environmental Engineering*, *148*(4), 04022011. DOI: 10.1061/(ASCE)EE.1943-7870.0001987

Chapter 4
Restoration Technique to Mitigate Plant Invasion

Kunal N. Odedra
https://orcid.org/0000-0002-5476-0342
M.D. Science College, Porbandar, India

Krishna B. Bhutiya
M.S. University, Vadodara, India

Nohil Kodiyatar
M.D. Science College, Porbandar, India

B. A. Jadeja
M.D. Science College, Porbandar, India

ABSTRACT

Plant invasions are a significant ecological challenge, threatening biodiversity, ecosystem functions, and human well-being. This chapter explores ecological restoration as a strategy to mitigate the impacts of invasive plant species. It discusses various restoration techniques, including mechanical, chemical, biological, cultural, and assisted regeneration methods, emphasizing their role in rebuilding native plant communities and restoring ecosystem functions. The chapter highlights the importance of integrating these methods into adaptive management frameworks, incorporating innovative approaches such as geospatial technologies and genetic engineering. Additionally, the role of stakeholder engagement and community-based initiatives in successful restoration is examined. By providing a comprehensive overview, this chapter aims to offer practical solutions for sustainable invasive species management and ecosystem recovery.

DOI: 10.4018/979-8-3693-8724-5.ch004

1. INTRODUCTION

Plant invasions represent a significant ecological challenge worldwide, affecting biodiversity, ecosystem function, and human well-being. The introduction of non-native species to new environments, whether intentionally or accidentally, can result in the establishment of invasive species that outcompete native flora, alter habitat structure, and disrupt ecosystem processes (Lockwood, Hoopes, & Marchetti, 2013). These invasions have been identified as one of the leading causes of biodiversity loss globally, alongside habitat destruction and climate change (Pyšek et al., 2020). As a consequence, plant invasions not only threaten the stability and resilience of ecosystems but also pose substantial economic and social costs due to their impact on agriculture, forestry, and water resources (Vilà et al., 2011).

The problem of invasive species is particularly acute in areas with high levels of biodiversity, such as islands, coastal regions, and biodiversity hotspots. These areas often have unique species that are highly specialized and vulnerable to the competitive pressures posed by invasive plants (Simberloff et al., 2013). For instance, the introduction of *Lantana camara* in India and Australia or *Pueraria lobata* in the United States has resulted in the widespread displacement of native species and significant alterations in ecosystem structure and function (Sharma et al., 2005; Forseth & Innis, 2004). Such invasions can modify fire regimes, soil chemistry, hydrological cycles, and nutrient availability, creating feedback loops that further facilitate their spread and hinder the recovery of native communities (D'Antonio & Vitousek, 1992).

Given these pervasive impacts, the management of invasive plant species has become a priority for conservationists, land managers, and policymakers. Traditional control measures such as mechanical removal, chemical treatments, and biological control have had varying degrees of success but are often limited by their high costs, unintended side effects, and the difficulty of complete eradication (Funk et. al., 2020). Moreover, these methods may not address the underlying ecological processes that facilitate invasions, such as disturbances, habitat fragmentation, or altered nutrient dynamics (Hobbs & Humphries, 1995). Consequently, there is a growing recognition of the need for a more holistic approach to invasive species management one that integrates ecological restoration principles to rebuild and maintain ecosystem resilience against invasions (Hulvey et al., 2013).

Ecological restoration, defined as the process of assisting the recovery of an ecosystem that has been degraded, damaged, or destroyed, offers a promising pathway for mitigating the impacts of invasive plant species (Society for Ecological Restoration, 2004). Restoration efforts aim to re-establish native plant communities, restore ecosystem functions, and enhance the resistance and resilience of ecosystems to future invasions (Funk et al., 2008). By focusing on restoring native biodiversity and ecological processes, restoration approaches can provide long-term, sustainable solutions to the challenges posed by invasive species (Clewell & Aronson, 2012).

This chapter aims to provide a comprehensive overview of restoration techniques that are used to mitigate plant invasions. It will explore the principles of ecological restoration, examine various methods for controlling invasive plants and restoring invaded ecosystems, and discuss the importance of integrated management strategies that combine traditional control methods with restoration practices. The chapter will also highlight key challenges and opportunities in restoration practice, emphasizing the need for adaptive management, stakeholder involvement, and policy support to enhance the effectiveness of restoration efforts in the face of ongoing invasions. In doing so, the chapter seeks to provide a practical framework for applying restoration techniques in diverse ecological contexts, offering valuable insights for practitioners, researchers, and policymakers engaged in the management of invasive species.

2. ECOLOGICAL IMPACT OF INVASIVE PLANTS

2.1 Effects on native biodiversity and ecosystem functions.

Invasive plants are recognized as a major threat to native biodiversity and ecosystem functions globally. They have the potential to alter the structure and function of ecosystems in various ways, leading to changes in species composition, nutrient cycling, and ecosystem productivity. The ecological impact of invasive plants is complex and multifaceted, often resulting in significant consequences for native flora, fauna, and the services ecosystems provide.

Invasive plants compete with native species for resources such as light, water, and nutrients. This competition often leads to the decline or local extinction of native plant species, thereby reducing species diversity. For example, studies have shown that invasive species like *Lantana camara* in India and *Fallopia japonica* (Japanese Knotweed) in Europe can outcompete native species by forming dense thickets that prevent sunlight from reaching the understory, leading to reduced regeneration of native plants (Sharma et al., 2005; Gerber et al., 2008).

Invasive plants can also alter ecosystem functions by changing nutrient cycling and soil properties. For instance, nitrogen-fixing invasive species, such as Acacia species in South Africa, can increase nitrogen levels in the soil, thereby disrupting the nutrient balance and favoring the growth of other invasive or non-native species adapted to higher nitrogen levels (Yelenik et al., 2004). This process, known as "Invasional Meltdown" can lead to a cascade of ecological changes, further exacerbating the decline of native biodiversity (Simberloff & Von Holle, 1999).

The impact of invasive plants extends beyond plant communities to affect entire ecosystems. Invasive plants can alter fire regimes by changing fuel loads and flammability. For example, the invasive grass *Bromus tectorum* (cheatgrass) in the western United States has been shown to increase the frequency and intensity of fires, which in turn prevents the recovery of native plant communities and promotes further invasion by fire-adapted species (D'Antonio & Vitousek, 1992). This alteration in fire regimes can have profound effects on ecosystem structure and function, as well as on the species that depend on these ecosystems.

Moreover, invasive plants can affect hydrological processes by altering water availability and quality. For example, Tamarix species (salt-cedar) in the southwestern United States consume large amounts of water compared to native vegetation, leading to reduced water availability in riparian ecosystems (Zavaleta, 2000). This not only affects plant communities but also has significant implications for wildlife that depend on these water resources.

In summary, the ecological impact of invasive plants is extensive and varied, affecting native biodiversity, ecosystem functions, and the provision of ecosystem services. The complex interactions between invasive species, native species, and ecosystem processes highlight the need for an integrated approach to managing invasive plants and restoring invaded ecosystems (Pyšek & Richardson, 2010). Effective restoration strategies must consider the specific ecological impacts of invasive plants in order to develop tailored management plans that promote the recovery of native biodiversity and ecosystem function.

2.2 Impacts on soil health, water availability, and nutrient cycling.

Invasive plants significantly alter soil health, water availability, and nutrient cycling in invaded ecosystems (Fig. 1). These impacts can result in profound changes to ecosystem structure and function, affecting both terrestrial and aquatic environments. Understanding these impacts is crucial for developing effective management and restoration strategies to mitigate the consequences of invasive plant species.

Figure 1. Ecological Impact of Invasive Plants

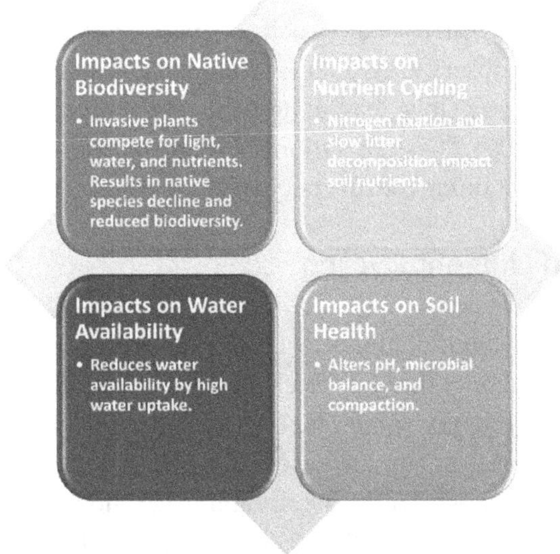

2.2.1 Impacts on Soil Health

Invasive plants can drastically modify soil properties, including soil structure, organic matter content, and microbial communities. These changes can have cascading effects on native vegetation and ecosystem processes. For example, the invasion of *Ailanthus altissima* (Tree of Heaven) in North America has been shown to alter soil pH and decrease microbial biomass, which in turn affects soil nutrient dynamics and the regeneration of native plant species (Gómez-Aparicio & Canham, 2008). The introduction of invasive species often leads to changes in soil compaction and porosity, which can affect water infiltration and retention.

Certain invasive plants, such as Eucalyptus species in Australia, release allelopathic chemicals that inhibit the growth of native plants by altering the soil's chemical properties (Abensperg-Traun et al., 1998). These chemicals can suppress the germination and growth of native flora, leading to shifts in plant community composition and a reduction in plant diversity. Additionally, invasive plants like *Spartina alterniflora* (smooth cordgrass) in salt marshes can modify the soil's physical structure, affecting sediment deposition rates and soil stability, which has implications for coastal erosion and habitat stability (Feagin et al., 2009).

2.2.2 Impacts on Water Availability

In tropical regions, invasive species such as *Mimosa pigra* (giant sensitive plant) in wetlands have been observed to alter hydrological dynamics by forming dense thickets that impede water flow, leading to water stagnation and changes in water quality (Morin et al., 2009). These alterations can disrupt the natural flooding regimes that are critical for maintaining the diversity and functioning of wetland ecosystems.

2.2.3 Impacts on Nutrient Cycling

Invasive plants can significantly affect nutrient cycling processes, particularly nitrogen and carbon cycling, through mechanisms such as altering litter decomposition rates and nutrient uptake patterns. Invasive nitrogen-fixing species, such as Acacia and Prosopis species, can increase soil nitrogen levels, leading to changes in plant community composition by favouring nitrogen-loving species (Ehrenfeld, 2003). These changes can disrupt the balance of nutrient cycling in ecosystems and create conditions that facilitate further invasions by other non-native species.

The decomposition of invasive plant litter can also influence soil nutrient dynamics. For instance, the invasive shrub *Berberis thunbergii* (Japanese barberry) in north-eastern North America has been shown to produce litter that decomposes more slowly than that of native species, leading to changes in soil organic matter accumulation and nutrient availability (Kourtev et al., 2002). These shifts can have long-term effects on soil fertility and the ability of native plants to regenerate, ultimately impacting ecosystem productivity and resilience.

Furthermore, invasive plants can alter the carbon cycle by changing carbon storage and fluxes within ecosystems. For example, invasive grasses such as *Phragmites australis* (common reed) in North American wetlands have been found to increase carbon sequestration rates due to their dense biomass and high growth rates (Were et al., 2019). While this might seem beneficial, the displacement of native species and alteration of habitat structure can have negative consequences for biodiversity and ecosystem functioning.

Overall, the impacts of invasive plants on soil health, water availability, and nutrient cycling underscore the need for comprehensive management strategies that address these changes to restore ecosystem health and function. Understanding the specific mechanisms through which invasive plants alter these ecological processes can guide the development of targeted restoration efforts and improve the effectiveness of invasive species control programs.

3. PRINCIPLES OF ECOLOGICAL RESTORATION

Ecological restoration involves the process of assisting the recovery of ecosystems that have been degraded, damaged, or destroyed. As an applied science, restoration aims to re-establish ecological integrity, enhance biodiversity, and rebuild ecosystem functions and services. The principles of ecological restoration provide a framework for understanding how to restore ecosystems in a way that is effective, sustainable, and aligned with both ecological and human goals.

3.1 Definition and Goals of Ecological Restoration

Ecological restoration is defined as the intentional process of initiating or accelerating the recovery of an ecosystem from a degraded state to its original condition (Clewell & Aronson, 2012). The goal is to return ecosystems to a state of health, integrity, and sustainability by re-establishing ecological processes, functions, and interactions that maintain resilience and resistance to disturbance (Society for Ecological Restoration [SER], 2004). Restoration efforts may include reforestation, removal of invasive species, reintroduction of native species, soil rehabilitation, and the restoration of hydrological systems (Hobbs & Harris, 2001).

Restoration is particularly important in the context of invasive plant management, as invasive species can significantly alter ecosystem structure and function. In this regard, the goals of ecological restoration are twofold: to remove or control invasive species and to re-establish native vegetation and ecological processes that support a healthy, functioning ecosystem (Allison, 2023).

3.2 Key Principles: Resilience, Resistance, and Recovery

The principles of resilience, resistance, and recovery are central to ecological restoration and guide the restoration process to ensure long-term sustainability and success (Fig. 2).

Figure 2. Principles of Ecological resistance

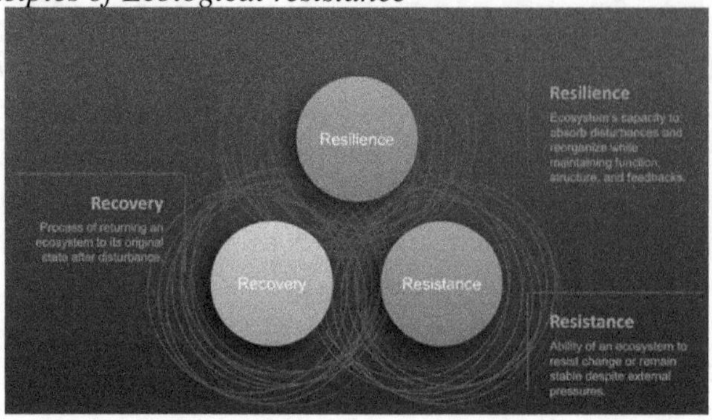

3.2.1 Resilience refers to an ecosystem's capacity to absorb disturbances and reorganize while undergoing change, so as to retain essentially the same function, structure, and feedbacks (Walker et al., 2004). In the context of invasive species management, resilience is the ability of an ecosystem to resist invasion or recover after invasive species have been removed. Restoration strategies aim to enhance the resilience of native communities by promoting species diversity, improving soil health, and maintaining ecological processes such as fire regimes and hydrological cycles (Seastedt et al., 2008).

3.2.2 Resistance is the ability of an ecosystem to resist change or remain stable in the face of external pressures, such as invasive species or environmental changes (D'Antonio & Chambers, 2006). Restoring ecosystem resistance may involve planting competitive native species that inhibit the growth of invasives, managing nutrient levels to favour natives, or restoring natural predators or herbivores that control invasive species (Bakker & Wilson, 2004). The goal is to create conditions that make it difficult for invasive plants to establish or spread.

3.2.3 Recovery refers to the process of returning an ecosystem to its original state after disturbance or degradation. Restoration practitioners often focus on restoring the ecological processes that facilitate natural recovery, such as seed dispersal, pollination, nutrient cycling, and water retention (Suding et al., 2004). Recovery is often measured by the establishment of target species, improvement in ecosystem functions, and the return of biodiversity and ecological interactions (Hobbs & Harris, 2001).

3.3 Role of Ecological Processes and Community Dynamics in Restoration

Successful restoration relies on the restoration of ecological processes and community dynamics that support ecosystem stability and function. Ecological processes such as nutrient cycling, water flow, and energy transfer are foundational to the health and resilience of ecosystems (Clewell & Aronson, 2012). The restoration of these processes often involves reintroducing keystone species, restoring hydrological regimes, and enhancing soil fertility. Community dynamics, including species interactions, competition, and facilitation, also play a crucial role in shaping restored ecosystems (Allison, 2023).

For example, in grassland restoration, the reintroduction of native herbivores like bison can help control invasive grasses and promote native plant diversity (Fuhlendorf et al., 2009). In riparian zones, restoring natural flooding regimes can facilitate the reestablishment of native vegetation and reduce the dominance of invasive species (Stromberg et al., 2007). Recognizing the importance of ecological processes and community dynamics allows for more holistic and effective restoration efforts that go beyond simple removal of invasive plants.

4. TECHNIQUES FOR RESTORING INVASION-AFFECTED ECOSYSTEMS

The restoration of ecosystems affected by invasive plants is a multifaceted challenge that requires a combination of techniques to effectively manage and restore native biodiversity and ecological functions. The following section explores various restoration techniques, focusing on mechanical control methods, chemical and biological control, and the use of integrated management approaches (Fig. 3). Each method has its specific applications, advantages, and limitations depending on the ecosystem context, type of invasive species, and restoration goals.

Figure 3. Technique for Restoring Invasion-Affected Ecosystems

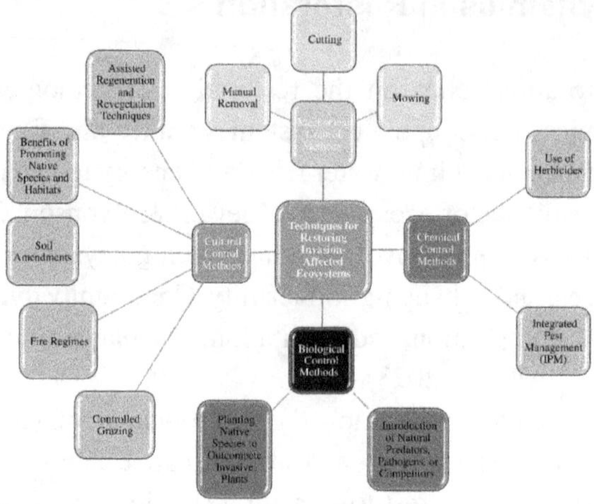

4.1 Mechanical Control Methods

Mechanical control involves the physical removal or destruction of invasive plants to reduce their abundance and impact on native ecosystems. This method is often considered one of the most direct approaches for managing invasive species, particularly in areas where chemical or biological controls may not be feasible or desirable. Mechanical control techniques include manual removal, mowing, cutting, and the use of machinery to manage invasive plant populations.

4.1.1 Manual Removal: This technique involves the physical extraction of invasive plants by hand or using hand tools such as shovels, spades, or hoes. Manual removal is particularly effective for small infestations or in sensitive areas where heavy machinery could cause further ecological damage (Skurski et al., 2014). It is often used in combination with other methods to ensure that root systems are removed or treated, preventing regrowth. While manual removal is labor-intensive and costly, it is highly targeted and reduces non-target impacts. However, it is not always feasible for large-scale infestations due to the time and labor required (D'Antonio & Meyerson, 2002).

4.1.2 Mowing: Mowing involves the use of mowing machines to cut invasive plants at ground level, which can reduce their biomass and prevent seed production. Mowing is commonly applied in grasslands, pastures, and roadside areas where invasive grasses or herbaceous species dominate (DiTomaso, 2000). Although mowing can be an effective tool for reducing plant height and spread, it does not eliminate root systems, allowing plants to regenerate. To improve its effectiveness, mowing is often combined with herbicide application or repeated multiple times during the growing season to deplete the plant's resources (DiTomaso et al., 2013).

4.1.3 Cutting: Cutting involves the selective removal of invasive woody plants or shrubs using saws, pruners, or chainsaws. This method is particularly useful for managing larger, woody invasive species such as shrubs and trees that are difficult to remove manually (Miller et al., 2010). Cutting can effectively reduce the aboveground biomass of invasive plants and limit seed production, but it does not address root systems or prevent resprouting. Follow-up treatments such as stump grinding, herbicide application, or controlled burns are often necessary to achieve long-term control (Richardson & Rejmanek, 2011).

4.1.4 Effectiveness and Limitations: Mechanical control methods are effective in reducing the immediate impact of invasive species and are often preferred for small-scale projects or where chemical use is restricted. However, these methods have several limitations. The most significant is their labor-intensive nature, which limits their applicability to larger landscapes. Additionally, mechanical methods can cause soil disturbance, potentially facilitating the spread of other invasive species or leading to erosion problems if not managed properly (Nielsen & Beglin, 2003). Furthermore, mechanical control may require repeated treatments to prevent regrowth and ensure the removal of all invasive propagules, increasing costs and labor demands over time (Skurski et al., 2014).

Mechanical control methods should be selected based on the specific characteristics of the target invasive species, the ecological context, and the availability of resources. Integrating these methods with other control strategies, such as chemical and biological approaches, often yields better outcomes and enhances the restoration of native plant communities (Richardson & Rejmanek, 2011).

4.2 Chemical Control Methods

Chemical control methods involve the use of herbicides to manage and suppress invasive plant populations. Herbicides are chemical agents designed to kill or inhibit the growth of unwanted plants and are widely used in agricultural, urban, and natural environments. This section provides an overview of various chemical control techniques, discusses their integration with pest management strategies, and highlights potential impacts on non-target species and the environment.

4.2.1 Use of Herbicides: Herbicides are a primary tool in chemical control and can be applied in several ways, such as foliar sprays, basal bark treatments, cut-stump treatments, and soil applications (Tu et al., 2001). Each application method is chosen based on the type of invasive species, its growth stage, and the characteristics of the site being treated. For example, foliar sprays are effective for treating large infestations of herbaceous weeds, while cut-stump treatments are suitable for managing woody invasive species (Owen et al., 2015).

Herbicides can be selective or non-selective. Selective herbicides target specific plant species or groups (e.g., broadleaf weeds) without harming the surrounding vegetation, while non-selective herbicides kill all plants they come in contact with (Radosevich *et al.*, 2007). For instance, glyphosate is a widely used non-selective herbicide effective against a broad spectrum of invasive plants, while 2,4-D is a selective herbicide targeting broadleaf weeds without harming grasses (Mahmood et. al., 2014). However, careful consideration must be given to the choice and application of herbicides to minimize environmental impact.

4.2.2 Integrated Pest Management (IPM): Chemical control methods are often integrated into a broader framework of Integrated Pest Management (IPM), which combines biological, cultural, mechanical, and chemical strategies to achieve effective and sustainable invasive plant control (Liebman & Gallandt, 1997). IPM approaches emphasize minimizing herbicide use and applying chemicals only when necessary, using targeted applications to reduce non-target effects and avoid the development of herbicide-resistant plant populations (Norris et al., 2003). For example, combining chemical control with manual removal or grazing can reduce the need for repeated herbicide applications and enhance restoration success (Florence et al., 2019).

Herbicides are particularly effective when used as part of an integrated approach in large-scale restoration projects or in situations where invasive plants dominate the landscape, and other control methods are insufficient (Tu et al., 2001). For example, herbicide application can precede seeding or planting of native species, reducing competition and giving native plants a better chance to establish (Sheley et al., 2010).

4.2.3 Impacts on Non-Target Species and Environment: While herbicides can effectively control invasive plants, they may have unintended consequences on non-target species and the broader environment. Non-target impacts can include damage to native plant communities, changes in soil microbial communities, and contamination of water bodies (Boutin et al., 2004). For instance, glyphosate, while effective, has been linked to declines in non-target plant diversity and can affect soil organisms, including beneficial mycorrhizal fungi and earthworms (Powles, 2008). Similarly, atrazine, a common herbicide, can persist in the environment, leading to contamination of water sources and affecting aquatic ecosystems (Gilliom et al., 2006).

There is also concern over the potential development of herbicide resistance in invasive plant populations, which can reduce the efficacy of chemical control measures over time. Repeated use of the same herbicide or reliance on chemical control as the sole method can lead to resistant biotypes, necessitating the use of higher doses or more toxic alternatives (Mahmood et. al., 2014). Therefore, chemical control must be used judiciously and in conjunction with other management practices to avoid these adverse outcomes and achieve long-term restoration goals.

Chemical control methods remain a valuable tool in the management of invasive plants, especially when other methods are not feasible. However, their application must be carefully planned, considering potential environmental impacts and the need for sustainable, integrated approaches to invasive species management.

4.3 Biological Control Methods

Biological control involves the intentional introduction of natural enemies, such as predators, pathogens, or competitors, to suppress invasive plant populations. This method harnesses ecological relationships to manage invasive species, offering a sustainable alternative to chemical and mechanical controls. Biological control is particularly effective for large-scale or difficult-to-access infestations, providing a long-term solution to invasive plant management when implemented correctly. This section will explore the principles of biological control, introduce different types of biological agents, and present case studies showcasing successful programs.

4.3.1 Introduction of Natural Predators, Pathogens, or Competitors: Biological control relies on the concept of re-establishing ecological balance by introducing natural enemies from the invasive plant's native range to control its spread in the invaded area (Van Driesche & Hoddle, 2009). These biological agents include herbivores (e.g., insects and mites), pathogens (e.g., fungi, bacteria, viruses), and competitive plants that target the invasive species without significantly impacting native flora and fauna.

The selection of biological control agents involves rigorous testing to ensure host specificity, meaning the agent should only target the invasive species without affecting non-target organisms. For instance, the introduction of *Rhinocyllus conicus*, a weevil that targets the invasive musk thistle (*Carduus nutans*), has demonstrated the importance of host specificity to prevent unintended ecological consequences (Louda et al., 2003). Effective biological control programs typically require substantial upfront research and monitoring to identify suitable agents and predict their impacts.

4.3.2 Case Studies of Successful Biological Control Programs: Several biological control programs have successfully managed invasive plants by leveraging natural enemies. One notable example is the use of *Cactoblastis cactorum*, a moth introduced from Argentina, to control invasive prickly pear cacti (Opuntia spp.) in Australia (Fig. 4). The introduction of Cactoblastis in 1925 led to the rapid collapse of prickly pear infestations across millions of hectares, transforming Australia's agricultural landscape and establishing biological control as a viable strategy (Dodd, 1940).

Figure 4. Cactoblastis cactorum to control invasive prickly pear cacti (Photo credit: Fundación para el Estudio de Especies Invasivas (FuEDEI), Buenos Aires, Argentina. Source: FuEDEI Cactus Moth article)

Another successful case is the management of invasive water hyacinth (*Eichhornia crassipes*) in Lake Victoria, East Africa. Water hyacinth, a highly invasive aquatic plant, forms dense mats that hinder water transport, reduce oxygen levels, and affect fish populations. The introduction of the weevils *Neochetina eichhorniae* and *Neochetina bruchi* has significantly reduced water hyacinth biomass (Fig. 5), demonstrating effective biological control in an aquatic ecosystem (Wilson et al., 2005).

Figure 5. Introduction of the weevils to reduced water hyacinth biomass (Photo credit Willey Durde and Katherine Parys, USDA-ARS, Bugwood.org).

Additionally, biological control has shown promise in managing Tamarix spp. (salt cedar), an invasive shrub in riparian ecosystems of the western United States. The release of the Diorhabda beetle, a leaf-feeding beetle, has resulted in significant reductions in salt cedar populations, which in turn supports the recovery of native vegetation and improves riparian habitat quality (Carruthers et al., 2008).

4.3.3 Challenges and Considerations in Biological Control: Despite its potential, biological control is not without challenges. The introduction of biological control agents can sometimes have unintended consequences, such as non-target effects, hybridization with native species, or disruption of existing ecological balances (Simberloff & Stiling, 1996). For instance, *Rhinocyllus conicus*, initially released to control musk thistle, has also attacked native thistles in North America, leading to negative impacts on native ecosystems (Louda et al., 2003). Therefore, careful assessment and long-term monitoring are crucial to avoid ecological harm.

Moreover, biological control programs require considerable time and financial investment, as they involve extensive research, testing, and post-release evaluation (Van Driesche & Hoddle, 2009). However, when successfully implemented, biological control can provide a cost-effective, sustainable, and environmentally friendly solution to invasive plant management, reducing reliance on chemical herbicides and promoting ecosystem restoration.

In summary, biological control remains an essential component of integrated invasive species management strategies. While offering significant advantages, such as cost-effectiveness and sustainability, the method also requires careful planning, research, and management to avoid potential risks and maximize benefits for native ecosystems.

4.4 Cultural Control Methods

Cultural control methods involve modifying environmental management practices to suppress invasive plant species while promoting the growth and recovery of native vegetation. These methods focus on creating conditions that favor native species and ecosystems, often using traditional land management practices adapted to modern ecological needs. Cultural control methods are essential in integrated invasive plant management, as they provide sustainable and cost-effective approaches that align with ecosystem restoration goals. This section discusses the various cultural control techniques, such as controlled grazing, fire regimes, and soil amendments, and explores their benefits in promoting native species and habitats.

4.4.1 Controlled Grazing: Controlled or targeted grazing is a cultural method that uses livestock to manage invasive plant species by grazing them selectively. Livestock, such as cattle, sheep, and goats, can be strategically introduced to graze on invasive plants, reducing their biomass and suppressing their growth. This method is particularly effective for managing invasive grasses and forbs that dominate rangelands and pastures (Davies & Sheley, 2011). Controlled grazing is often combined with rotational grazing systems to avoid overgrazing and ensure that native vegetation has sufficient time to recover between grazing periods. Studies have shown that targeted grazing can reduce invasive species like Medusahead (*Taeniatherum caput-medusae*) while promoting the growth of native grasses (DiTomaso, 2000).

However, the effectiveness of controlled grazing depends on various factors, including the timing, duration, and intensity of grazing, as well as the type of livestock used. Overgrazing can lead to soil degradation, erosion, and further spread of invasive species, highlighting the need for careful planning and monitoring to achieve desired ecological outcomes (Filazzola et. al., 2020). When managed properly, controlled grazing can help maintain ecosystem balance, enhance soil health, and support biodiversity conservation.

4.4.2 Fire Regimes: The use of prescribed or controlled burns is another cultural control method employed to manage invasive plants. Fire regimes can be manipulated to control the growth of invasive species that are less fire-tolerant than native plants. Prescribed burns can reduce the seed bank of invasive species, remove accumulated biomass, and create favourable conditions for native species to establish and grow (Brooks et al., 2004). For instance, prescribed burning has been used to manage invasive cheatgrass (*Bromus tectorum*) in sagebrush steppe ecosystems, thereby reducing its dominance and allowing native shrubs and grasses to recover (James et al., 2015).

Fire as a cultural control method requires careful consideration of the timing, frequency, and intensity of burns to prevent unintended ecological damage. Factors such as weather conditions, fuel loads, and the presence of fire-adapted invasive species must be accounted for to maximize the effectiveness of prescribed burns (Keeley, 2006). When used appropriately, fire regimes can be an effective tool for managing invasive plants, restoring natural fire cycles, and enhancing habitat quality for native flora and fauna.

4.4.3 Soil Amendments: Soil amendments involve adding materials to the soil to alter its properties, making it less favourable for invasive species and more conducive to native plant growth. Common soil amendments include organic matter (e.g., compost, mulch), nutrients, and other soil conditioners that can improve soil structure, fertility, and microbial communities (Durán et. al., 2018). For example, adding carbon-rich materials, such as wood chips or straw, can reduce soil nitrogen availability, limiting the growth of invasive plants that thrive in high-nitrogen environments while favouring native species that are adapted to lower nutrient conditions (Alpert & Maron, 2000).

Another approach involves using soil solarization, a technique that involves covering soil with clear plastic to increase soil temperature and reduce the seed bank of invasive plants. This method can be particularly effective in arid and semi-arid regions where soil temperatures can reach levels that are lethal to invasive plant seeds and pathogens (Stapleton, 2000). Soil amendments can also promote the establishment of native plant communities by enhancing soil microbial diversity, which plays a crucial role in nutrient cycling and plant growth (van der Heijden et al., 2008).

4.4.4 Benefits of Promoting Native Species and Habitats: Cultural control methods, such as controlled grazing, fire regimes, and soil amendments, offer several benefits for promoting native species and restoring habitats. By modifying land management practices to create conditions that favour native plants, these methods can help reduce the dominance of invasive species and support the recovery of degraded ecosystems. Moreover, cultural control methods often integrate well with other invasive plant management strategies, such as mechanical, chemical, and biological controls, providing a holistic and adaptive approach to ecological restoration (Hobbs & Humphries, 1995).

Implementing cultural control methods also supports long-term ecosystem resilience by maintaining ecological processes, such as nutrient cycling, soil formation, and fire dynamics, that are essential for ecosystem stability and function. Additionally, cultural control methods are often more sustainable and cost-effective than other approaches, particularly when they rely on locally available resources and traditional knowledge (Pywell et al., 2002). By promoting native species and habitats, cultural control methods contribute to biodiversity conservation, climate change mitigation, and the provision of ecosystem services critical for human well-being.

4.5 Assisted Regeneration and Revegetation Techniques

Assisted regeneration and revegetation techniques focus on actively restoring native plant communities by introducing or promoting the growth of native species that can outcompete invasive plants. These methods are essential components of ecological restoration efforts, as they help re-establish natural vegetation dynamics, improve ecosystem resilience, and support biodiversity. Assisted regeneration involves direct intervention to support the natural recovery of ecosystems affected by invasive species, whereas revegetation typically includes planting native species and managing soil conditions to create favourable environments for their growth.

4.5.1 Planting Native Species to Outcompete Invasive Plants: Planting native species is a widely used revegetation technique that involves introducing native plants back into areas that have been degraded by invasive species. The goal is to establish a dense cover of native vegetation that can suppress the regrowth of invasive plants through competition for resources such as light, water, and nutrients (Török et al., 2021). Native species are often chosen for their fast growth rates, high biomass production, and ability to form dense canopies or root systems that can effectively shade or crowd out invasive species. For example, studies have shown that planting native grasses, such as *Elymus canadensis* and *Panicum virgatum*, can help suppress invasive species like *Phragmites australis* in wetland restoration projects (Hazelton et al., 2014).

Successful planting programs require careful selection of native species that are well-adapted to local environmental conditions and have traits that enable them to compete effectively with invasive plants. It is also essential to consider the diversity and composition of native plant communities to ensure functional redundancy and resilience against future invasions (Funk et al., 2008). In some cases, revegetation efforts may need to be combined with other control methods, such as mechanical or chemical treatments, to reduce the existing populations of invasive species before planting native vegetation (Holl & Aide, 2011).

4.5.2 Soil Preparation and Seed Bank Management: Soil preparation is a critical step in assisted regeneration and revegetation, as it creates suitable conditions for the establishment and growth of native species. Soil preparation techniques may include mechanical treatments, such as tilling or ploughing, to break up compacted soils, remove invasive plant roots, and enhance soil aeration and water infiltration (Wortley et al., 2013). In some cases, soil amendments, such as organic matter or fertilizers, may be applied to improve soil fertility and microbial activity, which can benefit native plant establishment (Wortley et al., 2013). However, soil preparation must be carefully planned to avoid soil erosion, nutrient leaching, or other unintended ecological impacts.

Managing the seed bank is also a crucial aspect of assisted regeneration, particularly in areas where invasive plant seeds dominate the soil seed bank. Techniques such as soil solarization, which involves covering the soil with clear plastic to increase soil temperatures and kill weed seeds, or the application of carbon-based amendments to reduce soil nitrogen levels, can help manage the seed bank and reduce the likelihood of invasive plant regrowth (Blumenthal et al., 2003). Additionally, direct seeding of native species is often employed to introduce a diverse and abundant source of native plant propagules into the soil seed bank, further promoting the establishment of desired vegetation (Sheley et al., 2006).

Effective assisted regeneration and revegetation efforts require an adaptive management approach, where monitoring and feedback are used to refine methods and strategies over time. This ensures that restoration efforts remain responsive to changing conditions and ecological dynamics, ultimately increasing the success of restoration projects and the recovery of invasion-affected ecosystems (Zedler & Callaway, 1999).

5. RESTORATION PLANNING AND IMPLEMENTATION

Restoration planning and implementation are critical phases in ecological restoration projects, particularly when mitigating the impacts of invasive plant species. This process involves selecting appropriate sites, developing comprehensive restoration plans, and engaging stakeholders to ensure sustainable and effective restoration outcomes (Fig. 6). This chapter delves into the key steps involved in planning and executing restoration efforts, including site assessment, setting restoration goals, and fostering community participation.

Figure 6. Restoration Planning and Implementation

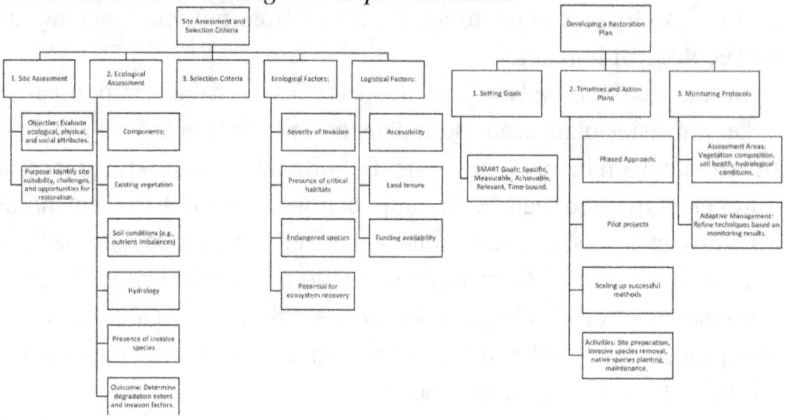

5.1 Site Assessment and Selection Criteria

Site assessment is the first step in restoration planning, involving a thorough evaluation of the ecological, physical, and social attributes of a potential restoration site. This process helps determine the suitability of a site for restoration efforts and identifies the specific challenges and opportunities present.

5.1.1 Ecological Assessment: This includes analysing the existing vegetation, soil conditions, hydrology, and presence of invasive species. Ecological assessments aim to understand the extent of degradation and the factors contributing to the invasion. For example, soil testing may reveal nutrient imbalances that favour invasive species, while vegetation surveys can identify native plant remnants that could aid in restoration efforts (Hobbs & Norton, 2004).

5.1.2 Selection Criteria: The selection of sites for restoration is based on criteria such as the severity of invasion, the presence of critical habitats or endangered species, potential for ecosystem recovery, and feasibility of restoration interventions. Sites with a high degree of ecological importance or those serving as critical wildlife corridors are often prioritized. Moreover, logistical factors such as accessibility, land tenure, and funding availability are also considered in the selection process (Clewell & Aronson, 2012).

5.2 Developing a Restoration Plan: Setting Goals, Timelines, and Monitoring Protocols

A comprehensive restoration plan is essential for guiding restoration activities and ensuring their success. Developing such a plan involves setting clear goals, establishing realistic timelines, and defining monitoring protocols.

5.2.1 Setting Goals: Restoration goals should be specific, measurable, achievable, relevant, and time-bound. Goals may include reducing invasive plant cover by a certain percentage, increasing native plant diversity, or improving soil health within a specified period (Suding, 2011). For instance, a goal might be to achieve a 70% reduction in invasive *Phragmites australis* within three years while restoring native wetland vegetation (Hazelton et al., 2014).

5.2.2 Timelines and Action Plans: Timelines are developed to outline the sequence of restoration activities, from site preparation and invasive species removal to native species planting and follow-up maintenance. An action plan may include a phased approach, starting with pilot projects to test restoration techniques, followed by scaling up successful methods across larger areas (Holl & Aide, 2011).

5.2.3 Monitoring Protocols: Monitoring is a critical component of restoration planning that involves regular assessment of progress toward the set goals. Monitoring protocols should be established to evaluate changes in vegetation composition, soil health, hydrological conditions, and other ecological parameters. Adaptive management approaches allow restoration practitioners to refine techniques and strategies based on monitoring results (Hobbs & Harris, 2001).

5.3 Stakeholder Involvement and Community-Based Restoration Initiatives

Effective restoration often requires collaboration among diverse stakeholders, including government agencies, local communities, NGOs, scientists, and private landowners. Engaging stakeholders in restoration efforts can enhance the sustainability and success of projects.

5.3.1 Community-Based Restoration: Involving local communities in restoration initiatives is crucial for building local capacity, fostering stewardship, and ensuring long-term maintenance of restored areas. Community-based restoration programs often provide training and employment opportunities for local residents, encouraging them to participate in invasive species removal, native planting, and monitoring activities (Berkes, 2004). For example, in the restoration of the South African fynbos, local community members were employed to remove invasive Australian Acacia species and replant native vegetation, leading to successful outcomes and socio-economic benefits (Pearson & Ortega, 2009).

5.3.2 Stakeholder Engagement Strategies: Effective stakeholder engagement involves transparent communication, inclusive decision-making, and the establishment of partnerships. Participatory approaches, such as workshops, public consultations, and co-management agreements, help align restoration goals with local values and interests (Reed, 2008). By involving stakeholders from the planning stage through to implementation and monitoring, restoration projects can build broader support and ensure more resilient outcomes.

6. MONITORING AND EVALUATION OF RESTORATION EFFORTS

Monitoring and evaluation are critical components of ecological restoration, especially when managing and mitigating plant invasions. Effective monitoring allows for the assessment of restoration progress and the adaptation of strategies to improve outcomes. This chapter focuses on the significance of long-term monitoring, the use of specific metrics and indicators to evaluate success, and real-world case studies that illustrate the effectiveness of various monitoring frameworks.

6.1 Importance of Long-Term Monitoring and Adaptive Management

Long-term monitoring is essential in ecological restoration to understand the dynamics of both invasive and native species, track ecosystem recovery, and identify any emerging threats or new invasions. It involves repeated observations and measurements over time to assess changes in ecological conditions and to determine whether the restoration objectives are being met (Lindenmayer & Likens, 2009). Adaptive management, a process of iterative decision-making in the face of uncertainty, is often used in conjunction with long-term monitoring to refine restoration practices based on empirical evidence (Holling, 1978).

6.1.1 Adaptive Management Approaches: Adaptive management emphasizes learning from restoration outcomes and adjusting methods as necessary to achieve desired goals. This approach is particularly useful in invasion ecology, where unexpected challenges often arise, such as the re-invasion of previously controlled species or shifts in ecosystem processes (Williams, Szaro, & Shapiro, 2009). By incorporating feedback from monitoring data, managers can modify techniques, reallocate resources, and develop new strategies to enhance restoration success (Walters & Holling, 1990).

6.1.2 Policy Implications: Monitoring data and adaptive management practices can inform policy decisions at local, regional, and national levels. By demonstrating the efficacy or shortcomings of particular restoration methods, data can help shape guidelines and regulations that promote sustainable land management and invasive species control (Allen et al., 2011).

6.2 Metrics and Indicators for Evaluating Restoration Success

To effectively evaluate restoration success, it is crucial to establish clear, measurable metrics and indicators that reflect the objectives of the restoration project. Metrics may vary based on the type of ecosystem, the extent of invasion, and the specific restoration goals (Table 1).

Table 1. Examples of Monitoring Indicators Used to Assess Restoration Success in Invasion-Affected Ecosystems

Indicator Type	Specific Indicator	What It Measures	Example Use Case	Reference
Ecological	Species richness (native/invasive)	Biodiversity recovery & invasion suppression	Assessing native plant recovery post-*Phragmites* removal	Palmer et al., 1997
Soil Health	Soil microbial biomass and activity	Soil ecosystem functioning and recovery	Recovery of microbial communities after invasive grass eradication	Harris, 2003
Hydrological	Water table depth, infiltration rate	Restoration of hydrological balance	Post-restoration in riparian zones previously invaded by *Tamarix*	Cleverly et al., 2006
Functional	Litter decomposition rate	Nutrient cycling and detritus processing	Wetland restoration following *Eichhornia* removal	Ehrenfeld, 2003
Socioeconomic	Community participation rate in restoration	Stakeholder engagement and long-term project sustainability	Monitoring participation in community-based Acacia removal projects	Datta et al., 2012

6.2.1 Ecological Indicators: Common ecological indicators include plant community composition, species richness, diversity indices, and the abundance of invasive versus native species (Ruiz-Jaen & Aide, 2005). For instance, the recovery of native species diversity and the decline in invasive species cover are frequently used to assess the progress of restoration efforts (Palmer et al., 1997). Soil health indicators, such as nutrient content, microbial activity, and soil structure, can also provide valuable insights into ecosystem recovery (Harris, 2003).

6.2.2 Functional Indicators: These indicators focus on ecosystem processes, such as nutrient cycling, hydrological dynamics, and primary productivity, which are essential for long-term ecosystem stability and resilience. Restoration efforts that aim to re-establish these processes may monitor indicators like leaf litter decomposition rates, soil respiration, or groundwater recharge levels (Young et al., 2005).

6.2.3 Socioeconomic Indicators: Restoration projects can also be evaluated based on their social and economic impacts, particularly when they involve community participation or provide ecosystem services. Metrics such as community engagement, employment generated through restoration activities, or improvements in local livelihoods are often considered (Aronson et al., 2006).

6.3 Case Studies of Monitoring Frameworks and Their Effectiveness

Several case studies demonstrate the effectiveness of monitoring frameworks in restoration projects aimed at managing invasive species and rehabilitating ecosystems. These examples illustrate different approaches to monitoring and the importance of tailoring frameworks to specific ecological and social contexts.

6.3.1 Prairie Restoration in North America: In a large-scale prairie restoration project in North America, a monitoring framework was established that included both short-term and long-term indicators, such as plant species diversity, soil organic matter, and water infiltration rates. The project used adaptive management to address unexpected challenges, such as the re-invasion of non-native grasses, by adjusting seeding rates and introducing controlled burns. The monitoring data revealed significant improvements in native species establishment and soil health over a 10-year period (Trowbridge et al., 2017).

6.3.2 Wetland Restoration in Australia: A wetland restoration initiative in Australia employed a comprehensive monitoring framework that integrated hydrological, biological, and geomorphological indicators. By assessing changes in water quality, vegetation cover, and sediment deposition rates, the project team was able to adaptively manage water flow regimes and invasive species control measures. The framework's effectiveness was demonstrated by the recovery of native wetland vegetation and the restoration of natural hydrological patterns within five years (Hayes et al., 2017).

6.3.3 Community-Based Mangrove Restoration in Southeast Asia: In Southeast Asia, community-based mangrove restoration efforts have implemented participatory monitoring frameworks involving local communities. These frameworks include both ecological indicators (such as mangrove seedling survival rates and canopy cover) and socioeconomic indicators (such as community income from sustainable fisheries). The involvement of local stakeholders in monitoring has fostered a sense of ownership and increased the sustainability of restoration efforts (Datta et al., 2012).

7. INNOVATIVE APPROACHES IN RESTORATION ECOLOGY

This section explores cutting-edge methods and advancements in restoration ecology. These innovative approaches enhance the effectiveness of restoration efforts by incorporating new technologies, scientific principles, and research methodologies. Understanding these advancements is crucial for developing more robust and adaptable restoration strategies.

7.1 Use of Geospatial Technologies for Site Assessment and Monitoring

Geospatial technologies have revolutionized the field of restoration ecology by providing advanced tools for site assessment, monitoring, and management. These technologies offer high-resolution data and spatial analysis capabilities, which are critical for understanding complex ecological processes and informing restoration decisions.

7.1.1 Geographic Information Systems (GIS): GIS enables the mapping and analysis of spatial data related to land use, vegetation cover, and habitat conditions. By integrating various data layers, GIS helps identify areas of high ecological value and priority for restoration (Longley et al., 2015). GIS also supports the development of detailed restoration plans by visualizing spatial patterns and changes over time (Alexander et. al., 2016).

7.1.2 Remote Sensing: Remote sensing techniques, including satellite imagery and aerial photography, provide valuable information on vegetation cover, land degradation, and ecological changes. These methods facilitate large-scale monitoring and assessment of restoration projects, offering insights into vegetation dynamics and environmental impacts (Turner et al., 2015). High-resolution imagery can detect subtle changes in plant communities and guide adaptive management efforts (Rose et al., 2015).

7.1.3 Drones and UAVs: Unmanned Aerial Vehicles (UAVs) offer a flexible and cost-effective means of collecting high-resolution aerial data. Drones can capture detailed images of restoration sites, allowing for precise mapping and monitoring of plant growth, habitat recovery, and invasive species spread (Anderson & Gaston, 2013). UAVs also enable real-time data collection and rapid response to emerging issues in restoration projects (Manfreda et al., 2018).

7.2 Genetic and Ecological Engineering Approaches

Genetic and ecological engineering approaches represent a frontier in restoration ecology, aiming to enhance the resilience and adaptability of ecosystems through targeted interventions.

7.2.1 Genetic Engineering: Genetic engineering techniques can be employed to develop plant varieties with improved traits for restoration. For example, genetically modified plants with enhanced resistance to pests, diseases, or environmental stressors can be used to restore degraded habitats more effectively (Zahoor et al., 2025). These approaches also include the development of plants with increased competitive abilities to outcompete invasive species (Breed et al., 2019).

7.2.2 Ecological Engineering: Ecological engineering involves designing and implementing systems that restore or create ecosystems to address specific environmental problems. Techniques such as the construction of artificial wetlands, bioengineered riparian buffers, and engineered soils are used to restore ecosystem functions and improve habitat quality (Mitsch & Jørgensen, 2004). These engineered systems can enhance water quality, stabilize soil, and support biodiversity (Lewis, 2005).

7.2.3 Synthetic Ecology: Synthetic ecology involves the creation of novel ecosystems through the introduction of new species combinations or synthetic organisms. This approach aims to accelerate ecosystem recovery and improve resilience by assembling functional communities that can withstand disturbances and support ecological processes (Mee & Wang, 2012). Synthetic ecology also explores the potential for using engineered microorganisms to remediate polluted environments or restore degraded soils (singh et al., 2016).

7.3 Future Directions in Restoration Ecology Research

Despite the comprehensive overview of restoration techniques provided in this chapter, several gaps and future research directions remain unaddressed. Restoration ecology is a rapidly evolving field, and its future will be shaped by emerging technologies, interdisciplinary collaboration, and the urgent need to address global environmental challenges. Below, key areas for future research and innovation are outlined to enhance the effectiveness and scalability of restoration efforts in mitigating plant invasions.

7.3.1 Integration of Climate Change Adaptation: Future research should emphasize integrating climate change considerations into restoration planning and practice. This includes developing strategies for restoring ecosystems in the face of changing climatic conditions, such as altered precipitation patterns, temperature increases, and extreme weather events (Chazdon, 2008). Research on climate-resilient species and adaptive management practices will be essential for ensuring long-term restoration success (Heller & Zavaleta, 2009).

7.3.2 Advancements in Restoration Science: Continued advancements in ecological science and technology will drive innovation in restoration practices. This includes improvements in ecological modelling, remote sensing technologies, and genetic tools. Research on the interactions between restoration interventions and ecosystem services, such as carbon sequestration and water regulation, will also enhance the effectiveness and sustainability of restoration efforts (Martin et al., 2018).

7.3.3 Community and Policy Engagement: Engaging communities and policymakers in restoration efforts is critical for achieving broader and more sustainable outcomes. Future research should explore strategies for involving local stakeholders in restoration planning, decision-making, and implementation. Building partnerships between scientists, practitioners, and communities can facilitate knowledge exchange, promote effective management practices, and ensure the long-term success of restoration initiatives (Paudyal et al., 2017).

8. CHALLENGES AND OPPORTUNITIES IN RESTORATION PRACTICE

The practice of ecological restoration faces numerous challenges but also presents significant opportunities for improving ecosystem health and resilience. Understanding these challenges and seizing the opportunities can help advance the effectiveness of restoration projects and integrate them with broader environmental and societal goals.

8.1 Common Challenges: Funding, Policy, and Technical Expertise

8.1.1 Funding: Securing adequate and sustained funding is a critical challenge in restoration projects. Restoration efforts often require significant financial resources for planning, implementation, and long-term maintenance. Funding challenges can be exacerbated by economic downturns or competing priorities. To address this issue, restoration projects can explore diverse funding sources, including government grants, private investments, and public-private partnerships (BenDor et al., 2015). Additionally, leveraging cost-effective techniques and demonstrating the long-term economic benefits of restoration can help attract and secure funding (TEEB, 2010).

8.1.2 Policy: Effective restoration often depends on supportive policies and regulatory frameworks. Inadequate or inconsistent policies can hinder the progress of restoration projects. Policies may need to address land use, conservation priorities, and environmental regulations. Advocating for and developing robust policies that support restoration goals, incorporate adaptive management strategies, and incentivize sustainable practices are essential for overcoming policy-related challenges (Clewell & Aronson, 2006). Engaging policymakers early in the planning process can also facilitate better alignment between restoration objectives and regulatory requirements (Suding et al., 2015).

8.1.3 Technical Expertise: Restoration projects require specialized knowledge and technical expertise in areas such as ecological assessment, site management, and species propagation. A shortage of skilled professionals or insufficient training can impact the success of restoration efforts. Addressing this challenge involves investing in education and training programs for restoration practitioners, promoting interdisciplinary collaboration, and developing best practice guidelines that are informed by the latest scientific research (Clewell et al., 2004). Establishing networks and communities of practice can also help share knowledge and resources among restoration professionals (Hobbs & Harris, 2001).

8.2 Opportunities for Integrating Restoration with Climate Adaptation Strategies Climate Adaptation: Restoration projects provide an opportunity to enhance ecosystem resilience and adapt to climate change. By incorporating climate adaptation strategies into restoration planning, projects can address anticipated changes in temperature, precipitation, and other climatic factors. For example, selecting plant species that are resilient to future climate conditions and designing restoration projects to include climate buffers can improve long-term success (Chazdon, 2008). Integrating restoration with climate adaptation goals can also help protect and enhance ecosystem services that support human well-being, such as water regulation and carbon sequestration (TEEB, 2010).

8.2.1 Ecosystem Services: Restoration efforts can be designed to maximize the provision of ecosystem services that contribute to climate adaptation. For example, restoring wetlands and riparian zones can enhance flood regulation, improve water quality, and provide habitat for wildlife. Incorporating ecosystem service valuation into restoration planning can help prioritize projects that deliver multiple benefits and align with broader environmental and societal goals (BenDor et al., 2015). Additionally, promoting green infrastructure and nature-based solutions as part of restoration projects can further integrate ecological and climate adaptation objectives (EEA, 2017).

8.3 Collaborations Between Researchers, Practitioners, and Policymakers Interdisciplinary Collaboration: Effective restoration requires collaboration among researchers, practitioners, and policymakers. Interdisciplinary approaches that combine scientific research, practical experience, and policy knowledge can enhance the design and implementation of restoration projects. Facilitating dialogue and partnerships between these stakeholders can lead to more integrated and innovative solutions (Suding et al., 2015). Collaborative research initiatives and joint projects can also help bridge the gap between science and practice, ensuring that restoration efforts are grounded in robust evidence and practical considerations (Hobbs & Harris, 2001).

8.3.1 Community Engagement: Engaging local communities in restoration efforts can improve project outcomes and build support for long-term maintenance. Community-based restoration initiatives that involve local stakeholders in planning and implementation can foster stewardship, increase project success rates, and enhance social and ecological resilience (Paudyal et al., 2017). Developing effective communication strategies and involving community members in monitoring and evaluation can also contribute to the sustainability and effectiveness of restoration projects (Fox & Cundill, 2018).

9. Recommendations for improving restoration practices to mitigate plant invasions integrate multidisciplinary approaches: combining ecological, technological, and community-based methods can enhance the effectiveness of restoration projects. Incorporating advanced technologies, such as geospatial tools and genetic engineering, into restoration planning can provide more accurate assessments and targeted interventions (Bendor et al., 2015).

9.1 Enhance Stakeholder Engagement: Engaging local communities, policymakers, and restoration practitioners is essential for successful restoration. Building partnerships and fostering collaboration can improve project outcomes and ensure sustained support (Fox & Cundill, 2018).

9.2 Focus on Adaptive Management: Implementing adaptive management practices allows for ongoing adjustments based on monitoring and evaluation results. This approach ensures that restoration efforts remain responsive to changing conditions and emerging challenges (Suding et al., 2015).

9.3 Secure Adequate Funding and Resources: Developing diverse funding strategies and leveraging economic incentives can help address financial challenges in restoration projects. Demonstrating the long-term benefits of restoration can attract and maintain investment (TEEB, 2010).

9.4 Promote Policy Development: Advocating for supportive policies and regulatory frameworks can facilitate restoration efforts and align them with broader environmental goals. Effective policy development should address land use, conservation priorities, and environmental regulations (Clewell & Aronson, 2006).

9.5 Call for Continued Research, Innovation, and Policy Support in Restoration Ecology: Ongoing research and innovation are vital for advancing restoration practices and addressing the complex challenges posed by invasive species. Continued exploration of new technologies, methods, and ecological insights will enhance our ability to restore and protect ecosystems effectively. Additionally, sustained policy support and collaboration between researchers, practitioners, and policymakers are crucial for scaling up restoration efforts and achieving long-term ecological resilience.

REFERENCES

Abensperg-Traun, M., Atkins, L., Hobbs, R., & Steven, D. (1998). Exotic plant invasion and understorey species richness: A comparison of two types of eucalypt woodland in agricultural Western Australia. *Pacific Conservation Biology*, *4*(1), 21–32. DOI: 10.1071/PC980021

Alexander, S., Aronson, J., Whaley, O., & Lamb, D. (2016). The relationship between ecological restoration and the ecosystem services concept. *Ecology and Society*, *21*(1), art34. DOI: 10.5751/ES-08288-210134

Allen, C. R., Fontaine, J. J., Pope, K. L., & Garmestani, A. S. (2011). Adaptive management for a turbulent future. *Journal of Environmental Management*, *92*(5), 1339–1345. DOI: 10.1016/j.jenvman.2010.11.019 PMID: 21168260

Allison, S. K. (2023). *Ecological restoration and environmental change: Renewing damaged ecosystems*. Routledge. Zavaleta, E. (2000). The economic value of controlling an invasive shrub. *Ambio*, *29*(8), 462–467. DOI: 10.1579/0044-7447-29.8.462

Alpert, P., & Maron, J. L. (2000). Carbon addition as a countermeasure against biological invasion by plants. *Biological Invasions*, *2*(1), 33–40. DOI: 10.1023/A:1010063611473

Anderson, K., & Gaston, K. J. (2013). Lightweight unmanned aerial vehicles will revolutionize spatial ecology. *Frontiers in Ecology and the Environment*, *11*(3), 138–146. DOI: 10.1890/120150

Aronson, J., Clewell, A. F., Blignaut, J. N., & Milton, S. J. (2006). Ecological restoration: A new frontier for nature conservation and economics. *Journal for Nature Conservation*, *14*(3-4), 135–139. DOI: 10.1016/j.jnc.2006.05.005

Bakker, J. D., & Wilson, S. D. (2004). Using ecological restoration to constrain biological invasion. *Journal of Applied Ecology*, *41*(6), 1058–1064. DOI: 10.1111/j.0021-8901.2004.00962.x

BenDor, T., Lester, T. W., Livengood, A., Yonavjak, L., & Davis, A. P. (2015). Estimating the Size and Impact of the Ecological Restoration Economy. *PLoS One*, *10*(6), e0128339. DOI: 10.1371/journal.pone.0128339 PMID: 26083034

Berkes, F. (2004). Rethinking community-based conservation. *Conservation Biology*, *18*(3), 621–630. DOI: 10.1111/j.1523-1739.2004.00077.x

Blumenthal, D. M., Jordan, N. R., & Svenson, E. L. (2003). Weed control as a rationale for restoration: The example of tallgrass prairie. *Conservation Ecology*, *7*(1), 6. DOI: 10.5751/ES-00480-070106

Boutin, C., Elmegaard, N., & Kjær, C. (2004). Toxicity testing of fifteen non-crop plant species with six herbicides in a greenhouse experiment: Implications for risk assessment. *Ecotoxicology (London, England)*, *13*(4), 349–369. DOI: 10.1023/B:ECTX.0000033092.82507.f3 PMID: 15344515

Breed, M. F., Harrison, P. A., Blyth, C., Byrne, M., Gaget, V., Gellie, N. J., Groom, S. V. C., Hodgson, R., Mills, J. G., Prowse, T. A. A., Steane, D. A., & Mohr, J. J. (2019). The potential of genomics for restoring ecosystems and biodiversity. *Nature Reviews. Genetics*, *20*(10), 615–628. DOI: 10.1038/s41576-019-0152-0 PMID: 31300751

Brooks, M. L., D'Antonio, C. M., Richardson, D. M., Grace, J. B., Keeley, J. E., DiTomaso, J. M., Hobbs, R. J., Pellant, M., & Pyke, D. (2004). Effects of invasive alien plants on fire regimes. *Bioscience*, *54*(7), 677–688. DOI: 10.1641/0006-3568(2004)054[0677:EOIAPO]2.0.CO;2

Carruthers, R. I., Jack Deloach, C., Herr, J. C., Anderson, G. L., & Knutson, A. E. (2008). Salt cedar areawide pest management in the western USA. In *Area-wide pest management: theory and implementation* (pp. 271–299). CABI. DOI: 10.1079/9781845933722.0271

Chazdon, R. L. (2008). Beyond deforestation: Restoring forests and ecosystem services. *Science*, *320*(5882), 1458–1460. DOI: 10.1126/science.1155365 PMID: 18556551

Cleverly, J. R., Smith, S. D., Sala, A., & Devitt, D. A. (2006). Invasive capacity of Tamarix ramosissima in a Mojave Desert floodplain: The role of drought. *Oecologia*, *147*(2), 296–306. DOI: 10.1007/s00442-005-0284-3 PMID: 28307496

Clewell, A. F., & Aronson, J. (2006). Motivations for the Restoration of Ecosystems. *Conservation Biology*, *20*(2), 420–428. DOI: 10.1111/j.1523-1739.2006.00340.x PMID: 16903103

Clewell, A. F., & Aronson, J. (2012). *Ecological restoration: Principles, values, and structure of an emerging profession*. Island Press.

Clewell, A. F., Rieger, J. P., & Munro, J. (2004). The Rights and Wrongs of Restoration. *Theoretical and Applied Genetics*, *109*(4), 733–742. DOI: 10.1007/s12031-003-1004-8 PMID: 15146319

D'Antonio, C. M., & Chambers, J. C. (2006). Using ecological theory to manage or restore ecosystems affected by invasive plant species. *Ecological Applications*, *16*(4), 1031–1038. DOI: 10.1890/1051-0761(2006)016[1031:UETTMO]2.0.CO;2

D'Antonio, C. M., & Meyerson, L. A. (2002). Exotic plant species as problems and solutions in ecological restoration: A synthesis. *Restoration Ecology*, *10*(4), 703–713. DOI: 10.1046/j.1526-100X.2002.01051.x

D'Antonio, C. M., & Vitousek, P. M. (1992). Biological invasions by exotic grasses, the grass/fire cycle, and global change. *Annual Review of Ecology and Systematics*, *23*(1), 63–87. DOI: 10.1146/annurev.es.23.110192.000431

Datta, D., Chattopadhyay, R. N., & Guha, P. (2012). Community based mangrove management: A review on status and sustainability. *Journal of Environmental Management*, *107*, 84–95. DOI: 10.1016/j.jenvman.2012.04.013 PMID: 22595074

Davies, K. W., & Sheley, R. L. (2011). Promoting native vegetation and diversity in exotic annual grass infestations. *Restoration Ecology*, *19*(2), 159–165. DOI: 10.1111/j.1526-100X.2009.00548.x

DiTomaso, J. M. (2000). Invasive weeds in rangelands: Species, impacts, and management. *Weed Science*, *48*(2), 255–265. DOI: 10.1614/0043-1745(2000)048[0255:IWIRSI]2.0.CO;2

DiTomaso, J. M., Kyser, G. B., & Pitcairn, M. J. (2013). Yellow star thistle management guide. California Invasive Plant Council. Retrieved from http://www.cal-ipc.org

Dodd, A. P. (1940). *The biological campaign against prickly pear*. Commonwealth Prickly Pear Board.

Durán, P., Tortella, G., Viscardi, S., Barra, P. J., Carrión, V. J., Mora, M. D. L. L., & Pozo, M. J. (2018). Microbial community composition in take-all suppressive soils. *Frontiers in Microbiology*, *9*, 2198. DOI: 10.3389/fmicb.2018.02198 PMID: 30283421

EEA (European Environment Agency). (2017). Green Infrastructure and Urban Planning. EEA Report No 7/2017. https://www.eea.europa.eu/publications/green-infrastructure-and-urban-planning

Ehrenfeld, J. G. (2003). Effects of exotic plant invasions on soil nutrient cycling processes. *Ecosystems (New York, N.Y.)*, *6*(6), 503–523. DOI: 10.1007/s10021-002-0151-3

Feagin, R. A., Lozada-Bernard, S. M., Ravens, T. M., Möller, I., Yeager, K. M., & Baird, A. H. (2009). Does vegetation prevent wave erosion of salt marsh edges? *Proceedings of the National Academy of Sciences of the United States of America*, *106*(25), 10109–10113. DOI: 10.1073/pnas.0901297106 PMID: 19509340

Filazzola, A., Brown, C., Dettlaff, M. A., Batbaatar, A., Grenke, J., Bao, T., Peetoom Heida, I., & Cahill, J. F.Jr. (2020). The effects of livestock grazing on biodiversity are multi-trophic: A meta-analysis. *Ecology Letters*, *23*(8), 1298–1309. DOI: 10.1111/ele.13527 PMID: 32369874

Florence, A. M., Higley, L. G., Drijber, R. A., Francis, C. A., & Lindquist, J. L. (2019). Cover crop mixture diversity, biomass productivity, weed suppression, and stability. *PLoS One*, *14*(3), e0206195. DOI: 10.1371/journal.pone.0206195 PMID: 30870424

Forseth, I. N., & Innis, A. F. (2004). Kudzu (Pueraria montana): History, physiology, and ecology combine to make a major ecosystem threat. *Critical Reviews in Plant Sciences*, *23*(5), 401–413. DOI: 10.1080/07352680490505150

Fox, H., & Cundill, G. (2018). Towards increased community-engaged ecological restoration: A review of current practice and future directions. *Ecological Restoration*, *36*(3), 208–218. DOI: 10.3368/er.36.3.208

Fuhlendorf, S. D., Engle, D. M., Kerby, J. A. Y., & Hamilton, R. (2009). Pyric herbivory: Rewilding landscapes through the recoupling of fire and grazing. *Conservation Biology*, *23*(3), 588–598. DOI: 10.1111/j.1523-1739.2008.01139.x PMID: 19183203

Funk, J. L., Cleland, E. E., Suding, K. N., & Zavaleta, E. S. (2008). Restoration through reassembly: Plant traits and invasion resistance. *Trends in Ecology & Evolution*, *23*(12), 695–703. DOI: 10.1016/j.tree.2008.07.013 PMID: 18951652

Funk, J. L., Parker, I. M., Matzek, V., Flory, S. L., Aschehoug, E. T., D'Antonio, C. M., Dawson, W., Thomson, D. M., & Valliere, J. (2020). Keys to enhancing the value of invasion ecology research for management. *Biological Invasions*, *22*(8), 2431–2445. DOI: 10.1007/s10530-020-02267-9

Gerber, E., Krebs, C., Murrell, C., Moretti, M., Rocklin, R., & Schaffner, U. (2008). Exotic invasive knotweeds (Fallopia spp.) negatively affect native plant and invertebrate assemblages in European riparian habitats. *Biological Conservation*, *141*(3), 646–654. DOI: 10.1016/j.biocon.2007.12.009

Gilliom, R. J., Barbash, J. E., Crawford, C. G., Hamilton, P. A., Martin, J. D., Nakagaki, N., Nowell, L. H., Scott, J. C., Stackelberg, P. E., Thelin, G. P., & Wolock, D. M. (2006). The quality of our nation's waters—Pesticides in the nation's streams and ground water, 1992–2001. U.S. Geological Survey Circular 1291. Retrieved from https://pubs.usgs.gov/circ/2005/1291/

Gómez-Aparicio, L., & Canham, C. D. (2008). Neighborhood models of the effects of invasive tree species on ecosystem processes. *Ecological Monographs*, *78*(1), 69–86. DOI: 10.1890/06-2036.1

Harris, J. A. (2003). Measurements of the soil microbial community for estimating the success of restoration. *European Journal of Soil Science*, *54*(4), 801–808. DOI: 10.1046/j.1351-0754.2003.0559.x

Hayes, M. A., Jesse, A., Hawke, B., Baldock, J., Tabet, B., Lockington, D., & Lovelock, C. E. (2017). Dynamics of sediment carbon stocks across intertidal wetland habitats of Moreton Bay, Australia. *Global Change Biology*, *23*(10), 4222–4234. DOI: 10.1111/gcb.13722 PMID: 28407457

Hazelton, E. L., Mozdzer, T. J., Burdick, D. M., Kettenring, K. M., & Whigham, D. F. (2014). Phragmites australis management in the United States: 40 years of methods and outcomes. *AoB Plants*, *6*, plu001. DOI: 10.1093/aobpla/plu001 PMID: 24790122

Heller, N. E., & Zavaleta, E. S. (2009). Biodiversity management in the face of climate change: A review of 22 years of recommendations. *Biological Conservation*, *142*(1), 14–32. DOI: 10.1016/j.biocon.2008.10.006

Hobbs, R. J., & Harris, J. A. (2001). Restoration ecology: Repairing the Earth's ecosystems in the new millennium. *Restoration Ecology*, *9*(2), 1–9. DOI: 10.1046/j.1526-100x.2001.009002239.x

Hobbs, R. J., & Humphries, S. E. (1995). An integrated approach to the ecology and management of plant invasions. *Conservation Biology*, *9*(4), 761–770. DOI: 10.1046/j.1523-1739.1995.09040761.x

Hobbs, R. J., & Norton, D. A. (2004). Ecological filters, thresholds, and gradients in resistance to ecosystem reassembly after invasion. *Ecology*, *85*(12), 3391–3397. DOI: 10.1890/04-0242

Holl, K. D., & Aide, T. M. (2011). When and where to actively restore ecosystems? *Forest Ecology and Management*, *261*(10), 1558–1563. DOI: 10.1016/j.foreco.2010.07.004

Holling, C. S. (1978). *Adaptive environmental assessment and management*. Wiley-Interscience.

Hulvey, K. B., Standish, R. J., Hallett, L. M., Starzomski, B. M., Murphy, S. D., Nelson, C. R., ... & Suding, K. N. (2013). Incorporating novel ecosystems into management frameworks. *Novel ecosystems: intervening in the new ecological world order*, 157-171.

James, J. J., Gornish, E. S., DiTomaso, J. M., Davy, J., Doran, M. P., Becchetti, T., Lile, D., Brownsey, P., & Laca, E. A. (2015). Managing medusahead (Taeniatherum caput-medusae) on rangeland: A meta-analysis of control effects and assessment of stakeholder needs. *Rangeland Ecology and Management*, *68*(3), 215–223. DOI: 10.1016/j.rama.2015.03.006

Keeley, J. E. (2006). Fire management impacts on invasive plants in the Western United States. *Conservation Biology*, *20*(2), 375–384. DOI: 10.1111/j.1523-1739.2006.00339.x PMID: 16903098

Kourtev, P. S., Ehrenfeld, J. G., & Häggblom, M. (2002). Exotic plant species alter the microbial community structure and function in the soil. *Ecology*, *83*(11), 3152–3166. DOI: 10.1890/0012-9658(2002)083[3152:EPSATM]2.0.CO;2

Lewis, R. R.III. (2005). Ecological engineering for successful management and restoration of mangrove forests. *Ecological Engineering*, *24*(4), 403–418. DOI: 10.1016/j.ecoleng.2004.10.003

Liebman, M., & Gallandt, E. R. (1997). Many little hammers: Ecological management of crop-weed interactions. *Ecological Applications*, *7*(1), 43–56. DOI: 10.1890/1051-0761(1997)007[0043:MLHEMO]2.0.CO;2

Lindenmayer, D. B., & Likens, G. E. (2009). Adaptive monitoring: A new paradigm for long-term research and monitoring. *Trends in Ecology & Evolution*, *24*(9), 482–486. DOI: 10.1016/j.tree.2009.03.005 PMID: 19409648

Lockwood, J. L., Hoopes, M. F., & Marchetti, M. P. (2013). *Invasion ecology*. John Wiley & Sons.

Longley, P. A., Goodchild, M. F., Maguire, D. J., & Rhind, D. W. (2015). *Geographic Information Systems and Science*. Wiley.

Louda, S. M., Kendall, D., Connor, J., & Simberloff, D. (2003). Ecological effects of an insect introduced for the biological control of weeds. *Science*, *277*(5329), 1088–1090. DOI: 10.1126/science.277.5329.1088

Mahmood, Q., Bilal, M., & Jan, S. (2014). Herbicides, pesticides, and plant tolerance: an overview. *Emerging technologies and management of crop stress tolerance*, 423-448.

Manfreda, S., McCabe, M. F., Miller, P. E., Lucas, R., Pajuelo Madrigal, V., Mallinis, G., Ben Dor, E., Helman, D., Estes, L., Ciraolo, G., Müllerová, J., Tauro, F., De Lima, M. I., De Lima, J. L. M. P., Maltese, A., Frances, F., Caylor, K., Kohv, M., Perks, M., & Toth, B. (2018). On the use of unmanned aerial systems for environmental monitoring. *Remote Sensing (Basel)*, *10*(4), 641. DOI: 10.3390/rs10040641

Martin, D. M., Mazzotta, M., & Bousquin, J. (2018). Combining ecosystem services assessment with structured decision making to support ecological restoration planning. *Environmental Management*, *62*(3), 608–618. DOI: 10.1007/s00267-018-1038-1 PMID: 29637276

Mee, M. T., & Wang, H. H. (2012). Engineering ecosystems and synthetic ecologies. *Molecular BioSystems*, *8*(10), 2470–2483. DOI: 10.1039/c2mb25133g PMID: 22722235

Miller, J. H., Manning, S. T., & Enloe, S. F. (2010). A management guide for invasive plants in southern forests. USDA Forest Service. Retrieved from https://www.srs.fs.usda.gov/pubs/gtr/gtr_srs131.pdf

Mitsch, W. J., & Jørgensen, S. E. (2004). *Ecological engineering and ecosystem restoration*. Wiley.

Morin, L., Reid, A. M., Sims-Chilton, N. M., Buckley, Y. M., Dhileepan, K., Hastwell, G. T., Nordblom, T. L., & Raghu, S. (2009). Review of approaches to evaluate the effectiveness of weed biological control agents. *Biological Control*, *51*(1), 1–15. DOI: 10.1016/j.biocontrol.2009.05.017

Norris, R. F., Caswell-Chen, E. P., & Kogan, M. (2003). *Concepts in integrated pest management*. Prentice Hall.

Owen, M. D., Beckie, H. J., Leeson, J. Y., Norsworthy, J. K., & Steckel, L. E. (2015). Integrated pest management and weed management in the United States and Canada. *Pest Management Science*, *71*(3), 357–376. DOI: 10.1002/ps.3928 PMID: 25346235

Palmer, M. A., Ambrose, R. F., & Poff, N. L. (1997). Ecological theory and community restoration ecology. *Restoration Ecology*, *5*(4), 291–300. DOI: 10.1046/j.1526-100X.1997.00543.x

Paudyal, K., Baral, H., Putzel, L., Bhandari, S., & Keenan, R. J. (2017). Change in land use and ecosystem services delivery from community-based forest landscape restoration in the Phewa Lake watershed, Nepal. *International Forestry Review*, *19*(4), 88–101. DOI: 10.1505/146554817822330524

Pearson, D., & Ortega, Y. (2009). Managing invasive plants in natural areas: moving beyond weed control. *In: Kingely, Rudolph V., ed. Weeds: Management, Economic Impacts and Biology. Hauppauge, NY: Nova Science Publishers, Inc. p. 1-21.*, 1-21.

Powles, S. B. (2008). Evolved glyphosate-resistant weeds around the world: Lessons to be learnt. *Pest Management Science*, *64*(4), 360–365. DOI: 10.1002/ps.1525 PMID: 18273881

Pyšek, P., Hulme, P. E., Simberloff, D., Bacher, S., Blackburn, T. M., Carlton, J. T., Dawson, W., Essl, F., Foxcroft, L. C., Genovesi, P., Jeschke, J. M., Kühn, I., Liebhold, A. M., Mandrak, N. E., Meyerson, L. A., Pauchard, A., Pergl, J., Roy, H. E., Seebens, H., & Richardson, D. M. (2020). Scientists' warning on invasive alien species. *Biological Reviews of the Cambridge Philosophical Society*, *95*(6), 1511–1534. DOI: 10.1111/brv.12627 PMID: 32588508

Pyšek, P., & Richardson, D. M. (2010). Invasive species, environmental change and management, and health. *Annual Review of Environment and Resources*, *35*(1), 25–55. DOI: 10.1146/annurev-environ-033009-095548

Pywell, R. F., Bullock, J. M., Hopkins, A., Walker, K. J., Sparks, T. H., Burke, M. J., & Peel, S. (2002). Restoration of species-rich grassland on arable land: Assessing the limiting processes using a multi-site experiment. *Journal of Applied Ecology*, *39*(2), 294–309. DOI: 10.1046/j.1365-2664.2002.00718.x

Reed, M. S. (2008). Stakeholder participation for environmental management: A literature review. *Biological Conservation*, *141*(10), 2417–2431. DOI: 10.1016/j.biocon.2008.07.014

Richardson, D. M., & Rejmánek, M. (2011). Trees and shrubs as invasive alien species–a global review. *Diversity & Distributions*, *17*(5), 788–809. DOI: 10.1111/j.1472-4642.2011.00782.x

Rose, R. A., Byler, D., Eastman, J. R., Fleishman, E., Geller, G., Goetz, S., Guild, L., Hamilton, H., Hansen, M., Headley, R., Hewson, J., Horning, N., Kaplin, B. A., Laporte, N., Leidner, A., Leimgruber, P., Morisette, J., Musinsky, J., Pintea, L., & Wilson, C. (2015). Ten ways remote sensing can contribute to conservation. *Conservation Biology*, *29*(2), 350–359. DOI: 10.1111/cobi.12397 PMID: 25319024

Ruiz-Jaen, M. C., & Aide, T. M. (2005). Restoration success: How is it being measured? *Restoration Ecology*, *13*(3), 569–577. DOI: 10.1111/j.1526-100X.2005.00072.x

Seastedt, T. R., Hobbs, R. J., & Suding, K. N. (2008). Management of novel ecosystems: Are novel approaches required? *Frontiers in Ecology and the Environment*, *6*(10), 547–553. DOI: 10.1890/070046

Sharma, G. P., Raghubanshi, A. S., & Singh, J. S. (2005). Lantana invasion: An overview. *Weed Biology and Management*, *5*(4), 157–167. DOI: 10.1111/j.1445-6664.2005.00178.x

Sheley, R. L., James, J. J., Smith, B. S., & Vasquez, E. A. (2010). Applying ecologically-based invasive plant management. *Rangeland Ecology and Management*, *63*(6), 605–613. DOI: 10.2111/REM-D-09-00187.1

Sheley, R. L., Mangold, J. M., & Anderson, J. L. (2006). Potential for successional theory to guide restoration of invasive plant-dominated rangeland. *Ecological Monographs*, *81*(2), 181–198. DOI: 10.1890/0012-9615(2006)076[0365:PFSTTG]2.0.CO;2

Simberloff, D., Martin, J. L., Genovesi, P., Maris, V., Wardle, D. A., Aronson, J., Courchamp, F., Galil, B., García-Berthou, E., Pascal, M., Pyšek, P., Sousa, R., Tabacchi, E., & Vila, M. (2013). Impacts of biological invasions: What's what and the way forward. *Trends in Ecology & Evolution*, *28*(1), 58–66. DOI: 10.1016/j.tree.2012.07.013 PMID: 22889499

Simberloff, D., & Stiling, P. (1996). How risky is biological control? *Ecology*, *77*(7), 1965–1974. DOI: 10.2307/2265693

Simberloff, D., & Von Holle, B. (1999). Positive interactions of nonindigenous species: Invasional meltdown? *Biological Invasions*, *1*(1), 21–32. DOI: 10.1023/A:1010086329619

Singh, A., Vaish, B., & Singh, R. P. (2016). Eco-restoration of degraded lands through microbial biomass: An ecological engineer. *Acta Biomedica Scientia*, *3*(1), 133–135.

Skurski, T. C., Rew, L. J., & Maxwell, B. D. (2014). Mechanisms underlying nonindigenous plant impacts: A review of recent experimental research. *Invasive Plant Science and Management*, *6*(1), 89–105. DOI: 10.1614/IPSM-D-13-00099.1

Society for Ecological Restoration (SER). (2004). The SER International Primer on Ecological Restoration. Society for Ecological Restoration International. Retrieved from http://c.ymcdn.com/sites/www.ser.org/resource/resmgr/custompages/publications/SER_Primer.pdf

Stapleton, J. J. (2000). Soil solarization in various agricultural production systems. *Crop Protection (Guildford, Surrey)*, *19*(8-10), 837–841. DOI: 10.1016/S0261-2194(00)00111-3

Stromberg, J. C., Beauchamp, V. B., Dixon, M. D., Lite, S. J., & Paradzick, C. (2007). Importance of low-flow and high-flow characteristics to restoration of riparian vegetation along rivers in arid southwestern United States. *Freshwater Biology*, *52*(4), 651–679. DOI: 10.1111/j.1365-2427.2006.01713.x

Suding, K., Higgs, E., Palmer, M., Callicott, J. B., Anderson, C. B., Baker, M., Gutrich, J. J., Hondula, K. L., LaFevor, M. C., Larson, B. M. H., Randall, A., Ruhl, J. B., & Schwartz, K. Z. (2015). Committing to ecological restoration. *Science*, *348*(6235), 638–640. DOI: 10.1126/science.aaa4216 PMID: 25953995

Suding, K. N. (2011). Toward an era of restoration in ecology: Successes, failures, and opportunities ahead. *Annual Review of Ecology, Evolution, and Systematics*, *42*(1), 465–487. DOI: 10.1146/annurev-ecolsys-102710-145115

Suding, K. N., Gross, K. L., & Houseman, G. R. (2004). Alternative states and positive feedbacks in restoration ecology. *Trends in Ecology & Evolution*, *19*(1), 46–53. DOI: 10.1016/j.tree.2003.10.005 PMID: 16701225

TEEB (The Economics of Ecosystems and Biodiversity). (2010). The Economics of Ecosystems and Biodiversity Ecological and Economic Foundations. Pushpam Kumar (Ed.). Earthscan.

Török, P., Brudvig, L. A., Kollmann, J. N., Price, J., & Tóthmérész, B. (2021). The present and future of grassland restoration. *Restoration Ecology*, *29*, e13378. DOI: 10.1111/rec.13378

Trowbridge, C. C., Stanley, A., Kaye, T. N., Dunwiddie, P. W., & Williams, J. L. (2017). Long-term effects of prairie restoration on plant community structure and native population dynamics. *Restoration Ecology*, *25*(4), 559–568. DOI: 10.1111/rec.12468

Tu, M., Hurd, C., & Randall, J. M. (2001). Weed control methods handbook: Tools and techniques for use in natural areas. The Nature Conservancy. Retrieved from https://tncweeds.ucdavis.edu/handbook.html

Turner, W., Rondinini, C., Pettorelli, N., Mora, B., Leidner, A. K., Szantoi, Z., Buchanan, G., Dech, S., Dwyer, J., Herold, M., Koh, L. P., Leimgruber, P., Taubenboeck, H., Wegmann, M., Wikelski, M., & Woodcock, C. (2015). Free and open-access satellite data are key to biodiversity conservation. *Biological Conservation*, *182*, 173–176. DOI: 10.1016/j.biocon.2014.11.048

Van der Heijden, M. G., Bardgett, R. D., & van Straalen, N. M. (2008). The unseen majority: Soil microbes as drivers of plant diversity and productivity in terrestrial ecosystems. *Ecology Letters*, *11*(3), 296–310. DOI: 10.1111/j.1461-0248.2007.01139.x PMID: 18047587

Van Driesche, R., & Hoddle, M. (2009). *Control of pests and weeds by natural enemies: an introduction to biological control.* John Wiley & Sons.

Vilà, M., Espinar, J. L., Hejda, M., Hulme, P. E., Jarošík, V., Maron, J. L., Pergl, J., Schaffner, U., Sun, Y., & Pyšek, P. (2011). Ecological impacts of invasive alien plants: A meta-analysis of their effects on species, communities and ecosystems. *Ecology Letters*, *14*(7), 702–708. DOI: 10.1111/j.1461-0248.2011.01628.x PMID: 21592274

Walker, B., Holling, C. S., Carpenter, S. R., & Kinzig, A. (2004). Resilience, adaptability, and transformability in social–ecological systems. *Ecology and Society*, *9*(2), 5. https://www.ecologyandsociety.org/vol9/iss2/art5/. DOI: 10.5751/ES-00650-090205

Walters, C. J., & Holling, C. S. (1990). Large-scale management experiments and learning by doing. *Ecology*, *71*(6), 2060–2068. DOI: 10.2307/1938620

Were, D., Kansiime, F., Fetahi, T., Cooper, A., & Jjuuko, C. (2019). Carbon sequestration by wetlands: A critical review of enhancement measures for climate change mitigation. *Earth Systems and Environment*, *3*(2), 327–340. DOI: 10.1007/s41748-019-00094-0

Williams, B. K., Szaro, R. C., & Shapiro, C. D. (2009). *Adaptive management: The U.S. Department of the Interior technical guide*. U.S. Department of the Interior.

Wilson, J. R. U., Holst, N., & Rees, M. (2005). Determinants and patterns of population growth in water hyacinth. *Weed Research*, *47*(5), 391–400. DOI: 10.1111/j.1365-3180.2007.00582.x

Wortley, L., Hero, J.-M., & Howes, M. (2013). Evaluating ecological restoration success: A review of the literature. *Restoration Ecology*, *21*(5), 537–543. DOI: 10.1111/rec.12028

Yelenik, S. G., Stock, W. D., & Richardson, D. M. (2004). Ecosystem level impacts of invasive Acacia saligna in the South African fynbos. *Restoration Ecology*, *12*(1), 44–51. DOI: 10.1111/j.1061-2971.2004.00289.x

Young, T. P., Petersen, D. A., & Clary, J. J. (2005). The ecology of restoration: Historical links, emerging issues, and unexplored realms. *Ecology Letters*, *8*(6), 662–673. DOI: 10.1111/j.1461-0248.2005.00764.x

Zahoor, M. A. A., Haidar, S. H., Rafique, S., Ahmed, U., Umber Rauf, M. A., Hanif, U., & Shehryar, M. (2025). Biotechnological Advances in Wildlife Conservation: Genetic Engineering, Cloning, Ecosystem Restoration, and Nanoparticle Applications. *Scholars Academic Journal of Biosciences*, *4*(4), 386–416. DOI: 10.36347/sajb.2025.v13i04.005

Zedler, J. B., & Callaway, J. C. (1999). Tracking wetland restoration: Do mitigation sites follow desired trajectories? *Restoration Ecology*, *7*(1), 69–73. DOI: 10.1046/j.1526-100X.1999.07108.x

Chapter 5
Restoration Techniques to Mitigate Plant Invasions Under Climate Change

P. Selvakumar
https://orcid.org/0000-0002-3650-4548
Department of Science and Humanities, Nehru Institute of Technology, India

A. Rohini
https://orcid.org/0000-0002-9779-6572
CVR College of Engineering, India

M. R. Arun
https://orcid.org/0000-0002-8291-2956
Vel Tech Rangarajan Dr. Sagunthala R&D Institute of Science and Technology, India

T. C. Manjunath
https://orcid.org/0000-0003-2545-9160
Rajarajeswari College of Engineering, India

ABSTRACT

In recent decades, the convergence of plant invasions and climate change has emerged as a critical concern for ecosystems and human societies worldwide. Plant invasions occur when non-native species spread into new areas, often outcompeting native flora and disrupting local ecological balances. Conversely, climate change can weaken native plants, making them more susceptible to invasion. Additionally, climate changes can disrupt plant-pollinator relationships, further complicating ecosystem

DOI: 10.4018/979-8-3693-8724-5.ch005

dynamics and potentially accelerating the spread of invasive species. Furthermore, climate change can facilitate the movement of invasive species across regions by altering habitat connectivity and allowing species to colonize new areas that were previously climatically unsuitable. This dynamic poses significant challenges for conservation efforts, as traditional management strategies may become less effective in the face of rapidly changing conditions. Addressing the dual challenges of plant invasions and climate change requires a multifaceted approach.

INTRODUCTION TO PLANT INVASIONS AND CLIMATE CHANGE IMPACTS

In recent decades, the convergence of plant invasions and climate change has emerged as a critical concern for ecosystems and human societies worldwide. Plant invasions occur when non-native species spread into new areas, often outcompeting native flora and disrupting local ecological balances. Conversely, climate change can weaken native plants, making them more susceptible to invasion. Additionally, climate changes can disrupt plant-pollinator relationships, further complicating ecosystem dynamics and potentially accelerating the spread of invasive species. Furthermore, climate change can facilitate the movement of invasive species across regions by altering habitat connectivity and allowing species to colonize new areas that were previously climatically unsuitable. This dynamic poses significant challenges for conservation efforts, as traditional management strategies may become less effective in the face of rapidly changing conditions. Addressing the dual challenges of plant invasions and climate change requires a multifaceted approach. Overall, the intersection of plant invasions and climate change presents a significant challenge for both ecological conservation and human well-being. By advancing our understanding of how climate change influences plant invasions and implementing comprehensive management strategies, we can better protect vulnerable ecosystems and ensure the sustainability of natural resources for future generations.

Assessing invasion risk through predictive models and mapping is an essential component of modern ecological management, offering valuable insights into how invasive species might spread and impact ecosystems. Predictive models use various algorithms and data sources to forecast the potential distribution of invasive species based on environmental variables, species traits, and historical invasion patterns. These models integrate data on climate, land use, soil types, and ecological interactions to estimate areas at high risk of invasion. They often employ ecological niche modeling (ENM), which evaluates the suitability of different environments for an invasive species by comparing the conditions of known invasion sites to those across potential new habitats (Mudereri B T., et al, 2020). Machine learning and statistical techniques are increasingly used to refine these models, enhancing their accuracy by incorporating large datasets and complex interactions. Mapping is a critical tool for visualizing and communicating the results of predictive models. Geographic Information Systems (GIS) are employed to create detailed maps that illustrate the potential spread of invasive species across different landscapes. These maps can highlight areas of high risk, inform resource allocation for monitoring and management, and guide policy decisions. By overlaying model predictions with existing spatial data, such as land cover and infrastructure, decision-makers can identify vulnerable regions and prioritize management actions. Additionally, mapping helps in tracking the actual spread of invasive species, allowing for the calibration and refinement of predictive models over time. The accuracy of predictive models and mapping depends on several factors, including the quality and resolution of input data, the choice of modeling techniques, and the inclusion of various ecological and socioeconomic variables (Ponti L., et al, 2021). Models must be regularly updated to reflect changes in environmental conditions and new information about species behavior. Moreover, while predictive models can provide valuable forecasts, they often involve uncertainties related to the variability of ecological interactions and future climate scenarios. Therefore, integrating model predictions with field observations and expert knowledge is crucial for improving their reliability and practical utility. Effective risk assessment also requires collaboration across disciplines, including ecology, geography, climate science, and data science. Multidisciplinary approaches can enhance model development and ensure that predictions are relevant and actionable. For instance, incorporating local ecological knowledge and stakeholder input can refine models to better address specific regional conditions and management needs. In summary, predictive models and mapping are vital for assessing invasion risk and guiding invasive species management (Pili A N., et al, 2020). They enable proactive identification of high-risk areas, support targeted interventions, and help allocate resources efficiently. However, their effectiveness relies on high-quality data, robust modeling techniques, and ongoing refinement based on real-world observations. As the field of predictive ecology advances, integrating new technologies and approaches will further enhance our ability to anticipate and manage the spread of invasive species, ultimately contributing to more effective conservation and ecosystem management strategies.

Restoration Goals and Objectives: Setting the Stage

The process begins with assessing the current state of the ecosystem and identifying the ecological, social, and economic values that need to be restored or preserved. This assessment helps in formulating clear, achievable goals that address both the ecological functions and the habitat quality necessary for supporting diverse plant and animal species (Lin X., et al, 2023). For instance, in wetlands, goals may involve reintroducing native plant species that enhance water filtration and support wildlife. In forests, objectives might focus on restoring native tree species to improve biodiversity and habitat complexity. Restoration goals should be specific and measurable, often framed in terms of quantitative targets such as the area of habitat restored, the number of species reintroduced, or the improvement in ecosystem services like carbon sequestration or flood regulation. Another key aspect of setting restoration goals involves addressing socioeconomic factors. Restoration projects often aim to balance ecological benefits with social and economic needs. This may involve engaging local communities, providing educational opportunities, or creating recreational spaces (Wade D T., 2020). Objectives might include enhancing local livelihoods through sustainable use of restored resources or involving community members in restoration activities to foster stewardship and support for the project. The formulation of restoration goals and objectives also requires consideration of the scale and timeframe of the project. Restoration efforts can vary widely in scale, from small-scale projects focusing on individual sites to large landscape-level initiatives. Goals should be realistic and achievable within the constraints of available resources and time. Long-term monitoring and adaptive management are essential to track progress, address unforeseen challenges, and make necessary adjustments to the restoration plan (Zhang Z., et al, 2023). Stakeholder engagement helps in identifying priorities, addressing potential conflicts, and securing support for the restoration effort. By establishing clear, measurable, and achievable goals, restoration projects can effectively guide efforts to rehabilitate ecosystems, enhance ecological functions, and provide socioeconomic benefits. Successful restoration depends on careful planning, stakeholder collaboration, and ongoing evaluation to ensure that objectives are met and that ecosystems can thrive in the future.

Native Plant Species Selection: Criteria and Considerations

Selecting native plant species for ecological restoration or landscaping projects involves a careful evaluation of various criteria to ensure that the chosen species will thrive in the local environment and contribute to the restoration goals. The process is guided by several key considerations, each aimed at promoting ecological health and resilience. Ecological Suitability: The primary criterion for selecting native plant species is their ecological suitability for the specific site conditions. For example, species that thrive in well-drained soils might be chosen for upland areas, while those adapted to wet conditions are more suitable for riparian zones (Diener & Mudu, 2021). It is important to choose a variety of species to create a diverse plant community that supports different types of organisms and ecological processes, such as pollination, seed dispersal, and soil enrichment. Historical and Regional Context:

Understanding the historical and regional context of the area helps in selecting species that were originally part of the local ecosystem. Historical records, such as botanical surveys and palaeoecological data, can provide insights into which species were native to the area before disturbance. This approach ensures that the restored plant community resembles the pre-disturbance ecosystem and supports the restoration of natural ecological processes. Invasiveness and Compatibility: Avoiding species that have the potential to become invasive is critical. Invasive plants can outcompete native species, disrupt ecological balance, and lead to long-term ecological damage. Selecting species that are noninvasive and have a low risk of becoming problematic in the new environment is essential for the success of restoration projects.

Growth Characteristics and Maintenance: Consideration of the growth characteristics of selected plant species, such as their size, growth rate, and maintenance requirements, is important for achieving project goals and ensuring long-term success. For instance, selecting species that are low maintenance and require minimal intervention can reduce the need for ongoing management and costs. Additionally, understanding the growth patterns and interactions of different species helps in designing plant communities that are stable and resilient (Akinola R., et al, 2020). Cultural and Aesthetic Values: In addition to ecological criteria, cultural and aesthetic values may influence plant selection. Native plants that hold cultural significance or provide aesthetic benefits, such as seasonal blooms or attractive foliage, can enhance the social and visual appeal of a restoration project. Incorporating culturally significant species can also strengthen community engagement and support for the project. Resilience to Climate Change: As climate change alters environmental conditions, selecting plant species that are resilient to changing climate patterns is increasingly important (Weidlich et al., 2020).

Soil Remediation: Preparing for Restoration

Soil remediation is a crucial process aimed at restoring contaminated soils to a state where they can support healthy ecosystems and human activities. The need for soil remediation often arises due to industrial activities, agricultural practices, or other sources of pollution that introduce harmful substances into the soil. Preparing for soil remediation involves a comprehensive approach that includes site assessment, planning, and the selection of appropriate remediation techniques. The first step in preparing for soil remediation is conducting a thorough site assessment. This involves investigating the extent and nature of the contamination, including the types and concentrations of pollutants present, the depth and spread of contamination, and the characteristics of the soil itself. Techniques such as soil sampling, chemical analysis, and geophysical surveys are employed to gather this information (Wang L., et al, 2021). Understanding the contamination's origin and its potential impacts on surrounding environments and human health is critical for determining the appropriate remediation strategy. Following the site assessment, the next phase is planning. This involves setting clear remediation goals based on the assessment findings. Goals may include reducing pollutant levels to meet regulatory standards, restoring soil fertility, or ensuring the safety of the land for future use. The planning process also includes developing a detailed remediation plan that outlines the selected methods, timeline, and budget. This plan must consider factors such as the volume of contaminated soil, the types of pollutants, and the potential impacts of remediation activities on the surrounding environment and communities. Choosing the right remediation technique is a key aspect of the preparation process. There are several methods available, each with its advantages and limitations (Lee S W., et al, 2020). Techniques can be broadly categorized into in-situ and ex-situ methods. In situ methods treat the contaminated soil at its original location, reducing the need for excavation and minimizing disturbance. Examples include bioremediation, where microorganisms are used to degrade pollutants, and phytoremediation, which involves the use of plants to absorb or transform contaminants. Ex-situ methods, on the other hand, involve excavating the contaminated soil and treating it elsewhere. These methods include soil washing, which uses water and chemical solutions to remove contaminants, and thermal treatment, where soil is heated to vaporize pollutants.

The selection of a remediation method depends on various factors, including the type of contaminants, soil properties, and environmental conditions. For instance, bioremediation may be effective for organic pollutants but less so for heavy metals. Thermal treatment might be suitable for highly contaminated soils but is often costly. Additionally, the chosen method must be feasible within the project's budget and timeline constraints. Another important aspect of preparation is stakeholder engagement and regulatory compliance (Bakina L G., et al, 2021). Monitoring and evaluating the remediation process is also a crucial component of preparation. This involves establishing baseline conditions, setting up monitoring systems to track the progress of remediation, and evaluating the effectiveness of the chosen methods. Regular monitoring helps to ensure that the remediation goals are being met and allows for adjustments to the plan if necessary. It also provides valuable data for future projects and helps in assessing the long-term success of the remediation efforts. In conclusion, preparing for soil remediation is a multifaceted process that requires careful planning, method selection, stakeholder engagement, and ongoing monitoring. By thoroughly assessing the site, setting clear goals, choosing appropriate remediation techniques, and ensuring regulatory compliance, it is possible to effectively restore contaminated soils and mitigate the impact of pollution on the environment and human health. This comprehensive approach not only addresses the immediate issues of contamination but also contributes to the sustainable management of soil resources for future generations.

Invasive Plant Control Methods: Eradication and Management

Invasive plant species pose significant challenges to ecosystems, agriculture, and human health. Effective control of these species involves both eradication and management strategies, aimed at reducing their impact and preventing further spread. Understanding and implementing these methods requires a thorough assessment of the invasive species, the affected environment, and the most appropriate interventions. Eradication Methods: Eradication aims to completely remove invasive plants from a specific area. For instance, hand-pulling or using tools like hoes and shovels can be effective for small infestations. Mechanical control is labor-intensive and may need to be repeated regularly to ensure all plants and their roots are removed (Green & Grosholz, 2021). It is often used in conjunction with other methods to improve effectiveness. Chemical control requires careful application to avoid environmental damage and must comply with regulations to ensure safety and efficacy. Biological Control: This method uses natural predators, pathogens, or competitors to control invasive plants. For example, introducing specific insects or diseases that naturally target the invasive species can help reduce its population. Biological control requires extensive research to ensure that the introduced agents do not become invasive themselves and that they effectively target the intended plant.

Management Methods: Management involves ongoing practices to control invasive plants and prevent their spread over larger areas. These strategies are used when eradication is not feasible or when dealing with larger infestations: Integrated Pest Management (IPM): IPM combines multiple control methods to manage invasive plants in a comprehensive approach. It includes mechanical, chemical, and biological controls, along with monitoring and prevention strategies. The goal is to use the most effective methods while minimizing environmental impact and costs. Preventive Measures: Preventing the introduction and spread of invasive plants is a key aspect of management. This includes measures such as cleaning equipment, monitoring for new infestations, and implementing quarantine regulations to restrict the movement of soil, plant materials, or other potential vectors of invasive species (SimberloffD., 2021). Restoration and Replanting: After controlling invasive plants, it is often necessary to restore the affected area by replanting native species. This helps to stabilize the soil, support local wildlife, and reduce the chances of invasive plants reestablishing themselves. Restoration efforts should focus on selecting native species that are suited to the local conditions and capable of outcompeting invasive plants. Education and Outreach: Raising awareness about invasive plants and their impacts is crucial for effective management. Educating landowners, stakeholders, and the public about identification, reporting, and prevention can help prevent the spread of invasive species and support community-based management efforts. Evaluation and Monitoring: Monitoring the effectiveness of control methods and assessing the success of management strategies are essential for adapting and improving efforts over time (Feng Q., et al, 2021). Regular surveys and data collection help track the progress of invasive plant control, identify new infestations, and determine the need for adjustments in the management plan. In conclusion, controlling invasive plants requires a combination of eradication and management strategies tailored to the specific context and scale of the infestation. By employing mechanical, chemical, biological, and preventive measures, and by focusing on education, restoration, and monitoring, it is possible to mitigate the impacts of invasive plants and protect native ecosystems and biodiversity.

Ecological Restoration Techniques: Revegetation and Habitat Reconstruction

Ecological restoration aims to return degraded or disturbed ecosystems to a state of health and functionality. Two primary techniques in this field are revegetation and habitat reconstruction, each playing a crucial role in restoring ecological balance and biodiversity. Revegetation: Revegetation is the process of replanting native vegetation to restore ecological functions and promote biodiversity in areas where vegetation has been lost or degraded. This technique involves several key steps: Site Assessment: Before revegetation begins, a thorough site assessment is necessary to understand the soil conditions, hydrology, and existing vegetation. This includes evaluating the types of plants that were originally present, current site conditions, and the specific goals of the restoration project (Zhang D., et al, 2020). Understanding these factors helps in selecting appropriate plant species and planting techniques. Selection of Native Species: Choosing native plant species is critical for successful revegetation. Native plants are adapted to local conditions and support local wildlife, which helps to maintain ecological balance. A diverse mix of species, including grasses, shrubs, and trees, is often selected to mimic the original plant community and provide various ecological functions. Soil Preparation: The soil must be prepared to support the growth of new plants. This may involve removing invasive species, amending soil with organic matter, and ensuring proper drainage. Soil health is vital for the establishment of plant roots and the overall success of the revegetation efforts (Li Z., et al, 2024). Planting and Maintenance: Revegetation involves planting native species either by sowing seeds or planting seedlings. Proper planting techniques and timing are crucial for ensuring successful establishment. Ongoing maintenance, such as watering, mulching, and monitoring for pests and diseases, is necessary to support plant growth and address any issues that arise. Monitoring and Evaluation: Post-planting monitoring helps assess the success of the revegetation effort. Regular evaluations track plant survival rates, growth, and the development of plant communities. This data is used to make adjustments to management practices and ensure that the restoration goals are being met.

Figure 1. Enhancing Climate Change Resilience of Ecological Restoration

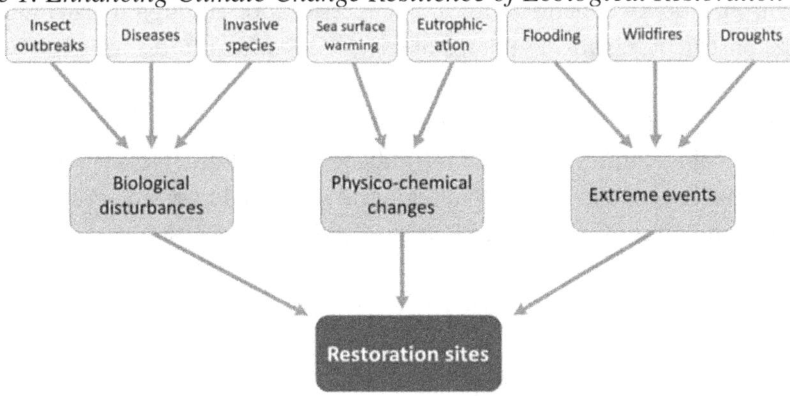

Habitat Reconstruction: Habitat reconstruction involves more extensive modifications to restore entire ecosystems or habitats that have been significantly altered or destroyed. This technique often includes Site Preparation and Analysis: Similar to revegetation, habitat reconstruction begins with a detailed site analysis. This involves understanding the historical and current conditions of the site, including soil types, hydrology, topography, and the presence of any remaining ecological features. Site preparation may involve significant alterations such as grading, excavation, or the reconfiguration of waterways. Re-establishment of Ecological Processes: Effective habitat reconstruction focuses on reestablishing key ecological processes that support ecosystem functionality. This might include restoring natural water flows, reintroducing natural fire regimes, or recreating specific landscape features like wetlands or floodplains. Reestablishing these processes is essential for creating a self-sustaining and resilient ecosystem. Construction of Habitat Features: Reconstructing habitat often involves creating or modifying physical features such as wetlands, stream channels, or forested areas. This might include constructing artificial wetlands, planting buffer zones along waterways, or creating nesting sites for wildlife. The goal is to recreate the essential elements of the habitat that support diverse plant and animal communities (Duque et al., 2021). Reintroduction of Species: In some cases, it may be necessary to reintroduce species that have been lost from the area. This can include planting specific native species or reintroducing wildlife that relies on the habitat. Careful planning is required to ensure that reintroduced species are compatible with the restored habitat and that they can thrive in their new environment. Long-Term Management and Monitoring: Successful habitat reconstruction requires ongoing management to ensure that the ecosystem remains functional and resilient. This includes monitoring ecological indicators, managing invasive species, and addressing any emerging issues. Long-term management ensures that the reconstructed habitat continues to provide ecological benefits and supports biodiversity over time. In conclusion, ecological restoration through revegetation and habitat reconstruction involves a range of techniques aimed at restoring ecosystems to their natural state. By carefully assessing sites, selecting appropriate species, and implementing effective management practices, these techniques contribute to the recovery of ecological functions and the enhancement of biodiversity. Both revegetation and habitat reconstruction play integral roles in addressing environmental degradation and promoting the health and resilience of ecosystems.

Assisted Migration: Facilitating Native Species Movement

This involves physically moving individuals or populations of a species from their current location to a new area where conditions are more suitable for their survival. Translocation can be used for both plants and animals and often requires careful planning to ensure the new habitat meets the species' ecological needs. For example, relocating a population of endangered amphibians to a higher elevation where temperatures are cooler can help them survive in the face of rising temperatures. Introduction to New Habitats: Sometimes, species are introduced to entirely new areas where they have not previously existed but where they might have a chance to thrive. This approach is typically used when natural migration corridors are blocked or when habitat conditions have changed significantly (Palik B J., et al, 2022). For example, planting a species of tree in a new region where could potentially establish a sustainable population due to favorable conditions. Assisted Migration of Ecosystem Components: This broader approach involves relocating entire ecological communities or key species to new areas. By moving multiple species that interact with each other, conservationists aim to recreate the complex ecological interactions necessary for ecosystem functioning. For instance, moving both the predators and prey of a specific habitat to a new location where conditions are more favorable can help reestablish a functioning ecosystem. Benefits of Assisted Migration: Enhanced Survival: Assisted migration can help species that are at risk of extinction due to environmental changes by providing them with new habitats that offer better conditions for survival and reproduction. This can be crucial for species that have limited dispersal capabilities or face barriers to natural migration. Preservation of Biodiversity: By facilitating the movement of species to suitable habitats, assisted migration helps preserve biodiversity and maintain ecosystem functions. This approach can also help protect species that play critical roles in their ecosystems, thereby supporting overall ecological health.

Adaptation to Climate Change: As climate change alters habitats and creates new environmental conditions, assisted migration can provide a proactive means of helping species adapt to these changes. It allows species to move to areas with climates similar to their historical ranges, potentially mitigating some of the impacts of climate change. Challenges and Considerations: Ecological Risks: Introducing species to new areas carries the risk of unintended ecological consequences. Nonnative species can become invasive, outcompeting local species and disrupting existing ecosystems. Assisted migration raises ethical questions about human intervention in natural processes. There may be concerns about the potential impact on local communities and ecosystems, as well as the long-term implications of relocating species. Engaging stakeholders and considering these concerns is important in developing and implementing assisted migration strategies. Monitoring and Evaluation: Effective monitoring and evaluation are essential components of assisted migration efforts (Charles & Stehlik, 2021). Tracking the success of relocated species involves assessing their survival, reproduction, and ecological impact in the new habitat. Regular monitoring helps to identify any emerging issues and allows for adaptive management to address potential problems. In conclusion, assisted migration is a valuable conservation tool that can help species adapt to changing environmental conditions and preserve biodiversity.

Ecosystem Engineering: Using Native Species to Restore Function

Ecosystem engineering involves using native species to restore or enhance the functionality of ecosystems, leveraging the natural roles these species play in shaping their environments. This approach capitalizes on the abilities of certain species to modify their surroundings in ways that promote ecological health and resilience. By focusing on native species, ecosystem engineers aim to restore natural processes and support biodiversity, ultimately leading to more sustainable and self-sustaining ecosystems. Key Concepts of Ecosystem Engineering: Roles of Native Species: Native species often play critical roles in their ecosystems by altering physical environments, influencing nutrient cycling, and affecting species interactions. Examples include beavers creating wetlands through dam building, trees providing shade and stabilizing soil, and certain plants enhancing soil fertility through nitrogen fixation. Restoration Goals: The primary goal of using native species in ecosystem engineering is to restore ecological processes and functions that have been disrupted or degraded (Emery-Butcher H E., et al, 2020). This can include improving habitat structure, enhancing biodiversity, and reestablishing natural nutrient cycles. Enhancing

Habitat Structure: Certain native species can be used to improve habitat structure and complexity. For instance, planting native trees and shrubs in riparian zones can stabilize stream banks, provide habitat for wildlife, and improve water quality through the filtration of runoff. Similarly, native grasses and sedges can be used to restore wetlands and floodplains, enhancing their ecological functions (Ning Z., et al, 2021). Promoting Soil Health: Native plants can enhance soil health through various mechanisms. For example, certain plants are capable of fixing nitrogen in the soil, enriching it, and supporting other vegetation. Others help in erosion control and soil stabilization by developing extensive root systems. Using native plants to restore these functions can improve soil fertility and reduce erosion. Restoring Natural Processes: Ecosystem engineers can help reestablish natural processes that have been disrupted. For example, reintroducing fire-adapted species and implementing controlled burns can help restore fire-dependent ecosystems, reduce the buildup of fuel, and promote the regeneration of fire-adapted plant species.

Benefits of Ecosystem Engineering with Native Species: Biodiversity Enhancement: Using native species for ecosystem engineering helps support and increase local biodiversity. Native species are adapted to local conditions and interactions, making them more likely to thrive and support a diverse range of other species. Ecological Resilience: Restoring native species and their functions can increase the resilience of ecosystems to disturbances such as climate change, invasive species, and pollution. By reestablishing natural processes, ecosystems can better withstand and recover from adverse conditions. Sustainable Management: Native species are often better suited to local environmental conditions and require fewer inputs such as water, fertilizer, and pesticides. This makes them a more sustainable choice for restoration projects and helps reduce the environmental impact of restoration efforts. Challenges and Considerations: Species Selection: Choosing the appropriate native species for ecosystem engineering requires careful consideration of their ecological roles, interactions, and compatibility with the target ecosystem. Misjudging these factors can lead to unintended consequences and hinder restoration success. Long-Term Monitoring: The success of ecosystem engineering efforts must be monitored over the long term to ensure that the desired ecological functions are being restored and maintained (Xu H., et al, 2023). Ongoing monitoring helps identify and address any emerging issues and adapt management practices as needed. Human and Ecological Impact: Ecosystem engineering projects should be designed to balance ecological benefits with potential impacts on local communities and existing ecosystems. Engaging stakeholders and considering the broader implications of restoration efforts are crucial for successful outcomes.

Monitoring and Adaptive Management: Tracking Restoration Success

These processes ensure that restoration efforts achieve their intended outcomes and can be adjusted based on observed conditions and feedback. By systematically tracking progress and being flexible in response to new information, restoration projects can adapt to changing circumstances and improve their effectiveness over time. Monitoring: Monitoring involves the systematic collection of data to assess the progress and effectiveness of restoration efforts. This process includes several key aspects: Setting Baselines: Establishing baseline conditions is essential for understanding the starting point of a restoration project. Baseline data provide a reference for evaluating changes over time and help in setting realistic goals and expectations. Defining Indicators: Indicators are specific metrics used to measure the success of restoration efforts. They can include ecological parameters such as species diversity, vegetation cover, soil health, and water quality. Choosing appropriate indicators depends on the goals of the restoration project and the ecological characteristics of the site (Camarretta N., et al, 2020). Data Collection: Regular and systematic data collection is necessary to track changes over time. This can involve field surveys, remote sensing, laboratory analyses, and other methods. The frequency and intensity of data collection should align with the restoration goals and the dynamics of the ecosystem. Data Analysis: Analyzing the collected data helps to assess the progress of restoration efforts and determine whether goals are being met. Statistical analyses, trend evaluations, and comparisons with baseline data are commonly used to interpret results. Reporting: Clear and transparent reporting of monitoring results is essential for communicating progress to stakeholders, funders, and the public. Reports should include findings, interpretations, and recommendations for future actions. Adaptive Management: Adaptive management is a systematic approach that involves adjusting management practices based on monitoring results and new information (Berkley & Beratan, 2021). It allows for flexibility and continuous improvement in restoration projects. Key components of adaptive management include: Setting Objectives and Hypotheses: At the outset of a restoration project, clear objectives and hypotheses are established. These provide a framework for monitoring and evaluating success. Implementing Management Actions: Based on the initial objectives, management actions are implemented to achieve the desired outcomes. These actions might include habitat modifications, species reintroductions, or changes in land use practices.

Monitoring and Feedback: Continuous monitoring provides feedback on the effectiveness of management actions. This feedback is used to assess whether the actions are achieving the desired results and to identify any unforeseen issues. Enhanced Effectiveness: By regularly assessing progress and making adjustments, monitoring and adaptive management improve the likelihood of achieving restoration goals. This iterative process helps to address emerging issues and adapt to changing conditions. Increased Resilience: Adaptive management fosters resilience by allowing projects to respond to environmental changes, unforeseen challenges, and new scientific knowledge. This flexibility helps to build more robust and sustainable ecosystems. Informed DecisionMaking: Monitoring provides valuable data that inform decision-making (Bahri T., et al, 2021). This ensures that management actions are based on empirical evidence rather than assumptions, leading to more effective and efficient restoration efforts. Stakeholder Engagement: Transparent monitoring and adaptive management processes help build trust and engage stakeholders. Regular updates and responsiveness to feedback enhance stakeholder support and involvement. Challenges and Considerations: Resource Constraints: Monitoring and adaptive management require time, expertise, and financial resources. Ensuring adequate resources are allocated for these activities is essential for their success. Complexity of Ecosystems: Ecosystems are complex and dynamic, which can make monitoring and management challenging. It is important to recognize the limitations of monitoring data and the need for flexible, nuanced approaches.

Community Engagement and Education: Building Support for Restoration

Community engagement and education are vital components in the success of ecological restoration projects. Involving local communities and fostering public understanding can significantly enhance project outcomes, ensure sustainability, and promote long-term support. Here's a comprehensive look at how community engagement and education contribute to building support for restoration initiatives. Community Engagement: Building Relationships: Establishing strong relationships with local communities is essential for gaining their support and cooperation. This involves reaching out to community members, understanding their needs and concerns, and incorporating their input into the restoration planning process (Li X., et al, 2020). Engaging communities in monitoring activities and providing feedback opportunities allows for ongoing involvement and helps address any emerging issues. Community-based monitoring programs can provide valuable data, enhance local investment in the project, and ensure that restoration efforts remain responsive to local conditions (Kim G., et al, 2020). Education: Raising Awareness: Educating the public about the importance of ecological restoration and the specific goals of a project helps build awareness and support. Educational campaigns can include informational brochures, social media content, public talks, and school programs. By increasing understanding of ecological issues and restoration benefits, education fosters a sense of stewardship and encourages community participation. Demonstrating Benefits: Highlighting the tangible benefits of restoration projects—such as improved water quality, enhanced biodiversity, and recreational opportunities—can make the value of these efforts more concrete and relatable to the public. Showcasing successful case studies and sharing stories of positive outcomes can further demonstrate the impact of restoration work.

Developing and distributing educational materials tailored to different audiences—such as fact sheets, guides, and interactive tools—helps disseminate information about restoration goals and methods. Accessible and engaging materials make it easier for people to understand complex concepts and become active participants in restoration efforts. Hosting Events: Organizing events such as restoration workshops, volunteer days, and public seminars provides opportunities for community members to learn about and engage with restoration projects. These events also serve as platforms for networking, sharing information, and celebrating successes (Miller-Rushing A J., et al, 2021). Benefits of Community Engagement and Education: Increased Support and Participation: Engaged and educated communities are more likely to support and participate in restoration projects. Their involvement can lead to increased volunteerism, advocacy, and contributions, enhancing the overall impact of the project. Sustainable Outcomes: Projects that actively involve and educate communities are more likely to achieve sustainable outcomes. Communities are diverse, and engaging with all relevant stakeholders can be challenging. It is important to address different perspectives and ensure that all voices are heard and considered in the planning and implementation processes (Akintobi T H., et al, 2020). Resource Allocation: Effective community engagement and education require resources, including time, funding, and expertise. Ensuring adequate resources are allocated for these activities is essential for their success. Building Trust: Establishing and maintaining trust with communities takes time and effort. It is important to be transparent, responsive, and accountable in all interactions to build and sustain trust.

Restoration in Different Ecosystems: Case Studies and Applications

Restoration projects vary widely across different ecosystems, each requiring tailored approaches to address unique environmental challenges and goals. Here are several case studies illustrating the diverse applications of ecological restoration in various ecosystems, highlighting methods used, challenges faced, and outcomes achieved. Wetland Restoration: The Florida Everglades: Background: The Florida Everglades, a vast and complex wetland ecosystem, has been severely impacted by land conversion, water management practices, and pollution. Restoration efforts aim to revive the natural flow of water, improve water quality, and restore habitat for wildlife. Approach: Water Flow Restoration: The Comprehensive Everglades

Restoration Plan (CERP) was initiated to restore natural water flow patterns by removing canals and levees that redirected water (Zhang X., et al, 2022). This effort also includes reestablishing the natural hydrological regime to support wetland habitats. Nutrient Reduction: Strategies to reduce nutrient runoff from agricultural areas were implemented to improve water quality. This includes constructing treatment wetlands to filter pollutants. Outcomes: Habitat Recovery: Improved water flow has led to the recovery of critical habitats for species such as the American crocodile and the endangered Florida panther. Water Quality Improvement: Reductions in nutrient levels have contributed to healthier aquatic plant communities and enhanced water quality. Restoration has led to the return of several native species and improved habitat connectivity for wildlife (Navarro A., et al, 2020). Restoration efforts aim to enhance coral resilience and recovery. Approach: Coral Nursery Programs: Coral nurseries are established to grow coral fragments in controlled conditions before transplanting them onto damaged reefs. This technique helps to restore coral populations more rapidly. Marine Protected Areas: Expanding and enforcing marine protected areas to reduce fishing pressure and pollution, allowing coral reefs to recover. Outcomes: Increased Coral Cover: Coral nursery programs have successfully increased coral cover in targeted areas, promoting reef recovery.

Grassland Restoration: The Prairie Restoration in the Midwest USA: Background: Historically, tallgrass prairies were widespread across the Midwest but have been largely converted to agriculture. Restoration aims to reestablish these grasslands and their associated biodiversity. Approach: Native Seed Planting: Restoration projects focus on planting a diverse array of native grass and wildflower species to recreate the structure and function of the original prairie. Fire Management: Implementing controlled burns to mimic natural fire regimes that maintain prairie ecosystems and prevent woody encroachment. Outcomes: Increased Biodiversity: Restored prairies have seen a resurgence of native plant species and pollinators, contributing to increased biodiversity. Ecosystem Function: Improved soil health and water infiltration have been observed as a result of restored grassland ecosystems (Feng Q., et al, 2020). Challenges: Soil and Seed Bank: Rebuilding soil health and reintroducing native seeds in areas with a depleted seed bank can be challenging. Land Use Conflicts: Balancing restoration with agricultural and land use pressures requires careful planning and negotiation. Desert Restoration: The Sonoran Desert in Arizona: Background: The Sonoran Desert has been impacted by overgrazing, land development, and invasive species. Restoration efforts aim to improve soil health, reestablish native vegetation, and enhance ecosystem resilience. Approach: Soil Restoration: Techniques such as adding organic matter and reducing erosion are employed to improve soil health and fertility. Native Plant Reintroduction: Planting native species and managing invasive plants to restore natural vegetation patterns and ecosystem functions. Outcomes: Vegetation Recovery: Increased cover of native plants has improved soil stability and habitat quality for wildlife. Enhanced Resilience: Restoration efforts have contributed to increased resilience against desertification and climate extremes (Ellison A M., et al, 2020). Challenges: Extreme Conditions: Harsh desert conditions pose challenges for plant establishment and survival.

Integrating Restoration with Climate Change Mitigation Strategies

Integrating ecological restoration with climate change mitigation strategies represents a holistic approach to addressing both environmental degradation and climate change. Restoration efforts, when strategically aligned with climate mitigation goals, can enhance carbon sequestration, improve ecosystem resilience, and contribute to the reduction of greenhouse gas emissions. For instance, reforestation and afforestation projects increase the number of trees that absorb CO2 through photosynthesis and store carbon in biomass and soils (Taillardat P., et al, 2020). By strengthening ecosystems' ability to absorb and buffer climatic impacts, restoration efforts reduce vulnerability and support sustainable adaptation strategies. Synergies with Renewable Energy and Sustainable Land Use: Integrating restoration with climate change mitigation involves aligning restoration projects with broader sustainability goals, including the promotion of renewable energy and sustainable land use. For instance, restoring degraded lands and wetlands can be paired with the development of solar farms or other renewable energy infrastructure, ensuring that both ecological and energy goals are met. Additionally, sustainable land management practices, such as agroforestry and conservation tillage, can be incorporated into restoration projects to enhance productivity while reducing greenhouse gas emissions. Restoration of Ecosystem Services: Restoration enhances ecosystem services that are critical for climate change mitigation. Services such as water filtration, soil fertility, and flood regulation are vital for maintaining ecological balance and supporting human communities (Smith P., et al, 2020). By restoring ecosystems that provide these services, we can ensure their continued function in mitigating climate-related risks. For example, riparian buffer zones restored along rivers and streams can improve water quality, reduce nutrient runoff, and stabilize banks, contributing to both environmental health and climate resilience.

Policies that promote the inclusion of restoration goals in climate strategies, provide funding and incentives for restoration projects, and facilitate collaboration between stakeholders are essential for achieving synergies (Wedding L M., et al, 2021). National and international agreements, such as the Paris Agreement, can incorporate restoration and conservation targets to align with climate goals. Effective governance ensures that restoration efforts are well-coordinated, transparent, and supported by appropriate legal and institutional mechanisms. Challenges and

Considerations: Despite the benefits, integrating restoration with climate change mitigation presents several challenges. These include securing adequate funding, managing competing land uses, addressing social and political barriers, and ensuring long-term sustainability. The restoration of the Loess Plateau in China has demonstrated how wetland restoration can enhance carbon sequestration and improve agricultural productivity. The project has led to increased carbon storage, reduced soil erosion, and improved water availability (Fawzy S., et al, 2020). By focusing on carbon sequestration, enhancing ecosystem resilience, promoting sustainable land use, and supporting economic and social benefits, this integration provides a holistic solution to climate change. Effective monitoring, adaptive management, and supportive policy frameworks are essential for overcoming challenges and achieving successful outcomes.

Policy and Management Frameworks: Supporting Restoration Efforts

Effective policy and management frameworks are crucial for supporting and guiding ecological restoration efforts. These frameworks provide the structure, resources, and coordination needed to implement successful restoration projects and ensure their sustainability. They address legal, financial, and administrative aspects, creating an environment that fosters collaborative and efficient restoration activities. Here's an overview of the key components and considerations for developing robust policy and management frameworks to support restoration efforts. This includes creating or amending environmental protection laws to include specific mandates for restoration, such as the Clean Water Act in the United States, which provides guidelines for wetland and waterway restoration (Di Sacco A., et al, 2021). Regulations and Standards: Developing clear guidelines and standards for restoration practices helps ensure consistency and quality. These regulations can cover aspects such as site assessment, project planning, and monitoring requirements. Funding and Financial Mechanisms: Funding Sources: Securing adequate funding is essential for the success of restoration projects. Policy frameworks should include mechanisms to attract and allocate financial resources, such as government grants, private sector investments, and international funding sources. Incentives and Market Mechanisms: Financial incentives, such as tax breaks, subsidies, or credits for conservation activities, can encourage participation in restoration efforts. Market-based mechanisms, like carbon credits, can also provide additional funding streams for restoration projects. Institutional Coordination and Governance: Coordinating Bodies: Establishing or designating institutions responsible for overseeing restoration efforts helps ensure coordinated and efficient implementation. This approach involves regularly monitoring and evaluating restoration progress and making adjustments as needed.

Monitoring and Evaluation: Performance Metrics: Establishing performance metrics and indicators helps track the success of restoration projects (Magableh G M., 2021). Metrics can include ecological indicators, such as species recovery and habitat quality, as well as socioeconomic outcomes, such as community engagement and economic benefits. Reporting and Accountability: Regular reporting on restoration outcomes and financial expenditures ensures transparency and accountability. It also provides valuable information for stakeholders and supports ongoing support and funding. Capacity Building and Training: Technical Training: Providing training and capacity-building opportunities for practitioners, local communities, and stakeholders enhances the effectiveness of restoration efforts. Training programs can cover technical skills, project management, and monitoring techniques. Knowledge Sharing: Promoting knowledge exchange and collaboration between experts, researchers, and practitioners helps disseminate best practices and innovations in restoration. This can be facilitated through workshops, conferences, and publications.

Challenges and Considerations: Policy Integration: Ensuring that restoration policies are integrated with other environmental and land use policies is crucial for achieving comprehensive and coherent outcomes. This requires cross-sectoral collaboration and alignment of objectives. Resource Allocation: Adequate and sustained funding is a common challenge in restoration projects (Kazancoglu & Sagnak, 2021). Outcomes: The NEP has successfully restored water quality, improved habitat, and enhanced community engagement in multiple estuarine areas across the country (Labadi S., et al, 2021). China's Sloping Land Conversion Program: Background: The Sloping Land Conversion Program is a large-scale restoration initiative aimed at addressing soil erosion and deforestation on steep lands in China. Approach: The program includes converting sloping farmland to forest or grassland, providing financial incentives to farmers, and integrating restoration with rural development goals. Outcomes: The program has significantly reduced soil erosion, improved vegetation cover, and contributed to poverty alleviation in rural areas. In summary, effective policy and management frameworks are essential for supporting ecological restoration efforts. By providing legal support, financial mechanisms, institutional coordination, strategic planning, and capacity building, these frameworks create an enabling environment for successful restoration projects. Addressing challenges such as policy integration, resource allocation, and local context ensures that restoration efforts are sustainable and impactful. Case studies from various regions illustrate the benefits and complexities of implementing comprehensive policy and management frameworks in different ecological and socioeconomic contexts.

Future Directions: Emerging Techniques and Research Frontiers

As ecological restoration evolves, emerging techniques and research frontiers are paving the way for more effective and innovative approaches to restoring ecosystems. These advancements address both the increasing complexity of environmental challenges and the need for enhanced restoration outcomes. These technologies allow for large-scale monitoring and assessment of restoration projects, enabling more precise and timely interventions (De Alwis C., et al, 2021). GIS Integration: Geographic Information Systems (GIS) facilitate spatial analysis and mapping of restoration sites, helping to identify priority areas, track progress, and model ecological processes. GIS can integrate various data layers, such as soil properties, land use, and species distribution, to support informed decision-making and adaptive management. Genetic and Genomic Approaches: Genetic Restoration: Advances in genetic and genomic technologies are improving our ability to restore genetic diversity in populations of endangered species and degraded ecosystems. Techniques such as genetic rescue and assisted gene flow help maintain or enhance genetic variability, which is crucial for species adaptation and resilience. Genomic Monitoring: Environmental DNA (eDNA) sampling and genomic monitoring provide new tools for assessing biodiversity, tracking species presence, and understanding ecological interactions. These methods offer noninvasive and highly sensitive ways to monitor the success of restoration efforts and detect changes in species composition. Restoration Ecology and Functional Diversity: Functional Diversity: Emphasizing functional diversity—the range of different ecological roles that species play—can enhance the resilience and functionality of restored ecosystems. Research is increasingly focusing on the functional traits of species, such as their roles in nutrient cycling, pollination, and seed dispersal, to guide restoration practices that support ecosystem functions and services. Ecosystem Function Models: Developing models that simulate ecosystem functions and dynamics helps predict the outcomes of restoration interventions and assess their effectiveness. These models can incorporate various factors, such as species interactions, climate change, and land use changes, to provide insights into the potential impacts of restoration strategies (Wang C L., 2021). This technique aims to ensure species survival and maintain ecological processes by moving them to areas where they are more likely to thrive under future climatic conditions. Trophic Rewilding: Reintroducing keystone species and restoring natural predator-prey relationships can enhance ecosystem structure and function. Trophic rewilding focuses on reestablishing ecological interactions that have been lost or diminished, leading to more resilient and self-sustaining ecosystems.

Socioecological Integration: Human Nature Interactions: Research is increasingly addressing the role of human activities and values in restoration projects. Understanding socio-ecological systems—how human and ecological systems interact—helps to design restoration efforts that are socially acceptable, economically viable, and ecologically effective. Community-Based Restoration: Engaging local communities in restoration efforts not only enhances project success but also builds social capital and resilience. Community-based restoration initiatives involve local knowledge, preferences, and practices, ensuring that restoration outcomes are beneficial for both people and ecosystems. Climate Change Adaptation and Mitigation: Climate Resilient Restoration: Incorporating climate projections into restoration planning helps design projects that are resilient to future climate conditions (Wang Y., et al, 2021). Implementing adaptive management frameworks allows for iterative learning and adjustment of restoration practices based on monitoring data and changing conditions. This approach ensures that restoration efforts remain responsive and effective over time. Policy and Governance Innovations: Integrated Policy Frameworks: Developing integrated policy frameworks that align restoration goals with broader environmental, economic, and social objectives is crucial for effective restoration. Policies should support cross-sectoral coordination, funding, and stakeholder engagement to facilitate comprehensive and successful restoration efforts. Governance Models: Innovative governance models, such as collaborative management and multistakeholder partnerships, enhance the effectiveness and inclusiveness of restoration projects (Tezel A., et al, 2020). These models facilitate joint decision-making, resource sharing, and knowledge exchange among diverse stakeholders. Conclusion: The future of ecological restoration is characterized by rapid advancements and emerging techniques that enhance our ability to restore ecosystems effectively and sustainably. By leveraging new technologies, incorporating genetic and genomic insights, exploring novel restoration techniques, and addressing socioecological dynamics, restoration efforts can achieve greater success and resilience. Continued research, innovation, and collaboration are essential for addressing the complex environmental challenges of the future and ensuring that restoration projects contribute to a more sustainable and resilient planet.

CONCLUSION

As climate change accelerates, plant invasions pose a growing threat to ecosystems worldwide, challenging traditional restoration techniques and demanding innovative approaches to mitigate their impacts. Effective restoration under these conditions requires a multifaceted strategy that combines proactive measures, adaptive management, and cutting-edge techniques. Climate change can alter temperature and precipitation patterns, creating new opportunities for invasive species to establish and spread. Restoration techniques must therefore include comprehensive assessments of potential invaders and their ecological impacts. This involves mapping invasion hotspots, monitoring species distributions, and analyzing how changing climatic conditions might influence invasion dynamics. By identifying high-risk areas and species, restoration efforts can be better targeted and more effective in preventing or managing invasions. Integrating Climate Projections: Incorporating climate change projections into restoration planning is essential for designing strategies that are robust against future uncertainties. Restoration projects should use climate models to predict potential changes in temperature, precipitation, and other environmental factors that may affect plant communities and invasion patterns. This foresight allows for the selection of plant species and restoration practices that are likely to remain effective under future conditions. For example, choosing native species with higher resilience to anticipated climate extremes or using climate-adapted cultivars can help ensure that restored ecosystems are better equipped to withstand invasions. Novel Restoration Techniques: The application of novel restoration techniques is increasingly important in addressing plant invasions. Techniques such as assisted migration, where native species are relocated to areas predicted to be suitable under future climate conditions, can enhance ecosystem resilience. Similarly, the use of bioengineering approaches, such as introducing mycorrhizal fungi or other beneficial microbes, can help strengthen native plants' competitive abilities against invasive species. Innovative methods, such as the deployment of biocontrol agents or genetic modifications, are also being explored to specifically target invasive plants and mitigate their impacts. Adaptive Management: Adaptive management is a cornerstone of effective restoration in the context of climate change and plant invasions. This approach involves continuously monitoring restoration sites, assessing the effectiveness of interventions, and making necessary adjustments based on new data and changing conditions. By adopting a flexible and iterative approach, restoration practitioners can respond to unexpected challenges and refine their strategies to better address ongoing invasion pressures. Regular monitoring and feedback mechanisms are crucial for identifying emerging threats and ensuring that restoration efforts remain relevant and effective over time.

Community Involvement and Education: Engaging local communities and stakeholders in restoration efforts is vital for addressing plant invasions. Community-based approaches can enhance awareness of invasive species and promote proactive management practices. Education and outreach programs can help residents and landowners recognize and report invasive species, participate in removal efforts, and adopt practices that reduce the risk of invasions. By fostering a sense of stewardship and collaboration, communities can play a crucial role in supporting and sustaining restoration initiatives. Policy and Governance: Effective policy frameworks and governance structures are essential for supporting restoration efforts aimed at mitigating plant invasions. Policies that promote early detection and rapid response to invasive species, provide funding for research and management, and encourage cross-sectoral coordination are critical for successful outcomes. Governance models that facilitate collaboration between governmental agencies, NGOs, scientists, and local communities can enhance the effectiveness of restoration efforts and ensure that resources are used efficiently. By integrating restoration objectives with broader environmental policies, we can create a supportive environment for addressing plant invasions. Future Research Directions: Continued research is needed to address knowledge gaps and develop new approaches for mitigating plant invasions under climate change. Research areas include understanding the interactions between climate change and invasive species, developing advanced monitoring technologies, and evaluating the long-term effectiveness of restoration techniques. Collaborations between researchers, practitioners, and policymakers can drive innovation and ensure that restoration strategies are based on the best available science. In conclusion, mitigating plant invasions under climate change requires a comprehensive and adaptive approach that integrates advanced techniques, proactive planning, and collaborative efforts. By understanding invasion dynamics, incorporating climate projections, employing novel restoration methods, and engaging communities, we can enhance the resilience of ecosystems and improve the effectiveness of restoration projects. The challenges posed by plant invasions in a changing climate are significant, but with continued innovation and commitment, we can develop strategies that not only address current threats but also anticipate and adapt to future changes. Through a combination of science, policy, and community engagement, we can build more resilient ecosystems and contribute to a sustainable future.

REFERENCES

Akinola, R., Pereira, L. M., Mabhaudhi, T., De Bruin, F. M., & Rusch, L. (2020). A review of indigenous food crops in Africa and the implications for more sustainable and healthy food systems. *Sustainability*, *12*(1), 1–15. DOI: 10.3390/su12083493 PMID: 33520291

Akintobi, T. H., Jacobs, T., & Sabbs, D. (2020). Community engagement of African Americans in the era of COVID-19: Considerations, challenges, implications, and recommendations for public health. *Preventing Chronic Disease*, •••, 2020. PMID: 32790605

Bahri, T., Vasconcellos, M., Welch, D. J., & Johnson, J. (2021). Adaptive management of fisheries in response to climate change. *FAO Fisheries and Aquaculture Technical Paper*, (667), 2021.

Bakina, L. G., Chugunova, M., & Polyak, Y. M. (2021). Bioaugmentation: Possible scenarios due to application of bacterial preparations for remediation of oil-contaminated soil. *Environmental Science and Pollution Research International*, *28*(1), 1–12. PMID: 33009614

Berkley, J., & Beratan, K. K. (2021). "Capturing practitioners'" how-to" knowledge in the form of recommendations for more effective planning of collaborative adaptive management projects. *Ecology and Society*, *26*(4), 2021. DOI: 10.5751/ES-12840-260424

Camarretta, N., Harrison, P. A., Bailey, T., Potts, B., Lucieer, A., Davidson, N., & Hunt, M. (2020). Monitoring forest structure to guide adaptive management of forest restoration: A review of remote sensing approaches. [– Springer]. *New Forests*, *51*(4), 2020. DOI: 10.1007/s11056-019-09754-5

Charles, K. M., & Stehlik, I. (2021). Assisted species migration and hybridization to conserve cold-adapted plants under climate change. [– Wiley Online Library]. *Conservation Biology*, *35*(2), 2021. DOI: 10.1111/cobi.13583 PMID: 32643822

De Alwis, C., Kalla, A., Pham, Q. V., & Kumar, P. (2021). Survey on 6G frontiers: Trends, applications, requirements, technologies and future research. *IEEE Open Journal of the Communications Society*, *2*, 2021. DOI: 10.1109/OJCOMS.2021.3071496

Di Sacco, A., Hardwick, K. A., Blakesley, D., Brancalion, P. H. S., Breman, E., Cecilio Rebola, L., Chomba, S., Dixon, K., Elliott, S., Ruyonga, G., Shaw, K., Smith, P., Smith, R. J., & Antonelli, A. (2021). Ten golden rules for reforestation to optimize carbon sequestration, biodiversity recovery and livelihood benefits. [– Wiley Online Library]. *Global Change Biology*, *27*(7), 2021. DOI: 10.1111/gcb.15498 PMID: 33494123

Diener, A., & Mudu, P. (2021). vegetation protects us from air pollution? A critical review on green spaces' mitigation abilities for air-borne particles from a public health perspective-with implications .... *The Science of the Total Environment*, *755*, 1–12. DOI: 10.1016/j.scitotenv.2021.148605

Duque, J. F. M., Zapico, I., Bugosh, N., & Tejedor, M. (2021). A Somolinos quarry land stewardship history: From ancient and recent land degradation to sensitive geomorphic-ecological restoration and its monitoring. [– Elsevier]. *Ecological Engineering*, •••, 2021.

Ellison, A. M., Felson, A. J., & Friess, D. A. (2020). Mangrove rehabilitation and restoration as experimental adaptive management. *Frontiers in Marine Science*, *7*, 2020. DOI: 10.3389/fmars.2020.00327

Emery-Butcher, H. E., Beatty, S. J., & Robson, B. J. (2020). The impacts of invasive ecosystem engineers in freshwaters: A review. [– Wiley Online Library]. *Freshwater Biology*, *65*(5), 2020. DOI: 10.1111/fwb.13479

Fawzy, S., Osman, A. I., Doran, J., & Rooney, D. W. (2020). Strategies for mitigation of climate change: A review. [– Springer]. *Environmental Chemistry Letters*, *18*(6), 2020. DOI: 10.1007/s10311-020-01059-w

Feng, Q., Wang, B., Chen, M., Wu, P., Lee, X., & Xing, Y. (2021). Invasive plants as potential sustainable feedstocks for biochar production and multiple applications: A review. [– Elsevier]. *Resources, Conservation and Recycling*, *164*, 2021. DOI: 10.1016/j.resconrec.2020.105204

Feng, Q., Zhao, W., Hu, X., Liu, Y., Daryanto, S., & Cherubini, F. (2020). Trading-off ecosystem services for better ecological restoration: A case study in the Loess Plateau of China. [– Elsevier]. *Journal of Cleaner Production*, *257*, 2020. DOI: 10.1016/j.jclepro.2020.120469

Green, S. J., & Grosholz, E. D. (2021). Functional eradication as a framework for invasive species control. *Frontiers in Ecology and the Environment*, *19*(1), 1–10. DOI: 10.1002/fee.2277

Kazancoglu I, & Sagnak M, (2021). "Circular economy and the policy: A framework for improving the corporate environmental management in supply chains." Journal of Strategy and the Environment, 2021 – Wiley Online Library

Kim, G., Newman, G., & Jiang, B. (2020). Urban regeneration: Community engagement process for vacant land in declining cities. [– Elsevier]. *Cities (London, England)*, *102*, 2020. DOI: 10.1016/j.cities.2020.102730 PMID: 32831449

Labadi, S., Giliberto, F., Rosetti, I., & Shetabi, L. (2021). Heritage and the sustainable development goals: Policy guidance for heritage and development actors. *International Journal of Heritage Studies*, •••, 2021.

Lee, S. W., Lee, W. C., Lee, S. H., & Kim, S. O. (2020). Changes of soil properties through the remediation processes and techniques for the restoration of remediated soils. *Economic and Environmental Geology*, *53*(1), 1–12. DOI: 10.9719/EEG.2024.57.1.1

Li, X., Zhang, F., Hui, E. C., & Lang, W. (2020). Collaborative workshop and community participation: A new approach to urban regeneration in China. [– Elsevier]. *Cities (London, England)*, *102*, 2020. DOI: 10.1016/j.cities.2020.102743

Li, Z., Shen, Q., Fan, W., Dong, S., Wang, Z., Xu, Y., Ma, T., & Cao, Y. (2024). … Solutions vs. Human-Induced Approaches for Alpine Grassland Ecosystem: "Climate-Help" Overwhelms "Human Act" to Promote Ecological Restoration in the Three …. *Remote Sensing (Basel)*, *16*(7), 2024. DOI: 10.3390/rs16071156

Lin X, He J, Chen Z, Lyu Z, Dai B, Yu F, (2023). "Diffbir: Towards blind image restoration with generative diffusion prior." arXiv preprint arXiv.

Magableh, G. M. (2021). "Supply chains and the COVID-19 pandemic: A comprehensive framework." European Management Review, 2021 – Wiley Online Library
Miller-Rushing A J, Athearn N, Blackford T, (2021). "COVID-19 pandemic impacts on conservation research, management, and public engagement in US national parks. [– Elsevier]. *Biological Conservation*, •••, 2021.

Mudereri, B. T., Abdel-Rahman, E. M., Dube, T., Landmann, T., Khan, Z., Kimathi, E., Owino, R., & Niassy, S. (2020). Multi-source spatial data-based invasion risk modeling of Striga (Striga asiatica) in Zimbabwe. *GIScience & Remote Sensing*, *57*(1), 1–18. DOI: 10.1080/15481603.2020.1744250

Navarro, A., Young, M., Allan, B., Carnell, P., Macreadie, P., & Ierodiaconou, D. (2020). The application of Unmanned Aerial Vehicles (UAVs) to estimate above-ground biomass of mangrove ecosystems. [– Elsevier]. *Remote Sensing of Environment*, *242*, 2020. DOI: 10.1016/j.rse.2020.111747

Ning, Z., Chen, C., Xie, T., Zhu, Z., Wang, Q., Cui, B., & Bai, J. (2021). Can the native faunal communities be restored from removal of invasive plants in coastal ecosystems? A global meta-analysis. [– Wiley Online Library]. *Global Change Biology*, *27*(19), 2021. DOI: 10.1111/gcb.15765 PMID: 34170600

Palik, B. J., Clark, P. W., D'Amato, A. W., Swanston, C., & Nagel, L. (2022). Operationalizing forest-assisted migration in the context of climate change adaptation: Examples from the eastern USA. [– Wiley Online Library]. *Ecosphere*, *13*(10), 2022. DOI: 10.1002/ecs2.4260

Pili, A. N., Tingley, R., Sy, E. Y., Diesmos, M. L. L., & Diesmos, A. C. (2020). Niche shifts and environmental non-equilibrium undermine the usefulness of ecological niche models for invasion risk assessments. *Scientific Reports*, *10*(1), 1–12. DOI: 10.1038/s41598-020-64568-2 PMID: 32409706

Ponti, L., Gutierrez, A. P., de Campos, M. R., Desneux, N., Biondi, A., & Neteler, M. (2021). Biological invasion risk assessment of Tuta absoluta: Mechanistic versus correlative methods. *Biological Invasions*, *23*(1), 1–15. DOI: 10.1007/s10530-021-02613-5 PMID: 33488272

Simberloff, D. (2021). Maintenance management and eradication of established aquatic invaders. *Hydrobiologia*, *848*(1), 1–12. DOI: 10.1007/s10750-020-04352-5 PMID: 32836349

Smith, P., Calvin, K., Nkem, J., Campbell, D., Cherubini, F., Grassi, G., Korotkov, V., Le Hoang, A., Lwasa, S., McElwee, P., Nkonya, E., Saigusa, N., Soussana, J.-F., Taboada, M. A., Manning, F. C., Nampanzira, D., Arias-Navarro, C., Vizzarri, M., House, J., & Arneth, A. (2020). Which practices co-deliver food security, climate change mitigation and adaptation, and combat land degradation and desertification? [– Wiley Online Library]. *Global Change Biology*, *26*(3), 2020. DOI: 10.1111/gcb.14878 PMID: 31637793

Taillardat, P., Thompson, B. S., Garneau, M., Trottier, K., & Friess, D. A. (2020). Climate change mitigation potential of wetlands and the cost-effectiveness of their restoration. *Interface Focus*, *10*(5), 2020. DOI: 10.1098/rsfs.2019.0129 PMID: 32832065

Tezel, A., Papadonikolaki, E., Yitmen, I., & Hilletofth, P. (2020). Preparing construction supply chains for blockchain technology: An investigation of its potential and future directions. [– Springer]. *Frontiers of Engineering Management*, *7*(4), 2020. DOI: 10.1007/s42524-020-0110-8

Wade, D. T. (2020). What is rehabilitation? An empirical investigation leading to an evidence-based description. *Clinical Rehabilitation*, *34*(1), 3–11. DOI: 10.1177/0269215515601176 PMID: 32037876

Wang, C. L. (2021). New frontiers and future directions in interactive marketing: Inaugural Editorial. *Journal of Research in Interactive Marketing*, *15*(1), 2021. DOI: 10.1108/JRIM-03-2021-270

Wang, L., Rinklebe, J., Tack, F. M. G., & Hou, D. (2021). A review of green remediation strategies for heavy metal contaminated soil. *Soil Use and Management*, *37*(1), 1–15. DOI: 10.1111/sum.12717

Wang, Y., Derakhshan, A., & Zhang, L. J. (2021). Researching and practicing positive psychology in second/foreign language learning and teaching: The past, current status and future directions. *Frontiers in Psychology*, *12*, 2021. DOI: 10.3389/fpsyg.2021.731721 PMID: 34489835

Wedding, L. M., Moritsch, M., Verutes, G., & Arkema, K. (2021). Incorporating blue carbon sequestration benefits into sub-national climate policies. [– Elsevier]. *Environmental Science & Policy*, •••, 2021.

Weidlich, E. W. A., Florido, F. G., Sorrini, T. B., & Brancalion, P. H. S. (2020). Controlling invasive plant species in ecological restoration: A global review. *Journal of Applied Ecology*, *57*(1), 1–12. DOI: 10.1111/1365-2664.13656 PMID: 31957374

Xu, H., Xu, F., Lin, T., Xu, Q., Yu, P., Wang, C., Aili, A., & Zhao, X. (2023). A systematic review and comprehensive analysis on ecological restoration of mining areas in the arid region of China: Challenge, capability and reconsideration. [– Elsevier]. *Ecological Engineering*, •••, 2023.

Zhang, D., Ge, W., & Zhang, Y. (2020). Evaluating the vegetation restoration sustainability of ecological projects: A case study of Wuqi County in China. [– Elsevier]. *Journal of Cleaner Production*, *264*, 2020. DOI: 10.1016/j.jclepro.2020.121751

Zhang, X., Liu, K., Wang, S., Wu, T., Li, X., Wang, J., Wang, D., Zhu, H., Tan, C., & Ji, Y. (2022). Spatiotemporal evolution of ecological vulnerability in the Yellow River Basin under ecological restoration initiatives. [– Elsevier]. *Ecological Indicators*, *135*, 2022. DOI: 10.1016/j.ecolind.2022.108586

Zhang, Z., Zou, Y., & Xiao, W. (2023). Exploration of a virtual restoration practice route for architectural heritage based on evidence-based design: A case study of the Bagong House. *Heritage Science*, *11*(1), 1–12. DOI: 10.1186/s40494-023-00878-8 PMID: 36844367

Chapter 6
Economic Perspectives on Managing Invasive Plant Species in Pakistan

Amina Farooq
https://orcid.org/0009-0008-5671-8982
University of Agriculture, Faisalabad, Pakistan

Usman Tufail
https://orcid.org/0009-0000-2798-3113
University of Agriculture, Faisalabad, Pakistan

Farooq Ahmad
University of Agriculture, Faisalabad, Pakistan

Jazab Shafqat
University of Agriculture, Faisalabad, Pakistan

Ansa Asghar
University of Agriculture, Faisalabad, Pakistan

Mishal Iqbal
University of Agriculture, Faisalabad, Pakistan

Sajid Aqeel
University of Agriculture, Faisalabad, Pakistan

Nida Hussain
University of Agriculture, Faisalabad, Pakistan

Anum Javaid
https://orcid.org/0009-0001-0294-924X
University of Agriculture, Faisalabad, Pakistan

Muhammad Sufyan
Agriculture University, Beijing, China

Zunaira Zulifqar
University of Agriculture, Faisalabad, Pakistan

Sana Maqsood
University of Agriculture, Faisalabad, Pakistan

DOI: 10.4018/979-8-3693-8724-5.ch006

ABSTRACT

Exotic species turns to invasive have negative impacts on native biota and ecosystem functioning. Globalization, development of active transport accelerates plants trade, create threat to global biodiversity. Developing countries like Pakistan invasive species pose economic loss like agricultural productivity, Eucalyptus camaldulensis intentionally introduced for green cover and soil reclamation, Parthenium hysterophorus L. for trade, Lantana camara for horticulture purposes, Broussonetia papyrifera (L.) for landscaping purpose. Invasive plants have adaptability against wide range of environmental conditions, reduced yield by competing with native biodiversity and have econo-medicinal properties fodder, oil, some other allelopathic effects that could pay their damage cost. Pakistan requires specific management program to control their introduction, spread and their impacts on native species. The control and management of invasive species can be implemented by assessment of their risk factors on biodiversity degradation and loss.

INTRODUCTION

Pakistan is home to a wide range of ecosystems including forests, grasslands wetlands and agricultural lands (Aziz, 2021). However, the introduction and spread of invasive plant species pose a serious to the country spread rapidly and disrupt local ecosystem by outcompeting native species. These species often thrive due to lack of natural predators and favorable climatic conditions leading to ecological and economic consequences (Adedibu, 2023). Invasive plants species pose a significant challenge to Pakistan economy agriculture and natural ecosystems (Aziz, 2021). These nonnative species spread aggressively outcompeting indigenous flora, reducing agriculture productivity and increasing the costs of land management. The economic impact of invasive plants is multifaceted affecting farmers, policymakers and environmental agencies by increasing expenditures on control measures reducing crop yields and disrupting local biodiversity (Shin et al., 2022). Pakistan diverse climate and extensive agricultural sector make it particularly vulnerable to biological invasions with species such as *Parthenium hysterophorus*, *Eichhornia crassipes* and *Lantana camara* spreading rapidly across different ecosystems (Mohammeda & Bahmudb, 2023). The unchecked growth of these species leads to significant losses in crop production increased irrigation costs and additional expenses for herbicides and manual removal (Gathala et al., 2021).

Invasive plant species represent one of the most pressing ecological threats to natural ecosystems worldwide. These species non-native to the regions they invade disrupt the balance of native biodiversity, alter habitat structure, degrade ecosystem services, and impair ecological resilience (Vilà & Hulme, 2017). In countries like Pakistan, where ecosystems range from mountainous regions and riverine plains to arid deserts and coastal belts, the vulnerability to plant invasions is particularly acute due to climatic variability, anthropogenic disturbances, and inadequate environmental regulation (Khan et al., 2015). The ecological implications of these invasions go far beyond the presence of an undesirable species; they encompass profound and often irreversible shifts in community composition, trophic dynamics, soil health, hydrological processes, and overall ecosystem functionality.

The increasing prevalence of invasive plants such as *Parthenium hysterophorus, Lantana camara, Prosopis juliflora, Ailanthus altissima, and Eichhornia crassipes* in Pakistan's landscapes signals an urgent need for science-based management frameworks that are rooted in ecological principles. Historically, responses to plant invasions have often focused on mechanical removal or chemical control, but such methods, while useful in certain contexts, often fail to address the root ecological drivers of invasion and sometimes lead to unintended environmental consequences. An ecological perspective provides a broader, systems-level understanding of invasive species dynamics encompassing factors such as species interactions, successional stages, disturbance regimes, and feedback mechanisms allowing for more effective, sustainable, and holistic management (Lugo, 2020).

From an ecological standpoint, plant invasions are not random phenomena but rather the outcome of specific environmental conditions, human interventions, and biological traits that favor the establishment and spread of non-native flora (Rai, 2015). Many invasive plants possess characteristics that give them competitive advantages over native species: rapid growth, prolific seed production, high phenotypic plasticity, allelopathy, and adaptability to disturbed habitats. These traits enable them to colonize new territories swiftly and dominate native plant communities. In many cases, invasions are facilitated by anthropogenic disturbances such as deforestation, overgrazing, infrastructure development, unregulated trade, and climate change, all of which weaken the natural resistance of ecosystems to biological invasions (Bogale & Tolossa, 2021). In turn, invasive species alter ecosystem processes, often creating feedback loops that further reinforce their dominance for example, by changing fire regimes, altering soil nutrient cycles, or modifying hydrology.

Managing invasive plants through an ecological lens begins with recognizing the importance of prevention and early intervention. Prevention is widely regarded as the most cost-effective and environmentally sound strategy in the management continuum (El-Halwagi, 2025). By identifying vulnerable ecosystems, monitoring high-risk pathways (such as ports, transportation networks, and disturbed lands), and applying ecological risk assessments, it becomes possible to reduce the likelihood of introduction and establishment. Once an invasive species has been introduced, early detection and rapid response (EDRR) are essential to contain its spread before it becomes unmanageable (Reaser et al., 2020). These approaches require robust ecological knowledge, including understanding species' life cycles, modes of reproduction, dispersal mechanisms, and habitat preferences.

Beyond prevention and containment, the ecological management of invasive plants often involves restoration-based approaches aimed at reestablishing native plant communities and ecosystem processes (Nuñez et al., 2017). Restoration ecology offers valuable tools for repairing invaded ecosystems by improving soil conditions, reintroducing native flora, enhancing habitat heterogeneity, and restoring ecological interactions such as pollination, seed dispersal, and herbivory. This process is not merely about removing an invader but about recreating the ecological conditions that promote native resilience and resistance to future invasions. In this regard, passive and active restoration strategies, assisted natural regeneration, and successional management are essential components of an ecologically informed response.

The aim of this chapter is to examine the economic implications of invasive plant species and evaluate cost-effective strategies for their management. It seeks to highlight the direct and indirect financial impacts on agriculture, ecosystems, and national economies, while emphasizing the importance of economic assessments in guiding policy decisions and prioritizing control measures. To analyze the economic burden caused by invasive plant species and to explore financially sustainable management approaches that minimize their impact on agriculture, natural ecosystems, and public infrastructure.

CAUSES OF SPREADING INVASIVE SPECIES IN PAKISTAN

Unregulated trade and transportation

Unregulated trade and transportation play a significant and increasingly alarming role in the spread of invasive species in Pakistan. As international trade has expanded over the years, the movement of goods, commodities, and plant materials across borders has increased substantially (Hulme, 2021). Unfortunately, this trade often takes place without adequate enforcement of phytosanitary regulations or the implementation of strict quarantine measures. A wide range of invasive organisms such as insect pests, weed seeds, fungal spores, and plant pathogens can easily hitchhike with imported goods, either as contaminants in agricultural produce or hidden in packing materials like wooden crates, pallets, or shipping containers. These materials, especially if untreated or not properly inspected, can harbor species that are not native to Pakistan and that pose a serious threat to local biodiversity, agriculture, and ecosystems.

The risk is particularly high in ports and land entry points where cargo is received and transported inland without sufficient biosecurity protocols. Due to limited awareness and under-resourced inspection systems, many potentially harmful organisms bypass detection during customs clearance (Kachigunda et al., 2023). For instance, ornamental plants and exotic seeds are often imported for horticultural and landscaping purposes without proper screening, resulting in the accidental introduction of fast-spreading species that later become invasive. Moreover, once these invasive species enter the country, the internal transportation infrastructure such as roads, railways, and local trade routes facilitates their rapid movement across different ecological zones. This unchecked distribution allows the invasive species to colonize new areas, often with devastating impacts on native species, natural habitats, and crop production (Simberloff, 2013).

Climate change

Climate change is an increasingly critical factor contributing to the spread and establishment of invasive species in Pakistan. As global temperatures rise and weather patterns shift, ecosystems across the country are undergoing significant transformations. These changes create new environmental conditions that can favor the survival, growth, and expansion of non-native species over indigenous ones. Invasive plants, insects, and pathogens often exhibit high adaptability to environmental stress and variability, allowing them to take advantage of the altered climates more effectively than native species, which may be more specialized and sensitive to temperature or moisture changes.

One of the key ways in which climate change facilitates biological invasions is through the expansion of suitable habitats. Warmer temperatures can allow tropical and subtropical invasive species to spread into previously cooler regions, including higher elevations and northern areas of Pakistan. Similarly, changes in precipitation patterns such as increased rainfall in some areas or prolonged droughts in others can disturb the balance of native plant communities, weakening their competitiveness and making room for fast-growing, opportunistic invaders (Ren et al., 2023). For instance, invasive weed species may quickly colonize areas where native vegetation has died back due to heat stress or water shortages.

Deforestation and land disturbance

Deforestation and land disturbance are among the most significant anthropogenic factors contributing to the spread of invasive species in Pakistan. These activities drastically alter the structure, composition, and ecological balance of natural ecosystems, creating ideal conditions for invasive species to establish and proliferate (Battisti et al., 2016). When forests are cleared for agriculture, urban development, or logging, the protective canopy is lost, soil is exposed, and native vegetation is removed, leaving the land vulnerable to colonization by opportunistic non-native species that thrive in disturbed environments.

In Pakistan, deforestation has been driven by multiple factors, including the growing demand for fuelwood, overgrazing, infrastructure development, and illegal logging (Faiza et al., 2017). These activities not only reduce forest cover but also fragment habitats, weakening the ecological integrity of the landscape. Fragmented and degraded ecosystems are less resilient to biological invasions because the natural checks and balances such as competition, predation, and mutualistic relationships are disrupted. Invasive plant species, in particular, are often highly competitive, fast-growing, and capable of exploiting disturbed soils and open spaces, allowing them to outcompete native plants and form dense monocultures.

Lack of natural predators

Invasive plants often thrive in Pakistan because their natural predators such as insects and herbivores are absent in the local ecosystem.From an economic perspective managing invasive plant species requires a cost effective and sustainable approach. Strategies such as biological control integrated weed management and community driven eradication programs can help mitigate the financial burden. Additionally, economic polices including subsidies for control measures and research funding play a crucial role in addressing the issue (Bradley et al., 2021).

Impact on environment

Invasive plant species in Pakistan have sever environmental consequences, disrupting ecosystems, reducing biodiversity, and altering natural landscapes. These species outcomes native vegetation, leading to habitat loss for indigenous flora and fauna. The unchecked spread of invasive plants such as *Parthenium hysterophorus,Prosopis juliflora* and *Eichhornia crassipes* has caused significant ecological damage by monopolizing resources like water, sunlight and nutrients. Additionally, invasive plants can contribute to climate change by altering carbon sequestration patterns (Finch et al., 2021). Some species such *Lantana camara* release allelopathic chemicals that inhibit the growth of native plants reducing the ability of forests and grasslands to act as carbon sinks, this impact can exacerbate the effects of climate change making ecosystems more vulnerable to extreme weather events (Clarke et al., 2022).

Invasive Species of Pakistan

Pakistan has a diverse range of flora, with over 5,700 species of vascular plants more than 400 species are endemic that is 7.1% of the total flora (Qazi et al., 2023). Besides the endemic plant species, a huge number of non-endemic species naturalized and outcompeted with endemic species. Alien species from all continents have invaded Pakistan but the primary donors of alien species are Temperate Asia (176) and Africa (143 species) (Jehangir et al., 2024). Alien taxa inhabit all six geographic regions of Pakistan and the largest number (117) is found on the Punjab fertile plains. Ruderal habitats support a wide array of alien plant species, hosting almost all the alien species irrespective of their invasion status. Based on field observations, published work, scientific reports and online databases, 400 alien taxa were identified (83 families, 265 genera), 180 of which were classified as casual alien, 117 as naturalized, and 103 as invasive (Seebens et al., 2023). Poaceae (61 taxa) and Asteraceae (46 taxa) were the richest families, while Ipomoea (9), Amaranthus (8) and Euphorbia (8) were the genera with the largest numbers of alien taxa (Khuroo et al., 2021). The greater majority of alien species were annuals (32%) irrespective of their stage in the process of invasion.

Some of the most significant invaders which vigorously impact biodiversity in Pakistan include *Parthenium hysterophorus*, *Lantana camara*, *Broussonetiapapyrifera*, *Eucalyptus camaldulensis*, *Xanthium strumarium*, *Prosopis juliflora*, *Eicchornia crassipes*, *Leucanealeucosephala*, and *Salvinia molesta*. Exotic invaders included *P. hysterophorus*, *L. camara*, *B. papyrifera*, and *X. strumarium* are considered the worst weeds (R. Ullah et al., 2022).

Parthenium hysterophorus

Parthenium hysterophorus L. belongs to Asteraceae family naturalized all across the globe due to its rapid growth rate and reproductive potential, adaptability to high temperature and drought tolerance, allelopathic effects declared as one of the worst weed. The *Parthenium* endemic range is limited to 20 countries (South America, North and Central America) and invasive range is limitless almost every region of the world (South Asia, East and south-east Asia, Middle East, Australia and Oceania, East Africa, Southern Africa, Northern Africa). Now Parthenium weed has been spread to over the 50 countries around the world.

Distribution in Pakistan

In 80s *Parthenium* was first reported in Gujarat, Punjab later on its invade various regions of Pakistan rapidly (Khan & Fahad, 2020), reported as invasive alien species in Punjab from lower to high mountainous regions and river fed fertile plains throughout the Kpk and Fata and Kashmir and Sindh province and irrigated regions of Baluchistan are also invade by this noxious weed. The southern Punjab and Sindh regions are less affected and are supposed to be invaded by this aggressive weed species in future due to extensive irrigation system and climatic conditions as predicted y.

Agricultural and Environmental Impact

Parthenium invades croplands and competes fiercely with cultivated plants for nutrients, water, sunlight, and space. It reduces crop yields, suppresses the growth of desirable species, and degrades pasture lands, thereby lowering the carrying capacity of rangelands and affecting livestock productivity. Its allelopathic chemicals inhibit the germination and growth of nearby plants, making it highly competitive and ecologically destructive. It is particularly problematic in the cultivation of cereals, vegetables, and pulses, where even low-density infestations can cause significant yield reductions. In addition to crop loss, Parthenium infestations increase the cost of weed management and land restoration, placing an economic burden on farmers and agricultural systems (Khan & Fahad, 2020).

From an ecological perspective, Parthenium reduces native biodiversity by displacing indigenous plants, altering soil characteristics, and disturbing ecological balances. It can dominate natural habitats, degrade wildlife areas, and interfere with ecological succession. The plant's adaptability to different climatic conditions and soil types, as well as its resistance to many common herbicides, make it a persistent and expanding threat to Pakistan's ecosystems (Ghaffar et al., 2022).

Lanta camara

Lantana camara belongs to the Verbenaceae family, basically native to subtropical and tropical American regions and some tropical areas of Asia and Africa (Negi et al., 2019). This introduced ornamental or hedge species escaped and naturalized in more than 60 countries of the world and enlisted as one of the 10 worst invasive species. The plant growth potential and allopathic effect are attributed for the quick establishment of *L. camara* in disturbed or regions with low source availability (Negi et al., 2019).

Distribution in Pakistan

The *lantana camara* known as red sage and blacksage, flowered sage, lantana, white sage or wild sage. In Pakistan, Panchphulli is a vernacular name commonly used (Nawaz et al., 2016). The introduction of lantana is dated back to 1800 with British invasion in the subcontinent. This alien species is widely distributed in Pakistan It is most common in northern Punjab and also found in KPK. It has invaded hilly areas of Islamabad, Margalla National Park, Rawal Lake and along Murree Road towards Bharako and almost every ecological zones extensively studied by and naturalized pothar region (Bashir et al.).

Ecological Impact

The ecological damage caused by Lantana camara is profound. It aggressively invades forests and grasslands, forming dense thickets that crowd out native flora. These monocultures disrupt the natural plant community structure, reduce biodiversity, and degrade wildlife habitats. Native species struggle to regenerate under its canopy due to competition for light, water, and nutrients. Furthermore, Lantana camara produces allelopathic compounds that inhibit the growth of surrounding plants, making it extremely difficult for native vegetation to recover once it has established dominance (Kato-Noguchi & Kurniadie, 2021). The result is the degradation of entire ecosystems and a decline in ecological stability.

Its spread also alters fire regimes. The dry, woody stems of Lantana are highly flammable, increasing the risk and intensity of forest fires. These fires not only destroy habitats but also promote the further spread of Lantana, which regenerates quickly after burning, often more aggressively than before (Debuse & Lewis, 2014). This creates a dangerous feedback loop, where fire leads to more Lantana, and Lantana leads to more fire.

Agricultural and Livestock Issues

In agricultural landscapes, *Lantana camara* invades crop fields, orchards, and grazing lands. It competes with cultivated plants for resources, reduces farm productivity, and is costly to control. It also encroaches on rangelands, reducing the availability of palatable grasses and forage for livestock. Unlike many weeds, Lantana is toxic to most grazing animals (Kaushik & Arora, 2024). Its leaves contain triterpenoids that can cause liver damage, photosensitivity, and even death in cattle, sheep, and goats if consumed in large quantities. This makes infested pastures not only unproductive but also dangerous to livestock, affecting the livelihoods of pastoral communities.

Prosopis juliflora

Prosopis is a genus with 44 species belonging to the family Fabaceae, Mesquite (*Prosopis juliflora* (Sw.) DC) is one of those evergreen leguminous tree species native to hot dry arid and semi-arid regions of the world (Balcha, 2022). Mesquite native to 129 countries of North America, Central America, Caribbean South America. The *P. juliflora* has been introduced into several ecosystems, causing economic and ecological biohazards. The *P. juliflora* invasion has negative impacts on native biodiversity, ecosystem structure and function, bulk soil, seed bank, and hydrological cycle. Management strategies are presented to control *P. juliflora* invasion (Dakhil et al., 2021).

Distribution of *Prosopis juliflora* in Pakistan

In Pakistan, it is mostly distributed in saharo-sindhianphyto-geoghraphical regions which includes vast tracts of areas of southern Balochistan, Sindh and lower Punjab. The *P. juliflora* like other species introduced in the 1800s to reduce sand dunes dispersion government and non-government organizations now established in every region of Pakistan from plain to seacost (Dakhil et al., 2021). The regions of Sindh, more notably, southern kirthar and coastal region of Karachi are highly affected. The areas of Suleiman range, sizeable areas of makran coast highways and canals riparian forest, salt range and pabbi hills area has densely naturalized by mesquite and southern regions of Punjab (Rashid et al., 2014) (areas adjacent to Multan) and Kpk (Bannu, LakkiMarwat, Tank and Dera Ismail Khan) also colonized by hardy and resilient Prosopis.

Ecological Impact

Prosopis juliflora is highly invasive due to its deep root system, fast growth, and prolific seed production. It forms dense thickets that outcompete native vegetation by monopolizing water, sunlight, and nutrients. The tree alters the structure and function of ecosystems by displacing indigenous plant species, reducing biodiversity, and degrading natural habitats. Its leaf litter changes soil chemistry and creates unsuitable conditions for the regeneration of native flora. Moreover, the dense canopy and thorny nature of Prosopis make it difficult for wildlife to forage or move through invaded areas, thereby reducing habitat quality for native fauna (Ravhuhali et al., 2021).

In wetland and rangeland ecosystems, its expansion disrupts water flow and reduces the availability of forage for livestock. The tree is capable of surviving in saline, arid, and degraded soils, which allows it to colonize even the most fragile ecosystems, further intensifying environmental degradation (Osman, 2014).

Impact on Agriculture and Livelihoods

While Prosopis juliflora has been used as a source of firewood, charcoal, and fencing material, its economic benefits are often outweighed by its negative impact on agriculture and rural livelihoods. The aggressive spread of Prosopis into farmlands and pastures reduces the area available for crop cultivation and grazing. Its deep roots extract large amounts of groundwater, affecting the water availability for nearby agricultural fields, especially in already water-scarce regions (Khair et al., 2019).

The thorns of Prosopis can injure livestock and humans, and its thick growth makes land access difficult. Moreover, its pods, though used as livestock feed in small quantities, can cause digestive issues and even toxicity if consumed in excess. Farmers and pastoralists report declining productivity of their land, reduced access to water sources, and higher labor and financial costs in managing the invasion of this tree.

PAPER MULBERRY

Broussonetiapapyrifera (L.) L'Herit ex Vent. commonly known as paper mulberry evergreen tree species of family Moraceae, native to East Asia, Taiwan and Japan, extensively cultivated within its natural range for its bark (The bark is used for manufacturing paper in China and Japan and Taps cloth in Polynesia) while introduced in Russia, Indo-Pakistan subcontinent, W. Asia, tropical Africa, Europe, Polynesia, Philippines, U.S.A. (CABI, 2020).

Ecological Impact

Paper mulberry aggressively colonizes disturbed areas, often forming dense stands that displace native vegetation (Olupot, 2022). Its rapid growth and high reproductive ability allow it to dominate landscapes, outcompeting indigenous plant species for light, water, and nutrients. This reduces plant biodiversity and alters the natural composition of ecosystems. Its ability to grow in a wide range of soil types and climates makes it especially dangerous to local habitats, including hillsides, roadsides, and urban green spaces. Additionally, its fallen leaves decompose slowly, altering soil chemistry and affecting the regeneration of native plant species.

Human Health Impact

One of the most serious concerns associated with paper mulberry in Pakistan is its effect on human health. The tree produces massive quantities of airborne pollen during its flowering season (typically from March to May), which is a potent allergen. In Islamabad and other cities, pollen from *Broussonetia papyrifera* is one of the leading causes of seasonal allergic rhinitis, commonly known as hay fever. Symptoms include sneezing, nasal congestion, runny nose, itchy eyes, and in severe cases, asthma attacks. Thousands of residents suffer each year from pollen-related allergies, with hospitals reporting a sharp increase in patients during the peak pollen season (Kay et al., 2020). The economic burden of treatment, medication, and reduced productivity during this time is significant.

Environmental and Urban Challenges

In urban settings, paper mulberry trees pose additional problems. Their extensive root systems can damage infrastructure such as sidewalks, drainage systems, and building foundations (Rohela et al., 2025). The tree's weak wood structure also makes it prone to breakage during storms or strong winds, posing risks to people and property. Despite its fast growth, the tree offers limited ecological benefits in terms of supporting wildlife or promoting soil health.

Distribution in Pakistan

According to estimation 60-70 percent of the trees in Islamabad are of Paper mulberry (Hussain & Shah, 2020) . The World Conservation union lists paper mulberry as one of the six worst plant invaders in Pakistan. It has been declared as a Highly Invasive plant due to its aggressive growth, outcompeting the endemic flora. Other than metropolitan area, himalayan foothill region especially areas of Punjab (Rawalpindi, Attock, Narowal, Sialkot, Gujrat, Jhelum, and Chakwal) and KPK (Abbottabad, Buner, Tor Ghar, and Peshawar) and Northern Kashmir are highly invaded by paper mulberry.

Eucalyptus

Eucalyptus is a perennial angiosperm tree species which belongs to the family Myrtaceae, subfamily Myrotideae and consists of some 800 species (Hussain & Shah, 2020). It is native to Australia and Tasmania with a small number of species also found in New Guinea, the Philippines and Indonesia and has become the most widely planted genus of tree in the world. Current plantations around the world are dominated by the "big nine" species (*E. camaldulensis, E. grandis, E. tereticornis, E. globulus, E. nitens, E. urophylla, E. saligna, E. dunnii, and E. pellita*) and their hybrids which together account for more than 90% of Eucalyptus planted forests (Varghese et al., 2024). These species around the globe are used for the purpose of afforestation and control of soil erosion but due to its deleterious effects on ecosystem high water and nutrient usage as compared to native species and containing volatile compounds, it's considered not as much valuable species for plantation.

Ecological Concerns

Although Eucalyptus grows quickly and provides a source of wood in a short period, its ecological impact has raised significant concern. One of the most prominent issues is its high water consumption. Eucalyptus trees are known for their deep root systems and voracious uptake of groundwater, which can deplete water tables in arid and semi-arid regions like those found in many parts of Pakistan (Khan et al., 2020). This is particularly problematic in areas where water scarcity is already a pressing issue. Their ability to alter local hydrology makes Eucalyptus plantations unsuitable for water-sensitive regions, leading to reduced water availability for crops and native vegetation.

Moreover, Eucalyptus leaves contain allelopathic chemicals that inhibit the germination and growth of other plant species around them. This results in poor undergrowth and minimal biodiversity in Eucalyptus-dominated plantations. Native flora and fauna are often displaced, leading to ecological imbalances and habitat degradation. The monoculture nature of Eucalyptus plantations provides limited habitat and food sources for native insects, birds, and mammals, further diminishing biodiversity (Bayle, 2019).

Soil Degradation and Fire Hazard

Eucalyptus trees can also contribute to soil degradation. The thick canopy reduces light penetration and the accumulation of leaf litter creates a layer that decomposes slowly, limiting the growth of understory plants and affecting soil microflora. Additionally, due to the high oil content in their leaves and bark, Eucalyptus trees are highly flammable, increasing the risk of wildfires, particularly in dry seasons. This is a concern for areas of Pakistan that are becoming more prone to drought and high temperatures due to climate change (Fahad & Wang, 2020).

Economic and Social Dimensions

Economically, Eucalyptus has played a positive role in meeting the demand for fuelwood, timber, and poles in rural and urban areas. It provides a quick source of income for farmers and forest departments. However, the short-term economic benefits must be weighed against the long-term environmental degradation and water scarcity it may cause. In some regions, farmers have experienced declining agricultural productivity in fields adjacent to Eucalyptus plantations due to reduced soil moisture and water availability (Desta et al., 2023). Moreover, communities dependent on traditional agriculture or pastoralism have reported adverse effects on their livelihoods.

Distribution in Pakistan

Pakistan has a long history of introduction of eucalyptus dated back to 1860s in the subcontinent (Khan et al., 2020). In 1911, again tried in the Botanical garden at Lahore. In May 1913, several species of Eucalyptus were planted (BUG, 2015). In 1962, seed of *E. camaldulensis* imported from Australia was used for afforestation and reclamation of high sub-soil water table areas in Pakistan. *E. camaldulensis* has established itself in Pakistan and can be seen everywhere from the mountains to the seacoast. This introduced invasive alien species has become a threat to the ecosystem due to its potential environmental constraints like allelopathy, loss of soil fertility, the substitution of conventional forests and causing various hydro-ecological imbalances of an ecosystem (Reid et al., 2019). Eucalyptus trees consume three times more water in arid and semi-arid environments. In Pakistan *E. camaldulensis* is a prime species in afforestation programs such as Billion Tree Afforestation Program (BTAP) and Green Pakistan Program. The million hectares of areas in all provinces (Malakand, Balochistan, Indus River's bank, Chakwal, Bhakkar, Mianwali, Sargodha and Sheikhupura M2 Motorway, Haiedrabad) were cultivated with eucalyptus under the project of Pakistan Forest Institute sponsored by USAID (F. Ullah et al., 2022). Moreover, the Potohar region of Pakistan is found to be highly invaded by eucalyptus.

Cockle bur

The origin of rough cocklebur is also considered to old world on the basis of Fossil records in Europe dating back to the last glacial period support the idea that the plant is native to the Old World (Manzo & Tomasello, 2025). *X. strumarium* is an annual herbaceous plant of Asteraceae family globally distributed to the temperate, subtropical and Mediterranean regions, usually found along sides of roads in cultivated field and coastal dunes (Müller-Kiefer & Tomasello, 2022). This species was unintentionally introduced in Pakistan and soon became an invasive alien species.

This invasive alien species significantly reduced the native plant diversity abundance in invaded areas up to 55% as it's out competed with indigenous plants species (Early et al., 2016). In 1980s xanthium was introduced in Pakistan due to migration of afghan nomad to Pakistan (in Kpk) with their massive livestock results into propagation of this aggressive weed. The clinging ability of its tiny spiny fruits results in its spread with the help of livestock (cling to sheep wool and cattle hair) (Reiss, 2025). Moreover, tough and long dormant seed and allello-chemicals are successive factors in progression of this noxious weed.

The progressive invasion of *X. strumarium* in Pakistan is expanding more likely in the areas of upper Punjab, central Punjab and Khyber Pakhtunkhwa, approximately 21% areas in Pakistan are invaded with this weed (Waheed et al., 2024). The future invasion risk assessment of xanthium in various regions has been done using climate model. The region with high risk in upper Punjab included Rawalpindi, Gujrat, Mianwali, Jhelum and Chakwal, while in central Punjab, areas such as Kasur, Sahiwal, Hafizabad, Nankana Sahib, Sheikhpura and Okara were considered suitable. The regions of southern Punjab included Layyah, Dera Ghazi Khan, Muzaffargarh, Khanewal, Lodhran and Vehari while in Balochistan and Khyber Pakhtunkhwa the regions of Nowshera, Peshawar, LakiMarwat, Bannu, Kohat, Quetta, Karak, Mastung, Sibi and Kech are estimated to be heavily invaded with this aggressive weed (Waheed et al., 2023).The impact of xanthium on native flora and agricultural land of Pakistan is widely assessed and reported. The pothwar region native vegetation diversity declined due to progressive growth of this noxious weed (Waheed et al., 2023). Weed in Kpk maize crop and in wheat crop. In Punjab crops.

Economic perspectives

A comprehensive economic analysis required to manage invasive species cost, trade off benefits in Pakistan. Financial stability of Ecosystem and agriculture significantly challenged by invasive species. Pakistan loses billions of dollars every year because of these problems, necessitating immediate ecological and economic solutions.

Economic impacts:

Invasive plants like Parthenium heterophorias, *Prosopis juliflora*, and Lantana camara cause economic losses, ecosystem degradation, and increased healthcare costs, costing Pakistan billions annually and requiring urgent interventions (Nawaz et al., 2025).

Agricultural losses

Invasive plant species pose a serious economic threat to Pakistan's agricultural sector, which contributes significantly to the country's GDP and employs a large portion of the population. These invasive weeds compete directly with crops for vital resources such as water, nutrients, sunlight, and space, leading to substantial reductions in crop yields (Kaur et al., 2018). For example, species like Parthenium hysterophorus and *Lantana camara* infest farmlands, suppressing crop growth and sometimes producing allelopathic chemicals that inhibit seed germination of native or cultivated plants. As a result, farmers are compelled to increase their use of chemical herbicides and fertilizers, which significantly raises the cost of cultivation (Haggblade et al., 2017). In many rural areas, where smallholder farming is dominant, the additional financial burden caused by invasive species reduces profit margins, pushes farmers into debt, and threatens food security. Moreover, invasive species reduce the availability and quality of fodder for livestock, indirectly affecting the dairy and meat industries.

Ecosystem degradation

Ecosystem degradation is one of the most serious and far-reaching consequences of invasive plant species in Pakistan. Invasive plants disrupt the natural structure, function, and biodiversity of ecosystems by outcompeting native species, altering soil properties, changing hydrological cycles, and affecting nutrient dynamics (Eviner et al., 2012). These species often exhibit rapid growth, high reproductive capacity, and strong adaptability to a range of environmental conditions, enabling them to establish dense monocultures that suppress the growth of native vegetation. As native plant communities decline, the entire food web is affected, leading to reduced habitat quality for insects, birds, and mammals that depend on indigenous flora for food and shelter (McCary et al., 2016). This loss of biodiversity weakens the overall resilience of the ecosystem, making it more susceptible to further degradation and environmental stressors such as climate change, drought, or disease outbreaks.

In Pakistan, ecosystems ranging from riverbanks and wetlands to arid rangelands and forest margins are increasingly under threat from invasive species like *Parthenium hysterophorus*, *Lantana camara*, *Prosopis juliflora*, and *Ailanthus altissima* (Nawaz et al., 2025). These plants degrade soil quality by depleting nutrients, altering pH levels, and releasing allelopathic chemicals that hinder the germination and growth of native plants. In riparian ecosystems, invasive species can disrupt water flow, increase sedimentation, and reduce water availability for both wildlife and human use (Bolds et al., 2021). For example, the spread of Typha and Eichhornia crassipes (water hyacinth) in lakes and canals has choked waterways, reduced aquatic oxygen levels, and caused a decline in fish populations, impacting both biodiversity and local livelihoods dependent on fishing.

Moreover, ecosystem degradation caused by invasive plants can lead to the loss of vital ecosystem services such as pollination, water purification, carbon sequestration, erosion control, and natural pest regulation. The economic and ecological value of these services is immense, and their decline has cascading effects on agricultural productivity, public health, and climate regulation (Huang et al., 2020). In degraded ecosystems, natural regeneration becomes increasingly difficult, requiring extensive and costly restoration efforts. Furthermore, once an invasive species becomes dominant, it can create a self-reinforcing feedback loop, where its presence further degrades the habitat, allowing it to spread even more aggressively.

The degradation of ecosystems also has profound social and cultural implications. Many rural communities in Pakistan rely on healthy ecosystems for their livelihoods, including agriculture, grazing, medicinal plant collection, and fuelwood harvesting (Ahmad et al., 2015). When invasive species take over, these resources become scarce or unusable, leading to economic hardship and potential displacement of communities. Traditional ecological knowledge, which is often centered around native species and sustainable land use, also diminishes as invasive species transform the landscape.

FUTURE PERSPECTIVES

The challenge of managing invasive plant species through ecological approaches is expected to intensify in the coming decades due to accelerating environmental changes, increasing globalization, and continued anthropogenic disturbances. While significant progress has been made in understanding the ecological mechanisms of invasions, future strategies must be more proactive, predictive, and integrated across disciplines and sectors. The future of invasive plant management lies in harnessing emerging technologies, strengthening community-based and ecosystem-based approaches, and aligning invasive species control efforts with broader goals of climate resilience, biodiversity conservation, and sustainable development.

Predictive Modeling and Ecological Forecasting

One of the most promising future directions is the application of predictive ecological modeling to anticipate potential invasions before they become unmanageable (Büyüktahtakın & Haight, 2018). Species distribution models (SDMs), ecological niche modeling, and climate envelope models can be used to predict the potential range expansion of invasive species under current and future climate scenarios. These tools allow land managers and policymakers to identify invasion hotspots, prioritize surveillance efforts, and develop early warning systems. Integrating real-time environmental data, remote sensing, and GIS-based mapping can further enhance the ability to monitor invasive spread dynamically and design spatially explicit management interventions (Halmy, 2024).

Integration of Climate Change Adaptation

Climate change is expected to exacerbate the risk of plant invasions by altering species ranges, modifying disturbance regimes (such as droughts and wildfires), and weakening ecosystem resistance. Future ecological management must adopt climate-smart invasive species strategies that emphasize adaptive capacity, resilience, and forward-looking planning. Restoration of native plant communities should involve the selection of climate-resilient species that are not only locally adapted but also capable of withstanding future environmental stresses (Simonson et al., 2021). Additionally, integrating invasive species control into national climate adaptation policies—such as National Adaptation Plans (NAPs)can improve funding opportunities and institutional support.

Restoration Ecology and Nature-Based Solutions

Ecological restoration will remain a central pillar of invasive species management, but future efforts must go beyond mere species removal to embrace nature-based solutions that enhance ecosystem services, biodiversity, and socio-ecological resilience. Restoration strategies should aim to restore ecological processes, promote natural regeneration, and re-establish native plant–animal interactions (Lamb, 2010). The use of functional traits in selecting restoration species, rewilding techniques, and promoting natural successional dynamics are expected to gain prominence. Moreover, restoring degraded areas as ecological buffers or green corridors can reduce the vulnerability of ecosystems to invasion and promote connectivity across fragmented landscapes.

Genomic and Biotechnological Tools

Advances in molecular ecology and biotechnology offer exciting new avenues for understanding and managing invasive plant species (Dehnen-Schmutz & Novoa, 2022). Genomic tools can be used to study the genetic diversity, adaptive evolution, and invasion pathways of non-native species, providing insights into their ecological behavior and management potential. Additionally, gene editing technologies such as CRISPR/Cas9 may, in the future, play a role in altering the reproductive capacity or invasive potential of certain species, although these approaches raise ethical, ecological, and regulatory questions. The development of species-specific biological control agents using advanced molecular screening will also become a focus in ensuring safety and effectiveness (Harms et al., 2021).

Community Involvement and Participatory Management

The future of ecological invasive species management depends heavily on the meaningful engagement of local communities, indigenous groups, farmers, and other stakeholders. Community-based monitoring, participatory mapping, and citizen science can significantly enhance early detection, reporting, and localized control efforts (Pecl et al., 2019). Educational campaigns and capacity-building programs will be essential in fostering environmental stewardship and shifting perceptions especially in areas where certain invasive plants are still valued for short-term economic benefits (e.g., *Prosopis juliflora* or *Eucalyptus*) (Brundu et al., 2020). Empowering local actors through incentives, training, and institutional support can ensure the long-term success of ecological management strategies.

Policy Integration and International Cooperation

Future success in managing invasive plant species will depend on the integration of invasive species policies into broader environmental and land-use frameworks. National and provincial governments in Pakistan and elsewhere must prioritize invasive species in biodiversity action plans, land-use regulations, and agricultural and forestry programs. Furthermore, international cooperation is crucial, as invasive species often cross political boundaries and affect transboundary ecosystems. Shared databases, early warning systems, and coordinated response plans among neighboring countries (such as within South Asia) will be increasingly important. Aligning invasive species management with the goals of the Convention on Biological Diversity (CBD), Sustainable Development Goals (SDGs), and other global frameworks will help secure funding and technical collaboration.

Research and Capacity Building

There is a pressing need for sustained investment in ecological research and institutional capacity building to meet future invasive species challenges. Research priorities should include long-term ecological monitoring, understanding plant–microbe interactions, studying invasive impacts on soil and water systems, and evaluating restoration outcomes. Universities, research institutions, and government agencies must work collaboratively to fill data gaps, develop decision-support tools, and establish national and regional invasive species databases. Moreover, training programs for forest officers, agricultural extension workers, urban planners, and ecologists are essential to translate ecological knowledge into actionable management practices.

REFERENCES

Adedibu, P. A. (2023). Ecological problems of agriculture: impacts and sustainable solutions. *ScienceOpen preprints.*

Ahmad, M., Bano, A., Zafar, M., Sultana, S., & Rashid, S. (2015). Interdependence of biodiversity, applied ethnobotany, and conservation in higher ecosystems of northern Pakistan under fast climatic changes. In *Climate change impacts on high-altitude ecosystems* (pp. 455–489). Springer. DOI: 10.1007/978-3-319-12859-7_18

Aziz, T. (2021). Changes in land use and ecosystem services values in Pakistan, 1950–2050. *Environmental Development, 37*, 100576. DOI: 10.1016/j.envdev.2020.100576

Balcha, L. D. (2022). Prosopis juliflora distribution, impacts, and control methods available in Ethiopia. *Int. J. Nat. Resour. Ecol. Manag, 7*, 132–144.

Bashir, F., Altaf, M., Batool, S., Khan, M. S. H., Narejo, A. H., Iftakhar, A., Bashir, S. M., Hadi, A., & Ali, S. Diversity and Ecology of Birds from selected Sites of Islamabad, Pakistan.

Battisti, C., Poeta, G., & Fanelli, G. (2016). *An introduction to disturbance ecology.* Springer. DOI: 10.1007/978-3-319-32476-0

Bayle, G. (2019). Ecological and social impacts of eucalyptus tree plantation on the environment. *Journal of Biodiversity Conservation and Bioresource Management, 5*(1), 93–104. DOI: 10.3329/jbcbm.v5i1.42189

Bogale, G. A., & Tolossa, T. T. (2021). Climate change intensification impacts and challenges of invasive species and adaptation measures in Eastern Ethiopia. *Sustainable Environment, 7*(1), 1875555. DOI: 10.1080/23311843.2021.1875555

Bolds, S. A., Lockaby, B. G., Ditchkoff, S. S., Smith, M. D., & VerCauteren, K. C. (2021). *Impacts of a large invasive mammal on water quality in riparian ecosystems* (0047-2425).

Bradley, S. W., Kim, P. H., Klein, P. G., McMullen, J. S., & Wennberg, K. (2021). Policy for innovative entrepreneurship: Institutions, interventions, and societal challenges. *Strategic Entrepreneurship Journal, 15*(2), 167–184. DOI: 10.1002/sej.1395

Brundu, G., Pauchard, A., Pyšek, P., Pergl, J., Bindewald, A., Brunori, A., Canavan, S., Campagnaro, T., Celesti-Grapow, L., Dechoum, M. S., Dufour-Dror, J.-M., Essl, F., Flory, S. L., Genovesi, P., Guarino, F., Guangzhe, L., Hulme, P. E., Jäger, H., Kettle, C. J., & Richardson, D. M. (2020). Global guidelines for the sustainable use of non-native trees to prevent tree invasions and mitigate their negative impacts. [In.]. *NeoBiota, 61*, 65–116. DOI: 10.3897/neobiota.61.58380

BUG. T. G. (2015). Pests and pest control. *Centre of excellence in marine biology, University of Karachi, Karachi, 126*.

Büyüktahtakın, I. E., & Haight, R. G. (2018). A review of operations research models in invasive species management: State of the art, challenges, and future directions. *Annals of Operations Research, 271*(2), 357–403. DOI: 10.1007/s10479-017-2670-5

Clarke, B., Otto, F., Stuart-Smith, R., & Harrington, L. (2022). Extreme weather impacts of climate change: An attribution perspective. *Environmental Research: Climate, 1*(1), 012001.

Dakhil, M. A., El-Keblawy, A., El-Sheikh, M. A., Halmy, M. W. A., Ksiksi, T., & Hassan, W. A. (2021). Global invasion risk assessment of Prosopis juliflora at biome level: Does soil matter? *Biology (Basel), 10*(3), 203. DOI: 10.3390/biology10030203 PMID: 33803081

Debuse, V. J., & Lewis, T. (2014). Long-term repeated burning reduces Lantana camara regeneration in a dry eucalypt forest. *Biological Invasions, 16*(12), 2697–2711. DOI: 10.1007/s10530-014-0697-y

Dehnen-Schmutz, K., & Novoa, A. (2022). Advances in the management of invasive plants. In *Global plant invasions* (pp. 317–330). Springer. DOI: 10.1007/978-3-030-89684-3_15

Desta, T. T., Teklemariam, H., & Mulugeta, T. (2023). Insights of smallholder farmers on the trade-offs of eucalyptus plantation. *Environmental Challenges, 10*, 100663. DOI: 10.1016/j.envc.2022.100663

Early, R., Bradley, B. A., Dukes, J. S., Lawler, J. J., Olden, J. D., Blumenthal, D. M., Gonzalez, P., Grosholz, E. D., Ibañez, I., Miller, L. P., Sorte, C. J. B., & Tatem, A. J. (2016). Global threats from invasive alien species in the twenty-first century and national response capacities. *Nature Communications, 7*(1), 12485. DOI: 10.1038/ncomms12485 PMID: 27549569

El-Halwagi, M. M. (2025). *Sustainable design through process integration: fundamentals and applications to industrial pollution prevention, resource conservation, and profitability enhancement*. Elsevier.

Eviner, V. T., Garbach, K., Baty, J. H., & Hoskinson, S. A. (2012). Measuring the effects of invasive plants on ecosystem services: Challenges and prospects. *Invasive Plant Science and Management, 5*(1), 125–136. DOI: 10.1614/IPSM-D-11-00095.1

Fahad, S., & Wang, J. (2020). Climate change, vulnerability, and its impacts in rural Pakistan: A review. *Environmental Science and Pollution Research International, 27*(2), 1334–1338. DOI: 10.1007/s11356-019-06878-1 PMID: 31802341

Faiza, N., Weiguo, J., Aijun, Y., & Wenxing, S. (2017). Giant deforestation leads to drastic eco-environmental devastating effects since 2000; A case study of Pakistan. *JAPS: Journal of Animal & Plant Sciences, 27*(4).

Finch, D. M., Butler, J. L., Runyon, J. B., Fettig, C. J., Kilkenny, F. F., Jose, S., Frankel, S. J., Cushman, S. A., Cobb, R. C., & Dukes, J. S. (2021). Effects of climate change on invasive species. *Invasive species in forests and rangelands of the United States: a comprehensive science synthesis for the United States forest sector*, 57-83.

Gathala, M. K., Laing, A. M., Tiwari, T. P., Timsina, J., Rola-Rubzen, F., Islam, S., Maharjan, S., Brown, P. R., Das, K. K., Pradhan, K., Chowdhury, A. K., Kumar, R., Datt, R., Anwar, M., Hossain, S., Kumar, U., Adhikari, S., Magar, D. B. T., Sapkota, B. K., & Gerard, B. (2021). Improving smallholder farmers' gross margins and labor-use efficiency across a range of cropping systems in the Eastern Gangetic Plains. *World Development, 138*, 105266. DOI: 10.1016/j.worlddev.2020.105266

Ghaffar, A., Rahman, M. H. U., Ahmed, S., Haider, G., Ahmad, I., Khan, M. A., Afzaal, M., Ahmed, S., Fahad, S., & Hussain, J. (2022). Adaptations in cropping system and pattern for sustainable crops production under climate change scenarios. In *Improvement of plant production in the era of climate change* (pp. 1–34). CRC Press. DOI: 10.1201/9781003286417-1

Haggblade, S., Smale, M., Kergna, A., Theriault, V., & Assima, A. (2017). Causes and consequences of increasing herbicide use in Mali. *European Journal of Development Research, 29*(3), 648–674. DOI: 10.1057/s41287-017-0087-2

Halmy, M. W. A. (2024). Importance of Geospatial Data for Monitoring and Managing Invasive Species for Sustainable Development: A Semantic Review. *Annals of Library and Information Studies, 71*(4), 417–434.

Harms, N. E., Knight, I. A., Pratt, P. D., Reddy, A. M., Mukherjee, A., Gong, P., Coetzee, J., Raghu, S., & Diaz, R. (2021). Climate mismatch between introduced biological control agents and their invasive host plants: Improving biological control of tropical weeds in temperate regions. *Insects, 12*(6), 549. DOI: 10.3390/insects12060549 PMID: 34204761

Huang, C., Li, N., Zhang, Z., Liu, Y., Chen, X., & Wang, F. (2020). Assessment of the economic cascading effect on future climate change in China: Evidence from agricultural direct damage. *Journal of Cleaner Production, 276*, 123951. DOI: 10.1016/j.jclepro.2020.123951

Hulme, P. E. (2021). Unwelcome exchange: International trade as a direct and indirect driver of biological invasions worldwide. *One Earth, 4*(5), 666–679. DOI: 10.1016/j.oneear.2021.04.015

Hussain, A., & Shah, S. U. H. (2020). Economic and Environmental Costs and Benefits of Paper Mulberry: A Case Study of Islamabad-Pakistan. *Journal of Health Research and Society, 2*, 5–12. DOI: 10.34002/jhrs.v2i0.20

Jehangir, S., Khan, S. M., Ejaz, U., Zahid, N., Rashid, N., Noshad, Q., Din, Z. U., & Shoukat, A. (2024). Alien flora of Pakistan: Taxonomic composition, invasion status, geographic origin, introduction pathways, and ecological patterns. *Biological Invasions, 26*(8), 2435–2451. DOI: 10.1007/s10530-024-03311-8

Kachigunda, B., Coupland, G. T., Mengersen, K., Perera, D. I., Van Der Merwe, J., & McKirdy, S. (2023). A systems approach to biosecurity pays off: A case study investigating introduction pathways and interceptions of non-indigenous species at a biosecurity border. *Biological Invasions, 25*(9), 2849–2871. DOI: 10.1007/s10530-023-03077-5

Kato-Noguchi, H., & Kurniadie, D. (2021). Allelopathy of Lantana camara as an invasive plant. *Plants, 10*(5), 1028. DOI: 10.3390/plants10051028 PMID: 34065417

Kaur, S., Kaur, R., & Chauhan, B. S. (2018). Understanding crop-weed-fertilizer-water interactions and their implications for weed management in agricultural systems. *Crop Protection (Guildford, Surrey), 103*, 65–72. DOI: 10.1016/j.cropro.2017.09.011

Kaushik, Y., & Arora, P. (2024). Investigating the sustainable energy generation potential of an invasive weed: Lantana camara. *Environmental Science and Pollution Research International, 31*(54), 62493–62509. DOI: 10.1007/s11356-024-35322-2 PMID: 39453581

Kay, D., Kelly, J., Schaeffer, T., Tuong, T., Leru, P., Addison, W., & Neamtu, R. (2020). Pollen Allergies in Romania.

Khair, S. M., Mushtaq, S., Reardon-Smith, K., & Ostini, J. (2019). Diverse drivers of unsustainable groundwater extraction behaviour operate in an unregulated water scarce region. *Journal of Environmental Management, 236*, 340–350. DOI: 10.1016/j.jenvman.2018.12.077 PMID: 30738305

Khan, M. A., Gul, B., & Khan, H. (2015). Impact of climatic change on flora of high altitudes in Pakistan. In *Climate Change Impacts on High-Altitude Ecosystems* (pp. 361–381). Springer. DOI: 10.1007/978-3-319-12859-7_14

Khan, N., & Fahad, S. (2020, January 12). Economic Review of Parthenium Hysterophorus L. Plant in the World. *Plant World*.

Khan, N., Fahad, S., Faisal, S., Akbar, A., & Naushad, M. (2020). Socio-economic and medicinal review of Eucalyptus tree in the world. *Available at SSRN* 3644215. DOI: 10.2139/ssrn.3644215

Khuroo, A. A., Ahmad, R., Hamid, M., Rather, Z. A., Malik, A. H., & Rashid, I. (2021). An annotated inventory of invasive alien flora of India. *Invasive alien species: observations and issues from around the world, 2*, 16-37.

Lamb, D. (2010). Ecological restoration. In *Regreening the bare hills: tropical forest restoration in the Asia-Pacific region* (pp. 325–358). Springer.

Lugo, A. E. (2020). Effects of extreme disturbance events: From ecesis to social–ecological–technological systems. *Ecosystems (New York, N.Y.)*, *23*(8), 1726–1747. DOI: 10.1007/s10021-020-00491-x

Manzo, E., & Tomasello, S. (2025). Recent Long-Distance-Dispersal Explains the Range Disjunction of the Old-Word Cockleburs (Xanthium strumarium). *Journal of Biogeography*, *52*(5), e15104. DOI: 10.1111/jbi.15104

McCary, M. A., Mores, R., Farfan, M. A., & Wise, D. H. (2016). Invasive plants have different effects on trophic structure of green and brown food webs in terrestrial ecosystems: A meta-analysis. *Ecology Letters*, *19*(3), 328–335. DOI: 10.1111/ele.12562 PMID: 26757702

Mohammeda, A., & Bahmudb, S. (2023). The Status, Ecological and Biodiversity Impacts of Invasive Plant Species in Ethiopia: A review Paper. *GSJ, 11*(11).

Müller-Kiefer, J., & Tomasello, S. (2022). Invasive or threatened? The decline of the old-world Xanthium strumarium (Asteraceae) in Italy based on herbarium records. *Flora Mediterranea*, *32*, 17–24.

Nawaz, A., Ayub, M. A., Nadeem, F., & Al-Sabahi, J. N. (2016). Lantana (Lantana camara): A medicinal plant having high therapeutic potentials–A comprehensive review. *International Journal of Chemical and Biochemical Science*, *10*, 52–59.

Nawaz, H., Ishaq, M. W., Qadeer, S., Mahmood, M. H., Adeel, M., Anwar, I., Mazahr, M., Faisal, M., Arshad, M. A., Umar, M., & Zaheer, M. (2025). Assessing the Threat: Parthenium Adverse Effects on Biodiversity, Human Communities, and Environmental Integrity. *Sch Bull, 11*(3), 21–41. DOI: 10.36348/sb.2025.v11i03.001

Negi, G. C., Sharma, S., Vishvakarma, S. C., Samant, S. S., Maikhuri, R. K., Prasad, R. C., & Palni, L. M. (2019). Ecology and use of Lantana camara in India. *Botanical Review*, *85*(2), 109–130. DOI: 10.1007/s12229-019-09209-8

Nuñez, M. A., Chiuffo, M. C., Torres, A., Paul, T., Dimarco, R. D., Raal, P., Policelli, N., Moyano, J., García, R. A., Van Wilgen, B. W., Pauchard, A., & Richardson, D. M. (2017). Ecology and management of invasive Pinaceae around the world: Progress and challenges. *Biological Invasions*, *19*(11), 3099–3120. DOI: 10.1007/s10530-017-1483-4

Olupot, W. (2022). The potential of natural succession to restore degraded areas of a Ugandan rainforest dominated by the exotic paper mulberry Broussonetia papyrifera (L.) L'Hér. ex Vent. *Forest Ecology and Management*, *504*, 119861. DOI: 10.1016/j.foreco.2021.119861

Osman, K. T. (2014). *Soil degradation, conservation and remediation* (Vol. 820). Springer. DOI: 10.1007/978-94-007-7590-9

Pecl, G. T., Stuart-Smith, J., Walsh, P., Bray, D. J., Kusetic, M., Burgess, M., Frusher, S. D., Gledhill, D. C., George, O., Jackson, G., Keane, J., Martin, V. Y., Nursey-Bray, M., Pender, A., Robinson, L. M., Rowling, K., Sheaves, M., & Moltschaniwskyj, N. (2019). Redmap Australia: Challenges and successes with a large-scale citizen science-based approach to ecological monitoring and community engagement on climate change. *Frontiers in Marine Science*, *6*, 349. DOI: 10.3389/fmars.2019.00349

Qazi, A. W., Saqib, Z., & Zaman-ul-Haq, M. (2023). Diversity and distribution of endemic flora in Pakistan. *Proceedings of the Pakistan Academy of Sciences: B. Life and Environmental Sciences, 60*(2), 165-179.

Rai, P. K. (2015). What makes the plant invasion possible? Paradigm of invasion mechanisms, theories and attributes. *Environmental Skeptics and Critics*, *4*(2), 36.

Rashid, M., Abbas, S. H., & Rehman, A. (2014). The status of highly alien invasive plants in Pakistan and their impact on the ecosystem: A review. *Innov J Agric Sci*, *2*, 1–4.

Ravhuhali, K. E., Mudau, H. S., Moyo, B., Hawu, O., & Msiza, N. H. (2021). Prosopis species—An invasive species and a potential source of browse for livestock in semi-arid areas of South Africa. *Sustainability (Basel), 13*(13), 7369. DOI: 10.3390/su13137369

Reaser, J. K., Burgiel, S. W., Kirkey, J., Brantley, K. A., Veatch, S. D., & Burgos-Rodríguez, J. (2020). The early detection of and rapid response (EDRR) to invasive species: A conceptual framework and federal capacities assessment. *Biological Invasions*, *22*(1), 1–19. DOI: 10.1007/s10530-019-02156-w PMID: 33488272

Reid, A. J., Carlson, A. K., Creed, I. F., Eliason, E. J., Gell, P. A., Johnson, P. T., Kidd, K. A., MacCormack, T. J., Olden, J. D., Ormerod, S. J., Smol, J. P., Taylor, W. W., Tockner, K., Vermaire, J. C., Dudgeon, D., & Cooke, S. J. (2019). Emerging threats and persistent conservation challenges for freshwater biodiversity. *Biological Reviews of the Cambridge Philosophical Society*, *94*(3), 849–873. DOI: 10.1111/brv.12480 PMID: 30467930

Reiss, B. F. (2025). *The Social Life of Plants: Botanical Representations in Moche Art*. The University of New Mexico.

Ren, G., Du, Y., Yang, B., Wang, J., Cui, M., Dai, Z., Adomako, M. O., Rutherford, S., & Du, D. (2023). Influence of precipitation dynamics on plant invasions: Response of alligator weed (Alternanthera philoxeroides) and co-occurring native species to varying water availability across plant communities. *Biological Invasions*, *25*(2), 519–532. DOI: 10.1007/s10530-022-02931-2

Rohela, G. K., Saini, P., Syam, S., Roy, P., Gonal, B., Aziz, D., Ayoub, O. B., & Maheswari, M. (2025). Mulberry Trees: A Sustainable Solution for Urban Forestry and Improved Air Quality. In *Sustainable Urban Environment and Waste Management: Theory and Practice* (pp. 223-246). Springer. DOI: 10.1007/978-981-96-1140-9_12

Seebens, H., Meyerson, L. A., Rahlao, S. J., Lenzner, B., Tricarico, E., Aleksanyan, A., Courchamp, F., Keskin, E., Saeedi, H., & Tawake, A. (2023). Trends and status of alien and invasive alien species.

Shin, Y. J., Midgley, G. F., Archer, E. R., Arneth, A., Barnes, D. K., Chan, L., Hashimoto, S., Hoegh-Guldberg, O., Insarov, G., Leadley, P., Levin, L. A., Ngo, H. T., Pandit, R., Pires, A. P. F., Pörtner, H.-O., Rogers, A. D., Scholes, R. J., Settele, J., & Smith, P. (2022). Actions to halt biodiversity loss generally benefit the climate. *Global Change Biology*, *28*(9), 2846–2874. DOI: 10.1111/gcb.16109 PMID: 35098619

Simberloff, D. (2013). *Invasive species: what everyone needs to know*. OUP Us. DOI: 10.1093/wentk/9780199922017.001.0001

Simonson, W. D., Miller, E., Jones, A., García-Rangel, S., Thornton, H., & McOwen, C. (2021). Enhancing climate change resilience of ecological restoration—A framework for action. *Perspectives in Ecology and Conservation, 19*(3), 300–310. DOI: 10.1016/j.pecon.2021.05.002

Ullah, F., Baloch, M., Ullah, S., Kabir, M., & Khan, N. H. (2022). Eucalyptus plantation in Pakistan: Holistic view of environmental socioeconomic and medicinal perspective. *Biometrical Letters, 8*(1), 41–49.

Ullah, R., Khan, N., Hewitt, N., Ali, K., Jones, D. A., & Khan, M. E. H. (2022). Invasive Species as Rivals: Invasive Potential and Distribution Pattern of Xanthium strumarium L. *Sustainability (Basel), 14*(12), 7141. DOI: 10.3390/su14127141

Varghese, M., Kamalakannan, R., & Suraj, P. (2024). Eucalyptus: Taxonomy, Geographic Distribution, Domestication, Breeding, Ecology and Economic Importance. In *Economically Important Trees: Origin, Evolution, Genetic Diversity and Ecology* (pp. 81-128). Springer.

Vilà, M., & Hulme, P. E. (2017). Non-native species, ecosystem services, and human well-being. In *Impact of biological invasions on ecosystem services* (pp. 1–14). Springer. DOI: 10.1007/978-3-319-45121-3_1

Waheed, M., Haq, S. M., Arshad, F., Jameel, M. A., Siddiqui, M. H., Bussmann, R. W., Manshoor, N., & Alamri, S. (2023). Where will threatened Aegle marmelos L., a tree of the semi-arid region, go under climate change? implications for the reintroduction of the species. *Land (Basel), 12*(7), 1433. DOI: 10.3390/land12071433

Waheed, M., Haq, S. M., Arshad, F., Vitasović-Kosić, I., Bussmann, R. W., Hashem, A., & Abd-Allah, E. F. (2024). Xanthium strumarium L., an invasive species in the subtropics: Prediction of potential distribution areas and climate adaptability in Pakistan. *BMC Ecology and Evolution, 24*(1), 124. DOI: 10.1186/s12862-024-02310-6 PMID: 39390368

Chapter 7
Effects of Invasive Plants on Soil Health and Nutrient Cycling

Laxmi Biban
https://orcid.org/0000-0002-9809-6287
Pt. C.L.S. Government College, Karnal, India

Shaista Md. Mumtaz Khan
https://orcid.org/0009-0004-5202-7608
Datta Meghe Institute of Management Studies, India

Anjum Parvez
https://orcid.org/0000-0002-4874-1372
Uttaranchal University, India

Garima Mishra
Datta Meghe Institute of Management Studies, India

S. Swathi
https://orcid.org/0009-0004-2536-1627
Dayananda Sagar Business Academy, Bangalore, India

P. Selvakumar
https://orcid.org/0000-0002-3650-4548
Nehru Institute of Technology, India

T. C. Manjunath
https://orcid.org/0000-0003-2545-9160
Rajarajeswari College of Engineering, India

ABSTRACT

Native plants, when to new environments, establish themselves rapidly and spread aggressively, often to the detriment of local ecosystems. Their ability to outcompete native flora and alter habitat structure can have profound effects on biodiversity, ecosystem functions, and ecological processes. Understanding the impacts on preserving ecological balance. Definition and Characteristics of Invasive Plants: characterized by their high growth rates, prolific reproduction, and adaptability to various environmental conditions. Unlike native species that have evolved specific

DOI: 10.4018/979-8-3693-8724-5.ch007

relationships with local fauna and flora, invasive plants often lack natural predators or controls in their new environments, allowing them to proliferate unchecked. Key characteristics of invasive plants include vigorous growth, high seed production, and the ability to outcompete native species for resources. Ecological Impacts: Biodiversity Loss: plants water, significant reductions in plant diversity.

INTRODUCTION TO INVASIVE PLANT SPECIES

Native plants, when to new environments, establish themselves rapidly and spread aggressively, often to the detriment of local ecosystems. Their ability to outcompete native flora and alter habitat structure can have profound effects on biodiversity, ecosystem functions, and ecological processes. Understanding the impacts on preserving ecological balance. Definition and Characteristics of Invasive Plants: characterized by their high growth rates, prolific reproduction, and adaptability to various environmental conditions. Unlike native species that have evolved specific relationships with local fauna and flora, invasive plants often lack natural predators or controls in their new environments, allowing them to proliferate unchecked. Key characteristics of invasive plants include vigorous growth, high seed production, and the ability to outcompete native species for resources. Ecological Impacts: Biodiversity Loss: plants water, significant reductions in plant diversity. The dominance of invasive species can disrupt the balance of plant communities, which in turn affects the animals that depend on native plants for food and habitat. This can result in cascading effects throughout the food web, leading to declines in native animal populations and overall biodiversity loss. Alteration of Habitat Structure: physical habitats. For example, dense mats of invasive vegetation can alter soil composition, moisture levels, and light availability. This can impact the growth and survival of native plants and animals, leading to changes in ecosystem dynamics and habitat quality. In aquatic environments, invasive species like water hyacinths can choke waterways, affecting water flow and oxygen levels, which can severely impact fish and other aquatic organisms. Disruption of Ecological Processes: Invasive plants can interfere with processes of erosion, and pollination. For instance, species chemistry availability or increasing erosion rates can affect native health ecosystems. Additionally, invasive plants can disrupt the relationships between native plants and their pollinators, potentially leading to reduced plant reproduction and ecosystem instability. Economic and Social Impacts: Economic Costs: Plants lead to significant economic costs.

Management and Control Strategies: Prevention and Early Detection: Effective management plant begins with This involves identifying potential invasive species before they become established. Public awareness campaigns and education programs can help plants by informing people about the risks and promoting responsible gardening and landscaping practices. Control and Eradication: Various control methods are used to manage invasive plant populations, including mechanical removal, chemical herbicides, and biological control. Kill or employ natural predators or pathogens to target invasive species. Each method has its advantages and limitations, and often a combination of approaches is used to achieve effective control. Restoration and Rehabilitation: Restoring invaded habitats involves reintroducing native plant species and improving ecosystem conditions to support their recovery. This may include soil amendments, controlled burns, or planting native species to reestablish natural plant communities. Successful restoration efforts require careful planning and monitoring to ensure that the interventions are effective and do not inadvertently contribute to others. Case Studies and Research: Case Studies: Numerous case studies highlight plant effectiveness and various management strategies. The Spread of kudzu southeastern led to significant ecological and economic challenges, prompting extensive research into control methods and restoration practices. Similarly, the invasion of garlic mustard in North American forests has led to a better understanding of its effects on soil ecosystems and native plant communities. Ongoing Research: Research into invasive plant species continues their biology and impacts. Studies on mechanisms of invasion, native species, effectiveness of different management techniques contribute to the development of more targeted and effective strategies.

Soil Health Indicators: Assessing Impacts of Invasive Plants

Ecosystem function productivity can be significantly influenced by the presence Assessing soil health invasive plant impacts involves examining various indicators that reflect changes in soil quality and ecosystem dynamics. Invasive plants, by altering soil properties and ecological interactions, profound health, impacting microbial communities, and overall soil structure. Impacts essential restoration efforts. Key Indicators of Soil Health: Soil Organic Matter (SOM): matter crucial and Invasive plants can alter SOM levels by changing for example, some invasive species produce litter with different chemical properties compared to native plants, affecting decomposition processes and SOM accumulation. Soil Nutrient Levels: essential (N), (P), (K), aspect health (Rai & Singh, 2020). Invasive plants can influence nutrient dynamics through their growth patterns and root exudates. Certain invasive species may alter nutrient cycling by modifying soil pH or disrupting microbial communities responsible for nutrient processing. Soil pH: affects Invasive plants can their decomposition products. Changes in soil pH caused by invasive species can impact native soil, potentially leading to further ecological imbalances. Soil Structure and Compaction: Soil structure, including aggregate formation and compaction, affects water infiltration, root growth, and aeration. Invasive plants can impact soil structure by altering root systems and affecting soil compaction. For instance, some invasive species have deep or extensive root systems that can change soil physical properties, potentially leading to changes in water movement and root penetration. Impacts of Invasive Plants on Soil Health: Alteration of Nutrient Cycling: Invasive plants can disrupt nutrient cycling by changing activity example, invasive species may alter the balance or lead to shifts in nutrient availability. These changes can affect the growth of native plants and overall soil health. Changes in Soil Organic Carbon (SOC): The introduction of invasive plants changes levels. Invasive species with high biomass production may initially increase SOC, but their effect can be complex. The decomposition of invasive plant material may differ from that of native species, affecting the stability and storage. Impact on Soil Microbial Communities: Soil maintaining health by facilitating and Invasive plants reduced microbial diversity and altered functions. Changes in microbial communities can influence nutrient availability and decomposition (Bhaduri D., et al, 2022). Soil Erosion and Sedimentation: The presence of invasive plants can affect soil erosion and sedimentation rates. Invasive densities may be reduced while others that lack extensive root networks can increase susceptibility to erosion.

Management and Restoration Considerations: Invasive Species Management: Effective management of invasive plant species is crucial for mitigating their impacts on soil health. Strategies such as mechanical removal, chemical control, and biological management can help reduce the prevalence of invasive plants and prevent further degradation of soil quality. Integrated management approaches that combine multiple methods are often more effective in addressing complex invasive species issues. Restoration and Rehabilitation: Restoring invaded areas involves reintroducing native plant species and improving soil conditions to support ecosystem recovery. Restoration efforts may include soil amendment, planting native vegetation, and enhancing soil structure. Monitoring soil health indicators before and after restoration activities helps evaluate the success of these efforts and guides adaptive management (Rai P K., 2022). Educational and Collaborative Efforts: Educating landowners, managers, and policymakers about soil health is essential for fostering effective management practices. Collaborative efforts among researchers, conservationists, and land managers can enhance the development of innovative solutions and promote the adoption of best practices for managing invasive plants and preserving soil health. Future Directions in Research: Long-Term Studies: cumulative and prolonged impacts of invasive plants on soil health. Research that tracks changes in soil properties over time can provide valuable invasion interactions and effectiveness of management strategies. Mechanistic Studies: Research focusing on the mechanisms by which invasive plants alter soil health is crucial for developing targeted management approaches. Understanding the specific processes involved in structure can inform more effective control and restoration strategies. Climate

Change Considerations: Future research should consider the potential interactions between invasive plants, soil health, and climate change (Menta C., et al, 2020).

Altered Nutrient Cycling: Mechanisms and Consequences

Ecological continuous availability of nutrients to organisms within an ecosystem? It encompasses the transformation and movement of nutrients through biological, chemical, and physical processes. When invasive plant species are introduced into an ecosystem, they can significantly alter nutrient cycling, impacting overall health. Invasive plants influence nutrient cycling and the resulting consequences are crucial for effective ecosystem management and restoration. Mechanisms of Altered Nutrient Cycling: Changes in Nutrient Input: Invasive plants can modify nutrient input into the soil through variations in litter quality and quantity. The decomposition of invasive plant litter may differ from that of native species in terms of nutrient content and decomposition rates. For instance, invasive species like garlic mustard produce litter that decomposes more slowly, potentially leading to nutrient imbalances and altered soil nutrient dynamics. Root Exudates: The root exudates of can chemistry availability (Mekonnen Z A., et al, 2021). These exudates may include organic acids, enzymes, and other compounds that influence soil pH and microbial activity. For example, some invasive species exude compounds that acidify the soil, which can affect others. Disruption of Microbial Communities: Invasive plants can disrupt crucial. Changes in microbial diversity and activity can affect nitrogen fixation, For example, invasive plants may outcompete native species for microbial symbionts, leading to reduced nutrient uptake efficiency and altered nutrient cycling. Alteration of Soil Properties: Invasive plants can influence moisture, and pH, which in turn affect nutrient cycling. For instance, invasive or extensive change structure rates, impact nutrient availability and distribution within the soil profile. Consequences of Altered Nutrient Cycling: Nutrient Imbalances: The alteration of nutrient cycling by invasive plants can lead to nutrient imbalances in the soil. Changes in the rates of nutrient input, transformation, and loss can result in deficiencies or excesses of essential nutrients. For example, an invasive plant that alters nitrogen cycling may lead to either nitrogen enrichment or depletion, affecting the growth of native plants and overall ecosystem productivity. Reduced Soil Fertility: Invasive plants that disrupt nutrient cycling can negatively impact soil fertility. For instance, if invasive species deplete essential nutrients or inhibit the activities of beneficial soil microorganisms, soil fertility may decline (Rädecker N., et al, 2021).

Kudzu can alter soil nitrogen levels by promoting nitrogen fixation and modifying soil structure. Its ability to outcompete native vegetation can lead to changes in nutrient distribution and soil fertility (Dai Z., et al, 2021). Japanese Knotweed (Fallopian japonica): Japanese knotweed is an invasive species that can impact nutrient cycling by altering soil structure and chemistry. Its extensive rhizome system can change soil properties and affect nutrient availability. The presence of Japanese knotweed plant alterations in nutrient cycling processes. Management and Mitigation Strategies: Monitoring and Assessment: assessment of nutrient cycling in invaded ecosystems is essential for understanding the impacts of invasive plants. Soil tests, plant tissue analysis, and microbial assessments can provide valuable information on changes in nutrient dynamics and guide management decisions. Integrated Management Approaches: Effective management of invasive plants involves an integrated approach that combines various Addressing the impacts on nutrient cycling requires strategies that not only control the invasive species but also restore nutrient balance and support ecosystem recovery. Restoration and Rehabilitation: Restoration efforts should focus on reestablishing native plant communities and improving soil health to counteract the effects of altered nutrient cycling. Techniques such as soil amendments, planting native species, and enhancing soil microbial diversity can help restore nutrient dynamics and support ecosystem recovery. Future Research Directions: Mechanistic Studies: Future research should go through which invasive plants alter nutrient cycling. Investigating the interactions between invasive species, soil microorganisms, and nutrient processes can provide insights into how to manage and mitigate their impacts. Long-Term Monitoring: Long-term monitoring of nutrient cycling in invaded and restored ecosystems is crucial for assessing the effectiveness of management and restoration efforts. Research should nutrient dynamics to evaluate the outcomes of different interventions and adapt strategies as needed (Zhang X., et al, 2020).

Changes in Soil Microbial Communities: Invasive Plant Effects

Communities facilitating essential processes of soil structure formation. Plants can significantly alter these microbial communities, leading to soil, plant growth, and overall function. Understanding how invasive plants impact soil microorganisms and the subsequent consequences is essential for effective ecosystem management and restoration. Mechanisms of Microbial Community Alteration: Changes in Organic Matter Inputs: Invasive plants can alter the quantity entering the type of root exudates produced by invasive species differs from those of native plants, affecting microbial communities. For example, invasive plants that produce high amounts of low-nutrient litter may lead to slower decomposition rates and shifts in microbial community composition, favoring decomposers adapted to these new conditions. Root Exudates and Soil Chemistry: The root exudates of invasive plants can influence soil chemistry by altering pH, nutrient availability, and the composition of dissolved organic carbon. These changes can impact microbial activity and community structure (Torres N., et al, 2021). For instance, some invasive species release organic acids that acidify the soil, potentially suppressing certain microbial groups while promoting others. Competition and Allelopathy: often including nutrients This competition can influence microbial communities by altering the availability of substrates for microbial growth. Additionally, some invasive species compounds potentially affect soil microbial populations by altering the availability. Disruption of Symbiotic Relationships: Many microorganisms plants, Invasive plants can disrupt these relationships by altering the availability of symbiotic partners or changing the conditions under which these interactions occur. For example, invasive species that do not form mycorrhizal associations may reduce nutrient uptake and soil microbial diversity(Shende, A. D., et al. 2025). Consequences of Altered Microbial Communities: Nutrient Cycling Disruption: Soil microbial communities are integral to nitrogen fixation, and phosphorus mobilization. Alterations in microbial community structure caused by invasive plants can disrupt these processes, leading to changes in nutrient availability and soil fertility. For example, shifts in microbial communities may affect nitrogen cycling, resulting in reduced nitrogen availability for plants or increased nitrogen loss through leaching. Impact on Soil Health: Changes in microbial communities can affect matter and moisture retention. Invasive plants that alter microbial communities may lead and changes in soil texture. These impacts can affect quality (Fahey & Flory, 2022).

The presence of garlic mustard can suppress and alter the composition, particularly those involved in nitrogen cycling. Japanese Knotweed (Fallopian japonica): Japanese knotweed is an invasive species that can alter soil microbial communities through its extensive rhizome system and high biomass production. Research has indicated that Japanese knotweed can affect microbial diversity and soil nutrient dynamics, leading to changes in soil health and plant community composition (Keet J H., et al, 2021). Management and Mitigation Strategies: Monitoring and Assessment: Regular soil microbial communities for understanding impacts of invasive plants. Soil microbial surveys, including molecular techniques, can change microbial diversity and function. Assessing microbial community structure and activity can help inform management decisions and track the effectiveness of control measures. Future Research Directions: Mechanistic Studies: Future research should through which alter Investigating the interactions between invasive species, soil microorganisms, and soil processes can provide insights into how to manage and mitigate their impacts effectively (Bell J K., et al, 2020). Long-Term Studies: assess cumulative and prolonged invasive plants on microbial communities. Research that tracks over time dynamics microbial-invasion interactions and the effectiveness of management strategies. Climate Change Considerations: consider interactions between invasive plants, and soil microbial communities, As conditions, the behavior impacts of invasive species may shift, affecting microbial communities and soil processes. Addressing these interactions is essential for developing adaptive management strategies in a changing environment. In conclusion, invasive plant species can significantly alter soil microbial communities through changes in organic matter inputs, root exudates, competition, and symbiotic relationships. These alterations can disrupt nutrient cycling, soil health, and functions. As a consequence of these changes and restoration, we can work towards maintaining healthy and resilient ecosystems that support biodiversity and ecosystem stability (Ranjitha P. K., et al. 2025).

Invasive Plant-Soil Feedback Loops: Understanding the Dynamics

Often have profound effects on the ecosystems they invade, and one of the most critical areas of impact is through soil feedback loops. These feedback loops, which involve complex interactions between invasive plants and soil properties, can reinforce the dominance of invasive species and lead to significant ecological changes. Understanding these dynamics is essential for developing plants. Mechanisms of Invasive Plant-Soil Feedback Loops: Altered Soil Nutrient Dynamics: Invasive plants can change soil nutrient availability through several mechanisms. For instance, some invasive species may alter soil pH or nutrient cycling processes, affecting the availability of essential nutrients. These changes can create conditions that favor the invasive species while disadvantaging acidification may reduce the availability of nutrients like phosphorus, which can further inhibit the growth of native species(Subramanian, M., et al. 2025).. Changes in Soil Microbial Communities: Invasive Plants Soil-altering invasive plants can affect decomposition and organic matter breakdown. This disruption can lead to a feedback loop where changes in microbial communities reinforce the dominance of invasive species and further alter soil conditions (Frouz J., 2023). Inhibition of Native Vegetation: Invasive plants often reduce the growth and survival of native plants, which may, in turn, affect them. For example, the decline of native plants can reduce the diversity of root exudates and organic matter inputs, leading to chemistry that favors invasive species. Allelopathic Effects: Some invasive plant compounds affect soil microbial communities. These compounds create conditions that further support the invasive species. For example, invasive plants like garlic mustard native disrupt changes in soil dynamics microbial communities. The alteration of soil properties and microbial communities by invasive plants can lead to declines in soil health. Changes in soil nutrient dynamics, pH, and organic matter content can affect These declines in health can further impact plant growth, reduce ecosystem productivity, and impair overall ecosystem function (Chen P., et al, 2021).

Examples of Invasive Plants and Their Feedback Loops: Kudzu (Pueraria Montana var. Lobata): Kudzu is a highly invasive vine that can alter soil properties and feedback loops. Its rapid growth and extensive root system can change soil nutrient dynamics and microbial communities. Kudzu's ability to fix nitrogen can alter soil nitrogen levels, creating conditions that favor its continued dominance while disadvantaging native species. Japanese Knotweed (Fallopian japonica): Japanese knotweed is an invasive species that can impact soil feedback loops through its extensive rhizome system and high biomass production. Its presence can alter soil structure, pH, and nutrient availability, leading to changes in microbial communities and reinforcing its dominance in invaded areas (Lowry C J., et al, 2024). Techniques such as soil amendments, planting native species, and fostering beneficial microbial interactions can help restore soil properties and microbial communities. Monitoring the effectiveness of restoration activities is important for evaluating success and guiding future efforts. Future Research Directions: Understanding Mechanisms: Future research should focus on specific plants to create feedback loops. Investigating the interactions between invasive species, can these feedback loops operate, and how to manage them effectively. Long-Term Studies: assess cumulative and prolonged invasive plants on soil feedback loops. Research that tracks properties, and plant health dynamics these feedback loops the effectiveness of management strategies (Wei W., et al, 2020).

Nitrogen Cycling: Impacts of Invasive Plants on Availability and Loss

Alteration of Nitrogen Inputs: amount and form nitrogen entering soil through their litter and root exudates. Some invasive species, like legumes, can fix atmospheric nitrogen in the soil, increasing nitrogen inputs and potentially leading to nutrient imbalances (Yu H., et al, 2021). Conversely, other invasive species may contribute less nitrogen through litter decomposition or disrupt nitrogen fixation processes, affecting overall nitrogen dynamics. Changes in Nitrogen Mineralization and Immobilization: The decomposition of invasive plant litter can influence nitrogen into usable by plants. Invasive plants that produce litter with low nitrogen content or high carbon-to-nitrogen (C: N) ratios can slow down nitrogen mineralization, leading to reduced nitrogen availability for other plants. On the other hand, invasive species that contribute high nitrogen inputs can lead to increased nitrogen immobilization by microbes, reducing the nitrogen available for plant uptake. Disruption of Nitrogen Cycling Microbes: Invasive plants can affect communities' nitrogen cycling, such as nitrifying and denitrifying bacteria. Can impact processes like altering nitrogen availability loss. For example, invasive species may suppress beneficial microbial populations or alter soil conditions in ways that reduce microbial activity and affect nitrogen transformations. Changes in Soil Properties: such as pH, moisture, and texture, influence nitrogen cycling processes. For instance, invasive species that acidify the soil can affect the activity of nitrifying bacteria and alter nitrogen availability. Changes in soil moisture or structure due to invasive plants can also impact nitrogen transformations and movement within the soil profile (Bowen J L., et al, 2024). These losses can contribute to environmental issues such as water pollution, eutrophication of aquatic systems, and greenhouse gas emissions.

Garlic Mustard (Alliaria petiolate): Garlic mustard is known for its ability to alter nitrogen cycling through its allelopathic effects and changes in litter quality. The slow decomposition of garlic mustard litter can affect nitrogen mineralization and reduce nitrogen availability for other plants. Additionally, garlic mustard can disrupt mycorrhizal fungi, which nitrogen plants (Lu Y., et al, 2024). Japanese Knotweed (Fallopian japonica): Japanese knotweed can impact nitrogen cycling through its extensive rhizome system and high biomass production. Its presence of moisture affects nitrogen transformations. Changes in nitrogen availability and soil conditions can favor the continued growth of Japanese knotweed and impact native plant species. Management and Mitigation Strategies: Monitoring and Assessment: Regular monitoring of nitrogen levels, soil properties, and plant health is essential for understanding nitrogen cycling. Soil tests, plant tissue analysis, and assessments of nitrogen transformations can provide valuable information for guiding management decisions and evaluating the effectiveness of control measures. Long-Term Studies: assess cumulative and prolonged invasive plants on nitrogen cycling (Liu Y., et al, 2023). Research that tracks changes in nitrogen availability, soil properties, and plant communities over time dynamics nitrogen cycling and effectiveness of management strategies. Climate Change Considerations: consider interactions with invasive plants, and nitrogen cycling as conditions, the behavior impacts of invasive species may shift, affecting nitrogen dynamics and ecosystem health.

Phosphorus Dynamics: Invasive Plant Effects on Uptake and Retention

Ecosystem functioning, such as cellular metabolism. The dynamics of phosphorus in soils involve complex interactions between soil chemistry, plant uptake, and microbial activity. Invasive plant species can significantly alter phosphorus dynamics, impacting both phosphorus uptake by plants and its retention in the soil. Understanding these impacts is crucial for managing invasive species and maintaining ecosystem health and productivity. Mechanisms of Phosphorus Dynamics Disruption by Invasive Plants: Altered Phosphorus Inputs: Invasive plants can influence the amount and form of phosphorus entering the soil through their litter and root exudates. The quality and decomposition rate of invasive plant litter can affect phosphorus availability. For instance, invasive species with high phosphorus concentrations in their tissues or those that decompose slowly can alter phosphorus dynamics by either increasing or decreasing the phosphorus available in the soil

(Wang C., et al, 2020). Changes in Phosphorus Uptake Efficiency: Invasive plants may exhibit different phosphorus uptake efficiencies compared to native species. Some invasive plants have adapted mechanisms to acquire phosphorus more effectively from low-phosphorus soils, such as through enhanced root structures or mycorrhizal associations. This can lead to increased phosphorus uptake and depletion of phosphorus in the soil, potentially disadvantaging native plants that are less efficient in phosphorus acquisition. Alteration of Soil Phosphorus Chemistry: Invasive plants can affect soil phosphorus chemistry by altering soil pH and moisture levels (Nikolakopoulou M., et al, 2020). Soil Phosphorus Retention and Loss: Invasive plants can impact soil phosphorus retention by influencing soil phosphorus binding and leaching. For example, invasive species can affect the retention of phosphorus matrix, potentially leading to increased phosphorus loss through leaching into groundwater or runoff into aquatic systems.

Garlic mustard can impact phosphorus dynamics through its litter decomposition. The slow decomposition of garlic mustard litter can affect phosphorus mineralization and availability. Additionally, garlic mustard's allelopathic effects on native plants and mycorrhizal fungi can influence phosphorus uptake and soil phosphorus dynamics (Watson G., 2024). Japanese Knotweed (Fallopian japonica): Japanese knotweed can alter phosphorus dynamics through its high biomass production and impact on soil properties. Its presence can change soil pH and moisture levels, affecting phosphorus solubility and retention. Japanese knotweed's extensive rhizome system can also influence phosphorus uptake and soil nutrient dynamics. Future Research Directions: Understanding Mechanisms: Future research should through which invasive plants alter phosphorus dynamics. Investigating the interactions between invasive species, can these processes operate, and how to manage them effectively. Long-Term Studies: assess cumulative and prolonged invasive plants on phosphorus dynamics (Hui Y., et al, 2021). Research that tracks changes in phosphorus availability, soil properties, and plant communities over time the dynamics of phosphorus cycling, and effectiveness of management strategies. Climate Change Considerations: consider interactions between invasive plants, and phosphorus dynamics As conditions, the behavior and impacts of invasive species may shift, affecting phosphorus dynamics and ecosystem health.

Carbon Sequestration: Invasive Plants' Impact on Soil Carbon Pools

In soils and vegetation, a critical component of Soils serves as reservoirs, and their ability to sequester carbon composition and management practices. Invasive plant species can profoundly impact soil carbon pools by altering carbon input, decomposition rates, and soil properties. Understanding these impacts is crucial for managing invasive species and enhancing soil carbon sequestration. Mechanisms of Carbon Sequestration Disruption by Invasive Plants: Altered Carbon Input: Invasive plants can influence carbon entering through their root biomass. Species with high biomass production or rapid growth rates can increase carbon inputs, potentially leading to greater soil carbon storage. However, if invasive plants produce litter that decomposes quickly or has low carbon content, it can be retained. The quality of plant litter, including its, affects decomposition rates and carbon sequestration. Changes in Decomposition Rates: The decomposition of plant material is a key process in carbon cycling, influencing how much carbon is retained in the soil. Invasive plants can alter decomposition rates by changing soil properties, litter chemistry, and microbial communities (Shimoda & Wagai, 2020). For instance, invasive species with high lignin content or low nutrient availability in their litter can slow down decomposition rates, leading to increased carbon accumulation in the soil. Conversely, invasive plants that produce easily decomposable litter can lead to faster carbon turnover and potentially less soil carbon sequestration(Pradeep, E., et al. 2025). However, species that accelerate decomposition or alter soil conditions may reduce soil carbon storage. These changes can impact capacity and contribute to atmospheric CO_2 concentrations (Nagy R C., et al, 2021). Impacts on Ecosystem Carbon Balance: carbon sequestration of Increased soil carbon storage due to invasive species may offset carbon emissions, while reduced soil carbon storage can contribute to higher atmospheric CO_2 levels.

The slow decomposition of garlic mustard litter and its allelopathic effects on native plants and mycorrhizal fungi can influence carbon cycling processes and soil carbon sequestration. Japanese Knotweed (Fallopian japonica): Japanese knotweed can alter soil carbon dynamics through its high biomass production and impact on soil properties. Changes in soil pH, moisture, and texture due to Japanese knotweed may affect carbon decomposition rates and soil carbon storage. Management and Mitigation Strategies: Monitoring and Assessment: Regular monitoring of soil carbon levels, plant community composition, and invasive plant distribution is essential for understanding the impacts of carbon sequestration. Soil carbon assessments, plant biomass measurements, and evaluations of decomposition rates can provide valuable information for guiding management decisions and evaluating the effectiveness of control measures (Liu J., et al, 2023). Integrated Management Approaches: Managing the impacts of carbon sequestration requires integration that combines various removal, and treatments, which should be used to manage invasive species while considering their effects on carbon dynamics. Additionally, restoring native vegetation and improving soil health can help mitigate the impacts of invasive plants and support carbon sequestration. Restoration and Rehabilitation: Restoration efforts should focus on reestablishing native plant communities and enhancing soil carbon dynamics to counteract the effects of invasive species. Techniques such as soil amendments, planting carbon-sequestering species, and fostering beneficial microbial interactions can help restore soil carbon pools and support ecosystem recovery. Monitoring the success of restoration activities is important for evaluating progress and guiding future efforts. Future Research Directions: Understanding Mechanisms: Future research should focus on the specific through which plants alter carbon dynamics. Investigating the interactions between invasive species, can these processes operate, and how to manage them effectively. Long-Term Studies: assess cumulative and prolonged invasive plants on soil carbon pools (Lin M., et al, 2023). Research that tracks changes in carbon inputs, decomposition rates, and soil properties over time can provide valuable information on dynamics and effectiveness management strategies.

Soil Water Relations: Invasive Plant Effects on Hydrology

Soil water relations encompass the interaction between soil properties, water availability, and plant uptake, which together influence hydrological processes such as infiltration, runoff, and water retention. Invasive plant species can significantly impact soil-water relations by altering soil structure, water availability, and the overall hydrological balance of ecosystems. (Berio Fortini L., et al, 2021). Some invasive species have high water demands and can deplete soil moisture more rapidly than native plants. Conversely, other invasive species may have adaptations that allow them to survive in dry conditions, potentially reducing soil moisture and affecting the overall hydrological balance of the area. Impact on Water Infiltration and Runoff: water relations can influence water infiltration and runoff patterns. Conversely, invasive species that alter soil moisture retention can lead to waterlogging or drought conditions, impacting vegetation and soil health (Sueltenfuss J P., et al, 2020). Impacts on Hydrological Processes: water relations can influence broader hydrological processes such as infiltration, runoff, and groundwater recharge. Changes in water infiltration rates and surface runoff patterns can affect streamflow, erosion, and sediment transport. For example, reduced infiltration due to hydrophobic soils can increase surface runoff and contribute to flooding, while increased infiltration can impact groundwater recharge and streamflow patterns.

Figure 1. Invasive plants accelerate nitrogen cycling

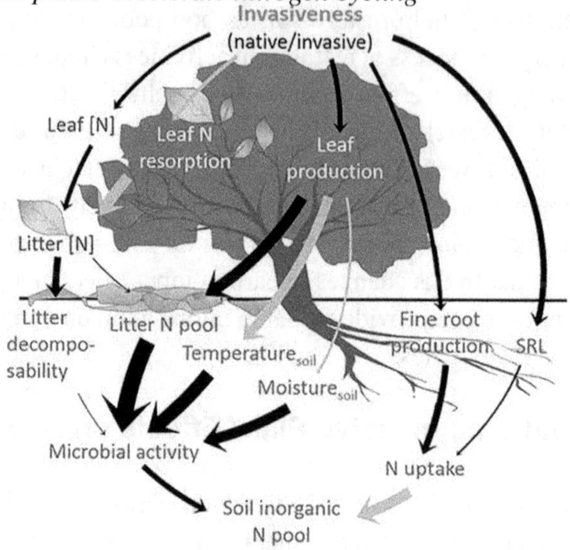

Japanese knotweed can also affect soil moisture availability by altering soil properties (Ning Z H., et al, 2023). Management and Mitigation Strategies: Monitoring and Assessment: Regular monitoring of soil moisture, vegetation health, and invasive plant distribution is essential for understanding the impacts on soil water relations. Soil moisture measurements, infiltration tests, and assessments of water runoff and erosion can provide valuable information for guiding management decisions and evaluating the effectiveness of control measures. Integrated Management Approaches: Managing soil water relations requires an integrated approach that combines various removal, treatments, and should used to manage invasive species while considering their effects on soil hydrology. Additionally, restoring native vegetation and improving soil health can help mitigate the impacts of invasive plants and support water balance. Restoration and Rehabilitation: Restoration efforts should focus on reestablishing native plant communities and enhancing soil-water relations to counteract the effects of invasive species. Techniques such as soil amendments, planting species with favorable water relations, and improving soil structure can help restore hydrological balance and support ecosystem recovery. Monitoring the success of restoration activities is important for evaluating progress and guiding future efforts. Future Research Directions: Understanding Mechanisms: Future research should through which invasive plants alter soil-water relations (Xie T., et al, 2021). Investigating the interactions between invasive species, soil properties, and hydrological processes can provide insights into how these processes operate and how to manage them effectively. Long-Term Studies: assess cumulative and prolonged invasive plants on soil-water relations. Research that tracks changes in soil moisture, infiltration rates, and hydrological balance over time dynamics soil water relations the effectiveness of management strategies. Climate Change Considerations: consider interactions with invasive plants, and soil water relations, As conditions, the behavior impacts of invasive species may shift, affecting soil moisture, infiltration rates, and overall hydrological balance.

Invasive Plant Impacts on Soil Structure and Erosion

The nature of an invasive plant's root system affects how it interacts with the soil matrix, influencing soil structure and stability. Organic Matter Input: input from invasive plants can influence soil structure. Invasive species that produce large amounts of which, in turn, enhance aggregation and improve soil structure (Scotton & Andreatta, 2021). However, if the litter is high in lignin or low in nutrients, it may decompose slowly, potentially impacting soil properties differently than native plant litter. Increased Soil Erosion: Disruptions in soil structure caused by invasive plants can lead to increased soil erosion. Invasive species that impair soil aggregation or cause soil compaction can reduce the soil's ability to resist erosion. Compacted soils are less permeable, leading to higher surface runoff and increased susceptibility to erosion. Additionally, invasive species that create dense, low-growing canopies may reduce soil protection, exposing it to erosion (Jamin J., et al, 2022). Loss of Soil Fertility: and organic matter Invasive plants that contribute to soil erosion can lead to decreased soil fertility, impacting land productivity and ecosystem health. The loss of fertile topsoil can reduce agricultural yields and affect the growth of both invasive and native vegetation. Decreased Water Infiltration: Altered soil structure and increased erosion can impact water infiltration rates. Compacted or degraded soils have reduced water infiltration capacity, leading to increased surface runoff and reduced exacerbated flooding, reduced plants, affect overall hydrological balance. Altered Vegetation Dynamics: Changes in soil structure and increased erosion can affect plant community dynamics by altering the physical environment. Invasive species that cause soil degradation may create conditions that favor their growth while disadvantaging native plants adapted to more stable soil conditions. This plant reduced biodiversity.

Garlic mustard's impact on soil structure and erosion may vary depending on local environmental conditions. Japanese Knotweed (Fallopian japonica): Japanese knotweed has an extensive rhizome system that can lead to soil compaction and altered soil structure (Mairaing W., et al, 2024). Its ability to rapidly colonize areas can also contribute to soil erosion by destabilizing soil and reducing vegetation cover. Japanese knotweed's impact on soil structure and erosion is particularly evident in riparian and disturbed areas. Management and Mitigation Strategies: Monitoring and Assessment: Regular monitoring of soil structure, erosion rates, and invasive plant distribution is essential for understanding the impacts on soil stability. Soil tests, erosion assessments, and vegetation surveys can provide valuable information for guiding management decisions and evaluating the effectiveness of control measures. Integrated Management Approaches: soil structure erosion requires an integrated approach that combines various removal, and treatments. It should be used to manage invasive species while considering their effects on soil stability. Additionally, it mitigates the impacts of invasive plants. Restoration and Rehabilitation: Restoration efforts should focus on reestablishing native vegetation and improving soil structure to counteract the effects of invasive species. Techniques such as planting erosion-resistant species, using ground cover plants, and applying soil amendments can help restore soil stability and reduce erosion. Monitoring the success of restoration activities is important for evaluating progress and guiding future efforts. Future Research Directions: Understanding Mechanisms: Future research should through which invasive plants alter soil structure and erosion. Investigating the interactions between invasive species, soil properties, and erosion processes can provide insights into how these factors operate and how to manage them effectively (Lázaro-Lobo A., et al, 2023). Long-Term Studies: assess cumulative and prolonged invasive plants on soil structure and erosion. Research that tracks changes in soil properties, erosion rates, and vegetation dynamics over time dynamics soil stability and effectiveness of management strategies.

Comparative Analysis: Native vs. Invasive Plant Effects on Soil Health

Their root structures often channels improve aeration, water infiltration, and aggregation. Native plants through root turnover can structure and reduce erosion. Invasive Plants: Invasive species often possess root systems that differ from native plants, which can disrupt soil health (Xu H., et al, 2022). For instance, some invasives have aggressive, deep, or dense root systems that reduce porosity and alter patterns. These changes can affect soil structure and function, potentially leading to increased erosion and reduced soil fertility. Invasive plants may also alter soil chemistry through their unique root exudates, impacting microbial communities and nutrient availability. Organic Matter Input and Decomposition: Native Plants: Native plants generally contribute to organic matter decay and plant residues. Decomposition of native plant materials typically occurs at a balanced rate, influenced by local climate and soil conditions. This process enhances soil organic matter content, supports soil aggregation, and contributes to nutrient cycling. Native plant litter often decomposes at a rate that is well-matched to the soil's microbial community, promoting healthy soil ecosystems. Invasive Plants: The organic matter input from invasive plants can differ significantly from that of native species. Invasive plants may produce litter with different chemical compositions, such as higher lignin content or lower nutrient availability, which can alter decomposition rates and soil organic matter dynamics. Some invasive species may lead to the accumulation of undecomposed plant material or change the microbial community structure, affecting nutrient cycling and soil health. The altered decomposition processes can disrupt soil organic matter balance and impact overall soil fertility. Soil Chemical Properties: Native Plants: Native plants are often well-adapted to local soil conditions and contribute to maintaining soil chemical stability (Xu Z., et al, 2021). They can enhance soil nutrient availability through symbiotic relationships with mycorrhizal fungi, which aid in nutrient uptake. Native species typically contribute to balanced soil pH and nutrient cycling, supporting a diverse and functional soil microbial community. Invasive Plants: Invasive species can significantly alter soil chemical properties, sometimes leading to adverse effects on soil health. For example, some invasives can acidify the soil or alter nutrient concentrations, affecting soil pH and nutrient availability.

Case Studies and Examples: Native Plants: Research has shown that native plants, such as prairie grasses and forest species, contribute to soil health by enhancing organic matter content, supporting diverse microbial communities, and maintaining stable soil structure. For example, native prairie grasses with aeration while reducing erosion. Forest balanced inputs and nutrient cycling (Teixeira L H., et al, 2020). Invasive Plants: Studies have documented the impacts of specific invasive species on soil health. For instance, kudzu (Pueraria Montana var. Lobata) can alter soil structure and nutrient dynamics through its aggressive growth and root systems, while garlic mustard (Alliaria petiolate) can disrupt soil microbial communities and nutrient cycling. Japanese knotweed (Fallopian japonica) can lead to soil compaction and erosion due to its extensive rhizome system. Management and Mitigation Strategies: Monitoring and Assessment: organic matter content, and diversity, are essential for assessing the impacts of both native and invasive plants. Soil tests, vegetation surveys, and erosion assessments can provide valuable information for guiding management decisions. Investigating the interactions between plant species, can these processes operate, and how to manage them effectively (Xia T., et al, 2020). Long-Term Studies: and prolonged native invasive plants on soil health. Research that tracks properties, vegetation dynamics soil health, and effectiveness of management strategies. Climate Change Considerations: Research should consider the interactions between plant species, soil health, and climate change. As climate conditions change, the behavior and impacts of both native and invasive species may shift, affecting soil health and ecosystem functionality. Addressing these interactions is essential for developing adaptive management strategies in a changing environment.

Management Strategies: Restoring Soil Health after Invasive Plant Removal

Restoring soil health after the removal of invasive plants is crucial for reestablishing ecological balance, enhancing soil fertility, and supporting native plant communities. Regular monitoring during and after restoration can ensure that interventions are effective and adjustments can be made as needed (Weidlich et al, 2020). Enhancing Soil Structure and Fertility: Soil Aeration and Compaction: Invasive plant removal can lead to soil compaction or altered soil structure. To restore soil health, it is important to improve soil aeration and reduce compaction. Techniques such as mechanical tillage, soil loosening, and subsoiling can help alleviate compaction and improve soil structure. Additionally, biochar enhances aggregation and improves aeration. The removal of invasive plants can disrupt soil microbial communities, affecting soil health and function (Moore E., et al, 2023). To support the restoration of beneficial microbial populations, microbial inoculants or biofertilizers can be applied. Bacteria or fungi promote plant growth and support soil health. Encouraging Natural Microbial Recovery: In addition to microbial inoculation, promoting natural microbial recovery plant cover can help restore soil microbial communities. Diverse plant communities and organic inputs provide soil, contributing to a balanced and functional soil ecosystem. For example, in a riparian restoration project, the application of organic amendments and planting of native grasses effectively improved soil structure and reduced erosion (Phang L Y., et al, 2024). Similarly, a prairie restoration project demonstrated the benefits of diverse native plantings and soil nutrient management in enhancing soil health and supporting native ecosystems. Exploring innovative techniques for soil restoration, such as advanced soil amendments or novel microbial inoculants, can enhance the effectiveness of restoration efforts. Research into the interactions between different restoration practices and their impacts on soil health can lead to the development of more efficient and sustainable approaches (Leger A M., et al, 2022).

Case Studies: Invasive Plant Impacts on Soil Health in Different Ecosystems

Invasive plant species pose significant threats to soil health across diverse ecosystems, manifesting through alterations in soil structure, nutrient dynamics, and microbial communities. Case studies reveal how these plants disrupt the balance of native systems and their associated soil functions. In temperate forests, for instance, the invasion of Japanese knotweed (Fallopian japonica) has profound impacts on soil health. This plant forms dense stands that shade out native vegetation and disrupt the natural regeneration processes (Milanović M., et al, 2020). Tamarisk, an aggressive invader in Western U.S. riparian areas, significantly alters soil salinity levels due to its high salt excretion. This change in soil salinity affects the native plant species' ability to thrive and alters such structure (Kihara J., et al, 2020). In tropical rainforests, the spread of the African tulip tree (Spathose campanulate) presents a different but equally concerning scenario. This species grows rapidly and forms a dense canopy, which significantly reduces light penetration to the forest floor. Consequently, the undergrowth diminishes, leading to reduced soil organic matter inputs and changes in soil microclimate. Studies have found that the invasion of the African tulip tree results in a decrease in soil nitrogen levels, impacting the entire nutrient cycling process. The alteration of nutrient dynamics affects microbial communities, which in turn influences soil fertility and plant diversity. In alpine ecosystems, the invasion of non-native grasses like Agrostis capillaries has been documented to impact soil properties significantly (Rai R K., et al, 2022). These seaweeds alter nutrient concentrations in the sediment, impacting the health of coral reefs and associated marine life. The accumulation of seaweed can lead to changes in sediment composition, reducing the habitat quality for native marine species and affecting the overall ecological balance (Müller & Bahn, 2022). These case studies illustrate that invasive plants affect soil health in multifaceted ways across various ecosystems.

Modeling Invasive Plant Effects on Soil Health: Tools and Applications

These models allow researchers to explore different scenarios and management strategies, helping to identify potential outcomes and guide decision-making. Geographic Information Systems (GIS) provide spatial analysis capabilities that are modeling spread and their impacts on health (Bzdęga K., et al, 2021). GIS tools can map invasive species distributions, correlate these with soil properties, and visualize how different scenarios might unfold. By integrating spatial data with soil health metrics, GIS models can identify areas at high risk of invasion and prioritize monitoring and management efforts. Remote Sensing offers valuable data for modeling invasive plant impacts on soil health. These tools can capture large-scale vegetation cover changes, and monitor and assess soil conditions indirectly. For example, satellite data can reveal shifts in vegetation patterns and canopy cover, which can be linked to changes in soil moisture and nutrient levels. Combining remote sensing data with ground-truthing efforts enhances the accuracy of models and helps in assessing the effectiveness of management interventions (Pouyat R V., et al, 2020). Ecological Modeling Frameworks Ecological modeling frameworks, such as the and useful for understanding the interactions between invasive plants, soil health, and other ecological factors. IBMs simulate the behavior and interactions of individual plants, while aims focus on the actions of autonomous agents (e.g., invasive plants) and their effects on the environment. These models can capture complex dynamics and feedback loops, providing insights into how invasive species influence soil processes and interactions with native flora. Statistical and Machine Learning Models Statistical models and machine learning techniques are increasingly used to analyze large datasets related to invasive plants and soil health. Regression analysis, classification models, and neural networks can identify patterns and predict soil changes based on invasive properties, these models can help refine predictions, identify key drivers of soil health changes, and optimize management strategies (Tataridas A., et al, 2022). Applications of Modeling The applications of these modeling tools are diverse and impactful. For instance, models can inform land management practices by predicting how invasive species control measures will affect soil health (Divya K L., et al, 2021). They can also guide restoration efforts by identifying the most effective strategies for soil rehabilitation in invaded areas.

Future Directions: Research and Management Priorities for Mitigating Invasive Plant Impacts

Future research should focus on integrating ecological, climatic, and socio-economic data to improve the accuracy of models (Sueltenfuss J P., 2020). Incorporating genetic and behavioral traits of invasive species into models can enhance predictions about their potential impact. Additionally, machine learning and artificial intelligence employed datasets through traditional methods. Research should focus on identifying native species and ecological processes that can restore invaded areas to their natural state (Ning Z H., et al, 2023). This includes experimenting with various replanting techniques, soil amendments, and habitat enhancements to support recovery. Evaluating the effectiveness of different restoration approaches and understanding their impacts on soil health and ecosystem resilience will be essential for successful rehabilitation efforts. Socio-Economic Implications and Management: Understanding the socio-economic impacts of species comprehensive research should explore plant management and infrastructure. Additionally, studies should assess the effectiveness of different policy measures and management interventions. Engaging stakeholders, including landowners, policymakers, and local communities, in collaborative management efforts, will be vital for ensuring the success of invasive species control programs(Athavale, V. A., et al. 2025). Genetic and Biological Control Methods: Exploring genetic and biological control methods offers promising avenues for managing invasive plants (Funk J L., et al, 2020). Research should focus on developing and testing biocontrol agents, such as pathogens or herbivores, that target invasive species without harming native plants. Advances in genetic engineering could also provide tools for modifying invasive plants or their interactions with ecosystems. However, these methods must be carefully assessed for potential risks and ecological consequences. Climate Change Considerations: Integrating climate change projections into invasive plant management strategies is increasingly important. Future research should examine how changing climate conditions may. This includes studying how shifts affect invasive plant growth and interactions with native species. Developing adaptive management strategies that account for these changes will be essential for effective long-term control (Meyerson L A., et al, 2022).

CONCLUSION

Health and cycling are profoundly multifaceted, underscoring the urgent need for comprehensive understanding and management strategies. Invasive species often disrupt soil structure, alter nutrient dynamics, and impact microbial communities, leading to significant ecological consequences. These disruptions can manifest in various ways, such as changes in soil pH, reductions in soil fertility, and shifts in microbial community composition. Impacts invasive plants on their influence physical properties. Many invasive species, such as Japanese knotweed and tamarisk, can lead to erosion. This alteration in soil structure affects water infiltration and retention, which in turn impacts plant growth and soil stability. Invasive plants can also change soil texture and density, further compounding the challenges for native vegetation and ecosystem functions. Nutrient cycling is another critical area affected by invasive species. Invasive plants can alter soil nutrient levels through various mechanisms, including changes in nutrient uptake, release of allelopathic compounds, and modifications to the soil organic matter. For example, invasive species like the African tulip tree can decrease soil nitrogen levels, impacting the availability of essential nutrients for other plants. Similarly, the high salt excretion by tamarisk alters soil salinity, affecting nutrient availability and plant health in arid regions. These changes disrupt the natural nutrient cycling processes, leading to reduced soil fertility and compromised ecosystem productivity. Microbial communities in the soil are also significantly impacted by introduction can alter microbiota, crucial roles in Invasive plants changing soil chemistry, providing novel substrates, or outcompeting native plant species that support beneficial microbes. The resulting further soil nutrient dynamics, create a feedback loop that exacerbates the impact of invasive species. The cascading effects of these disruptions can lead to broader ecological consequences. Changes can affect plant community composition, reduce biodiversity, and alter the ecosystem. Invasive plants often outcompete native species, leading to reduced plant diversity and changes in habitat structure. This, in turn, impacts wildlife and other organism's dependent on native plant communities, contributing to declines in biodiversity and ecosystem resilience. Invasive plants are both preventative measures and active management. Effective strategies involve to well as targeted control measures to manage existing infestations. Restoration efforts should focus on rehabilitating affected soils. Additionally, research is needed to further understand the specific mechanisms through which invasive species impact soil health and nutrient cycling, as well as to develop innovative solutions for mitigating these effects. Public education and engagement are also crucial components of a comprehensive management strategy. Raising awareness about the impacts of invasive plants and promoting citizen science initiatives can help in early detection and management efforts. Collaborative approaches involving researchers address challenges posed by invasive species. In conclusion, invasive plants pose significant threats with widespread implications for ecosystem function and biodiversity. Their impacts on soil physical properties and nutrients highlight the need for a holistic approach to management and research. By integrating advanced monitoring techniques, predictive models, restoration efforts, and public engagement, we can better understand and mitigate the effects of invasive species, ultimately protecting and preserving soil health and ecosystem integrity. Addressing these challenges will be critical for maintaining ecological balance and ensuring the sustainability of our natural environments in the face of ongoing and future invasions.

REFERENCES

Athavale, V. A., Selvakumar, P., Poorani, S., Anand, T., & Rathore, D. S. (2025). Designing Chatbots for Educational Leadership. In Musundire, A. (Ed.), *Chatbots in Educational Leadership and Management* (pp. 181–204). IGI Global Scientific Publishing., DOI: 10.4018/979-8-3693-8734-4.ch008

Bell, J. K., Siciliano, S. D., & Lamb, E. G. (2020). A survey of invasive plants on grassland soil microbial communities and ecosystem services. *Scientific Data*, *7*(1), 1–11. DOI: 10.1038/s41597-020-0422-x PMID: 32152302

Berio Fortini, L., Leopold, C. R., Perkins, K. S., Chadwick, O. A., Yelenik, S. G., Jacobi, J. D., Bishaw, K.II, & Gregg, M. (2021). Landscape-level effects of invasive plants and animals on water infiltration through Hawaiian tropical forests. *Biological Invasions*, *23*(5), 1421–1437. DOI: 10.1007/s10530-021-02494-8

Bhaduri, D., Sihi, D., Bhowmik, A., & Verma, B. C. (2022). A review on effective soil health bio-indicators for ecosystem restoration and sustainability. *Frontiers in Environmental Science*, •••, 10. PMID: 36060788

Bowen, J. L., Spivak, A. C., Bernhard, A. E., & Fulweiler, R. W. (2024). Salt marsh nitrogen cycling: Where land meets sea. *Trends in Ecology & Evolution*, *39*(1), 53–65. PMID: 37827901

Bzdęga, K., Zarychta, A., & Urbisz, A. (2021). Geostatistical models with the use of hyperspectral data and seasonal variation–A new approach for evaluating the risk posed by invasive plants. *Ecological Informatics*, *61*, 101295.

Chen, P., Huang, Q., Zhuge, Y., Li, C., Zhu, P., & Hou, Y. (2021). The effects of plant–soil feedback on invasion resistance are soil context dependent. *Oecologia*, *197*(2), 257–269. DOI: 10.1007/s00442-021-05004-8 PMID: 34328557

Dai, Z., Xiong, X., Zhu, H., Xu, H., Leng, P., Li, J., Tang, C., & Xu, J. (2021). Association of biochar properties with changes in soil bacterial, fungal and fauna communities and nutrient cycling processes. *Biochar*, *3*(1), 1–15. DOI: 10.1007/s42773-021-00099-x

Divya, K. L., Mhatre, P. H., Venkatasalam, E. P., & Sudha, R. (2021). Crop simulation models as decision-supporting tools for sustainable potato production: A review. *Potato Research*, *64*(2), 149–164. DOI: 10.1007/s11540-020-09483-9

Fahey, C., & Flory, S. L. (2022). Soil microbes alter competition between native and invasive plants. *Journal of Ecology*, *110*(1), 141–155. DOI: 10.1111/1365-2745.13807

Frouz, J. (2023). Plant-soil feedback across spatiotemporal scales from immediate effects to legacy. *Soil Biology & Biochemistry*, *177*, 108790.

Funk, J. L., Parker, I. M., Matzek, V., Flory, S. L., Aschehoug, E. T., D'Antonio, C. M., Dawson, W., Thomson, D. M., & Valliere, J. (2020). Keys to enhancing the value of invasion ecology research for management. *Biological Invasions*, *22*(5), 1475–1490. DOI: 10.1007/s10530-020-02267-9

Hui, Y., Atkinson, J. F., Zhu, Z., Schlea, D., & Redder, T. (2021). Invasive dreissenid mussel effects on phosphorus dynamics in Lake Ontario: Insights from integrated hydrodynamic–ecological modeling. *Canadian Journal of Fisheries and Aquatic Sciences*, *78*(1), 14–26. DOI: 10.1139/cjfas-2020-0164

Jamin, J., Diehl, D., Meyer, M., David, J., Schaumann, G. E., & Buchmann, C. (2022). Physico-chemical soil properties affected by invasive plants in Southwest Germany (Rhineland-Palatinate)—A case study. *Soil Systems*, *6*(2), 35. DOI: 10.3390/soilsystems6040093

Keet, J. H., Ellis, A. G., Hui, C., Novoa, A., & Le Roux, J. J. (2021). Impacts of invasive Australian acacias on soil bacterial community composition, microbial enzymatic activities, and nutrient availability in fynbos soils. *Microbial Ecology*, *81*(2), 257–269. DOI: 10.1007/s00248-021-01683-1 PMID: 33515051

Kihara, J., Bolo, P., Kinyua, M., Nyawira, S. S., & Sommer, R. (2020). Soil health and ecosystem services: Lessons from sub-Sahara Africa (SSA). *Geoderma*, *361*, 114094. DOI: 10.1016/j.geoderma.2020.114342

Lázaro-Lobo A, Alonso Á, Fernández R D, (2023). "Impacts of plant invasions on ecosystem functionality: A perspective for ecosystem health and ecosystem services." Plant Invasions and Ecosystem Health, 241-262.

Leger, A. M., Ball, K. R., Rathke, S. J., & Blankinship, J. C. (2022). Mulch more so than compost improves soil health to reestablish vegetation in a semiarid rangeland. *Restoration Ecology*, *30*(2), e13546. DOI: 10.1111/rec.13698

Lin, M., Chen, Y., Cheng, L., Zheng, Y., Wang, W., Sardans, J., Song, Z., Guggenberger, G., Zou, Y., Ding, X., Tariq, A., Zeng, F., Fahad Alrefaei, A., & Peñuelas, J. (2023). Response of topsoil Fe-bound organic carbon pool and microbial community to Spartina alterniflora invasion in coastal wetlands. *Catena*, *220*, 106695. DOI: 10.1016/j.catena.2023.107414

Liu, J., Duan, X., Li, G., Cai, Z., Wei, S., Song, Q., & Zheng, Z. (2023). Changes in Bacterial Communities and Their Effects on Soil Carbon Storage in Spartina alterniflora Invasion Areas, Coastal Wetland Bare Flats, and Sueada salsa .... *International Journal of Environmental Research and Public Health, 20*(11), 5769. DOI: 10.3390/ijerph20054308 PMID: 36901319

Liu, Y., Xu, F., Ding, L., Zhang, G., Bai, B., Han, Y., Xiao, L., Song, Y., Li, Y., Wan, S., & Li, G. (2023). Microplastics reduce nitrogen uptake in peanut plants by damaging root cells and impairing soil nitrogen cycling. *Journal of Hazardous Materials, 443*, 130173. DOI: 10.1016/j.jhazmat.2022.130384 PMID: 36444071

Lowry, C. J., Blocklove, K. M., Eckert, K. E., & Koirala, N. (2024). If you can't beat them, join them: Using invasive plants to understand rhizodeposits as drivers of plant-soil feedbacks. *Frontiers in Environmental Science*, •••, 12.

Lu, Y., Kronzucker, H. J., Yu, M., Shabala, S., & Shi, W. (2024). Nitrogen-loss and carbon-footprint reduction by plant-rhizosphere exudates. *Trends in Plant Science, 29*(1), 53–65. DOI: 10.1016/j.tplants.2023.09.007 PMID: 37802692

Mairaing, W., Jotisankasa, A., Leksungnoen, N., Hossain, M., Ngernsaengsaruay, C., Rangsiwanichpong, P., Pilumwong, J., Pramusandi, S., Semmad, S., & Ahmmed, A. N. F. (2024). A Biomechanical Study of Potential Plants for Erosion Control and Slope Stabilization of Highland in Thailand. *Sustainability (Basel), 16*(4), 2445. DOI: 10.3390/su16156374

Mekonnen, Z. A., Riley, W. J., Berner, L. T., Bouskill, N. J., Torn, M. S., Iwahana, G., Breen, A. L., Myers-Smith, I. H., Criado, M. G., Liu, Y., Euskirchen, E. S., Goetz, S. J., Mack, M. C., & Grant, R. F. (2021). Arctic tundra shrubification: A review of mechanisms and impacts on ecosystem carbon balance. *Environmental Research Letters, 16*(4), 1–13. DOI: 10.1088/1748-9326/abf28b PMID: 33897808

Meyerson L A, Pauchard A, Brundu G, Carlton J T, (2022). "Moving toward global strategies for managing invasive alien species." Global Plant Invasions, 1-14.

Milanović, M., Knapp, S., Pyšek, P., & Kühn, I. (2020). Linking traits of invasive plants with ecosystem services and disservices. *Ecosystem Services, 45*, 101194. DOI: 10.1016/j.ecoser.2020.101072

Moore, E., D'Amico, V., & Trammell, T. L. E. (2023). Plant community dynamics following non-native shrub removal depend on invasion intensity and forest site characteristics. *Ecosphere, 14*(1), e02323. DOI: 10.1002/ecs2.4351

Müller, L. M., & Bahn, M. (2022). Drought legacies and ecosystem responses to subsequent drought. *Global Change Biology*, *28*(1), 211–223. DOI: 10.1111/gcb.16270 PMID: 35607942

Nagy, R. C., Fusco, E. J., Balch, J. K., Finn, J. T., Mahood, A., Allen, J. M., & Bradley, B. A. (2021). A synthesis of the effects of cheatgrass invasion on US Great Basin carbon storage. *Journal of Applied Ecology*, *58*(10), 2415–2426. DOI: 10.1111/1365-2664.13770

Nikolakopoulou, M., Argerich, A., Bernal, S., Gacia, E., Ribot, M., Martí, E., Sorolla, A., & Sabater, F. (2020). Effect of three emergent macrophyte species on nutrient retention in aquatic environments under excess nutrient loading. *Environmental Science & Technology*, *54*(19), 11753–11762. DOI: 10.1021/acs.est.0c03216 PMID: 33175518

Ning, Z. H., Chen, C., Zhang, S. Y., & Wang, A. D. (2023). Lateral Hydrological Connectivity Driven by Tidal Flooding Regulates Range-Expansion of Invasive Spartina alterniflora in Tidal Channel-Salt Marsh Systems. *Journal of Coastal Research*, *39*(3), 539–548.

Phang, L. Y., Mingyuan, L., Mohammadi, M., Tee, C. S., Yuswan, M. H., Cheng, W.-H., & Lai, K.-S. (2024). Phytoremediation as a viable ecological and socioeconomic management strategy. *Environmental Science and Pollution Research International*, *31*(1), 253–265. DOI: 10.1007/s11356-024-34585-z PMID: 39103580

Pouyat R V, Page-Dumroese D S, Patel-Weynand T, (2020). "Forest and rangeland soils of the United States under changing conditions: a comprehensive science synthesis."

Pradeep, E., Mukherjee, S. K., Selvakumar, P., Koppaka, A. K., Chandel, P. S., & Vats, C. (2025). Empowering Social Media Users With Ethical AI. In Chakraborty, S. (Ed.), *Ethical AI Solutions for Addressing Social Media Influence and Hate Speech* (pp. 91–118). IGI Global Scientific Publishing., DOI: 10.4018/979-8-3693-9904-0.ch005

Rädecker, N., Pogoreutz, C., Gegner, H. M., Cárdenas, A., Roth, F., Bougoure, J., Guagliardo, P., Wild, C., Pernice, M., Raina, J.-B., Meibom, A., & Voolstra, C. R. (2021). Heat stress destabilizes symbiotic nutrient cycling in corals. *Proceedings of the National Academy of Sciences of the United States of America*, *118*(10), 1–8. DOI: 10.1073/pnas.2022653118 PMID: 33500354

Rai, P. K. (2022). Environmental degradation by invasive alien plants in the anthropocene: Challenges and prospects for sustainable restoration. *Anthropocene Science*, *2*(1), 1–13.

Rai, P. K., & Singh, J. S. (2020). Invasive alien plant species: Their impact on environment, ecosystem services and human health. *Ecological Indicators*, *117*, 106613.

Rai R K, Shrestha L, Joshi S, Clements D R, (2022). "Biotic and economic impacts of plant invasions." Global Plant Invasions, 147-164.

Ranjitha, P. K., Selvakumar, P., Anute, N., Manjunath, T., Sureshkrishna, G., & Pougajendy, S. (2025). Transportation Challenges In Islands Tourism. In Hassan, V. (Ed.), *Navigating Mass Tourism to Island Destinations: Preservation and Cultural Heritage Challenges* (pp. 287–312). IGI Global Scientific Publishing., DOI: 10.4018/979-8-3693-9107-5.ch011

Scotton, M., & Andreatta, D. (2021). Anti-erosion rehabilitation: Effects of revegetation method and site traits on introduced and native plant cover and richness. *The Science of the Total Environment*, *793*, 148714. DOI: 10.1016/j.scitotenv.2021.145915 PMID: 33647653

Shende, A. D., Tekade, S. A., Deshmukh, A. A., Tembhurkar, S. P., & Selvakumar, P. (2025). Water Quality Monitoring and Control With AI. In Ruiz-Vanoye, J., & Díaz-Parra, O. (Eds.), *Smart Water Technology for Sustainable Management in Modern Cities* (pp. 51–80). IGI Global Scientific Publishing., DOI: 10.4018/979-8-3693-8074-1.ch003

Shimoda, S., & Wagai, R. (2020). Ecosystem dynamics after abandonment of rice paddy fields: Does alien plant invasion enhance carbon storage? *Ecosystems (New York, N.Y.)*, *23*(5), 931–944. DOI: 10.1007/s10021-019-00426-1

Subramanian, M., Portia, R., Mishra, B. R., & Selvakumar, P. P. V., R., & T. C., M. (2025). Assessment and Evaluation in Personalized STEAM Education. In N. Son (Ed.), *Integrating Personalized Learning Methods Into STEAM Education* (pp. 103-122). IGI Global Scientific Publishing. https://doi.org/DOI: 10.4018/979-8-3693-7718-5.ch005

Sueltenfuss, J. P., Ocheltree, T. W., & Cooper, D. J. (2020). Evaluating the realized niche and plant–water relations of wetland species using experimental transplants. *Plant Ecology*, *221*(1), 53–65. DOI: 10.1007/s11258-020-01015-2

Tataridas, A., Jabran, K., Kanatas, P., Oliveira, R. S., Freitas, H., & Travlos, I. (2022). Early detection, herbicide resistance screening, and integrated management of invasive plant species: A review. *Pest Management Science*, *78*(1), 13–24. DOI: 10.1002/ps.6963 PMID: 35510308

Teixeira, L. H., Yannelli, F. A., Ganade, G., & Kollmann, J. (2020). Functional diversity and invasive species influence soil fertility in experimental grasslands. *Plants*, *9*(11), 1511. DOI: 10.3390/plants9010053 PMID: 31906387

Torres, N., Herrera, I., Fajardo, L., & Bustamante, R. O. (2021). Meta-analysis of the impact of plant invasions on soil microbial communities. *BMC Ecology and Evolution*, *21*(1), 1–13. DOI: 10.1186/s12862-021-01899-2 PMID: 34496752

Wang, C., Wang, W., Sardans, J., Ouyang, L., Tong, C., Asensio, D., Gargallo-Garriga, A., Wiesmeier, M., & Peñuelas, J. (2020). Higher fluxes of C, N and P in plant/soil cycles associated with plant invasion in a subtropical estuarine wetland in China. *The Science of the Total Environment*, *703*, 135027. DOI: 10.1016/j.scitotenv.2020.139124 PMID: 32388112

Watson G, (2024). "Phosphorus Dynamics in the Sediment of a Lake Erie Coastal Wetland."

Wei, W., Zhu, P., Chen, P., Huang, Q., Bai, X., Ni, G., & Hou, Y. (2020). Mixed evidence for plant–soil feedbacks in forest invasions. *Oecologia*, *192*(3), 651–663. DOI: 10.1007/s00442-020-04703-y PMID: 32642792

Weidlich, E. W. A., Flórido, F. G., Sorrini, T. B., & Brancalion, P. H. S. (2020). Controlling invasive plant species in ecological restoration: A global review. *Journal of Applied Ecology*, *57*(5), 931–942. DOI: 10.1111/1365-2664.13656

Xia, T., Wang, Y., He, Y., Wu, C., Shen, K., Tan, Q., Kang, L., Guo, Y., Wu, B., & Han, X. (2020). Invasive plant experiences greater benefits of root morphology from enhancing nutrient competition associated with arbuscular mycorrhizae in karst soil than a native …. *PLoS One*, *15*(10), e0239778. DOI: 10.1371/journal.pone.0234410 PMID: 32516341

Xie, T., Wang, Q., Ning, Z., Chen, C., Cui, B., Bai, J., Shi, W., & Pang, B. (2021). Artificial modification on lateral hydrological connectivity promotes range expansion of invasive Spartina alterniflora in salt marshes of the Yellow River delta, China. *The Science of the Total Environment*, *751*, 141673. DOI: 10.1016/j.scitotenv.2020.144476 PMID: 33460837

Xu, H., Liu, Q., Wang, S., Yang, G., & Xue, S. (2022). A global meta-analysis of the impacts of exotic plant species invasion on plant diversity and soil properties. *The Science of the Total Environment*, *807*, 150786. DOI: 10.1016/j.scitotenv.2021.152286

Xu, Z., Guo, X., Caplan, J. S., Li, M., & Guo, W. (2021). Novel plant-soil feedbacks drive adaption of invasive plants to soil legacies of native plants under nitrogen deposition. *Plant and Soil*, *467*(1), 347–362. DOI: 10.1007/s11104-021-05057-x

Yu, H., Le Roux, J. J., Jiang, Z., Sun, F., Peng, C., & Li, W. (2021). Soil nitrogen dynamics and competition during plant invasion: Insights from Mikania micrantha invasions in China. *The New Phytologist, 229*(2), 841–854. DOI: 10.1111/nph.17125 PMID: 33259063

Zhang, X., Ward, B. B., & Sigman, D. M. (2020). Global nitrogen cycle: Critical enzymes, organisms, and processes for nitrogen budgets and dynamics. *Chemical Reviews, 120*(2), 987–1016. DOI: 10.1021/acs.chemrev.9b00613 PMID: 31549814

Chapter 8

Role of Aerial and Edaphic Factors in the Invasive Success of Justicia adhatoda L. in Arid Environments

Zunaira Naeem
https://orcid.org/0009-0005-8092-088X
University of Agriculture, Faisalabad, Pakistan

Farooq Ahmad
University of Agriculture, Faisalabad, Pakistan

Muhammad Sajid aqeel Ahmad
University of Agriculture, Faisalabad, Pakistan

Safura Bibi
https://orcid.org/0000-0002-1715-1871
University of Agriculture, Faisalabad, Pakistan

Mansoor Hameed
University of Agriculture, Faisalabad, Pakistan

Sadia Faryad
University of Agriculture, Faisalabad, Pakistan

ABSTRACT

The invasive plant species are non-native and rapidly grow in heterogeneous environments may affect native species. The invasive plant, Justicia adhatoda L,shows complex growth patterns controlled by various above-ground and below-ground factors. The aerial factors include; environmental constraints and biotic relationships that determine the growth patterns, physiological processes and invasion of Justicia adhatoda L. Whereas, the edaphic factors also play a crucial role in root growth and plant health. An interaction of these two factors will help to understand the basic

DOI: 10.4018/979-8-3693-8724-5.ch008

mechanism of survival of Justicia adhatoda L. under different environmental constraints. This chapter will provide a complete review of the most challenging abiotic factors i.e., the drought that influence J. adhatoda growth patterns by combining current research and their practical applications. It emphasizes the importance of interactive strategies to manage invasive species and restoration ecology, while also providing insights into sustainable practices and novel research tools.

1. INTRODUCTION

1.1 *Justicia adhatoda*

Justicia adhatoda is a perennial, small evergreen, shrub that belongs to the family Acanthaceae, it has a wide distribution especially in India, Malaysia and Sri Lanka, but in Pakistan, Northern latitudes of Punjab is enriched with diversity of *Justicia adhatoda*. It was previously known as vasaka, later redefined as *Justicia adhatoda*. It mostly grows to a considerable height of about 1.5 m and possesses long and wide leaves (Figure 1). The leaves of *J. adhatoda* look dark green, but its most distinctive morphological feature is a white flower-forming spike (Isha et al. 2025). *Justicia adhatoda* is an erect shrub, leaves are simple, opposite and decussate. Flowers are bracteolate, white and sessile. The most outstanding features is the widespread occurrence of variously shaped cystoliths, acicular fibres and crystals in stem and leaf. The economic products obtained a yellow dye from the leaves of *Justicia adhatoda* L. (Kumar and Singh 2025).

Figure 1. Pictorial view of the Morphology of Justicia adhatoda L. and its distribution in Punjab, Pakistan

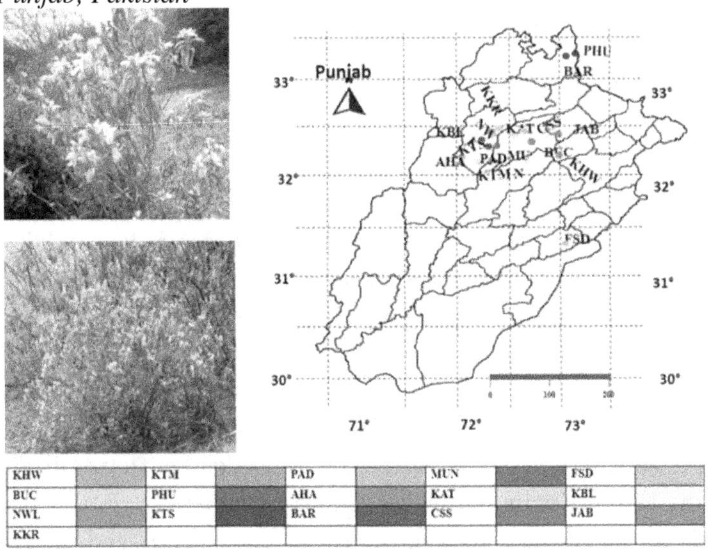

Justici adhatoda is a strong invader, capable of dominating local flora and altering ecosystem dynamics. It exhibits allopathic effects, which can inhibit the growth of surrounding plant species, leading to reduced biodiversity. Despite its invasive nature, *J. adhatoda* contributes positively to ecosystems by stabilizing soil to enhance nitrogen content in compost and aiding in ecological restoration (Isha et al. 2025, Kumar and Singh 2025).

The invasive potential of *J.adhatoda* is significantly supported by its robust root system, adaptive leaf morphology, and notable phenotypic plasticity. These traits enable the plant to survive in diverse environments, contributing to its invasive success.

- It exhibits a strong root system characterized by increased root length and weight, particularly in xeric environments, which enhances its ability to access water and nutrients (Naeem et al. 2024a).
- The leaves of *J. adhatoda* are adapted for various climatic conditions, which aids in its survival and spread. The leaf structure also supports efficient photosynthesis and resource utilization, crucial for growth in resource-limited environments. The leaves also contribute to its allelopathic effects, suppressing the growth of competing species, thereby facilitating its dominance in invaded ecosystems (Isha et al. 2025).

- *J.adhatoda* demonstrates significant phenotypic plasticity, allowing it to adjust its growth and resource acquisition strategies in response to environmental changes(Matloob and Hameed 2024).

It is considered as most important medicinal plant, leaves comprise of various alkaloids, but mainly they contain vasicine. The leaves, roots and stem of *J. adhatoda* also contain various chemical constituents including essential oils, fats, phenols, amino acids, flavonoids, proteins and vitamin 'C' that show its antioxidative properties (Sarada and Rao 2008). This plant species has traditionally been used for treatment of various diseases specifically asthma, bronchitis and leprosy. In addition to its therapeutic significance, it also has insecticidal and parasiticidal properties (Lone et al. 2013). *Justicia adhatoda* L. is used for food, spices, colours, religious and cultural relevance, economic importance, and horticultural or floricultural significance (Singh and Singh 2013). A. vasica contains several bioactive compounds different parts of the plant (root, stem, leaves, flower, bark, etc.) i.e., vasicine, 5-hydroxy vasicine, vasicine glycoside, deoxyvasicine, vasicinol, vasicinone, vasicolinone, adhavasicinone, adhatodine, vasnetine and anisotine (Ali et al. 2020). The leaves are high in vitamin C and carotene and are produced as an essential oil, sugar, resin, gums, proteins and amino acids. The plant has a pungent taste because of the presence of phytochemicals i.e., phenols, tannins, alkaloids, anthraquinone, saponins, flavonoids, (Basit et al. 2022). Additionally, alkaloids, phytosterols, polyphenolics, and glycosides are among the most common types of phytochemicals (Hossain et al. 2023) that are used to treat infantile convulsions by grinding the leaves with heated water.

Due to industrialization, urbanization and numerous cultivation practices, the habitat of Malabar nut has shrunk. There is a dire need to reintroduce and restore the potential habitats of *Justicia adhatoda* (Yang et al. 2013).

Importance of Justicia adhtoda

Justicia adhatoda showed multifaceted properties, ranging from its medicinal values to commercial cultivation. This plant, recognized for its therapeutic benefits, plays a significant role in the pharmaceutical industry, influencing both local economies and healthcare systems. The following highlights show key aspects of its economic importance in various aspects.

a) Medicinal importance: *Justicia adhatoda* is valued for its alkaloids, particularly vasicine which is effective in treating respiratory conditions such as cough and asthma(Akbar and Akbar 2020). The increasing demands for herbal medicines has led to rise in the commercial cultivation of *Justicia adhatoda* contributing agriculture economics (Saravanan and Karthi 2015). The use of *Justicia adhatoda* in traditional medicine can potentially reduce healthcare costs by providing affordable treatment options for respiratory ailments(Akbar and Akbar 2020). Its effectiveness in treating conditions like tuberculosis may alleviate the financial burden on healthcare systems, especially in regions with limited access to modern medicine (Saravanan and Karthi 2015).

b) Environmental and agricultural values: The cultivation of *Justicia adhatoda* under elevated CO_2 conditions has shown enhanced growth and biochemical properties, indicating its adaptability and potential for increased yield medicine (Saravanan and Karthi 2015). This adaptability can lead to sustainable agricultural practices, promoting economic stability in rural areas (Gilani et al. 2011, Yang et al. 2013).

c) *Justicia adhatoda* offers, reliance on a single medicinal plant. It can pose risks, such as market volatility and overexploitation, which may threaten biodiversity and long-term sustainability in herbal medicine sectors(Khan et al. 2018)

1.2 Importance of studying invasive species

Studying invasive species in arid environments is crucial due to their significant ecological impacts, interactions with native species and implications for biodiversity and ecosystem health. Understanding these dynamics can help to formulate management strategies to mitigate negative effects and preserve native communities. Ecologically, invasive species can alter community structure and dynamics, often outcompeting native flora for resources (Khan et al. 2023). For instance, invasive annual grasses like *Bromus rubens* and *B. tectorum* have been shown to increase fire frequency and intensity, further threatening native species post-fire (Underwood et al. 2019). There is a need of creating positive interactions, some native shrubs may facilitate the growth of invasive species, creating a complex interaction that can reduce native plant abundance and diversity (Lucero et al. 2022). The presence of certain invasive species can also enhance soil nutrient availability, potentially benefiting both invaders and some native species (Elsheikh et al. 2021). However, for ecological risk care, such benefits are often balanced by long-term disruptions to community structure and nutrient cycling (Simberloff, 2013; Pysek, 2020). For example, nitrogen-fixing invasive plants can lead to nutrient imbalances, creating competitive dynamics and facilitating further invasions (Xu, H., et al. 2022). Risk assessments emphasise the importance of evaluating ecosystem-wide consequences, rather than focusing solely on short-term productivity gains, to prevent unintended ecological degradation. Moreover, the hyperspectral remote sensing tools are valuable for monitoring invasive species and assessing their distribution, which aids in understanding their ecological roles in arid ecosystems (Nagler et al. 2018). While the focus on invasive species often highlights their detrimental effects, it is also essential to consider their potential roles in ecosystem processes, such as nutrient cycling and habitat modification, which can sometimes lead to unexpected ecological benefits.

1.3 Objectives

- The main objectives of studying the role of aerial and edaphic factors in the invasive success of *Justicia adhatoda* L. in arid environments encompass understanding its adaptability, ecological distribution and potential for management strategies.
- This study will help to elucidate how these factors contribute to the plant's invasive characteristics and resilience in challenging habitats.
- It will explore the invasive potential of *Justicia adhatoda* and consider broader ecological implications of invasive species, including their impact on native biodiversity and ecosystem health. .

2. AERIAL FACTORS AFFECTING INVASIVENESS

2.1 Effect of Temperature:

The most influencing environmental factor is the temperature, affecting germination, seedling establishment, species interactions, growth rates, competitive dynamics and overall fitness. An increase in temperature provides a competitive advantage to invasive species over native species, altering community dynamics and ecological balance. It also enhances plants ability to establish and spread in new environments. Temperature impacts invasiveness in various ways:

2.1.1 Plant growth and seed germination

Temperature significantly affect plant growth and productivity from seedling stage to harvest (Hellmers and Warrington 2018). High temperature induces heat stress that negatively affect plant development specifically reproductive stages, causing reduction in overall yield (Bhattacharya 2019). Invasive species respond differently to temperature changes. Optimal temperature ranges promote high germination rates. Some species are tolerant showing high germination percentages in a broad spectrum range of temperatures, while others show avoidance against temperature extremes by showing dormancy under unfavorable conditions (Bhatt et al. 2023). Extreme temperatures can limit germination; however, nonnative invasive species are frequently more resistant to high temperatures than native species, allowing them to better grow under warm climates (Hou et al. 2014).

2.1.2 Climate Change and Competitive Dynamics: Climate change increases the risk of global warming that exacerbates plant invasiveness risk. High temperature promotes invasiveness and low temperature favors native species, establishing a complex relationship between temperature extremes and invasiveness (Hou et al. 2014). Climate change is expected to increase global temperatures further, potentially facilitating the spread of invasive species while challenging native populations. This shift may lead to alter geographic ranges and community compositions species adapt to new thermal environments (Finch et al. 2021). It further influence the distribution of invasive species within communities, indicating, invasive species are adapted to heterogeneous climatic variations that lead to their increase distributions and invasiveness (Chen et al. 2023, Silveira et al. 2023).

2.1.3 Species Interaction

Change in temperature affects the species interaction i.e., Intra- and interspecific competition among invasive species. Rising temperatures can alter interactions among invasive species, potentially increasing competition or facilitation depending on plant density and neighbor identity. High temperatures favored the invasive species *Chilo partellus* over native *Noctuids* in maize fields, enhancing its growth rates (Ntiri et al. 2016). However, lower temperatures benefited native species. These temperature fluctuations can shift competitive hierarchies within ecosystems (Ntiri et al. 2016). Moreover, temperature affects herbivorous arthropod development and fertility, potentially altering plant resistance mechanisms (Nechols et al. 2020). Warmer temperatures may increase insect populations, increasing more pressure on plants and influencing their invasiveness through changing defense mechanisms (Nechols et al. 2020).

2.1.4 Effect of Temperature on *Justicia adhatoda* L. invasiveness

Increased temperatures can facilitate root growth, which is crucial for the establishment and spread of invasive plants like *Justicia adhatoda.* It was studied that optimal growth occurs at mid-range temperatures, but higher temperatures can enhance root depth and width, providing competitive advantages over native species(Ni et al. 2023).

Temperature changes causes adaptations that modify plants architecture, affecting root growth and physiology (metabolic pathways) and reduced competitiveness against invasive species (González-García et al. 2022). Long and robust root system provide successive environment to compete against abiotic stress by increasing surface area in surrounding environment. Moreover, root structural changes provide better resilience against increasing temperature ultimately lead to dry conditions (NAEEM et al. 2024b, Waseem et al. 2024). Moreover, anatomical modifications i.e., thick epidermal and endodermal tissues promote growth of *Justicia adhatoda* under dry conditions. Though temperature is a crucial determinant, other factors such as soil conditions and biotic interactions are also important in determining the overall invasiveness of *Justicia adhatoda*, implying that an integrated strategy is required for effective management of *Justicia adhatoda* in new environment (Figure 2).

Figure 2. Aerial and edaphic factors affecting the invasiveness of Justicia adhatoda L in arid environments

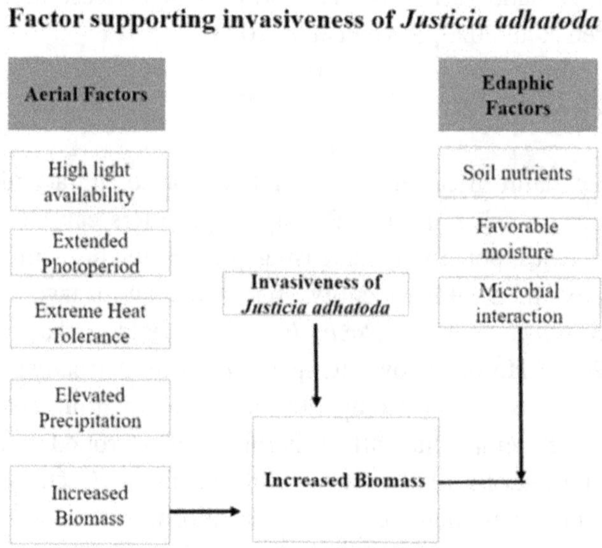

2.2 Effect of Precipitation

The effect of precipitation on plant invasiveness is significant, as varying moisture levels enhance or inhibit the growth of invasive species. It was studied earlier, that extreme precipitation events, particularly increased wet conditions, tend to promote the abundance of invasive plant species. The presence of exceptionally rainy months is associated with higher invasive species abundance, particularly in managed grasslands (Ratcliff et al. 2023). While drought conditions are expected to reduce invasiveness. It was suggested that they may not be sufficient to counteract the effects of increasing wetness (Ratcliff et al. 2023). Moreover, the seasonal changes particularly include winter precipitation, enhanced winter rainfall has been associated with enhanced growth rates of invasive species such as *Bromus tectorum*, allowing them to survive when native species competition is minimal (Prevéy and Seastedt 2015). Other factors include, nutrient interactions, the combined effect of rainfall and soil nutrients significantly affects the invasiveness of plant species. The nutrient-rich environment provides space to occupy in a new environment than nutrient-poor ones (Eskelinen and Harrison 2014). Varied precipitation patterns alter the demographic features of hybrid invasive species, potentially increasing survival and reproductive success as compared to non-hybrid populations(Westbrook and DiTommaso 2023).

2.2.1 Effect of Precipitation on *Justicia adhatoda* L. invasiveness

Precipitation has a multidimensional effect on the invasiveness of *Justicia adhatoda*, regulating growth, reproductive success, and ecological interactions. Understanding these dynamics is critical for controlling spread of invasive species, especially in areas where it could become invasive.

a) **Growth:** Increased precipitation promotes growth rate of *Justicia adhatoda*, as moisture is essential for its physiological processes. It was previously studied that optimal water availability correlates with increased biomass production, promotes its spread in suitable environments (Berry et al. 2020).
b) **Reproductive Success:** The reproductive success of *Justicia adhatoda's* directly affected by precipitation level. Increased rainfall boosts seed yield and viability. The plants have spatial and temporal adaptations that help to tolerate flooding environmental conditions contributing to their invasiveness by allowing them to compete with native species for resources. The presence of *Justicia adhatoda* can disrupt local ecosystems, especially in wetland areas, by affecting herbivores and other plant populations (Shivanna 2009).
c) **Ecological Interactions:** The presence of *Justicia adhatoda* can disrupt local ecosystems, especially in wetland areas, by affecting herbivores and other plant populations. Leaf extract of *Justicia adhatoda* has the potential to lower the fitness of specific pests, perhaps influencing its competitive advantage in a moist environment (Thanigaivel et al. 2017).

In contrast, significant precipitation can cause waterlogging, thereby affecting the growth of *Justicia adhatoda* and limiting its invasiveness in some regions **(Figure 3)**.

Figure 3. Management and control strategies to promote native vegetation and sustainable environment

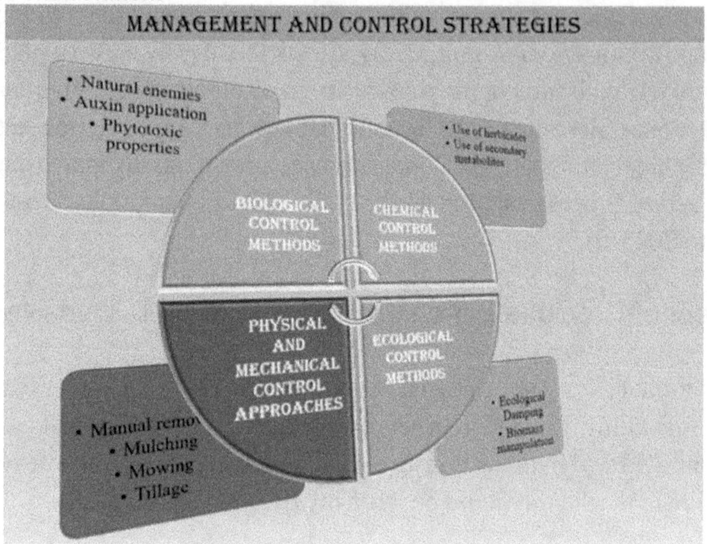

2.3 Effect of Wind Patterns on Invasiveness

Wind patterns have a considerable impact on the invasiveness of specific plant species, particularly those that rely on wind for seed dispersion. Wind not only helps the long-distance movement of invasive species but also influences their establishment and spread throughout different biomes.

a) **Wind as a dispersal mechanism:** Wind-dispersed invasive plants, such as those in the Asteraceae family, exhibiting some unique features that enhance their dispersal capabilities, including pappus length and weight of seeds, which correlate with horizontal dispersal distance and invasiveness (Jian-Hua et al. 2010). Atmospheric pathways, particularly Lagrangian coherent structures, have been identified as critical in the long-distance aerial dispersal of invasive species, allowing them to travel along predictable routes (Pretorius et al. 2023).
b) **Biome specific effects:** Wind affects the dispersal of invasive plants differently throughout biomes. It has a detrimental impact in tropical and subtropical moist biomes but a favorable impact in temperate and boreal forests (Wan et al. 2017). In Mediterranean wetlands, high winds have a considerable impact on the dispersal patterns of species such as *Cortaderia selloana*, resulting in

a concentration of seeds and recruits in specific orientations (Saura-Mas and Lloret 2005).

c) **Disturbance and invasion:** Windstorms can cause disturbances that aid invasions by changing microhabitats and resource availability, allowing invasive species to establish themselves in newly produced gaps (Bonanomi et al. 2018).

Though, wind patterns are important for invasive species distribution and establishment, other factors such as soil conditions and competition with native species must also be considered, as these might influence invasion success.

2.3.1 Effect of Wind Patterns on *Justicia adhatoda* L. invasiveness

Wind patterns have a considerable impact on the invasiveness of *Justicia adhatoda*, especially through its dispersal mechanisms. Understanding these dynamics is critical for controlling its spread because wind can aid in the spreading of its seeds, increasing its invasive potential.

Like other invasive species, *Justicia adhatoda* produces wind-dispersed seeds. Wind dispersal allows for more horizontal movement of seeds, which can lead to the colonization in new places and expand the plant's invasive range (Jian-Hua et al. 2010).

Seed morphology, such as size and weight, can influence dispersal distance and settling velocity, both of which are important for invasiveness. *Justicia adhatoda* possesses lighter seeds, and wind dispersal modifications help invade new environments sharply (Jian-Hua et al. 2010).

The growth of *Justicia adhatoda* can affect local ecosystems by outcompeting native vegetation and modifying habitat structures(Akbar and Akbar 2020). Its medicinal qualities may lead to increased cultivation, accelerating its spread and possibly invasiveness in non-native areas (Basit et al. 2022). Though wind patterns help in the dispersal of *Justicia adhatoda*, other environmental and topographic factors such as the soil composition and structure and moisture availability play important roles in the growth of *Jutsicia adhatoda* into new environments (Figure 2).

2.3.2 Effect of light availability and photoperiod on invasiveness

The availability of light, quality, intensity and photoperiod have significant impact on the invasiveness of plant species, affecting their growth, germinations, fitness and competition. Variations in light conditions can either enhance or limit the invasive potential of certain species, shaping their ecological success. It was studied earlier that light availability and photoperiod help in invasiveness of species, while dim light can limit their growth and promote native species. Light affect growth and biomass, for example; invasive species such as *Commelina benghalensis* and *Bromus madritensis* exhibit lower growth and biomass in low-light circumstances, potentially limiting their competitive advantage over native species (Riar et al. 2016, Pik et al. 2020). Moreover, in *Alternanthera philoxeroides* light intensity improves branching and biomass (Xu KaiYang et al. 2005), while the germination of *Eupatorium adenophorum* depend on high light intensity (Yong et al. 2012). Another light factor, shade of light also plays a significant role in species invasiveness, such as *Orbea variegata*, which grow better in shaded habitats (low light) (Lenz and Facelli 2003). The duration of light show influence on invasiveness of plant species. Photoperiod affect reproductive timings, short photoperiods promotes flowering in invasive species while reducing vegetative growth, as studied in *C. benghalensis* (Chenxi 2016). This reproductive adaptability help them to survive in their environments. Different species respond differently to photoperiod, with some requiring longer days for optimal reproduction, which can impact their invasiveness (Funk and McDaniel 2010). Quality of light also affect invasiveness of plant species, light of different wavelength affect i.e., red and far red light promote germination pattern and photosynthetic efficiency of plants in different environments (Hu et al. 2009).

2.3.3 Effect of light availability and photoperiod on invasiveness of *Justicia adhatoda*

The invasiveness of *Justicia adhatoda* greatly affected by light availability, light quality, intensity and photoperiod. These factors influence the growth and biochemical properties of *Justicia adhatoda*. High light helps *Justicia adhatoda* by increasing leaf area and allocating more biomass to leaves. Light intensity, quality and duration affect growth and development and competitive abilities of *J. adhatoda*. The quality of light affect chlorophyll contents affect photosynthetic efficiency of *Justicia adhatoda* (Saravanan and Karthi 2015). Moreover, increased photosynthetic efficiency leads to faster growth rates. The root growth and leaf expansion, crucial factors of invasiveness promoted by light intensity (Saravanan and Karthi 2015). The ecological success of *Justicia adhatoda* in different environmental conditions also depend on it biochemical and medicinal properties specifically alkoids (vascine) production help to compete native flora and pest attack(Saravanan and Karthi 2015). It was studied that invasive species *J. adhatoda* often survive in new environments in sufficient light, due to its adaptability potential related to biomass allocation and leaf morphology to optimize light capture. These modifications help to improve growth rates and competitive edge over native species (WANG and FENG 2004). Moreover, longer photoperiods lead to intensify growth and reproductive abilities, potentially leading to increased interactions with other species (competitive interaction), which may further influence the dynamics of invasion (Kehoe et al. 2020).

While low light, *Justicia adhatoda* exhibited characteristics such as larger, thinner leaves to maximize light interception. The most importantly, phenotypic plasticity allow *Justicia adhatoda* to survive in diverse environmental conditions, potentially increasing invasive potential (Figure 2).

3. EDAPHIC (SOIL) FACTORS INFLUENCING INVASION

The edaphic factors i.e., the soil is crucial in establishing and succeeding invasive plants Figure 3. Soil properties i.e., soil composition, structure, texture, and nutrient availability are critically important to identify community structure and stable ecosystem. The growth, competitive abilities and survival success of invasive plants occur in nutrient-rich soil. In arid environments, increased carbon contents correlate with increasing diversity of invasive plant species (Vitti et al. 2020). Invasive species alter nutrients in soil, improving their growth and suppressing the growth of native flora. They can exploit nutrients nitrogen (N) and phosphorus (P) resources more effectively than native flora (Gibbons et al. 2017, Afreen et al. 2023).

Soil texture is a fundamental property of soil to determine the type of flora in a specific environment. It influences water retention and aeration, affecting plant growth and seed germination (Vitti et al. 2020, Trotta 2024). Coarse texture has better drainage, promoting the growth of *Justicia adhatoda* while fine textured soil retains more water, affecting competition with native flora (Inderjit and Cahill 2015). The sandy soil is suitable for invasive species near coastal dunes, while saline water of the ocean limits the growth of these species. Increased salinity and organic matter contents limit the growth of *Justicia adhatoda*, while promoting native plant cover (Trotta 2024). Moreover, the condition of the soil also influences the invasion of species. The soil type enhances or inhibits invasiveness (Gao et al. 2017). Invasiveness of plant species in soil rich in microbial organisms, depending on soil type and condition. Soil microbes play an important role in making soil healthy and shaping the community by facilitating the spread of invasive species (Mahla and Mlambo 2019). Though soil physico-chemical properties play an important role in plant species invasion, other factors such as climate change and human exploitation significantly influence the invasion of *Justicia adhatoda*. Therefore ecological interactions are necessary to consider effective management strategies.

Soil moisture is a crucial factor determining the overall structure of soil. The growth of plant species in different environments depends on the moisture content in the soil. Though it is a complex interplay between moisture levels and species dynamics. Some plants can better survive and germinate under unfavorable environmental conditions due to better water use efficiency (Chadha et al. 2019). Increased water-holding capacity promotes the growth of exotic as well as native flora e.g.,

Justicia adhatoda flourishes with increasing soil moisture contents. Soil with maximum water holding capacity leading to more organic matter content (humus), thereby promoting growth of invasive plant *Justcia adhatoda* (Hartzell et al. 2017, Acharya et al. 2020). Conversely, high precipitation promotes invasions, hence disturbing the native flora (McKinney and Cleland 2014, Ratcliffe et al. 2024). During drought, native flora utilize maximum available water, thereby limiting open resources for exotic invaders (Duca et al. 2022).

Soil pH and ECe are critical factors in controlling the spread of invasive species. These are essential to study biotic and abiotic interactions that affect plant performance in different environments. The invasiveness of Justicia adhatoda is significantly influenced by soil pH and electrical conductivity (ECe). It was previously studied that acid and high saline soils can restrict the growth of invasive species. Low soil pH creates phytotoxicity that limit the growth of *Justcia adhatoda*. It was previously observed that acidification leads to biomass cover of invasive plants by 71% in Cerrado region (Lira Junior et al. 2017). The less acidic and nutrient- rich soils provide a suitable environment to establish a dynamic community of *Justica adhatoda* (Soti et al. 2020).

Ece value shows salinity level in soil that negatively affect the growth of invasive plants. It was negative correlation of soil Ece and pH. High Ece, less acidic soil and irrigation practices are effective management strategies to limit the growth of *Justicia adhatoda* and establishment community of native flora(Cherie 2022, Pereira et al. 2022).

Soil microbes play a significant role in growth and establishment of *Justicia adhatoda* in various environments. Microbes promotes in invasion by altering soil microbial communities and soil microbial feedback mechanism. *Justicia adhatoda* have ability to manipulate soil microbes by increasing pathogenicity and altering biotic interactions. In this way they can spread easily in new environment. On the other hand, soil microbes also negatively effects growth of *Justicia adhatoda*, when microbes become acclimatize and stabilize according to environment and positive soil plant relationship and feedback mechanisms. *Justicia adhatoda* can change the nutrient cycles by affecting enzymatic activity that affects the ecosystem stability, significantly limiting the invasiveness of other species over time (Ravichandran and Thangavelu 2017).

However, not all microbial interactions are beneficial. Some native soil microbial communities may exert inhibitory effects on invaders species like *J.adhatoda* have allopathic effect that limit growth and seedling establishment (Van der Putten 2010). Therefore, the net outcome of microbial interaction depends on the balance between facilitative mutualisms and antagonistic biotic **resistance** from the native soil biota.

4. INTERACTION BETWEEN AERIAL AND EDAPHIC FACTORS

The growth of *Justicia adhatoda* is significantly influenced by the interaction between aerial and edaphic factors (Figure 2). These factors include soil composition, moisture levels, and atmospheric conditions, which together create a conducive environment for the plant's adaptability and growth. The presence of essential nutrients such as nitrogen, phosphorus, and potassium are important for the growth of *J. adhatoda*. Moreover, soil organic manures, moisture have been shown to enhance growth parameters significantly, including height and leaf area (Rahman et al. 2021), but, the application of organic fertilizers can counteract these effects (Al-Rekaby and Merza 2015). Plant's response to aerial factors i.e., air pollutants is influenced by soil water status, which affects stomatal behaviour. Increased soil moisture can reduce sensitivity to contaminants like sulfur dioxide (Al-Rekaby and Merza 2015). However, edaphic factors play a critical role in plant growth, extreme climatic conditions, such as prolonged drought or high salinity, can severely limit the adaptability and survival of *Justicia adhatoda*, highlighting the delicate balance required for optimal growth. Salinity levels in the soil can interact with aerial factors, impacting overall plant health and growth. High salinity can exacerbate stress and promote certain adaptive traits (Sabbagh et al. 2014).Salinity reduces chlorophyll content, impairing photosynthesis and stomatal conductance, limiting CO_2 uptake and further inhibiting photosynthetic efficiency. Other climatic variables, including temperature and precipitation, are essential in determining the species' distribution and invasiveness, as they influence growth and reproductive success.

5. STRESS FACTORS AND ADAPTIVE MECHANISMS

The prevalence of stress in relation to *Justicia adhatoda* highlights the plant's adaptability and therapeutic potential. It was indicated that *Justicia adhatoda* exhibits significant morphological and physiological adaptations to drought stress, enhancing its survival in xeric environments. This adaptability is linked to various root modifications i.e., thick root epidermis and enlarged vascular tissues(xylem and phloem) that allow the plant to conserve water and maintain structural integrity under adverse conditions (NAEEM et al. 2024b). Additionally, the plant's leaf extracts demonstrate notable anti-inflammatory and antioxidant properties, suggesting its potential to mitigate oxidative stress(Basit et al. 2022).These properties suggested that *J. adhatoda* are beneficial in treating stress, potentially serving as a natural remedy (Basit et al. 2022). Conversely, while Justicia adhatoda shows promise in stress adaptation and therapeutic applications, its efficacy may vary based on environmental conditions and extraction methods.

5.1 Adaptive mechanism

J.adhatoda exhibits remarkable adaptive mechanisms that enable it to survive in diverse environmental conditions, particularly against native vegetation. This adaptability is primarily manifested through morphological, physiological, and biochemical modifications that enhance its survival in xeric environments. The following sections detail these adaptive strategies.

5.1.1 Morphological adaptations:

a) **Root Modifications:** In severely dry habitats, *J. adhatoda* shows increased root fresh and dry weight, along with enhanced root length, which aids in water absorption (NAEEM et al. 2024b).
b) **Morphotype variation:** Two distinct morphotypes have been identified, with one adapted to drier regions exhibiting specific phenotypic traits that confer resilience against drought (Berry et al. 2020).

5.1.2 Physio-chemical adaptations:

The plant's ability to manage ionic content, particularly sodium, calcium, and potassium, is crucial for maintaining cellular function and osmotic adjustment under stress (NAEEM et al. 2024b). Changes in phytochemical composition, such as increased flavonoids and phenols, are observed in response to varying altitudinal conditions, enhancing its medicinal properties and stress resistance (Jamwal et al. 2023)

5.1.3 Ecological Implications:

The adaptability of *J. adhatoda* not only allows it to survive but also to compete effectively with native vegetation, potentially altering local ecosystems. However, this competitive edge may also lead to concern regarding its invasiveness in certain regions, highlighting the need for careful management and distribution (Sobia et al. 2018).

Table 1. Key adaptations that support the invasion success of Justicia adhatoda in arid and disturbed environments

Adaptation Type	Description
Physiological	High drought and temperature tolerance, High water use effeciency (Zhang et al. 2017)
Phenological	Flexibility in germination and growth timing across seasons(Kumar and Singh 2025)
Morpho-anatomical	Thick cuticle, dense trichomes, deep root systems (Lombardini and Rossi 2019)
Allelopathic Potential	Releases compounds that inhibit native plant growth (GULZAR)
Soil Microbe Interaction	Exploits beneficial microbes (e.g., mycorrhizae) while resisting pathogens (Kushwaha et al. 2020)
Shade Tolerance	Capable of surviving under reduced light; modulates photosynthetic rates(Cheah and Teh 2020)

6. MANAGEMENT AND CONTROL STRATEGIES

6.1 Biological Control Methods:

The biological control of *Justicia adhatoda* growth can be effectively managed through various methods, including the use of natural enemies, auxins application, phytotoxic properties of plants' organic manures, and innovative propagation techniques. These methods help to mitigate the ecological impact of this species in non-native environments, but also ensure sustainable management of *Justicia adhatoda* **(Figure 4)**.

6.1.1 Natural enemies: The herbivores, insects or fungal pathogens from the native range of *Justicia adhatoda* effectively suppress its growth. This method is beneficial in sensitive ecosystems where chemical control is unsuitable (Van Driesche and Center 2013).

6.1.2 Auxin application: The application of auxins such as IBA (Indole-3-butyric acid) at 1000 ppm significantly increases root number and biomass in stem cuttings (Uddin and Rashid 2023). IAA (Indole-3-acetic acid) and NAA (Naphthaleneacetic acid) also promote root length and biomass, indicating that auxin treatment is crucial for effective propagation (Uddin and Rashid 2023). The use of organic manures, particularly water hyacinth enhances growth metrics such as height and leaf area (Rahman et al. 2021). Incorporating organic manures improves plant growth and contributes to soil health, making it a sustainable approach for managing *J. adhatoda* cultivation (Rahman et al. 2021). Though these biological control methods are effective, it is essential to consider potential challenges such as the variability in growth responses to different treatments.

6.1.3 Phytotoxic properties: The methanolic extract from the roots of *Justicia adhatoda* showed significant phytotoxic properties. This extract shows an inhibitory effect on seed germination and root growth of *Justicia adhatoda*. This suggests the potential for using as eco-friendly herbicides (Nasir et al. 2024).

6.2 Chemical Control Approaches

Controlling the growth of *Justicia adhatoda*, an invasive species, can be effectively achieved through various chemical methods. These methods focus on the application of herbicides and natural pesticides, which can selectively target the invasive plant while minimizing harm to surrounding ecosystems. Utilizing a mixture of herbicides with C_2-C_5 mono- or dicarboxylic acid additives has been shown to enhance the speed and effectiveness of weed control compared to herbicides alone (Anderson and Mao 2008). The use of natural pesticides (endocides) derived from other invasive species can inhibit the growth of *Justicia adhatoda*. These endocides act specifically on the target species without affecting the producing species (Li et al. 2018). Moreover, secondary metabolites like terpenoids have demonstrated phytotoxic activity, offering an ecological alternative to synthetic herbicides for sustainable management (de Oliveira et al. 2022). Though chemical methods are effective, there is a growing concern regarding their environmental impact and potential harm to non-target species. It led to increased interest in integrating biological control methods and promoting the use of natural products in invasive species management (Figure 3).

6.3 Physical and Mechanical Control Techniques

Controlling the growth of *Justicia adhatoda* an invasive species, can be effectively achieved through various physical and mechanical control techniques. These methods focus on reducing its spread and minimizing its ecological impact. The following are the most effective strategies based on recent research findings.

6.3.1 Manual Removal: Regularly uprooting the plant can significantly reduce its population. This method is labour-intensive but effective in small areas(Isha et al. 2025).

6.3.2 Mulching: Applying organic or synthetic mulch can suppress the growth of *Justicia adhatoda* by blocking sunlight and preventing seed germination (Nasir et al. 2024).

6.3.3 Mowing: Frequent mowing can prevent flowering and seed production, thereby limiting the plant's ability to spread. This technique should be combined with other methods for effectiveness (Hossain et al. 2022).

6.3.4 Tillage: Cultivating the soil can disrupt root systems and reduce the plant's ability to regenerate. This is particularly effective in agricultural settings (NAEEM et al. 2024b).

While these control techniques are effective, it is essential to consider the potential ecological impacts of removing an invasive species. It was studied that *Justicia adhatoda* may have beneficial uses in traditional medicine, which complicates its management (Ravali et al., 2024).

6.4 Ecological stratagies

The ecological control methods offer promising alternatives to traditional approaches to mitigate invasiveness of *Justicia adhatoda*. These stratagies help to focus on enhancing ecosystem resilience and utilizing natural processes to suppress invasive species without the adverse effects associated with chemical treatments or the introduction of non-native predators.

6.4.1 Ecologicl Damping: It method involves concept and implementation of methamatical models to dampen the population growth of invasive species like *Justicia adhatoda*, preventing exponential growth of invasive species that lead to ecological disruption. Moreover, this method avoid the use of chemical treatments, making a safer option for ecosystem stability (Parshad et al. 2016).

6.4.2 Biomass Manipulation: It involves practices including selective graing and prescribed burrning. It can maintain ecosystem stability by suppressing the growth of invasive species like Justicia adhatoda (Guo et al. 2018).

7. RESTORATION OF NATIVE VEGETATION

The restoration of native vegetation in areas where *Justicia adhatoda* is prevalent can be effectively approached through various strategies that emphasize the selection and management of native species. Justicia adhatoda, while beneficial, can influence local ecosystems' dynamics, necessitating careful restoration planning.

Selecting appropriate native species is crucial for reducing the invasibility of exotic plants. For instance, early successional forbs have been shown to significantly decrease the biomass of invasive species in desert ecosystems (Abella et al. 2012). Moreover,defining target species assemblages based on environmental conditions can enhance restoration outcomes. It was demonstrated that modeling of species can also help to identify which native plants are best suited for specific restoration sites (Siles et al. 2010).

On the other hand, the application of organic amendments, such as sewage sludge, has been shown to boost the growth of native vegetation. In a semi-arid environment, the addition of biowaste significantly increased plant biomass and canopy cover (Walter and Calvo 2009).While biowaste can enhance growth, it may also reduce native species richness if applied excessively, indicating the need for careful management of amendment rates(Walter and Calvo 2009).Conversely, the introduction of non-native species, including *Justicia adhatoda*, can sometimes lead to drastic ecological consequences, such as increased competition for resources among native plants.

Moreover, the use of biowaste materials such as sewage sludge, compost and green manure has potential benefits in enhancing soil fertility, supporting vegetation recovery and water retention in degraded arid lands that stimulate microbial activity, fostering a more favourable environment for native plant reestablishment(Adhikari et al. 2009, Chibuike and Obiora 2014).

However, the use of untreated or poorly managed biowaste can result in unintended consequences such as nutrient runoff and pathogen spread, posing risks to both soil and ecosystem health. There is a need to adopt site-specific application rates and rotational strategies to reduce contaminant build-up.

Future Research Directions

- Improve monitoring of invasive species like *Justicia adhatoda* by utilizing an unmanned aerial system or remote sensing techniques.
- Restore ecosystem hydrology by actively removing invasive species
- Design integrated spectral–structural workflow to classify (native and non-native flora) and precise mapping of *Justicia adhatoda* population over large areas.
- Identify invasive species and their impact on the local ecosystem can be checked by ecological assessment and remote sensing techniques.
- Implement continuous effective management strategies to prevent the spread of *Justicia adhatoda* in semiarid regions.
- Impose dynamic monitoring techniques that can detect the distribution of *Justicia adhatoda* over time.
- Focus on restoration and introduction of native flora that can improve ecosystem resilience against invasiveness in semiarid environments.
- Publicly aware and educate the local residents about the risk to native flora associated with the invasiveness of *Justicia adhatoda* that can help to prevent its spread.

- Establish a balanced ecosystem by monitoring and effective management strategies including various chemical, physical and biological control approaches to control the spread of *Justicia adhatoda.*
- Promote the growth of native flora *Ziziphus mauritiana, Cenchrus biflorus and Senegalia modesta* that can outcompete and limit the growth of *Justicia adhatoda.*
- Focus on the interaction of aerial and edaphic factors to develop targeted control measures of the invasiveness of *Justicia adhatoda.*

Though, these future perspectives are necessary, but, they require significant resources and long-term commitment, which can be challenging. Balancing economic considerations with ecological needs is essential for sustainable management.

CONCLUSION

The ecological success of *Justicia adhatoda* L. in arid and semiarid environments is significantly affected by both aerial and edaphic factors. It processed remarkable adaptability potential through various physiological and morphological attributes that help to compete with aridity. *Justicia adhatoda* shows phenotypic plasticity to adjust growth patterns in response to varying water availability to tolerate drought. It can modify its leaf structure and stomatal density which can improve water use efficiency which is crucial for survival in arid environments. Moreover, the adaptability of *Justicia adhatoda* is also linked to its root morpho-physiological attributes which are optimized for nutrient and water uptake in arid environments. While, edaphic factors i.e., soil physicochemical attributes i.e., soil pH, and moisture play a critical role in the distribution of *Justicia adhatoda.* It can also significantly influence the plant's growth and invasive potential in semiarid regions. Though *Justicia adhatoda* can survive in arid environments, other invasive species can compete due to less effective adaptations to similar conditions, highlighting the better resilience of *Justicia adhatoda.*

REFERENCES

Abella, S. R., Craig, D. J., Smith, S. D., & Newton, A. C. (2012). Identifying native vegetation for reducing exotic species during the restoration of desert ecosystems. *Restoration Ecology*, *20*(6), 781–787. DOI: 10.1111/j.1526-100X.2011.00848.x

Acharya, S., McLaughlin, D., Kaplan, D., & Cohen, M. J. (2020). A proposed method for estimating interception from near-surface soil moisture response. *Hydrology and Earth System Sciences*, *24*(4), 1859–1870. DOI: 10.5194/hess-24-1859-2020

Adhikari BK, Barrington S, Martinez J, King S (2009) Effectiveness of three bulking agents for food waste composting. Waste management 29(1): 197-203

Afreen, T., Kumari, S., Bhadouria, R., Devi, R. S., Singh, S., & Tripathi, S. (2023). Plant Invasion and Soil Processes: A Mechanistic Understanding. In *Plant Invasions and Global Climate Change* (pp. 227–246). Springer. DOI: 10.1007/978-981-99-5910-5_10

Akbar, S., & Akbar, S. (2020) Justicia adhatoda L.(Acanthaceae) (Syns.: Adhatoda vasica Nees.; A. pubescens Moench.; A. zeylanica Medik.; Dianthera latifolia Salisb). Handbook of 200 Medicinal Plants: A Comprehensive Review of Their Traditional Medical Uses and Scientific Justifications: 1059-1066

Al-Rekaby LS, Merza TK (2015) Effect of water stress and organic fertilizer (Azomin) on some growth parameters of Adhatoda vasica L.(Nees). Kufa Journal for Agricultural Sciences 7(2)

Ali M, Hakeem KR, Ali M, Hakeem KR (2020) Phyto-pharmaceutical Potential and the Isolation of Novel Compounds from Adhatoda vasica L. Nees. Scientific Explorations of Adhatoda vasica: An Asian Health Remedy: 37-60

Anderson, T., & Mao, J. (2008). Herbicidal compositions and methods of controlling weed growth. Google Patents, Basit A, Shutian T, Khan A, Khan SM, Shahzad R, Khan A, Khan S, Khan M (2022) Anti-inflammatory and analgesic potential of leaf extract of Justicia adhatoda L.(Acanthaceae) in Carrageenan and Formalin-induced models by targeting oxidative stress. *Biomedicine and Pharmacotherapy*, *153*, 113322.

Berry, E., Choudhary, A. K., Mishra, G., Tandon, R., & Geeta, R. (2020). Justicia adhatoda reveals two morphotypes with possible functional significance. *Journal of Plant Research*, *133*(6), 783–805. DOI: 10.1007/s10265-020-01224-w PMID: 32979146

Bhatt, A., Chen, X., Pompelli, M. F., Jamal, A., Mancinelli, R., & Radicetti, E. (2023). Characterization of invasiveness, thermotolerance and light requirement of nine invasive species in China. *Plants*, *12*(5), 1192. DOI: 10.3390/plants12051192 PMID: 36904052

Bhattacharya A (2019) Effect of high-temperature stress on crop productivity. Effect of high temperature on crop productivity and metabolism of macro molecules

Bonanomi G, Incerti G, Abd El-Gawad AM, Sarker TC, Stinca A, Motti R, Cesarano G, Teobaldelli M, Saulino L, Cona F (2018) Windstorm disturbance triggers multiple species invasion in an urban Mediterranean forest. iForest-Biogeosciences and Forestry 11(1): 64

Chadha, A., Florentine, S. K., Chauhan, B. S., Long, B., & Jayasundera, M. (2019). Influence of soil moisture regimes on growth, photosynthetic capacity, leaf biochemistry and reproductive capabilities of the invasive agronomic weed; Lactuca serriola. *PLoS One*, *14*(6), e0218191. DOI: 10.1371/journal.pone.0218191 PMID: 31251746

Cheah, S., & Teh, C. (2020). Parameterization of the Farquhar-von Caemmerer-Berry C3 photosynthesis model for oil palm. *Photosynthetica*, *58*(3), 769–779. DOI: 10.32615/ps.2020.020

Chen, S., Wang, Z., Zhang, M., Shi, X., Wang, L., An, W., Li, Z., Pan, F., & Yang, L. (2023). Practical evaluation of prelithiation strategies for next-generation lithium-ion batteries. *Carbon Energy*, *5*(8), e323. DOI: 10.1002/cey2.323

Chenxi Z (2016) Environmental Influences on Growth and Reproduction of Invasive Commelina benghalensis.

Cherie, D. A. (2022). Determination and Correlation of pH and Electrical Conductivity of Assosa Agricultural Research Center Sites Soil. *Asian Soil Research Journal*, *6*(3), 6–10. DOI: 10.9734/asrj/2022/v6i330131

Chibuike, G. U., & Obiora, S. C. (2014). Heavy metal polluted soils: Effect on plants and bioremediation methods. *Applied and Environmental Soil Science*, *2014*(1), 752708. DOI: 10.1155/2014/752708

de Oliveira MS, Cruz JN, de Aguiar EH (2022) Potential Use of Terpenoids in Weed Management. Terpenoids: Recent Advances in Extraction, Biochemistry and Biotechnology: 200

Duca, M., Boian, I., & Domenco, R. (2022). The impact of droughts on sunflower production in the Republic of Moldova. *Notulae Botanicae Horti Agrobotanici Cluj-Napoca*, *50*(4), 1–16. DOI: 10.15835/nbha50413040

Elsheikh, E. A., El-Keblawy, A., Mosa, K. A., Okoh, A. I., & Saadoun, I. (2021). Role of endophytes and rhizosphere microbes in promoting the invasion of exotic plants in arid and semi-arid areas: A review. *Sustainability (Basel)*, *13*(23), 13081. DOI: 10.3390/su132313081

Eskelinen, A., & Harrison, S. (2014). Exotic plant invasions under enhanced rainfall are constrained by soil nutrients and competition. *Ecology*, *95*(3), 682–692. DOI: 10.1890/13-0288.1 PMID: 24804452

Finch DM, Butler JL, Runyon JB, Fettig CJ, Kilkenny FF, Jose S, Frankel SJ, Cushman SA, Cobb RC, Dukes JS (2021) Effects of climate change on invasive species. Invasive species in forests and rangelands of the United States: a comprehensive science synthesis for the United States forest sector: 57-83

Funk, J. L., & McDaniel, S. (2010). Altering light availability to restore invaded forest: The predictive role of plant traits. *Restoration Ecology*, *18*(6), 865–872. DOI: 10.1111/j.1526-100X.2008.00515.x

Gao, L., Wang, R., Shen, G., Zhang, J., Meng, G., & Zhang, J. (2017). Effects of biochar on nutrients and the microbial community structure of tobacco-planting soils. *Journal of Soil Science and Plant Nutrition*, *17*(4), 884–896. DOI: 10.4067/S0718-95162017000400004

Gibbons, S. M., Lekberg, Y., Mummey, D. L., Sangwan, N., Ramsey, P. W., & Gilbert, J. A. (2017). Invasive plants rapidly reshape soil properties in a grassland ecosystem. *mSystems*, *2*(2), 00178–00116. DOI: 10.1128/mSystems.00178-16 PMID: 28289729

Gilani, S. A., Fujii, Y., Kikuchi, A., Shinwari, Z. K., & Watanabe, K. N. (2011). Ecological consequences, genetic and chemical variations in fragmented populations of a medicinal plant, Justicia adhatoda and implications for its conservation. *Pakistan Journal of Botany*, *43*, 29–37.

González-García, A., Palomo, I., González, J. A., García-Díez, V., García-Llorente, M., & Montes, C. (2022). Biodiversity and ecosystem services mapping: Can it reconcile urban and protected area planning? *The Science of the Total Environment*, *803*, 150048. DOI: 10.1016/j.scitotenv.2021.150048 PMID: 34500265

Guo, Q., Brockway, D. G., Larson, D. L., Wang, D., & Ren, H. (2018). Improving ecological restoration to curb biotic invasion—A practical guide. *Invasive Plant Science and Management*, *11*(4), 163–174. DOI: 10.1017/inp.2018.29

Hartzell, S., Bartlett, M. S., & Porporato, A. (2017). The role of plant water storage and hydraulic strategies in relation to soil moisture availability. *Plant and Soil*, *419*(1-2), 503–521. DOI: 10.1007/s11104-017-3341-7

Hellmers, H., & Warrington, I. (2018). Temperature and plant productivity. In *Handbook of Agricultural Productivity* (pp. 11–22). CRC Press.

Hossain, M. A., Akhter, S., Sohrab, M. H., Afroz, F., Begum, M. N., & Rony, S. R. (2023). Applying an optimization technique for the extraction of antioxidant components from Justicia adhatoda leaves. *Engineering and Science*, *24*(2), 913. DOI: 10.30919/es913

Hossain, R., Islam, M. S., Dey, D., & Islam, M. T. (2022). Vinca rosea (Madagascar Periwinkle) and Adhatoda vesica (Malabar Nut). In *Herbs, Shrubs, and Trees of Potential Medicinal Benefits* (pp. 301–338). CRC Press. DOI: 10.1201/9781003205067-16

Hou, Q.-Q., Chen, B.-M., Peng, S.-L., & Chen, L.-Y. (2014). Effects of extreme temperature on seedling establishment of nonnative invasive plants. *Biological Invasions*, *16*(10), 2049–2061. DOI: 10.1007/s10530-014-0647-8

Hu, Y., Jiang, S., Li, J., Gao, J., Gao, Y., & Gu, S. (2009). Effects of light intensity and quality on plant growth and development. [Nat Sci Ed]. *J Inn Mongolia Agri Univ*, *30*, 296.

Inderjit, & Cahill JF (2015) Linkages of plant–soil feedbacks and underlying invasion mechanisms. AoB plants 7: plv022

Isha, K. P., Kumar, P., & Singh, A. N. (2025). An Overview of Justicia adhatoda: A Medicinal Plant but Native Invader in India. *Conservation*, *5*(1), 2. DOI: 10.3390/conservation5010002

Jamwal, M., Puri, S., Sharma, N., Prakash, S., & Pundir, A. (2023). Altitudinal variation in phytochemical, physicochemical, and morphological aspects of Justicia adhatoda L. plant growing wildly in Western Himalayas. *Journal of Applied Biology and Biotechnology*, *11*(3), 85–96.

Jian-Hua, H., Sheng, Q., Kang-Ning, D., & Yuan-Xing, G. (2010). Wind-dispersed traits of cypselas in ten Asteraceae alien invasive species. *Acta Phytoecologica Sinica*, *34*(8), 957.

Kehoe, R., Sanders, D., Cruse, D., Silk, M., Gaston, K. J., Bridle, J. R., & van Veen, F. (2020). Longer photoperiods through range shifts and artificial light lead to a destabilizing increase in host–parasitoid interaction strength. *Journal of Animal Ecology*, *89*(11), 2508–2516. DOI: 10.1111/1365-2656.13328 PMID: 32858779

Khan, I., Ahmad, B., Azam, S., Hassan, F., Aziz, A., Rehman, N., Ullah, F., & Liaqat, Z. (2018). Pharmacological activities of Justicia adhatoda. *Pakistan Journal of Pharmaceutical Sciences*, *31*(2). PMID: 29618423

Khan, N., Ullah, R., Okla, M. K., Abdel-Maksoud, M. A., Saleh, I. A., Abu-Harirah, H. A., AlRamadneh, T. N., & AbdElgawad, H. (2023). Spatial distribution of the four invasive plants and their impact on natural communities' dynamics across the arid and semi-arid environments in northwest Pakistan. *Frontiers in Plant Science*, *14*, 1207222. DOI: 10.3389/fpls.2023.1207222 PMID: 37692447

Kumar P, Singh AN (2025) An Overview of Justicia adhatoda: A Medicinal Plant but Native Invader in India. Conservation (2673-7159) 5(1)

Kushwaha, R. K., Rodrigues, V., Kumar, V., Patel, H., Raina, M., & Kumar, D. (2020). Soil microbes-medicinal plants interactions: ecological diversity and future prospect. In *Plant microbe symbiosis* (pp. 263–286). Springer. DOI: 10.1007/978-3-030-36248-5_14

Lenz, T. I., & Facelli, J. M. (2003). Shade facilitates an invasive stem succulent in a chenopod shrubland in South Australia. *Austral Ecology*, *28*(5), 480–490. DOI: 10.1046/j.1442-9993.2003.01304.x

Li S, Wang P, Yuan W, Zushang S, Bullard SH (2018) Compositions and methods to selectively control invasive species. Google Patents, Lira Junior MA, Fracetto GGM, Araújo ASF, Fracetto FJC, Nóbrega RSA, Silva K, Galdino AC (2017) Rhizobial diversity for tropical pulses and forage and tree legumes in Brazil. Microbes for Legume Improvement: 135-151

Lombardini, L., & Rossi, L. (2019). Ecophysiology of plants in dry environments. In *Dryland ecohydrology* (pp. 71–100). Springer. DOI: 10.1007/978-3-030-23269-6_4

Lone, S., Yadav, A., Sharma, A., Tafazul, M., Badkhane, Y., & Raghuwanshi, D. (2013). A review on Adhatoda vasica Nees-An important and high demanded medicinal plant. *Indo American Journal of Pharm Research*, *3*, 2600.

Lucero, J. E., Faist, A. M., Lortie, C. J., & Callaway, R. M. (2022). Risk of facilitated invasion depends upon invader identity, not environmental severity, along an aridity gradient. *Frontiers in Ecology and Evolution*, *10*, 886690. DOI: 10.3389/fevo.2022.886690

Mahla, N., & Mlambo, D. (2019). Influence of two co-occurring invasive plant species on resident woody species and surface soil properties in Chipinge Safari Area, Zimbabwe. *Tropical Ecology*, *60*(1), 129–139. DOI: 10.1007/s42965-019-00016-1

Matloob, H., & Hameed, M. (2024). Osmoregulation facilitated by leaf structural and functional traits in Malabar nut (Justicia Adhatoda L.) along elevation gradient. *Pakistan Journal of Botany*, *56*(4), 1419–1429. DOI: 10.30848/PJB2024-4(2)

McKinney, J., & Cleland, E. E. (2014). Root inputs influence soil water holding capacity and differentially influence the growth of native versus exotic annual species in an arid ecosystem. *Restoration Ecology*, *22*(6), 766–773. DOI: 10.1111/rec.12139

Naeem, Z., Ahmad, M., Hameed, M., Bibi, S., Rani, W., Ratnasekera, D., Soufan, W., Adnan, M., & El Sabagh, A. (2024). Adaptability potential of *Justicia Adhatoda L.* To xeric environments through root morphophysiological and architectural modulations. *Applied Ecology and Environmental Research*, *22*(4), 2977–2998. DOI: 10.15666/aeer/2204_29772998

Nagler, P. L., Sridhar, B. M., Olsson, A. D., van Leeuwen, W. J., & Glenn, E. P. (2018). Hyperspectral remote sensing tools for quantifying plant litter and invasive species in arid ecosystems. In *Advanced Applications in Remote Sensing of Agricultural Crops and Natural Vegetation* (pp. 97–129). CRC Press. DOI: 10.1201/9780429431166-6

Nasir, M., Ramash, R., Fatima, H., Ashraf, S., Munir, I., Asghar, S., Adnan, M., Masood, A., & Chaudhari, S. K. (2024). Phytochemical Characterization and Assessment of Crude Extracts from Justicia adhatoda for Phytotoxic and Cytotoxic Activity. *Scientifica*, *2024*(1), 1374346. DOI: 10.1155/2024/1374346 PMID: 39359907

Nechols, J. R., Hough, A. R., Margolies, D. C., Ruberson, J. R., McCornack, B. P., Sandercock, B. K., & Murray, L. (2020). Effect of temperature on plant resistance to arthropod pests. *Environmental Entomology*, *49*(3), 537–545. DOI: 10.1093/ee/nvaa033 PMID: 32280953

Ni, M., Luo, H., Xu, H., Chu, C., & Fang, S. (2023). High temperature can improve the performance of invasive plants by facilitating root growth. *American Journal of Botany*, *110*(10), e16227. DOI: 10.1002/ajb2.16227 PMID: 37561668

Ntiri, E. S., Calatayud, P.-A., Van Den Berg, J., Schulthess, F., & Le Ru, B. P. (2016). Influence of temperature on intra-and interspecific resource utilization within a community of lepidopteran maize stemborers. *PLoS One*, *11*(2), e0148735. DOI: 10.1371/journal.pone.0148735 PMID: 26859748

Parshad, R. D., Quansah, E., Black, K., & Beauregard, M. (2016). Biological control via "ecological" damping: An approach that attenuates non-target effects. *Mathematical Biosciences*, *273*, 23–44. DOI: 10.1016/j.mbs.2015.12.010 PMID: 26767801

Pereira, G. W., Valente, D. S. M., de Queiroz, D. M., Santos, N. T., & Fernandes-Filho, E. I. (2022). Soil mapping for precision agriculture using support vector machines combined with inverse distance weighting. *Precision Agriculture*, *23*(4), 1189–1204. DOI: 10.1007/s11119-022-09880-9

Pik, D., Lucero, J. E., Lortie, C. J., & Braun, J. (2020). Light intensity and seed density differentially affect the establishment, survival, and biomass of an exotic invader and three species of native competitors. *Community Ecology*, *21*(3), 259–272. DOI: 10.1007/s42974-020-00027-2

Pretorius, I., Schou, W. C., Richardson, B., Ross, S. D., Withers, T. M., Schmale, D. G.III, & Strand, T. M. (2023). In the wind: Invasive species travel along predictable atmospheric pathways. *Ecological Applications*, *33*(3), e2806. DOI: 10.1002/eap.2806 PMID: 36660794

Prevéy, J. S., & Seastedt, T. R. (2015). Increased winter precipitation benefits the native plant pathogen Ustilago bullata that infects an invasive grass. *Biological Invasions*, *17*(10), 3041–3047. DOI: 10.1007/s10530-015-0934-z

Rahman, S., Nishi, K., Muslim, M., & Rahman, M. (2021). Effects of organic manures on the growth and yield of malabar nut (Justicia adhatoda L.). *Dhaka University Journal of Biological Sciences*, *30*(1), 79–86. DOI: 10.3329/dujbs.v30i1.51811

Ratcliff, J. J., Hall, A. J., Porto, E., Saville, B. R., Lewis, R. J., Allen, J. W., Frankel, M., Wright, D. W., Barrow, D. L., & Pradilla, G. (2023). Early Minimally Invasive Removal of Intracerebral Hemorrhage (ENRICH): Study protocol for a multi-centered two-arm randomized adaptive trial. *Frontiers in Neurology*, *14*, 1126958. DOI: 10.3389/fneur.2023.1126958 PMID: 37006503

Ratcliffe, H., Kendig, A., Vacek, S., Carlson, D., Ahlering, M., & Dee, L. E. (2024). Extreme precipitation promotes invasion in managed grasslands. *Ecology*, *105*(1), e4190. DOI: 10.1002/ecy.4190 PMID: 37877294

Ravali, K., Sarathchandra, G., Kumary, S. U., Sujatha, P. L., Kannan, T. A., & Parthiban, M. (2024, July 18). An in silico Evaluation of Anti-diabetic Potential of Vasicine, A Quinazoline Alkaloid of Justicia adhatoda Linnaeus. *Indian Journal of Animal Research*, *1*(Of), 4. DOI: 10.18805/IJAR.B-5278

Ravichandran, K. R., & Thangavelu, M. (2017). Role and influence of soil microbial communities on plant invasion. *Ecological Questions*, *27*, 9–23. DOI: 10.12775/EQ.2017.024

Riar, M. K., Carley, D. S., Zhang, C., Schroeder-Moreno, M. S., Jordan, D. L., Webster, T. M., & Rufty, T. W. (2016). Environmental influences on growth and reproduction of invasive Commelina benghalensis. *International Journal of Agronomy*, *2016*(1), 5679249.

Sabbagh, E., Lakzayi, M., Keshtehgar, A., & Rigi, K. (2014). The effect of salt stress on respiration, PSII function, chlorophyll, carbohydrate and nitrogen content in crop plants. *Intl J Farm & Alli Sci*, *3*(9), 988–993.

Sarada, K., & Rao, G. (2008). Antimicrobial activity of essential oils of Adhatoda vasica (Basaka) fresh and shade dried leaves and roots from Anantapur District, Andhra Pradesh. *Indian J Microb World*, *10*, 86–90.

Saravanan S, Karthi S (2015) distinction in morphological and biochemical changes in valuable medicinal herb Justicia Adhatoda under elevated CO2.

Saura-Mas, S., & Lloret, F. (2005). Wind effects on dispersal patterns of the invasive alien Cortaderia selloana in Mediterranean wetlands. *Acta Oecologica*, *27*(2), 129–133. DOI: 10.1016/j.actao.2004.12.001

Shivanna, K. (2009). Pollination biology, breeding system and reproductive success of Adhatoda vasica, an important medicinal plant. *Current Science*, •••, 408–412.

Siles, G., Alcántara, J. M., Rey, P. J., & Bastida, J. M. (2010). Defining a target map of native species assemblages for restoration. *Restoration Ecology*, *18*(4), 439–448. DOI: 10.1111/j.1526-100X.2008.00473.x

Silveira, M. J., Florêncio, F. M., de Carvalho Harthman, V., & Thiébaut, G. (2023). Responses of three invasive alien aquatic plant species to climate warming and plant density. *Journal of Plant Research*, *136*(6), 817–826. DOI: 10.1007/s10265-023-01482-4 PMID: 37505305

Simberloff, D., Martin, J.-L., Genovesi, P., Maris, V., Wardle, D. A., Aronson, J., Courchamp, F., Galil, B., García-Berthou, E., Pascal, M., Pyšek, P., Sousa, R., Tabacchi, E., & Vilà, M. (2013). Impacts of biological invasions: What's what and the way forward. *Trends in Ecology & Evolution*, *28*(1), 58–66. DOI: 10.1016/j.tree.2012.07.013 PMID: 22889499

Singh, L. R., & Singh, O. M. (2013). Datura stramonium: An overview of its phytochemistry and pharmacognosy. *Research Journal of Pharmacognosy and Phytochemistry*, *5*(3), 143–148.

Sobia, H. N., Khan, S., & Nadeem, F. (2018). Use of malabar nut (Justicia adhatoda L.) from traditional medicine to current pharmacopeia–A review study. *International Journal of Chemical and Biochemical Science*, *13*, 46–51.

Soti, P. G., Purcell, M., & Jayachandran, K. (2020). Soil biotic and abiotic conditions negate invasive species performance in native habitat. *Ecological Processes*, *9*(1), 1–9. DOI: 10.1186/s13717-020-00220-1

Thanigaivel, A., Senthil-Nathan, S., Vasantha-Srinivasan, P., Edwin, E. S., Ponsankar, A., Selin-Rani, S., Pradeepa, V., Chellappandian, M., Kalaivani, K., Abdel-Megeed, A., Narayanan, R., & Murugan, K. (2017). Chemicals isolated from Justicia adhatoda Linn reduce fitness of the mosquito, Aedes aegypti L. *Archives of Insect Biochemistry and Physiology*, *94*(4), e21384. DOI: 10.1002/arch.21384 PMID: 28266058

Trotta, A. (2024). Environmental impact bonds: Review, challenges, and perspectives. *Current Opinion in Environmental Sustainability*, *66*, 101396. DOI: 10.1016/j.cosust.2023.101396

Uddin, M. S., & Rashid, M. H. U. (2023). Effects of auxins on the stem cuttings of Justicia Adhatoda L. *Bangladesh Journal of Botany*, *52*(3), 861–865. DOI: 10.3329/bjb.v52i3.68947

Underwood, E. C., Klinger, R. C., & Brooks, M. L. (2019). Effects of invasive plants on fire regimes and postfire vegetation diversity in an arid ecosystem. *Ecology and Evolution*, *9*(22), 12421–12435. DOI: 10.1002/ece3.5650 PMID: 31788187

Van der Putten, W. H. (2010). Impacts of soil microbial communities on exotic plant invasions. *Trends in Ecology & Evolution*, *25*(9), 512–519. DOI: 10.1016/j.tree.2010.06.006 PMID: 20638747

Van Driesche R, Center T (2013) Biological control of invasive plants in protected areas. Plant invasions in protected areas: patterns, problems and challenges: 561-597

Vitti, S., Pellegrini, E., Casolo, V., Trotta, G., & Boscutti, F. (2020). Contrasting responses of native and alien plant species to soil properties shed new light on the invasion of dune systems. *Journal of Plant Ecology*, *13*(6), 667–675. DOI: 10.1093/jpe/rtaa052

Walter, I., & Calvo, R. (2009). Biomass production and development of native vegetation following biowaste amendment of a degraded, semi-arid soil. *Arid Land Research and Management*, *23*(4), 297–310. DOI: 10.1080/15324980903231827

Wan, J.-Z., Wang, C.-J., & Yu, F.-H. (2017). Wind effects on habitat distributions of wind-dispersed invasive plants across different biomes on a global scale: Assessment using six species. *Ecological Informatics*, *42*, 38–45. DOI: 10.1016/j.ecoinf.2017.09.002

Waseem, M., Yao, G.-Q., Hasan, M. M., Li, Y.-R., Qi, S.-H., Cai, L., Yang, D., & Fang, X.-W. (2024). Divergent hydraulic and gas-exchange strategies in two closely related Salix species. *Journal of Plant Ecology*, *17*(6), rtae075. DOI: 10.1093/jpe/rtae075

Westbrook, A. S., & DiTommaso, A. (2023). Hybridization in agricultural weeds: A review from ecological, evolutionary, and management perspectives. *American Journal of Botany, 110*(12), e16258. DOI: 10.1002/ajb2.16258 PMID: 38031455

Xu, H.. (2022). A global meta-analysis of the impacts of exotic plant invasions on soil: Increased nutrient content and microbial activity but decreased plant diversity. *The Science of the Total Environment.*

Kaiyang, X. U., Wanhui, Y. E., Guomin, L. I., & Jing, L. (2005). Phenotypic plasticity in response to light intensity in the invasive species {\sl Alternanthera philoxeroides}. Wuhan zhi wu xue yan jiu= Wuhan Botanical Research, 23(6), 560-563.

Yang, X.-Q., Kushwaha, S., Saran, S., Xu, J., & Roy, P. (2013). Maxent modeling for predicting the potential distribution of medicinal plant, Justicia adhatoda L. in Lesser Himalayan foothills. *Ecological Engineering, 51*, 83–87. DOI: 10.1016/j.ecoleng.2012.12.004

Yong, J., Wen-Jie, W., Yan-Hong, L., & Shong-Lian, B. (2012). Influence of light quality and quantity on germination and seedling status of invasive species Eupatorium adenophorum. *Bulletin of Botanical Research, 32*(4), 415.

Zhang, Y., Wang, X., Hu, R., & Pan, Y. (2017). Stemflow volume per unit rainfall as a good variable to determine the relationship between stemflow amount and morphological metrics of shrubs. *Journal of Arid Environments, 141*, 1–6. DOI: 10.1016/j.jaridenv.2017.02.002

Chapter 9
Mechanism of Invasion: How Non-Native Species Establish and Spread

Anum Javaid
https://orcid.org/0009-0001-0294-924X
University of Agriculture, Faisalabad, Pakistan

Ansa Asghar
University of Agriculture, Faisalabad, Pakistan

Muhammad Sufyan
College of Biological Sciences, China Agricultural University, Beijing, China

Usman Tufail
https://orcid.org/0009-0000-2798-3113
University of Agriculture, Faisalabad, Pakistan

Ahmed Raza
https://orcid.org/0009-0002-5916-1899
University of Agriculture, Faisalabad, Pakistan

Zahida Parveen
University of Agriculture, Faisalabad, Pakistan

ABSTRACT

Invasive species, known as non-native species, are a significant global challenge due to their influences on ecosystems, economies, and human health. The mechanisms enabling these species to establish, spread, and disrupt ecological balance, emphasizing the importance of understanding their pathways, stages of invasion, impacts and provides a foundational understanding of invasive species. A detailed examination of the mechanisms facilitating invasion and give invasive species a viable edge over native flora. It also highlights the significant impacts like invasive plants modifying natural habitats. Recent studies highlight the importance in combating specific invasions and concludes that proper management and control measures, advocating for a proactive approach that combines prevention, early detection, and rapid response. Integrated strategies, such as biological control, habitat restoration, and public education, are emphasized as vital tools for mitigating the impacts of

DOI: 10.4018/979-8-3693-8724-5.ch009

invasive species and protect global biodiversity in an increasingly interconnected and changing world.

1. INTRODUCTION

As opposed to native species, alien species are those whose presence in a particular location may be ascribed to deliberate or inadvertent human actions that enabled them to transcend biogeographical boundaries (Álvarez-Tostado et al., 2024). The basis for the categorization of invasive species is the rapid spread of a subset of alien species across large distances from introduction sites. Aliens that procreate often to create self-replacing populations are said to establish themselves. A second definition supported by the (IUCN), the World Trade Organization, and the Convention on Biological Diversity states that an alien species is only thought "invasive" if it has a detrimental effect on the environment, the economy, or human health (Wang, 2025).

North American grasslands have been altered by invading exotic species and other human changes to past processes and landscapes. By altering and decreasing the resources available to native species in affected habitats, exotic-invasive species may have an effect on biodiversity (Deeksha et al., 2024). Furthermore, invading species have the ability to exploit existing environmental changes and modify historical processes, endangering species that rely on them. In many grasslands, the mix of plant species has altered due to dominant alien grass species in particular (Świerszcz et al., 2024). Alterations in the nitrogen cycle, favorable growth circumstances, and modified grassland disturbances required to sustain plant variety are the means by which these changes are accomplished. Global biodiversity is being threatened by invasive species (Roy et al., 2024). In addition to causing indirect environmental consequences such altered litter characteristics, water and nutrient cycling, and climate regulatory services, invasive species may cause the extinction of whole species (Zhao & Riaz, 2024). Systems are affected by invasive grasses through changes in community composition and structure, competitive relationships between species, disturbance regimes (e.g., fire frequency and intensity), and ecosystem processes. These invasive species frequently occur in nature preserves, where they change fire regimes, decrease native plant biomass, alter the microhabitat structure, and negatively impact native plant communities (Dezotti et al., 2024). Due to their ability to adapt to a variety of environmental circumstances, invasive plants are also predicted to benefit from climate change.

The associations with blooming forb species, may have indirect interactions with pollination insects like bees. For instance, decreased native plant species diversity has been linked to abundances of the invasive, exotic, *Poa pratensis* L. and smooth brome (Holden et al., 2024). It is unclear what processes provide Kentucky bluegrass and smooth brome their competitive edge. They are thought to have replaced native plant species cover, nevertheless, due to their favorable growth in the early and late growing seasons, strong production of tillers (new grass shoots), and capacity to take advantage of moderate-to-high nitrogen levels (Snorek et al., 2024). Because grassland pollinators, like bees, depend on native floral resources throughout their life cycles and have varying pollen diet needs, native plant declines are relevant to them. Even though invasive grasses have an influence on the Neotropics, little is known about their present and potential future distribution patterns (Hardy et al., 2024). Given that the majority of alien species are native to tropical and subtropical regions of Africa and Asia, they are likely to have a high tolerance for high temperatures, which might lead to an extension of their range under the projected global warming scenarios (Hardy et al., 2024). The examined invasive species' wide present range in the Neotropics indicates that they may effectively colonize a variety of soil types in forested savannas, shrublands, and dry forests (Moreira et al., 2024).

The invasive success of alien grasses may be attributed to their great adaptability, quick development, and high rates of reproduction (Khattak et al., 2024). However, areas with a weather more like to the species' origin have better suitability. The majority of invading grasses, for instance, demonstrated poor climatic suitability for the Amazon Rainforest. Moreover, unlike the fire-prone grassland vegetation of the Cerrado, the closed canopy of the Amazon Rainforest may serve as a barrier for invading species since shade makes the ecosystem unsuitable for grasses (Harris, 2018). However, particularly in the Amazon Arc of Deforestation zone, pastures with exotic grasses have been planted. The increasing spread of pastures into the Amazon Forest via deforestation zones, where major fires are common, is problematic because it increases the likelihood that invasive species would permeate the remaining native flora in this area (Cosenza & Carvalho, 2024). (Among invasive species, *M. maximus* L. and *U. brizantha* L. are particularly notable).

Natural highland grasslands dominated by Merxmuellera (C3), (a genus in the Poaceae family) and *T. triandra* (Forssk.) Hack (C4) have been shown to be invaded by Nassella species (C3) (Mapaura, 2022). Since C4 plants are already functioning at a rate close to maximum, their marginal reactions to the anticipated rise in temperature and atmospheric CO2 will be negated by the frequent droughts, but C3 plants are anticipated to grow more vigorous. Factors including herbivory pressure, management, and intra- and inter-species competition influence each species' response. According to (Magliozzi et al., 2024), this enabled them to transcend biogeographical borders. Certain alien species launch themselves (i.e., reproduce often to build self-replacing populations); a subset of these spread rapidly across considerable distances from the sites of introduction, which is the basis for the idea of invasive species (Daniels, 2024). Understanding how exotic-invasive species affect grassland biodiversity is crucial since many animal species rely on the richness and structure of plant communities to mediate their resources.

The objective of this chapter is to explore how non-native species establish, spread, and have an impact ecosystems, economies, and human health. It aims to define invasive species, analyze natural and human-mediated pathways of introduction, and explain the stages of invasion, from introduction to establishment and spread. The chapter also examines the traits and mechanisms that facilitate invasions, highlighting their competitive advantages over native species. By analyzing ecological, economic, and public health impacts, supported by real-world case studies, the chapter seeks to provide actionable insights into effective management and control strategies to mitigate the threats posed by invasive species and promote ecosystem resilience.

1.1. Impact of non-native species

Sage et al. (2024) stated that invasive grasses might spread so far and become so successful that they could change both regional and global elements of ecosystem function. In several habitats in the Americas, Australia, and the Pacific islands, grasses from Africa and Eurasia have proliferated, endangering local biodiversity and changing different facets of the fire regime. These grasses are now acknowledged as some of the most significant species that affect ecosystems worldwide. Notable examples are cogongrass (*Imperata cylindrical* L.) in the US, cheatgrass (*Bromus tectorum* L.), gamba grass in Australia, and buffel grass in Mexico, the US Southwest, and Australia. Among the effects these species may have include habitat degradation, decreased carbon storage, and loss of native biodiversity (Kerns et al., 2020).

1.2 Biodiversity losses

The ecological processes and population dynamics of creatures occupying invaded environments may be significantly impacted by buffel grass invasion. Usually, invasion results in buffel grass plants capturing significant amounts of light, water, and nutrients (Karban et al., 2024). Native vegetation is competitively displaced as a consequence, and overall plant diversity is decreased. According to (Wright et al., 2021), allelopathic substances in buffel grass vegetation decrease the viability of *Acacia victoriae* seeds and the germination rates of *Acacia tetragonophylla* (F. Muell.) F. Muell., often known as dead finish. Growing *Bromus tectorum* L. (cheatgrass) among native, perennial grasses from western North America has an impact on community composition and production (Molvar et al., 2024). Because it may totally replace native vegetation, *Bromus tectorum*, which is widespread in both Canada and the United States, is seen as a serious threat, especially in northwest North America. Since dry *B. tectorum* L. is very combustible, it is a pest of rangelands and agriculture that raises the danger of wildfires (Molvar et al., 2024). Over 20 million hectares of current plant ecosystems in the western United States have been replaced by cheatgrass since its introduction. Human health and biodiversity are closely related, both positively and negatively (Molvar et al., 2024). Applications in vector-borne control and ethno-medical applications are two benefits of IAPS. For example, *Lantana camara* L. extract is used as a mosquito repellent. Shugart and White (2024) demonstrated that the durable root arrangement of young rainbow eucalyptus trees in Costa Rica limits the lateral expansion of the root system. Regenerating from belowground rhizomes after fire disturbance, the plant is fire resistant and burns easily. If fires are more frequent or more severe and result in higher native species mortality, invader-driven changes in fire regimes may have a detrimental effect on native species via decreased post-fire recovery. While fire increases the germination of certain species, intense fires can decrease seedling emergence and prevent resprouting (Tomat-Kelly & Flory, 2023).

1.3 Human health

While *Sorghum bicolor* L., *Arundo donax* L., *Brachiaria ramose* (L.) Stapf., and *Chrysopogon zizanioides* (L.) Roberty, are widely used to treat microbial infections in cattle (Umair et al., 2024), *Cynodon dactylon* L., *Eleusine indica* L., *Bromus japonicus* (Thunb), *Phragmites australis* (Cav.), and *Desmostachya bipinnata* L., *Eragrostis minor* are said to treat a variety of stomach issues (Umair et al., 2024). Yadav et al. (2024) reported that Animal inflammation and fungal illnesses have been effectively treated by, *Sacchrum spontaneum* L., *Imperata cylindrica* L., and *Cynodon dactylon* L.. Additionally, *Heteropogon contortus* L. possesses anti-microbial, anti-cancer, and anti-inflammatory qualities, and it also boosts livestock's producing milk. *Chloris barbata* (Sw.) act as a disinfectant (Majeed et al., 2020).

1.4 Impacts on ecosystem

Invasive Alien Plant Species causes more disruptions to ecosystem functioning on the Islands than on the mainland. These species has been shown to impact ecosystem functioning through three main mechanisms: (a) a decrease in native plant and animal diversity, (b) notable alterations in soil physio-chemical properties (primarily due to allelopathy), and (c) an improvement in ecosystems' ability to adapt to changing fire regimes (Haq et al., 2024). Reduced native plant biodiversity is one well-documented effect of IAPS, this might be detrimental to global climate change, ecosystem services, and environmental functioning. It is also recognized that alien invaders have a negative impact on nature (Rai & Singh, 2020).The IAPS may enter aquatic systems and impede water flow by exhibiting distinct physiological characteristics (such as large biomass, deep roots, and high evapo-transpiration), making the water unfit for drinking and irrigation (Ababa, 2024).

Economic losses

Invasive species impose significant economic burdens globally, affecting multiple sectors, including agriculture, forestry, fisheries, infrastructure, and public health (Zhyvko, 2024). The economic losses are categorized into direct costs, such as damage to crops and infrastructure, and indirect costs, including loss of ecosystem services and expenses associated with control measure (Aller et al., 2024). Because of the shade tolerance and lanky sprawling growth, *U. maxima* (Jacq.) has become a problematic specie for citrus farmers in Texas and Florida. Dry tillers in shrubs can form ladders that transmit fire from the ground to the canopies of shrubs and trees during drought situations (Hardy et al., 2024). Ecosystem changes in Hawaii's dry tropical forests are attributed to variations in fire behavior and return intervals, which eliminate native forest species and permit the invasion of trees and shrubs (Hardy et al., 2024). Many alien species that were implemented for the benefit of humans have a history of wreaking havoc on the environment and the economy. Because it alters the water quality, *Acacia mangium* (Willd.), an IAPS in the northern Brazilian Amazon, is known to have detrimental consequences on the ecology, the economy, and native people (Nahla Zakariya).

Using the participatory rural appraisal (PRA) approach, it was determined that *Opuntia stricta* (Haw.), a high-risk IAPS in Africa, causes an annual economic loss of \$500 to \$1,000 per family (Hallstein & Iseman, 2021). Eschen et al. (2021) reported that foreign invaders harm crops in African nations' agricultural sectors, incurring an annual economic loss of \$1 billion USD. In Southeast Asia, disease-carrying foreign invaders caused a financial loss of US \$1.85 billion to the human health sector alone (Quarles, 2007). The alien invasive species caused a combined economic loss of US\$33.5 billion to the agriculture and health sectors. As a result, the agricultural sector suffered a greater economic loss from invaders (about 90% of total losses) than the human health sector.

Figure 1. **Impacts of invasive Species on ecology, economy, culture and human health**

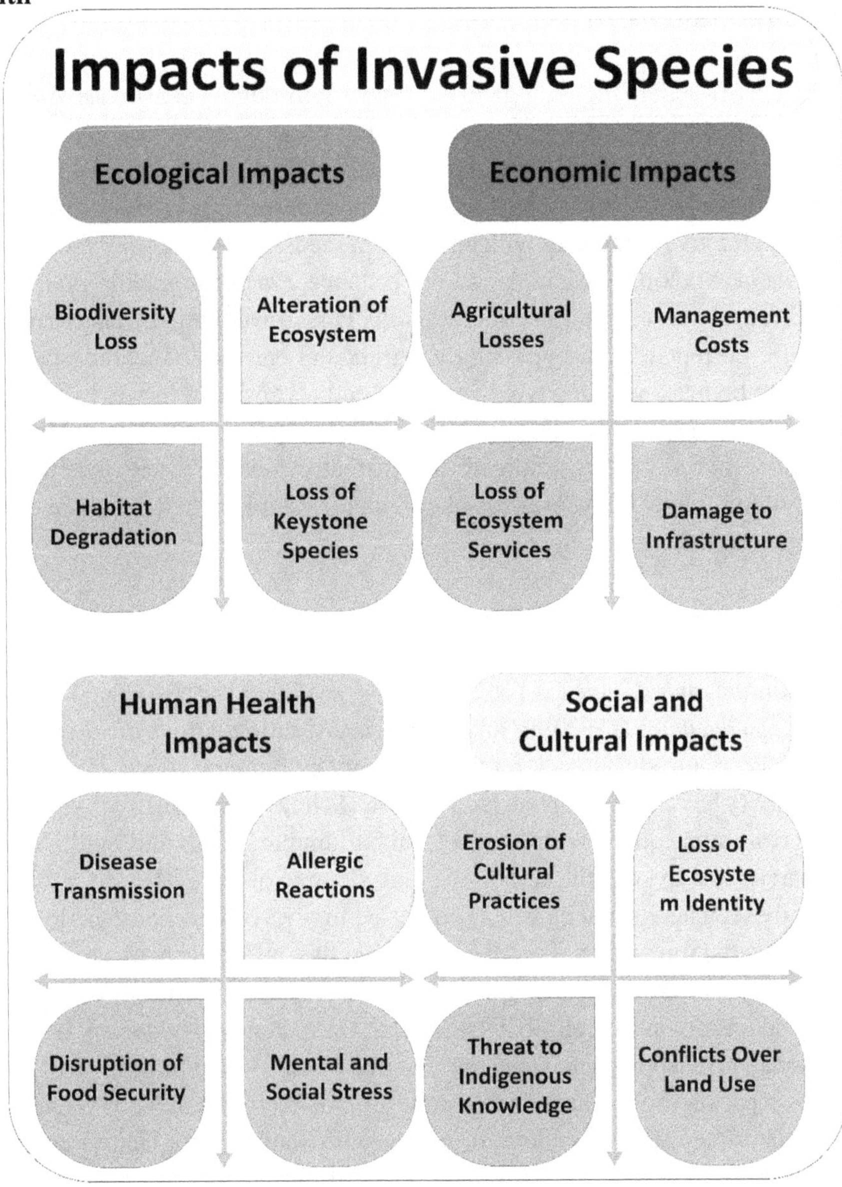

2. PATHWAYS OF INTRODUCTION

2.1 Natural pathways

Natural pathways are the ecological mechanisms through which invasive species disperse and establish themselves in new habitats without direct human involvement. These pathways utilize natural forces such as wind, water, and animal movements to transport organisms across varying distances (Chanda et al., 2024). Wind dispersal is particularly effective for species with lightweight seeds, spores, or other propagules, allowing them to colonize distant areas. For instance, *Parthenium hysterophorus* L. seeds and fungal spores like *Puccinia graminis* Pers. travel long distances through atmospheric currents, often bypassing geographical barriers (Chakrabarty et al., 2023). Water bodies, including rivers, streams, and oceans, also serve as significant conduits for the spread of aquatic and semi-aquatic species. The Water Hyacinth (*Eichhornia crassipes* (Mart.)) spreads downstream, creating dense mats that disrupt ecosystems, while buoyant seeds like coconuts establish themselves on distant shores (Staeck, 2022).

2.2 Human mediated pathways

Dichanthium annulatum (Forssk.) Stapf is accidentally introduced invasive species. *B. ischaemum* (L.) Keng and Caucasian bluestem (*Bothriochloa bladhii* (Retz.) has been introduced as a forage producers by humans (Wied et al., 2020). Human activity has caused *Cortaderia selloana* (Schult. & Schult.f.) to spread. It is delivered from stone quarries with gravel and building materials and is utilized as a natural barrier and to keep the soil on the road slope (Coleby, 2021). As a result, it is frequently seen near stone quarries and beside highways, new construction sites, or gravel-covered tiny walkways. When people gather inflorescences and use them as decorations in restaurants, shops, or residences, they may potentially accidentally spread the species. (Cavanaugh, 2019) stated that *Microstegium vimineum* (Trin.) was unintentionally transported to Tennessee, a state in the United States, when it was utilized as a packaging substance in ceramic shipments from China. It is currently present in 26 states after spreading throughout the Southeastern United States. In the 1820s, *Arundo donax* L. was imported to California from the Mediterranean to be used as roof covering and to prevent erosion in waterways near Los Angeles (Aghajani et al., 2021). It has grown naturalized across warm coastal freshwaters of North America by dispersion and later plantings as decorative plants and for use as reeds in woodwind instruments, and its range is still expanding. Another typical introduction channel is ornamental cultivation, and invasion is widespread in the wildland–urban border, where natural vegetation and built areas coexist (Sandenbergh et al.).

2.3 By ships and import

It seems that hurricane grass and *Dichanthium aristatum* (Poir.) were accidentally introduced to the Western Hemisphere by Caribbean Islands. The first known instance of *B. ischaemum* (L.) in the United States was a shipment to the California Agriculture Experiment Station in Berkeley from the U.S. embassy in Amoy (Wied et al., 2020). Similar species were brought to Angleton in 1914 by the U.S. Bureau of Plant Industry for the Texas Agriculture Experiment Station (Waller, 2024).

During one of Christopher Columbus's expeditions to the Caribbean Islands, *Cynodon dactylon* (L.) made its first appearance in the Western Hemisphere. In 174 A botanist Robert Miller carried material to Savannah, Georgia, which is thought to have been the first introduction to North America. Within 50 years, it was also documented in South Carolina (Langlois, 2024). Early in the 20th century, cogon grass was shipped in boxes to the Southeastern United States (Overholt et al., 2016). According to Bhandari (2020), distinct introduction episodes of *B. tectorum* through several entrance ports in eastern and western Canada are indicated by numerous genotypes of this species.

2.4 Import of plants and seeds

Originating in the wetter tropical and subtropical parts of Africa, guinea grass has been extensively spread into other countries as a fodder species and by accident since the seventeenth century (Shabbir et al., 2024). Guinea grass has been widely grown and used in tropical and subtropical places worldwide since it was first imported, either for range development or accidentally in chaff. Because *Cortaderia jubata* (Lemoine) seeds adhere to the hairy surface of kiwi fruits (*Actinidia deliciosa* (A.Chev.), they may be transmitted as food contaminants (Montagnani et al., 2022). Unintentionally, imported plants or seeds may potentially harbor non-native species. Invasive weed seeds might cling to plant material or mix with imported seed stocks. Similarly, when a plant is transplanted to a new area, the soil that is linked to its roots may include bacteria, fungi, or pests (Chakrabarty et al., 2023).

Table 1. Classification of Propagule Release Based on Human Involvement and Intent.

Deliberate Release of Propagules	Presence or Absence of Human Motivation as Underlying Driver in the Transport and Release of Propagules	Accidental Release of Propagules	Reference
Introduction and subsequent propagation	Target courses were purposefully relocated. Introducing new species to improve the environment. Ornamentals deposited as garden garbage	Transport losses in crop seeds by spilling from vehicles or trains	(Sych, 2024)
Release of propagules with solid liquid	Unintentionally relocated nontarget species. Manure made from wool factory waste that contains propagules. Propagules deposited on automobiles, animals, or human footwear. Release of propagules associated with products via sanitation methods (e.g., trains)	Transport losses of weed seeds associated with crop seeds	(Sych, 2024)
mostly clustered invasion foci caused by irregular propagule discharge	The spatial patterns that result	mostly linear invasion foci by constant propagule discharge	(Micael & Gíslason, 2024)

2.5. Land Transport

One of the most important routes for the introduction and dispersal of non-native species over different areas is land movement. Inadvertently, invasive species spread to new areas via the movement of automobiles, trucks, and train (Montagnani et al., 2022). It is possible for seeds, spores, and even tiny creatures to stick themselves to cars, especially on the tires, underbody, and cargo. Seeds lodged in tire tracks or entangled in mud on agricultural or construction machinery may travel great distances and deposit themselves in unfamiliar habitats (Frick-Ruppert, 2014). Trains serve as conduits for the transportation of products and animals. Because they are often the focus of ongoing human activity and upkeep, roadside areas and highways are ideal places for invasive species to establish themselves. Hardy non-native plants may flourish in these locations due to ongoing disturbance, exposure to runoff, pollutants from vehicles, and waste products created by humans (Hazen & Sutton, 2025). Roads also fragment ecosystems, making native species more vulnerable and opening up niches for invaders. Trucks and lorries transporting raw materials such as timber, soil, and agricultural products also can bring with them invasive pests, pathogens, or seeds (Aller et al., 2024). For instance, invasive plant seeds might be introduced into previously unaffected regions by imported dirt or gravel used in construction. The danger of non-native species spreading via land transportation has increased due to the acceleration of commerce and infrastructure development on a worldwide scale (Adeniran-Obey & Osagie, 2024).

Table 2. A measure of the success of accidentally vs purposely introduced alien plant species in the Montpellier region in southern France is the naturalization rate of foreign plants.

Pathway of Introduction	Total Number of Alien Species	Subset of Casual Species	% Casual Species	Subset of Naturalized Species	% Naturalized Species	Reference
Total Species Number	800	693	86.6%	107	13.4%	(Fried et al., 2014)
a) International Introductions	148	87	58.8%	61	41.2%	(Shang et al., 2024)
b) Accidental Introductions	621	575	92.6%	46	7.4%	(Kowarik & Lippe, 2007)
– Wool	526	507	96.4%	19	3.6%	(Kowarik & Lippe, 2007)
– Seed and Feed Grains	40	31	77.5%	9	22.5%	(Kalsa, 2019)
– Grain Crops	18	18	100.0%	0	0%	(Olugbenga & Chongs, 2024)
– Ballast Materials	19	10	52.6%	9	47.4%	(Tobbala et al., 2024)
– Transportation Vehicles	18	9	50.0%	9	50.0%	(Balcioglu et al., 2024)
c) Resulting from Hybridization	31	31	100.0%	0	0%	(Kudlu et al., 2025)

2.6 Horticulture and agriculture

Agriculture and horticulture are two important sectors that unintentionally bring and disperse non-native species. Many of the plant species that are purposefully imported by these industries for agricultural production, landscaping, or cultivation end up becoming invasive (Lokare & Pandya, 2024). Certain non-native species that are employed in horticulture, such trees and decorative plants, have the potential to escape from greenhouses, nurseries, and gardens and into neighboring areas (Tartaglia & Aronson, 2024). These plants have the ability to colonize natural regions by seed dispersal caused by wind, water, or animals. *Fallopia japonica* (Houtt.), the invasive Japanese knotweed, was once brought in as an attractive plant but has since spread over Europe and North America and become one of the most hazardous invasive species. Because of agriculture, damaged landscapes are more vulnerable to invasion. Non-native species may sometimes cross-pollinate with native species to create new, invasive hybrids. The natural biodiversity may be further threatened by these hybrids if they thrive or spread more aggressively (Adeniran-Obey & Osagie, 2024).

Figure 2. **Key modes of invasive species spreading through biological, natural and anthropogenic activities**

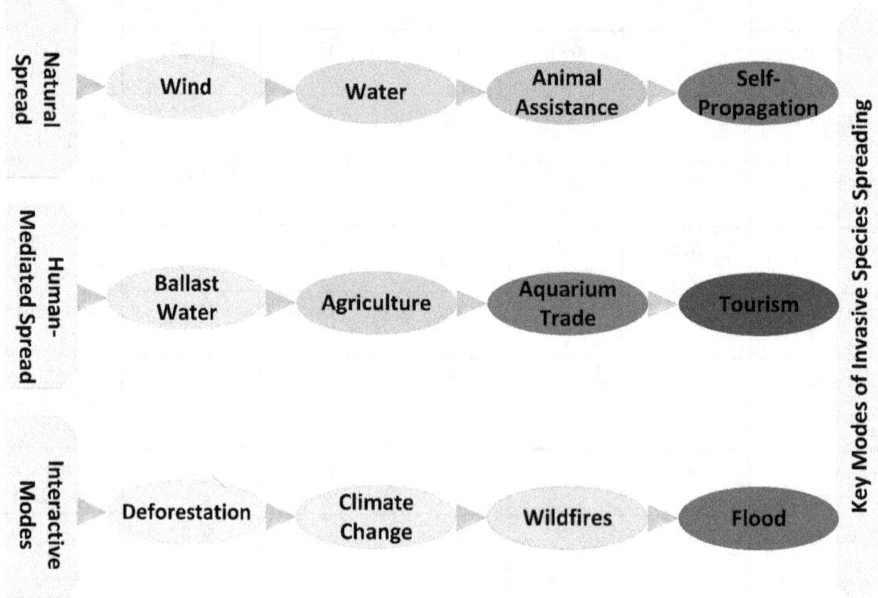

3. STAGES OF INVASION

Before becoming an annoyance in their new environments, several species were introduced for agronomic and erosion control purposes., some species are unintentional imports (Coulter et al., 2024). High productivity and drought tolerance are characteristics that encourage invasiveness and make plants desirable candidates for introduction. Intentional or unintentional human activity frequently facilitates the introduction of exotic grasses into new areas (Hardy et al., 2024). Although grasses can be introduced for agricultural, landscaping, or soil stabilization purposes, they may propagate beyond their introduced areas due to natural dispersal mechanisms along man-made corridors like roads and railroads, or transportation (ships, airplanes, and vehicles). These species can quickly spread to new areas once they're established, particularly if they have a wide range of ecological tolerances (Gillmann et al., 2024). By changing the composition of the soil, reducing biodiversity, and flourishing in environments like climate change, invasive grasses can outcompete native species and accelerate their spread (Khattak et al., 2024).

3.2. Survival at initial stage

The capacity of grasses to adapt to local conditions and outcompete native species is frequently what determines their survival in a new habitat during the early phases of invasion. Typical characteristics of invasive grasses include quick germination, adaptability to a broad range of temperatures and moisture content, and the capacity to change the dynamics of soil nutrients (Khattak et al., 2024). They also benefit from the new environment's lack of diseases or natural predators. Additionally, by altering microbial populations to their benefit, they can create beneficial plant-soil feedbacks that aid in their establishment and survival (Timmis & Ramos, 2021).

3.3. Establishment

Plant fitness is impacted by certain environmental conditions found in urban areas. By changing the air quality *(Johnson et al., 2024)*, impermeable surfaces, artificial light, nutrient subsidies, water quality, and heavy metal concentrations, anthropogenic activity affects plant establishment and growth (Miller et al., 2024). The competition between *Microstegium vimineum* (Trin.) litter with its associated pathogens and the natural grass *Elymus virginicus* L. in a greenhouse experiment was assessed (Benitez et al., 2022). *M. vimineum* (Trin.) litter inhibited both species' establishment and biomass while promoting illness in *E. virginicus* L. Compared to *M. vimineum*, *E. virginicus* was more negatively impacted by litter, which increased *M. vimineum's* (Trin.) relative biomass (Cunard, 2016). Both species' biomass was decreased by live plant competition, while *E. virginicus* L. illness incidence was raised by live *M. vimineum* (Trin.). Both species were inhibited overall by invading grass litter, which eventually favored the invasive species in competition and raised the prevalence of illness in the native species (Ramachandran et al., 2024).

(Akbar et al., 2024) concluded that invasive litter mediated competition, pathogens that may infect living plants and reside in litter are seldom considered. Invasive plants have the potential to develop new pathogens over time, the probability of apparent rivalry with native species that is, indirect competition via a common adversary that may be mediated by its litter increases with the length of time an invading species has been established. By more effectively suppressing the establishment of *E. virginicus* L. than *M. vimineum* (Trin.), and without the aid of disease, *M. vimineum* litter enhanced the relative biomass of M. vimineum (Franks, 2024). As a result, invading plant litter may eventually help non-native species in the field by causing interference competition and disease in native plants (Ullah et al., 2024).

Wied et al. (2020) stated that the aboveground biomass production of sandy and loamy-textured soils is comparable, and *P. ciliare* L. becomes a dominant species and spreads more easily on loams and sandy clays. According to (Ullah et al., 2024) establishment happens more easily on more alkaline soils than on acidic ones. Hard freezes can harm established plants, while wet winters can ruin seed sown during the growing season. If seed vigor is great, seedlings can swiftly reestablish themselves once older plants are removed from a place. Compared to natural species, *Eragrostis lehmanniana* Nees recovers faster and is thought to be a fire-tolerant plant. Higher seed production is possible within a year following a fire occurrence because to the speedier recovery over native grasses, and the removal of litter whether by fire or mechanical means improves seedling emergence by boosting temperature variations at ground level and red-light penetration. According to Lauman (2024) *E. lehmanniana* Nees stands had a greater amount of litter buildup, which might change the intensity of fire and promote the species' continued establishment.

3.4. Reproduction and growth

The ability of non-native species to reproduce and flourish is a critical factor in determining their survival in new settings. Both sexual and asexual reproduction are common very successful reproductive mechanisms shown by invasive organisms (Croft, 2024). Their capacity to reproduce twice allows them to increase their chances of establishing and proliferating. While asexual reproduction, such as vegetative propagation, enables these species to reproduce quickly even in the absence of partners or pollinators, sexual reproduction promotes genetic variety, which improves tolerance to various environmental situations. The water hyacinth (*Eichhornia crassipes* Mart.) has a fast population growth rate because it may reproduce asexually by creating offshoots and sexually by dispersing seeds (Croft, 2024). Furthermore, invasive species often have high fecundity, meaning they may produce a lot of seeds, spores, or progeny, which boosts their establishment potential (Xue et al., 2024). Because they can produce a large number of offspring, invading species may outcompete local species and swiftly take over new regions. In addition, invasive species can develop quickly, giving them the advantage over native species when it comes to resources like sunshine, water, and nutrients. They may monopolize resources and establish themselves more rapidly than slower-growing native species because to their fast growth (Gutgesell et al., 2024).

Heteropogon contortus L. is a C4 perennial bunchgrass belonging to the Andropogoneae tribe. Its erect culms usually reach a height of one meter. Culms have leaves along their length, and they terminate with spiklate raceme inflorescences. Each upper floret produces a long, twisted awn. The grass gets its popular name from the way these awns all wind together along the raceme. The stiff awns cling to clothing, fur, and vehicles, which carry the seeds and aid in their spread, but the florets are readily detached. Although sexual reproduction is known to occur, apomictic reproduction is the predominant mode of reproduction (Hörandl, 2024). Although vegetative reproduction by stolon's occurs during dispersion, apomictic reproduction is prevalent within both genera and among their hybrids. With a deep root structure, the ability to quickly recover after losing leaves, and a receptiveness to soil nitrogen addition, the species is fire-adapted (Zachman, 2024). Unlike most natural grass species, *P. ciliare* L. regrows from nodes along lower stems rather than the crown following defoliation. *Pennisetum ciliare* L. has been confirmed to significantly contribute to aboveground carbon losses in the Sonoran Desert and to change soil carbon and nitrogen in a variety of climatic areas throughout Mexico (Gamiz, 2024). (Lauman, 2024) discovered that *P. ciliare* L. responded better to natural plant cover in test plots when nitrogen was replaced with fertilizer supplementation. In *Urochloa maxima* (Jacq.), sexual or apomictical reproduction is possible. Diploid individuals reproduce sexually, but polyploid individuals undergo apomixis.

Although diploid individuals (2n = 2x = 20) produce seed sexually, polyploid individuals reproduce asexually by apomixis (Hörandl et al., 2024). Its native African range and the southwestern United States are dominated by apomictic individuals. *E. lehmanniana* Nees has a high degree of morphological flexibility that enables it to persist in a variety of ecological environments, even if nonsexual reproduction results in less genetic variability. A substantial yield in early summer and a smaller yield in late summer to early fall are the two seed harvests that established stands of *E. lehmanniana* Nees annually (Buerdsell, 2021). Higher clay fractions in soil textures seem to reduce seedling emergence, and germination is impeded when seeds are buried less than 5 m below the soil profile. Since then, several seeded cultivars and hybrids that are basically sterile and need to be propagated via sprigging have been created (Magni et al., 2023). Invasive cogongrass is a pyrogenic plant that mostly spreads vegetatively through rhizomes (Miller, 2007). Small (0.8–1.0 mm) and many (85 000 seeds per m2 for *Sporobolus pyramidalis*, or gigantic rat's tail grass) plants are produced by the *Sporobolus* genus. These seeds germinate all year long, grow robust root systems fast, and may last for up to ten years. In a wide range of conditions, guinea grass, a tufted perennial C4 bunchgrass, reproduces both vegetatively via rhizomes and stolon and by seed. High germination rates of 88% and germination across a wide variety of environmental circumstances were observed by (Dagar & Gupta, 2024) in her study of the invasive Sporobolus indicus's germination success in America.

3.5. Spread

Depending on the circumstances and their modes of dispersion, non-native species may spread quickly once they are established in a new area. Invasive species may spread locally or regionally, and they often begin in disturbed or altered ecosystems like urban areas, highway edges, or agricultural fields (Carlon & Dominoni, 2024). The species could have less rivals in these areas, which would enable them to spread swiftly. Invasive species have the potential to expand from these early establishing sites into more natural habitats, where their effects may be far more severe (Gallardo et al., 2024). Numerous invasive plants propagate by means of seeds or other vegetative elements that are dispersed by wind, water, animals, or human activity. Due to people's inadvertent long-distance transportation of seeds, spores, or microscopic creatures, invasive species often spread due to human activity. Non-native animals may spread by means of cars, ships, aircraft, and even clothes (Elsensohn et al., 2024). Invasive species thrive in areas that have been altered by humans, such as urban landscapes, agricultural fields, and traffic corridors. Road networks may serve as conduits for invasive species to spread over areas, while airports and ports can help bring in species from distant continents. Because they foster an environment that is conducive to the development and reproduction of invasive species, environmental disturbances whether caused by man-made activities like clearing land or natural occurrences like storms also aid in the spread of these species (Sharma & Naik, 2024).

3.6. Dispersal mechanism

Different dispersion mechanisms have been developed by non-native species to enhance their capacity to colonize new places. Many invasive plants have seeds that can be dispersed by wind because they are designed to be transported great distances by air currents (Shiba et al., 2024). For instance, dandelions (*Taraxacum officinale*) generate seeds with delicate, hair-like characteristics that enable the wind to carry them great distances. Another useful strategy is water dispersion, particularly for riparian or aquatic species. Because seeds and other vegetative portions of some species, like water hyacinth, float on water, they may spread out across rivers and lakes and colonize new areas that are distant from their original source (Abba & Sankarannair, 2024). Additionally, animals are important in the spread of invasive species. Certain non-native plants provide fruits or seeds that animals ingest, spreading them across large distances via their excrement. Some species may contain seeds that stick to an animal's hair or feathers and enable the animal to travel to other areas (Mudie, 2024). For instance, burdock (*Arctium minus*) generates spiky seeds that stick to animals' coats and help them go to new places. One of the most successful methods used by invasive species to propagate is human-mediated dispersion. Unintentionally, seeds, spores, or tiny creatures are carried by humans via their clothes, cars, or equipment, which helps these species spread to new places (Bunyard, 2024).

Cheatgrass's large propagule numbers (one plant may yield 25–5,000 seeds) and high rates of dissemination (e.g., by wind, human, or animal movement) are the reasons for its success, and effective soil resource acquisition (Mahood, 2021). According to, fire can increase *H. contortus* (L.) seedling production by up to 1,500 seedlings m^{-2}. *Pennisetum ciliare* (L.) seeds dispersed via wind, runoff, vehicles, and attachment to animal hair (Wied et al., 2020). The main methods of *Urochloa maxima* (Jacq.) seeds dispersed by animals, wind, and water. In a tropical dry forest in Bolivia, (SIS, 2024) showed that motor vehicles spread the seeds that grew on damaged logging sites. Many invasive grasses produced panicles with developed seeds that fall over a period of weeks. Arthropods or seed-eating birds are likely to spread seeds, and seeds will stick to damp surfaces to spread mechanically. Once established, stoloniferous growth from existing plants or local seed distribution by animals or birds are the methods by which short-range dispersal might happen (Rhodes, Plowes, et al., 2021).

4. MECHANISM FACILITATING INVASION

A thorough examination of the processes allowing non-native species to invade shows a complicated and comprehensive process. A combination of biological characteristics, ecological relationships, and outside influences enable invasive species to thrive and eventually take over new habitats.

4.1. Reproductive strategies and dispersal strategies:

High reproduction rates are a common feature of invasive species, both in terms of yielding huge numbers of offspring and reproducing regularly all year long (Griffen et al., 2024). They may swiftly establish in new environments thanks to their fast population growth, outcompeting local species for resources. For example, a lot of invasive plants spread their seeds widely, making it more likely that they will successfully colonize a variety of settings (Akbar et al., 2024). Similar to humans, certain invasive species have the ability to multiply many times a year, which causes population growth to explode (Newsome et al., 2024). Another important component in the success of invasive species is effective dissemination. While certain species benefit from human-mediated dispersion, such as via international commerce, agriculture, or transportation, many species may disseminate naturally by wind, water, or animal vectors.

4.2. Environmental tolerance and adaptability:

The ability of invasive species to flourish in a range of habitats and climatic circumstances is one of their defining characteristics (Nyssen et al., 2024). They can establish themselves in both pure and damaged habitats because to their versatility. Many invasive species have physiological characteristics that enable them to withstand high or low temperatures, droughts, salinities, or a lack of nutrients (Mishra et al., 2024). Their ability to adapt to changing environments offers them a competitive edge over native species, which may be less flexible and more specialized. Invasive species are often controlled by illnesses, parasites, or predators in their natural habitats. They could, however, be able to elude their natural enemies when they are brought to other areas, which would result in unchecked population increase. The "enemy release hypothesis," which describes this behavior, explains how invasive organisms may spread quickly in unfamiliar surroundings (Flickinger & Dukes, 2024). Because there are no herbivores or predators to control their population, they are able to proliferate uncontrolled and eventually outcompete and displace local species (Khattak et al., 2024). Certain invasive plants have the ability to emit substances into the soil known as allelopathic qualities, which prevent neighboring plants from growing or germinating. Because they are able to monopolize resources like water and nutrients and restrict local plant species, they have a huge competitive advantage (Song et al., 2025).

4.3. Hybridization

The degree of invasiveness may be increased by hybridization between invasive and native species. Hybrid Vigor is the phenomena whereby hybrid animals exhibit enhanced Vigor, adaptability, or reproductive success by combining the qualities of both parent species (Khorramnejad et al., 2024). Certain hybrid species have the potential to proliferate more quickly than their parent species because they are sometimes more resilient to environmental stresses like pollution and drought. Moreover, hybridization boosts genetic variety, which helps invading populations become more resilient to environmental change. New mutualistic connections between invasive species and other creatures inside the invaded habitat may develop (Khattak et al., 2024). To aid in their spread, invasive plants, for example, may form partnerships with regional seed distributors or pollinators (Shabbir et al., 2024).

4.4. Competitive Exclusion

For vital resources like light, nutrition, water, and space, invasive species usually outcompete native species. Invasive species have the ability to dominate resources via better resource acquisition tactics, which may result in the extinction or displacement of native species (Iqbal et al., 2024). By feeding on or competing with native species at different trophic levels, they may also change the structure of food webs, having an ecosystem wide cascade impact.

4.5. Genetic and phenotypic plasticity

Numerous invasive species have significant degrees of phenotypic and genetic plasticity, which allows them to modify their morphology, physiology, and behavior in response to changing environmental factors (Mai, 2024). Because of their adaptability, they can live in a variety of environments and deal with new stresses including changing soil conditions, water availability, and temperatures. The capacity of invading populations to adapt to local circumstances is further enhanced by genetic variety, which raises the likelihood of their survival and expansion (Lamb et al., 2024).

4.6. Other mechanism facilitating invasion

Fire timing can be altered by the presence of invasive species in a region, which increases their dominance over native plants (Suárez-Ronay et al., 2024). A positive feedback loop between the invader and the fire regime may arise when these changes are advantageous for the invader's establishment. Due to seasonal fluctuations in the quantity of available resources, this grass-fire cycle may cause changes in the fire season, which may alter the competitive interactions between invasive and native plants (Hart-Fredeluces et al., 2024). Allelopathic substances produced by *Urochloa* species aid in its invasions to new regions. The extensive usage of commercial lineages that have undergone genetic modification and artificial selection to become more durable, competitive, and aggressive further accelerates their rate of invasion (Climent et al., 2024). A highly flammable essential metabolite is produced in large amounts by *Melinis minutiflora P. Beauv.* (Damasceno & Fidelis, 2020). Compared to fires that solely burn native vegetation, both species have comparable effects on Cerrado fires, encouraging larger flames and hotter, more intense burns (da Silva Arruda et al., 2024). *Urochloa brizantha's Beauv.* recovery after fire is due to its increased invasiveness in post-fire conditions (Rosalem et al., 2024). Its rapid and effective resprouting following disturbances, the persistence of its fire-resistant seeds in the soil seed bank, the propagule pressure from nearby invaded areas and the microclimatic buffer produced by its tussock, are the reasons to raise fire temperatures and cause a “kill thy neighbor effect” (Damasceno & Fidelis, 2020).

Figure 3. **Stages of invasion into new land**

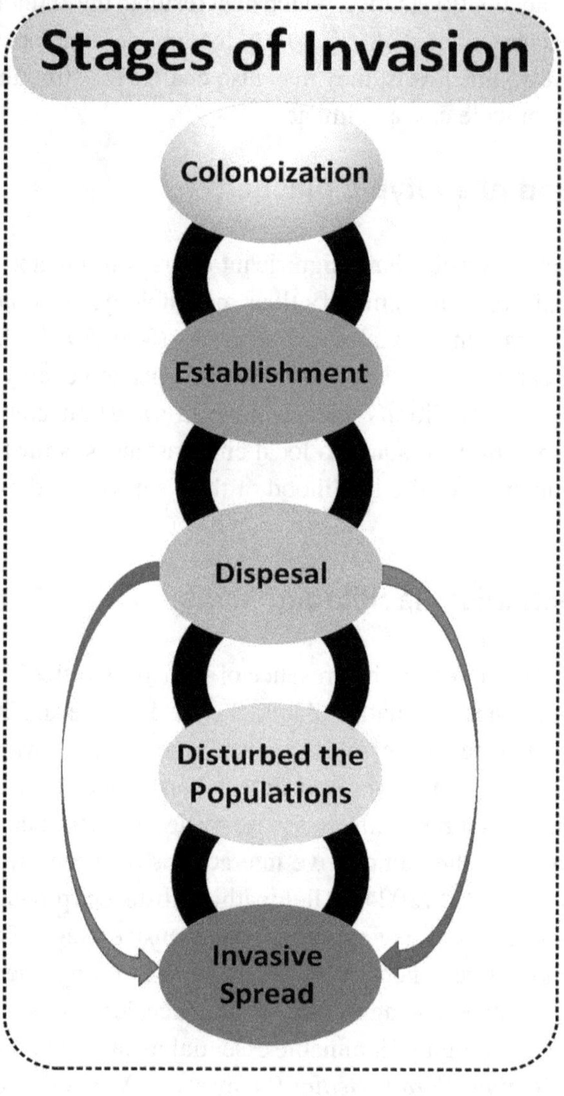

5. CASE STUDY

5.1 Successful Invasion

According to (Khattak et al., 2024), a complex combination of characteristics, such as the capacity to alter the environment for their persistence, contribute to the success of grass invasion on a worldwide scale. In South Africa, there are no native *Nassella* species (Mapaura et al., 2024). According to (Mapaura, 2022) three species: *N. neesiana* (Trin. & Rupr.), *N. tenuissima (Trin.)*, and *N. trichotoma* (Nees) have naturalized in the whole country. The Anglo-Boer War (1899–1902) is believed to have brought all three species through the ports of East London and Port Elizabeth, maybe as seed in hay from Argentina for use as horse feed (Mapaura et al., 2020). The National Environmental Management: Biodiversity Act's category 1b presently lists only *N. trichotoma* (Nees) and *N. tenuissima* (Trin. & Rupr.) as invasive species (NEM:BA, 10/ 2004): Alien and Invasive Species (AIS) Regulations of 1 October 2014 (Mapaura, 2022). According to these rules, these species have to be removed and destroyed whenever feasible and cannot be planted or exchanged in any way.

The most often utilized species for planted pastures in Brazil nowadays are *Urochloa* species, which have been introduced as cow pastures on several occasions since the early 20th century (Merloti, 2024). In order to improve ecosystem services, such as erosion control and livestock forage production, two warm-season and cool-season non-native perennial grasses, *Bothriochloa ischaemum* (L.) Keng (warm-season) and *Bromus inermis Leyss.* (cool-season), have been widely planted in the Great Plains of North America (Duell et al., 2021). According to (Singh et al., 2024) reports of *B. ischaemum* and *B. inermis* in 17 and 48 of the contiguous 48 states, respectively, these two species are linked to wide spread distribution and detrimental effects on invaded ecosystems. The invasion of neighboring grasslands, meadows, and roadsides by both of these non-native species sometimes results in dense, monotypic stands that severely affect native biodiversity at various trophic levels by outcompeting native flora (Duell et al., 2021). In contrast to native species, non-native invasive grasses may be able to withstand a larger range of environmental conditions. In the 18th century, the alien annual species *Bromus tectorum* was brought to North America from Europe, and it has since spread to much of the continent (Carpenter et al., 1999). In the early 1900s, the invasive grass known as "stiltgrass," or *Microstegium vimineum* (Trin.) A. Camus, was brought to the United States from Asia (Goldsmith et al., 2023). Since its introduction to the United States, the invasive grass *Microstegium vimineum* (Trin.) has acquired novel fungal leaf spot infections and reduces native populations, partly through litter production (Lane, 2021).

The capacity of *H. contortus* (L.) to flourish in a variety of environments makes it a successful species. It is possible for *H. contortus* to expand into regions with less productive soil (Wied et al., 2020). Established stands of *H. contortus* (L.) growing on more nutrient-rich soils can also endure longer than other herbaceous species that rapidly deplete soil nutrients due to slow absorption (Stehn, 2020). Because of this, *H. contortus* (L.) is linked to monoculture growth of closed canopy in areas where it has become invasive, which may reduce the amount of bare land and light available for other plants. It is assumed that invasive population of *U. maxima* escaped from an illegal *U. maxima* planting in the Rio Grande Valley using seeds from the Weslaco, Texas, agriculture experiment station (Barton, 2024). Now this species is quickly spreading within the San Antonio municipal boundaries along the southern San Antonio River (KG Lyons, personal communication). In southern Texas, the subtropical *U. maxima* population has spread to rangelands, croplands, and urban environments, becoming invasive (Wied et al., 2020). Many locations in the Rio Grande River corridor have a dominant herbaceous layer made up of *Urochloa maxima* (L.) and, to a lesser extent, *P. ciliare* L., which are now impossible to eradicate (Wied et al., 2020). According to a survey of seven locations along the Rio Grande, *U. maxima* dominated the ground layer, especially in locations with a high canopy cover of trees and shrubs (Flores, 2019). The invasion of *M. maximus* has caused problems for Texas (Massigoge et al., 2024). Furthermore, it faces competition with partial shade beneath shrubs from the endangered Tamaulipan kidney petal.

(Ryan-Colton et al., 2024) stated that the eradication of *P. ciliare* L. enhanced the cover of native herbaceous plants. According to (Solves, 2020) in southeastern Arizona, *E. lehmanniana* Nees produced three to four times as much green fodder per year as native grasses, Cattle prefer natural perennial grasses in the summer, and *E. lehmanniana* Nees has a low palatability. Because *E. lehmanniana* Nees produces green fodder for longer periods of time, usage rises in comparison to native grasses during other seasons. (Jenkins, 2024) noted a shift in rodent variety and postulated that *C. dactylon's* L. monoculture practice reduces small animals' access to food. Invasion by the perennial rhizomatous cogongrass (*Imperata cylindrica* (L.) can enhance fuel loading and create more powerful flames than those in native vegetation, even though pine savanna ecosystems are suited to regular, low-intensity fires (Butler et al., 2021). Originating in Asia, cogongrass is invasive in the Southeast United States, including over 500,000 hectares in Florida, particularly in areas that have been disturbed. This contradiction is best illustrated by guinea grass (*Megathyrsus maximus* (Jacq.), which is both a highly successful invasive species worldwide and a valuable pasture grass known for high livestock yields through increased forage production, palatability, and herbivory (Tanner, 2023). According to (Rhodes et al., 2022), guinea grass's great potential as an invasive plant is correlated with its favorable features for cattle, such as its high biomass production and resistance to herbivory and changing environmental circumstances. In some areas of southern Texas and northeast Mexico, the short form has grown to be a serious invasive danger. Although it is regarded as an invasive species, it has been a major forage grass in Queensland, Australia (Hardy et al., 2024).

One of the most harmful invasive plant species in the southwestern United States is buffelgrass (*Cenchrus ciliaris*). It has thrived in dry environments since being brought from east Africa to the arid parts of North America in the 1930s Hardy et al. (2024). Between 1949 and 1985, seed companies sold 7 million kg of t-4464 buffelgrass seed, which helped the plant spread across 4 million hectares in Texas and more than 6 million hectares in Mexico (Morrison et al., 2023). Buffelgrass became one of the most prevalent pastoral grasses on the continent after being imported into Australia for cattle grazing starting in the 1920s. Unfortunately, it also escaped from animal pastures and has now spread over the continent, taking up large amounts of natural space in grasslands, eucalyptus woods, and other woodlands (French et al., 2024). As a result, many rivers red gum trees (*Eucalyptus camaldulensis* Dehnh.), a keystone species of arid Australian environments, were lost as a result of buffelgrass. It is increasing the frequency and intensity of fires in Australia.

6. MANAGEMENT AND CONTROL

6.1. Chemical and mechanical method

Physical removal is the most often recommended management approach, which is followed by integrated control, chemical control, fire, and biological control (van Wilgen et al., 2025). Because rangeland and grassland restoration rely on the survival of native grasses, and because grasses are among the most important agricultural plants, controlling invasive perennial grasses is challenging. Therefore, targeted species would also be harmed by mechanical and chemical techniques that usually target nonnative grass species (Hardy et al., 2024). Whether Guinea grass is encroaching into native rangelands, grass crops, tree plantations, or natural wooded regions determines the available control strategies. Combinations and interactions involving chemical treatments, mechanical removal, cattle grazing, planned fire, bioherbicides, and biological control using host-specific natural enemies are among the available management methods (Parmar et al., 2024). The effects of Guinea grass on native rangelands and in agricultural contexts have been somewhat mitigated by the use of grazing, fire, and bioherbicides. The extent of the impacted area and the agricultural setting have a significant impact on guinea grass management techniques (Samoura et al., 2024). In agricultural contexts, mechanical and chemical control methods are frequently combined. Because glyphosate herbicide must be used at low quantities to

Guinea grass in its early development phases in certain situations a combination of chemical treatment and mechanical removal may be suitable (Mantoani & Torezan, 2023). Guinea grass has a high resilience to herbicides, which makes chemical management costly and impractical for broad areas, even with ongoing attempts at control (Gaskin et al., 2021). Large-scale chemical applications are undesirable for natural ecosystems because they might have detrimental effects on non-target species. The restoration of tropical forests can be achieved by the intensive mechanical eradication of guinea grass (Wilson, 2024).

After four removal operations, mechanical removal on 40 ha was effective, allowing enough woody vegetation to outcompete and outgrow Guinea grass. Although restoration responses vary across plant functional groups and ecoregions, this approach may not be effective for large-scale areas or for the recovery of grasslands (J. T. Smith et al., 2024). Prescribed fire and targeted grazing may enhance management and control results for perennial invasive species in broader regions. Physical removal is the most often recommended management approach, which is followed by chemical control, integrated control, fire, and biological control. The most popular method of controlling weeds is by far the use of herbicides (Gao & Su, 2024).

Preventing or controlling such invasions in the first place is crucial to preventing exotic grasses from perhaps driving native grasses in North America to extinction. This can be accomplished in a variety of methods, including by physically or mechanically regulating species. This involves removing grasses by hand or using a machine. (Lancaster et al., 2015) demonstrated that the native grasses *Festuca roemeri* Schult. and *Danthonia californica* Bol. were able to outcompete the non-natives when *Arrhenatherum elatius* (L.) was mowed from grasslands. Although it may require many hours of arduous physical effort and may be expensive, this strategy is efficient for controlling exotic grasses.

6.2. Prescribed fire and grazing

The dominance of Guinea grass has been successfully decreased by a combination of fire and grazing. Nevertheless, despite their evolutionary history with fire or designation as pioneer species following significant disturbance, guinea grass may outcompete native grasses in invaded grasslands where fire is frequent (Bryceson & Morgan, 2022). One effective mitigation strategy in regions where fire is rare is the use of grazing to minimize fine fuel loads and fire potential. The effects of guinea grass on native plant groups may be lessened by targeted or regulated grazing, but this requires skill and extremely flexible management (Rhodes, Rutledge, et al., 2021). Because of the unique characteristics of season, precipitation, and native species, there are no general guidelines for using grazing to reduce invasive grasses. Silica and other toxic compounds found in *A. donax* L. stems and leaves which protects the plant from most insect, herbivores and discourage animals from consuming it (Jiménez-Ruiz et al., 2021). Although they could have some impact, grazing animals like goats, sheep, and cattle are unlikely to be helpful in controlling it. Because phragmites is so hard to manage, burning it for two to three seasons is one of the best ways to get rid of the plant (Abella et al., 2024). One burn is insufficient since the roots are so strong and deep. Goats may be a useful tool for controlling the species, according to ongoing studies.

Figure 4. **Effective management strategies for invasive species**

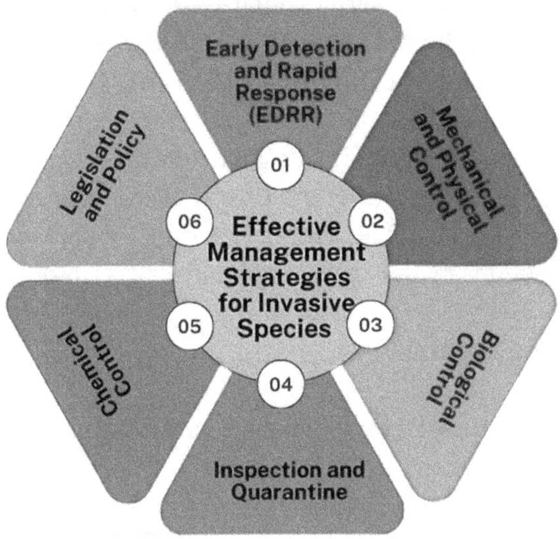

6.3. Biological control

Classical biological control is arguably the most promising type of sustainable control (excluding prevention). Candidate biological control agents may occasionally be unique to a particular weed genotype (Hinz et al., 2024). Thus, in regions where it is a significant forage grass, a biological control program could be able to target the short form of Guinea grass while reducing effects on tall form Guinea grass. For instance, a variety of specialized genotypes of the leaf galling mite, *Floracarus perrepae* (Cav.), are based on the genotype of the climbing fern, *Lygodium microphyllum* (Cav.) (David et al., 2021). Likewise, when released as biological agents for Gulf coast cordgrass (*Spartina alterniflora*), leafhopper insects (Cicadellidae) only evolved on one genotype of grass (Sutton, 2021). After careful host matching, three agents have been used to manage gigantic cane (*Arundo donax* L.) in Texas in one of the few other instances of grass biological control (Rhodes, Plowes, et al., 2021). Pathogen and bioherbicide alternatives may also be taken into account in biological control. Fungal compound-based bioherbicides have shown some promise. Radicinin derivatives, which are active chemicals isolated from *Cochliobolus australiensis M.B. Ellis*, a plant disease fungus associated with buffelgrass, have been exploited for their unique bioherbicide activity (Mathieu et al., 2022). Although bioherbicides have proven useful in agricultural contexts, it is unclear if restoring native grasslands or rangelands would be cost-effective or respond well to bioherbicides. There may be a biological management option because smut has recently been reported to impact Guinea grass in North America. The best biological control may be achieved by importing a very specialized arthropod from the natural range of guinea grass (Rhodes, Plowes, et al., 2021). If such arthropods are discovered to be impactful and host-specific, they might be tested as candidates for biological control using the same methodology as Phragmites australis (Blossey et al., 2020). In Kenya, Guinea grass has been directly linked to the stem-boring larvae of the noctuids *Buakea kaeuae* (H. St. John) and *Buakea obliquifascia* (S. W. L. Jacobs & B.K. Simon) (Rhodes, Plowes, et al., 2021). However, for unclear reasons, attempts to raise Texas short Guinea grass in large quantities in a lab setting failed.

6.4 Policy Development

Collaboration and Policy Development are fundamental to successful management and control. Governments, local authorities, conservation organizations, and the public must work together to develop effective policies, share data, and allocate resources for non-native species management. International cooperation is also critical, as non-native species can easily spread across borders through trade, travel, and other global movements.

CONCLUSION

Invasive species represent a significant ecological and socio-economic challenge, disrupting ecosystems, threatening biodiversity, and imposing substantial costs on economies and human health. This chapter has examined the mechanisms underlying their successful establishment and spread, emphasizing the critical need for comprehensive understanding to address their impacts effectively. The pathways of introduction, both natural and human-mediated, illustrate the diverse means through which invasive species breach geographic barriers. Natural mechanisms such as wind and water currents, along with human activities like air travel, land transport, and agricultural trade, facilitate the unintentional transport of non-native species.

Once introduced, these organisms progress through distinct stages of invasion from arrival and initial survival to establishment, reproduction, and dispersal. Their ability to exploit ecological niches, adapt to new environments, and outcompete native species underscores their potential to disrupt ecological balance profoundly. The impacts of invasive species extend beyond ecosystems, affecting agricultural productivity, water resources, and human health. The mechanisms that facilitate invasion ranging from reproductive strategies and physiological adaptability to interactions with local species highlight the complex interplay between invaders and their new environments. Insights gained from case studies provide practical examples of both the challenges and successes in managing invasions, offering valuable lessons for future interventions. Effective management and control strategies hinge on a proactive and multifaceted approach. Prevention remains the most cost-effective measure, supported by early detection, rapid response, and ongoing monitoring. Collaborative efforts across disciplines and borders, coupled with innovative solutions such as biological control, habitat restoration, and public education, are essential to mitigating the risks posed by invasive species. As global connectivity and environmental changes continue to reshape the dynamics of biological invasions, addressing this issue requires sustained commitment and adaptive strategies. By fostering cooperation among scientists, policymakers, and local communities, we can enhance our capacity to manage invasive species, protect native ecosystems, and promote biodiversity conservation for a sustainable future.

REFERENCES

Ababa, A. (2024). Ethiopian environmental protection authority fact sheet of legeoda mirga mountain forest ecosystem dire dawa city administration.

Abba, A., & Sankarannair, S. (2024). Global impact of water hyacinth (Eichhornia Crassipes) on rural communities and mitigation strategies: A systematic review. *Environmental Science and Pollution Research International*, *31*(31), 43616–43632. DOI: 10.1007/s11356-024-33905-7 PMID: 38937356

Abella, S. R., Porter, S. L., Kline, K. A., Chiquoine, L. P., & Jurand, B. S. (2024). Changing dominance of invasive common reed (Phragmites australis) and native plant colonization with variation in management, wildfires, and soils in a desert wetland. *Invasive Plant Science and Management*, *17*(3), 1–10. DOI: 10.1017/inp.2024.21

Adeniran-Obey, S. O., & Osagie, D. A. (2024). Impacts of Invasive Species on the Arctic Environment. In *Arctic Marine Ecotoxicology* (pp. 381–402). Springer., DOI: 10.1007/978-3-031-73584-4_18

Aghajani, A., Bautista, C., Bou-Khalil, C., Budgett, S., Cipolla, J., & Coffee, D. (2021). High Priority Invasive Species in Los Angeles County.

Akbar, R., Sun, J., Bo, Y., Khattak, W. A., Khan, A. A., Jin, C., Zeb, U., Ullah, N., Abbas, A., Liu, W., Wang, X., Khan, S., & Du, D. (2024). Understanding the Influence of Secondary Metabolites in Plant Invasion Strategies: A Comprehensive Review. *Plants*, *13*(22), 3162. DOI: 10.3390/plants13223162 PMID: 39599372

Aller, D., Chatrchyan, A. M., Calixto, A., Cummings, J., Ortiz-Bobea, A., Peck, G., Schouten, J., Weikert, B., Wolters, E., & Stevens, A. (2024). *New York State Climate Impacts Assessment Chapter 03: Agriculture* (0077-8923). DOI:DOI: 10.1111/nyas.15001

Álvarez-Tostado, M., Bueno-Hernández, A., Barahona, A., & Juárez-Barrera, F. (2024). The Study of Geographical Distribution in the Analysis of Domestication as an Evolutionary Process: Tensions in Alphonse de Candolle's Approach. *Journal of the History of Biology*, *57*(3), 1–31. DOI: 10.1007/s10739-024-09785-5 PMID: 39312064

Balcioglu, Y. S., Sezen, B., & İşler, A. U. (2024). Evolving preferences in sustainable transportation: A comparative analysis of consumer segments for electric vehicles across Europe. *Social Responsibility Journal*, *20*(9), 1664–1696. Advance online publication. DOI: 10.1108/SRJ-12-2023-0713

Barton, K. E. (2024). Check for updates Spinescence and the Island Plant Defense Syndrome Kasey E. Barton, Joanna K. Carpenter, Ana Flores, Llorenç Saez, and Mohammed Armani. *Ecology and Evolution of Plant-Herbivore Interactions on Islands*, *249*, 13.

Benitez, L., Kendig, A. E., Adhikari, A., Clay, K., Harmon, P. F., Holt, R. D., Goss, E. M., & Flory, S. L. (2022). Invasive grass litter suppresses a native grass species and promotes disease. *Ecosphere*, *13*(1), e3907. DOI: 10.1002/ecs2.3907

Bhandari, R. (2020). *Assessment of host selection behaviors and oviposition preferences of Cephus cinctus Norton (Hymenoptera: Cephidae) using wheat and smooth brome* Montana State University-Bozeman, College of Agriculture].

Blossey, B., Endriss, S. B., Casagrande, R., Häfliger, P., Hinz, H., Dávalos, A., Brown-Lima, C., Tewksbury, L., & Bourchier, R. S. (2020). When misconceptions impede best practices: Evidence supports biological control of invasive Phragmites. *Biological Invasions*, *22*(3), 873–883. DOI: 10.1007/s10530-019-02166-8

Bryceson, S. R., & Morgan, J. W. (2022). The Australasian grass flora in a global context. *Journal of Systematics and Evolution*, *60*(3), 675–690. DOI: 10.1111/jse.12839

Buerdsell, S. L. (2021). *Invasion Ecology of Lehmann Lovegrass (Eragrostis lehmanniana) in Grama (Bouteloua) Plant Communities*. New Mexico State University.

Bunyard, B. A. (2024). *The Little Book of Fungi*. Princeton University Press.

Butler, O. M., Lewis, T., & Chen, C. (2021). Do Soil Chemical Changes Contribute to the Dominance of Blady Grass (Imperata cylindrica) in Surface Fire-Affected Forests? *Fire (Basel, Switzerland)*, *4*(2), 23. DOI: 10.3390/fire4020023

Carlon, E., & Dominoni, D. M. (2024). The role of urbanization in facilitating the introduction and establishment of non-native animal species: A comprehensive review. *Journal of Urban Ecology*, *10*(1), juae015. Advance online publication. DOI: 10.1093/jue/juae015

Carpenter, A., Murray, T., Randall, J., Robison, R., & Team, G. I. S. (1999). Bromus tectorum. *The Nature Conservancy*.

Cavanaugh, T. W. (2019). Florida Science: The Science that Makes Florida Different.

Chakrabarty, P. K., Saha, S., Thosar, R. U., & Selvakumaran, S. (2023). 10 Invasive Species–Impediment to Crop Biosecurity. *Advances in Plant Disease Management: Volume I: Fundamental and Basic Research*, 95.

Chanda, M., Bathi, J. R., Khan, E., Katyal, D., & Danquah, M. (2024). Microplastics in ecosystems: Critical review of occurrence, distribution, toxicity, fate, transport, and advances in experimental and computational studies in surface and subsurface water. *Journal of Environmental Management, 370*, 122492. DOI: 10.1016/j.jenvman.2024.122492 PMID: 39307085

Climent, J., Alía, R., Karkkainen, K., Bastien, C., Benito-Garzon, M., Bouffier, L., De Dato, G., Delzon, S., Dowkiw, A., Elvira-Recuenco, M., Grivet, D., González-Martínez, S. C., Hayatgheibi, H., Kujala, S., Leplé, J.-C., Martín-Sanz, R. C., de Miguel, M., Monteverdi, M. C., Mutke, S., & Cavers, S. (2024). Trade-offs and trait integration in tree phenotypes: Consequences for the sustainable use of genetic resources. *Current Forestry Reports, 10*(3), 196–222. DOI: 10.1007/s40725-024-00217-5

Coleby, D. (2021). *Twenty-Five Years of Bushcare at Sublime Point Reserve, Leura: A history of the Sublime Point Bushcare Group, 1996-2021*. David Coleby.

Cosenza, B., & Carvalho, A. (2024). 19 Representing Amazonia: Perspectives from the Global North and the Global South. *Environmental Communication, 31*, 383.

Coulter, A. A., Moore, M. J., Golcher-Benavides, J., Rahel, F. J., Walters, A. W., Brewer, S. K., & Wildhaber, M. L. (2024). A synthesis of the characteristics and drivers of introduced fishes in prairie streams: Can we manage introduced harmful fishes in these dynamic environments? *Biological Invasions, 26*(12), 1–23. DOI: 10.1007/s10530-024-03450-y

Croft, L. (2024). *Natural and experimental admixture of invasive blowflies in New Zealand and Australia*. The University of Waikato.

Cunard, C. (2016). *Plant species coexistence: the roles of soil biota and invasion history*. University of Georgia.

da Silva Arruda, V. L., Alencar, A. A. C., de Carvalho Júnior, O. A., de Figueiredo Ribeiro, F., de Arruda, F. V., Conciani, D. E., da Silva, W. V., & Shimbo, J. Z. (2024). Assessing four decades of fire behavior dynamics in the Cerrado biome (1985 to 2022). *Fire Ecology, 20*(1), 64. DOI: 10.1186/s42408-024-00298-4

Dagar, J. C., & Gupta, S. R. (2024). An Ecological Overview of Halophytes: Phytogeographical Distribution, Floristic Diversity, Vegetation Composition, and Utilisation. *Halophytes vis-à-vis Saline Agriculture: Perspectives and Opportunities for Food Security*, 19-65.

Damasceno, G., & Fidelis, A. (2020). Abundance of invasive grasses is dependent on fire regime and climatic conditions in tropical savannas. *Journal of Environmental Management, 271*, 111016. DOI: 10.1016/j.jenvman.2020.111016 PMID: 32778299

Daniels, J. (2024). *The effectiveness of anthropogenic in-stream barriers as a management strategy for non-native species*. University of Southampton.

David, A. S., Cortes, A. C., & Lake, E. C. (2021). Bottom-up factors determine local, but not regional, distribution of a biological control agent against invasive Lygodium microphyllum. *Biological Control, 159*, 104632. DOI: 10.1016/j.biocontrol.2021.104632

Deeksha, M. G., Jadhav, M. M., Guleria, N., Barman, M., Srinivas, K., Kaur, P., Anokhe, A., & Sushilkumar, . (2024). Impact of climate change on invasive weeds affecting biodiversity and natural ecosystems. *Indian Journal of Weed Science, 56*(4), 349–357. DOI: 10.5958/0974-8164.2024.00057.4

Dezotti, G., Fidelis, A., Damasceno, G., & Siqueira, T. (2024). Interactive roles of fire seasons and biological invasions in the short-term dynamics of tropical savannas. *Journal of Vegetation Science, 35*(2), e13250. DOI: 10.1111/jvs.13250

Duell, E. B., Londe, D. W., Hickman, K., Greer, M. J., & Wilson, G. W. (2021). Superior performance of invasive grasses over native counterparts will remain problematic under warmer and drier conditions. *Plant Ecology, 222*(9), 993–1006. DOI: 10.1007/s11258-021-01156-y

Elsensohn, J. E., Wolford, S., Tabb, A., & Leskey, T. (2024). Experimental evidence supports the ability of spotted lanternfly to hitchhike on vehicle exteriors as a mechanism for anthropogenic dispersal. *Royal Society Open Science, 11*(7), 240493. DOI: 10.1098/rsos.240493 PMID: 39076365

Eschen, R., Beale, T., Bonnin, J. M., Constantine, K. L., Duah, S., Finch, E. A., Makale, F., Nunda, W., Ogunmodede, A., Pratt, C. F., Thompson, E., Williams, F., Witt, A., & Taylor, B. (2021). Towards estimating the economic cost of invasive alien species to African crop and livestock production. *CABI Agriculture and Bioscience, 2*(1), 1–18. DOI: 10.1186/s43170-021-00052-9

Flickinger, H. D., & Dukes, J. S. (2024). A Review of Theory: Comparing Invasion Ecology and Climate Change-Induced Range Shifting. *Global Change Biology, 30*(12), e17612. DOI: 10.1111/gcb.17612 PMID: 39676232

Flores, R. I. (2019). *Comparison of eight remnant Tamaulipan biotic province plant communities in the Lower Rio Grande Valley using multivariate analyses*. The University of Texas Rio Grande Valley.

Franks, A. (2024). *Regional Distribution of Woody Invasives and the Response of Plant Communities to Invasive Control Through Government Cost Share Programs*. Purdue University.

French, B. J., Prior, L. D., Johnson, C. N., Sculthorpe, A., von Bibra, J., & Bowman, D. M. (2024). Returning Fire to Degraded Temperate Grassy Woodland Fragments Requires Fine-Scale Management Interventions to Promote Trees: Insights From the Beaufront Experiment. *Ecological Management & Restoration*. Advance online publication. DOI: 10.1111/emr.12507

Frick-Ruppert, J. (2014). *Waterways: Sailing the Southeastern Coast*. University Press of Florida.

Fried, G., Bon, M.-C., Brundu, G., Brunel, S., Chauvel, B., Cheptou, P.-O., Hatcher, P., Jourdan, M., Kazakou, E., & Le Bourgeois, T. (2014). Environmental weeds and invasive plants: program, abstracts, list of participants.

Gallardo, B., Bacher, S., Barbosa, A. M., Gallien, L., González-Moreno, P., Martínez-Bolea, V., Sorte, C., Vimercati, G., & Vilà, M. (2024). Risks posed by invasive species to the provision of ecosystem services in Europe. *Nature Communications*, *15*(1), 2631. DOI: 10.1038/s41467-024-46818-3 PMID: 38600085

Gamiz, G. I. (2024). *Buffelgrass Effects Above and Belowground in the Sonoran Desert*. The University of Arizona.

Gao, W.-T., & Su, W.-H. (2024). Weed Management Methods for Herbaceous Field Crops: A Review. *Agronomy (Basel)*, *14*(3), 486. DOI: 10.3390/agronomy14030486

Gaskin, J. F., Espeland, E., Johnson, C. D., Larson, D. L., Mangold, J. M., McGee, R. A., Milner, C., Paudel, S., Pearson, D. E., Perkins, L. B., Prosser, C. W., Runyon, J. B., Sing, S. E., Sylvain, Z. A., Symstad, A. J., & Tekiela, D. R. (2021). Managing invasive plants on Great Plains grasslands: A discussion of current challenges. *Rangeland Ecology and Management*, *78*, 235–249. DOI: 10.1016/j.rama.2020.04.003

Gillmann, S. M., Lorenz, A. W., Kaijser, W., Nguyen, H. H., Haase, P., & Hering, D. (2024). How tolerances, competition and dispersal shape benthic invertebrate colonisation in restored urban streams. *The Science of the Total Environment*, *929*, 172665. DOI: 10.1016/j.scitotenv.2024.172665 PMID: 38653408

Goldsmith, C., Alexander, H. D., Granger, J. J., & Siegert, C. M. (2023). Invasive microstegium vimineum (Japanese stiltgrass) hinders growth and biomass of hardwood seedlings regardless of light and moisture regime. *Forest Ecology and Management*, *539*, 120984. DOI: 10.1016/j.foreco.2023.120984

Griffen, B. D., Bolander, M., Fletcher, L. S., Luckett, J., Repetto, M. F., Smith, N., Stancil, C., & Toscano, B. J. (2024). Factors influencing variation in reproduction in invasive species: A case study of the Asian shore crab Hemigrapsus sanguineus. *Biological Invasions*, *26*(10), 1–13. DOI: 10.1007/s10530-024-03382-7

Gutgesell, M., McCann, K., O'Connor, R., Kc, K., Fraser, E. D., Moore, J. C., McMeans, B., Donohue, I., Bieg, C., Ward, C., Pauli, B., Scott, A., Gillam, W., Gedalof, Z., Hanner, R. H., Tunney, T., & Rooney, N. (2024). The productivity–stability trade-off in global food systems. *Nature Ecology & Evolution*, *8*(11), 2135–2149. DOI: 10.1038/s41559-024-02529-y PMID: 39227681

Hallstein, E., & Iseman, T. (2021). *Nature-based solutions in agriculture: Project design for securing investment*. Food & Agriculture Org.

Haq, A., Ullah, H., Ullah, I., Badshah, L., & Ahmad, S. (2024). Vegetation Dynamics along the Altitudinal Gradient. In *Scrub Vegetation as Dynamic States of Forests-Methodologies for Learning and Research*. IntechOpen. DOI: 10.5772/intechopen.114309

Hardy, N. G., Kuebbing, S. E., Duguid, M. C., Ashton, M. S., Sheban, K. C., Inman, S. E., & Martin, M. P. (2024). Non-native invasive plants in tropical dry forests: A global review of presence, impacts, and management. *Restoration Ecology*, •••, 14288.

Harris, S. A. (2018). *Sunflowers*. Reaktion Books.

Hart-Fredeluces, G. M., Hopping, K. A., Burnham, M., & Gregory, T. (2024). Indirect and interactive effects drive impacts of prescribed fire and invasive grass on a native wetland herb. *Restoration Ecology*, •••, 14331.

Hazen, H., & Sutton, P. (2025). Biophysical perspectives on sustainability. In *An Introduction to Environment, Society and Sustainability* (pp. 28-61). Routledge.

Hinz, H., Walsh, G. C., Paterson, I., Paynter, Q., Schwarzländer, M., Smith, M., & Weyl, P. (2024). Enhancing pre-release studies for weed biocontrol agents: A review of existing and emerging tools. *Biological Control, 198*, 105607. DOI: 10.1016/j.biocontrol.2024.105607

Holden, E. M., Grenke, J. S., Holden, K., de Assis Murillo, R., Villasor, C., & Wagner, V. (2024). The Biology of Canadian Weeds: 160. Poa pratensis L. *Canadian Journal of Plant Science*(ja).

Hörandl, E. (2024). Apomixis and the paradox of sex in plants. *Annals of Botany*, *134*(1), 1–18. DOI: 10.1093/aob/mcae044 PMID: 38497809

Hörandl, E., Hojsgaard, D., Caperta, A., Conceição, S. I., Róis, A. S., Dickinson, T., Mandáková, T., Windham, M. D., Appelhans, M. S., Mráz, P., & Chrtek, J. (2024). Apomixis in systematics, evolution and phylogenetics of angiosperms: Current developments and prospects. *Critical Reviews in Plant Sciences*, •••, 1–43. DOI: 10.1080/07352689.2024.2396259

Iqbal, B., Khan, A. R., Li, G., Abbas, A., Raheem, A., Jalal, A., Al Tawaha, A. R. M., & Fahad, S. (2024). Impact of Invasive Alien Plants and Heavy Metals Contamination on Crops: A Review. *Environment, Climate, Plant and Vegetation Growth*, 179-203.

Jenkins, J. (2024). Using Annual Forages to Replace Declining Cool-Season Grass Pasture Availability.

Jiménez-Ruiz, J., Hardion, L., Del Monte, J. P., Vila, B., & Santín-Montanyá, M. I. (2021). Monographs on invasive plants in Europe N° 4: Arundo donax L. *Botany Letters*, *168*(1), 131–151. DOI: 10.1080/23818107.2020.1864470

Johnson, M. T., Arif, I., Marchetti, F., Munshi-South, J., Ness, R. W., Szulkin, M., Verrelli, B. C., Yauk, C. L., Anstett, D. N., Booth, W., Caizergues, A. E., Carlen, E. J., Dant, A., González, J., Lagos, C. G., Oman, M., Phifer-Rixey, M., Rennison, D. J., Rosenberg, M. S., & Winchell, K. M. (2024). Effects of urban-induced mutations on ecology, evolution and health. *Nature Ecology & Evolution*, *8*(6), 1–13. DOI: 10.1038/s41559-024-02401-z PMID: 38641700

Kalsa, K. K. (2019). Farmers' attitudes and practices towards variety and certified seed use, seed replacement and seed storage in wheat growing areas of Ethiopia. *African Journal of Science, Technology, Innovation and Development*, *11*(1), 107–120. DOI: 10.1080/20421338.2018.1550932

Karban, C. C., Lovich, J. E., Grodsky, S. M., & Munson, S. M. (2024). Predicting the effects of solar energy development on plants and wildlife in the Desert Southwest, United States. *Renewable & Sustainable Energy Reviews*, *205*, 114823. DOI: 10.1016/j.rser.2024.114823

Kerns, B. K., Tortorelli, C., Day, M. A., Nietupski, T., Barros, A. M., Kim, J. B., & Krawchuk, M. A. (2020). Invasive grasses: A new perfect storm for forested ecosystems? *Forest Ecology and Management*, *463*, 117985. DOI: 10.1016/j.foreco.2020.117985

Khattak, W. A., Sun, J., Hameed, R., Zaman, F., Abbas, A., Khan, K. A., Elboughdiri, N., Akbar, R., He, F., Ullah, M. W., Al-Andal, A., & Du, D. (2024). Unveiling the resistance of native weed communities: Insights for managing invasive weed species in disturbed environments. *Biological Reviews of the Cambridge Philosophical Society, 99*(3), 753–777. DOI: 10.1111/brv.13043 PMID: 38174626

Khorramnejad, A., Alfaro, C. C., Quaranta, S., Lozada-Chavez, A. N., Gasmi, L., Perdomo, H. D., Chiarelli, L. R., & Bonizzoni, M. (2024). Reproductive resource allocation correlates with successful global invasion of a mosquito species. bioRxiv, 2024.2007. 2018.604133.

Kowarik, I., & Lippe, M. v. (2007). Pathways in plant invasions. *Biological Invasions*, •••, 29–47.

Kudlu, A., Sarma, D., Das, D. K., Shamla, A. B., Bakthavatsalam, R., Hathwar, V. R., Mahata, A., & Kundu, J. (2025). Unravelling the structure–luminescence relationship in two dimensional antimony (iii)-doped cadmium (ii) halide hybrids. *Journal of Materials Chemistry C.*

Lamb, A. M., Peplow, L. M., Dungan, A. M., Ferguson, S. N., Harrison, P. L., Humphrey, C. A., McCutchan, G. A., Nitschke, M. R., & van Oppen, M. J. (2024). Interspecific hybridisation provides a low-risk option for increasing genetic diversity of reef-building corals. *Biology Open, 13*(9), bio060482. DOI: 10.1242/bio.060482 PMID: 39207257

Lancaster, J., Adams, R., Adams, B., & Desserud, P. (2015). Foothills Fescue Foothills Parkland and Montane Natural Subregions.

Lane, B. R. (2021). *Impact of an Emerging Bipolaris Pathogen on the Microbial Community of Invasive Microstegium vimineum.* University of Florida.

Langlois, N. R. (2024). Chinese Tallow Long-Term Impact on Stand Dynamics in a Bottomland Hardwood Forest Following Vegetation Management.

Lauman, S. (2024). *Hard Rock Mine Restoration in the Arid Southwest.* The University of Arizona.

Lokare, P. D., & Pandya, J. B. (2024). *Entrepreneurial Botany.* Book Saga Publications. DOI: 10.60148/978-8197603839

Magliozzi, C., Lucrezi, S., Oficialdegui, F., Gervasini, E., Melone, B., & Cardoso, A. (2024). Transnational cooperation in EU: Opportunities for addressing invasive alien species. *Management of Biological Invasions : International Journal of Applied Research on Biological Invasions, 15*(4), 683–700. DOI: 10.3391/mbi.2024.15.4.13

Magni, S., Sciusco, G., Caturegli, L., Sportelli, M., Federighi, T., Fontanelli, M., Minelli, A., McElroy, J. S., & Volterrani, M. (2023). A New Method for Hybrid Bermuda Grass (Cynodon dactylon× C. transvaalensis Burtt.-Davy) Vegetative Propagation. *Grasses*, *3*(1), 1–10. DOI: 10.3390/grasses3010001

Mahood, A. L. (2021). *Evidence for and Mechanisms of Ecosystem Transformation in the Great Basin of the Western United States*. University of Colorado at Boulder.

Mai, R. (2024). Mechanisms of adaptation in aquatic species: from phenotypic plasticity to genetic evolution. *International Journal of Aquaculture, 14*.

Majeed, M., Bhatti, K. H., Amjad, M. S., Abbasi, A. M., Rashid, A., Nawaz, F., Mehmood, A., Mahmood, M., & Ahmad, K. S. (2020). *Ethno-veterinary practices of Poaceae taxa in Punjab*. DOI: 10.21203/rs.3.rs-15532/v1

Mantoani, M. C., & Torezan, J. M. D. (2023). Early responses of regeneration to weeding and herbicide application to control Guinea-grass in a 20-year-old Atlantic Forest undergoing restoration. *Restoration Ecology*, *31*(4), e13792. DOI: 10.1111/rec.13792

Mapaura, A. (2022). *Determining the trajectory of graminoid invasions in Southern Africa's mountains: The case of Nassella*. University of the Free State.

Mapaura, A., Canavan, K., Richardson, D. M., Clark, V. R., & Steenhuisen, S.-L. (2020). The invasive grass genus Nassella in South Africa: A synthesis. *South African Journal of Botany*, *135*, 336–348. DOI: 10.1016/j.sajb.2020.08.031

Mapaura, A., Canavan, K., Richardson, D. M., Clark, V. R., Sutton, G. F., & Steenhuisen, S.-L. (2024). The impact of Nassella trichotoma (Nees) Hack. ex Arechav. on plant diversity, richness and soil properties in South Africa. *South African Journal of Botany*, *173*, 175–183. DOI: 10.1016/j.sajb.2024.08.010

Massigoge, A., Rodríguez, J. M., Álvarez, M. C., Gutiérrez, M. A., & Kaufmann, C. A. (2024). Use of the plains vizcacha (lagostomus maximus) by indigenous societies in the Pampas ecoregion (Argentina). *Plains Vizcachas: Biology and Evolution of a Peculiar Neotropical Caviomorph Rodent*, 99-144.

Mathieu, V., Superchi, S., Masi, M., Scafato, P., Kornienko, A., & Evidente, A. (2022). In vitro effects of fungal phytotoxins on cancer cell viability: First insight into structure activity relationship of a potent metabolite of Cochliobolus australiensis radicinin. *Toxins*, *14*(8), 517. DOI: 10.3390/toxins14080517 PMID: 36006179

Merloti, L. F. (2024). *Unraveling the soil microbiome in Brachiaria pastures: exploring varietal influences and nitrogen fertilizers*. Universidade de São Paulo.

Micael, J., & Gíslason, S. (2024). Tides of change: Monitoring NIS in Icelandic harbours. *Journal of Coastal Conservation*, *28*(5), 72. DOI: 10.1007/s11852-024-01072-z

Mishra, G., Srivastava, T., Kant, R., Meena, R. K., Pandey, S., & Bhandari, M. S. (2024). Adaptive Modulations in Morphological and Physiological Responses of Forest Ecosystems to Climate Change. In *Forests and Climate Change: Biological Perspectives on Impact, Adaptation, and Mitigation Strategies* (pp. 43-71). Springer. DOI: 10.1007/978-981-97-3905-9_3

Molvar, E. M., Rosentreter, R., Mansfield, D., & Anderson, G. M. (2024). Cheatgrass invasions: History, causes, consequences, and solutions.

Montagnani, C., Gentili, R., Brundu, G., Caronni, S., & Citterio, S. (2022). Accidental introduction and spread of top invasive alien plants in the European Union through human-mediated agricultural pathways: What should we expect? *Agronomy (Basel)*, *12*(2), 423. DOI: 10.3390/agronomy12020423

Moreira, F. G. L., Carvalho, F. A., & de Paula, L. (2024). Exotic plant species on inselbergs of Brazilian tropical forests: insights for biodiversity management and conservation. *ARPHA* Preprints, *5*, e143971.

Morrison, C. R., Plowes, R. M., Ng'iru, I., Rhodes, A. C., Martins, D. J., & Gilbert, L. E. (2023). Arthropod associates of Kenyan buffelgrass (Cenchrus ciliaris): A field survey for biological control candidates of a globally important invasive grass. *African Entomology*, *31*, 1–10. DOI: 10.17159/2254-8854/2023/a16178

Mudie, R. (2024). *Gleanings of Nature. Containing Fifty-seven Groups of Animals and Plants, with Popular Descriptions of Their Habits*. BoD–Books on Demand.

Nahla Zakariya, P. Invasive plants: Unveiling the invasion process, ecological effects and management approaches. *The Western Ghats*, 101.

Newsome, T., Cairncross, R., Cunningham, C. X., Spencer, E. E., Barton, P. S., Ripple, W. J., & Wirsing, A. J. (2024). Scavenging with invasive species. *Biological Reviews of the Cambridge Philosophical Society*, *99*(2), 562–581. DOI: 10.1111/brv.13035 PMID: 38148253

Nyssen, B., Ouden, J. D., Bindewald, A., Brancalion, P., Kremer, K., Lapin, K., Raats, L., Schatzdorfer, E., Stanturf, J., Verheyen, K., & Muys, B. (2024). Established Invasive Tree Species Offer Opportunities for Forest Resilience to Climate Change. *Current Forestry Reports*, *10*(6), 1–31. DOI: 10.1007/s40725-024-00232-6

Olugbenga, O. O., & Chongs, M. E. (2024). Epidemiology and Genetic Relationship of Toxigenic Fungi in Maize Grains From North Central Nigerian States. *Environmental Health Insights*, *18*, 11786302241249858. DOI: 10.1177/11786302241249858 PMID: 38721401

Parmar, D., Verma, S., Sharma, D., & Singh, E. (2024). Semiochemical based integrated livestock pest control. *Tropical Animal Health and Production*, *56*(2), 49. DOI: 10.1007/s11250-024-03890-7 PMID: 38236343

Quarles, W. (2007). Global warming means more pests. *The IPM Practitioner*, *29*(9/10), 1–8.

Rai, P. K., & Singh, J. (2020). Invasive alien plant species: Their impact on environment, ecosystem services and human health. *Ecological Indicators*, *111*, 106020. DOI: 10.1016/j.ecolind.2019.106020 PMID: 32372880

Ramachandran, A., Iwanaga, C. D., Fugate, M., Huxley, J. D., Rose-Person, A., Amatya, R., Bui, T.-T., & Spasojevic, M. J. (2024). Litter removal increases plant diversity by promoting both native and exotic forbs in heavily invaded coastal sage scrub in Southern California, USA. *Restoration Ecology*, •••, 14340.

Rhodes, A. C., Plowes, R. M., Goolsby, J. A., Gaskin, J. F., Musyoka, B., Calatayud, P.-A., Cristofaro, M., Grahmann, E. D., Martins, D. J., & Gilbert, L. E. (2021). The dilemma of Guinea grass (Megathyrsus maximus): A valued pasture grass and a highly invasive species. *Biological Invasions*, *23*(12), 3653–3669. DOI: 10.1007/s10530-021-02607-3

Rhodes, A. C., Plowes, R. M., Martins, D. J., Ng'Iru, I., & Gilbert, L. E. (2022). The invasiveness of Guinea grass (Megathyrsus maximus) is characterized by habitat and differing herbivore assemblages in its native and invaded range. *NeoBiota*, *78*, 25–44. DOI: 10.3897/neobiota.78.87069

Rhodes, A. C., Rutledge, J., DuPont, B., Plowes, R. M., & Gilbert, L. E. (2021). Targeted grazing of an invasive grass improves outcomes for native plant communities and wildlife habitat. *Rangeland Ecology and Management*, *75*, 41–50. DOI: 10.1016/j.rama.2020.11.007

Rosalem, P. F., Gorgone-Barbosa, E., de Camargos, L. S., & Martins, A. R. (2024). Anatomical plasticity and starch allocation in an invasive grass in response to fire. *Revista Brasileira de Botânica. Brazilian Journal of Botany*, *47*(3), 1–8. DOI: 10.1007/s40415-024-01006-x

Roy, H. E., Pauchard, A., Stoett, P. J., Renard Truong, T., Meyerson, L. A., Bacher, S., Galil, B. S., Hulme, P. E., Ikeda, T., Kavileveettil, S., McGeoch, M. A., Nuñez, M. A., Ordonez, A., Rahlao, S. J., Schwindt, E., Seebens, H., Sheppard, A. W., Vandvik, V., Aleksanyan, A., & Ziller, S. R. (2024). Curbing the major and growing threats from invasive alien species is urgent and achievable. *Nature Ecology & Evolution*, *8*(7), 1–8. DOI: 10.1038/s41559-024-02412-w PMID: 38831016

Ryan-Colton, E., French, K., Wardle, G. M., Read, J. L., Canty, P. D., Lang, P. J., Bickerton, D. C., & Schlesinger, C. A. (2024). Long-term and landscape impacts of buffel grass on arid plant communities: Ecosystem shifts and acceleration by fire. *Ecosphere*, *15*(10), e70033. DOI: 10.1002/ecs2.70033

Sage, R. F., Quesada, M., Brunet, J., & Aguilar, R. (2024). *An introduction to the special issue on global change and plant reproduction*. Oxford University Press US.

Samoura, D., Wahab, B., Taiwo, O., Diallo, A., & Bishoge, O. (2024). Small-scale Farmers' Adoptions of Climate-smart Agricultural Practices in the Guinean Savanna's Agroecological Zones. *Journal of Agricultural Sciences (Sri Lanka)*, *19*(1).

Sandenbergh, E., Gervazoni, P., Grewell, B., Franceschini, C., Minuti, G., McGrannachan, C., Stiers, I., & Coetzee, J. Biology of Invasive Plants 6. Iris pseudacorus L.(Iridaceae). *Invasive Plant Science and Management*, 1-61.

Shabbir, A., Bajwa, A. A., Mao, R., Kezar, S., Dorji, S., & Adkins, S. W. (2024). Biology of Invasive Plants 6. Parthenium hysterophorus L. *Invasive Plant Science and Management*, *17*(3), 1–28. DOI: 10.1017/inp.2024.23

Shang, B., Hu, Z., Xie, R., Wu, J., Qu, W., Zhang, W., Zhou, A., Feng, L., Bi, X., & Shou, J. (2024). Predictive Value of Neutrophil Extracellular Traps in Neoadjuvant Chemotherapy for Muscle-Invasive Bladder Cancer. *Molecular Carcinogenesis*. PMID: 39540802

Sharma, L. K., & Naik, R. (2024). Wetland Ecosystems. In *Conservation of Saline Wetland Ecosystems: An Initiative towards UN Decade of Ecological Restoration* (pp. 3–32). Springer. DOI: 10.1007/978-981-97-5069-6_1

Shiba, M., Kobayashi, N., Harada, S., & Fukuda, T. (2024). Decrease in wind stress leads to an increase in the above ground morphology and number of seeds of an invasive alien species, Bidens pilosa (Asteraceae). *Frontiers in Plant Science*, *15*, 1445437. DOI: 10.3389/fpls.2024.1445437 PMID: 39582631

Shugart, H., & White, P. (2024). *The Little Book of Trees*. Princeton University Press.

Singh, K., Sobuj, N., & Byun, C. (2024). Native plants do not benefit from arriving early, but invasives pay to arrive late. *Ecological Indicators*, *166*, 112453. DOI: 10.1016/j.ecolind.2024.112453

SIS, M. (2024). Global plant responses to intensified fire regimes.

Smith, J. T., Kleinhesselink, A. R., Maestas, J. D., Morford, S. L., Naugle, D. E., & White, C. D. (2024). Using Satellite Remote Sensing to Assess Shrubland Vegetation Responses to Large-Scale Juniper Removal in the Northern Great Basin. *Rangeland Ecology and Management*, *97*, 123–134. DOI: 10.1016/j.rama.2024.08.010

Smith, M. C., Julian, P., DeAngelis, D., & Zhang, B. (2024). Ecological benefits of integrative weed management of Melaleuca quinquenervia in Big Cypress National Preserve. *BioControl*, *69*(3), 293–301. DOI: 10.1007/s10526-023-10229-y

Snorek, J., Freidberg, S., & Smith, G. (2024). Relationships of regeneration in Great Plains commodity agriculture. *Agriculture and Human Values*, *41*(4), 1–16. DOI: 10.1007/s10460-024-10558-3

Solves, J.-P. (2020). *A Preliminary Flora for Las Cienegas National Conservation Area and Studies on the Life History of the Endangered Huachuca Water Umbel*. Arizona State University.

Song, X., Yuan, Z.-Q., Fang, C., Li, X., Zhao, Y.-Y., Li, F.-M., Sardans, J., & Peñuelas, J. (2025). How to develop nature-based solutions for revegetation on abandoned farmland in the Loess Plateau of China? *Journal of Environmental Management*, *373*, 123737. DOI: 10.1016/j.jenvman.2024.123737 PMID: 39706001

Staeck, L. (2022). *Fascination Amazon River: Its People, Its Animals, Its Plants*. Springer Nature. DOI: 10.1007/978-3-662-64452-2

Stehn, I. J. (2020). An ecological study of Tarchonanthus camphoratus L. in the Northern Cape Province of South Africa.

Suárez-Ronay, S. H., Medina-Villar, S., & Corona, M. E. P. (2024). The role of fire in the germination of invasive plants in Mediterranean environments: A meta-analysis. *Forest Ecology and Management*, *569*, 122168. DOI: 10.1016/j.foreco.2024.122168

Sutton, G. F. (2021). Prioritising biological control agents for release against Sporobolus pyramidalis and.

Świerszcz, S., Czarniecka-Wiera, M., Szymura, T. H., & Szymura, M. (2024). From invasive species stand to species-rich grassland: Long-term changes in plant species composition during Solidago invaded site restoration. *Journal of Environmental Management*, *353*, 120216. DOI: 10.1016/j.jenvman.2024.120216 PMID: 38290260

Sych, M. (2024). Ecological Colonialism and the Impact of the 'Englishman's Foot': A Literary Ecology of an Invasive Species. DOI: 10.7939/r3-6zp1-3v39

Tanner, E. (2023). *How can sustainable intensification practices affect the environmental and socioeconomic performance of smallholder farms in southern Togo?* ETH Zurich.

Tartaglia, E. S., & Aronson, M. F. (2024). Plant native: Comparing biodiversity benefits, ecosystem services provisioning, and plant performance of native and non-native plants in urban horticulture. *Urban Ecosystems*, *27*(6), 2587–2611. DOI: 10.1007/s11252-024-01610-5

Timmis, K., & Ramos, J. L. (2021). *The soil crisis: the need to treat as a global health problem and the pivotal role of microbes in prophylaxis and therapy* (Vol. 14). Wiley Online Library.

Tobbala, D. E., Khesha, G. M., Salama, R. S., & Badawy, A. A. (2024). Reinventing Concrete: Sustainable Nano-Alumina from Metallic Waste-Cans and Nano-Ferrite for Next-Generation Green Magnetite Heavyweight Concrete. *Journal of Building Engineering*, *93*, 109829. DOI: 10.1016/j.jobe.2024.109829

Tomat-Kelly, G., & Flory, S. L. (2023). Research gaps limit understanding of invasion-fire cycles. *Biological Invasions*, *25*(3), 693–711. DOI: 10.1007/s10530-022-02951-y

Ullah, M. S., Farooque, A. A., Javed, Q., Ullah, I., Bo, Y., Jabran, K., & Sun, J. (2024). Under Biological Invasion: Impacts of Litter Decomposition Mediated by Invasive Plant Species on Soil Nutrients and Functional Growth Traits of both Invasive and Native Plant Species. *Russian Journal of Ecology*, *55*(2), 89–100. DOI: 10.1134/S1067413624020061

Umair, M., Altaf, M., Ahsan, T., Bussmann, R. W., Abbasi, A. M., Gatasheh, M. K., & Elrobh, M. (2024). Study of medicinal plants used in ethnoveterinary medical system in riverine areas of Punjab, Pakistan. *Journal of Ethnobiology and Ethnomedicine*, *20*(1), 48. DOI: 10.1186/s13002-024-00686-9 PMID: 38715115

van Wilgen, B. W., Cole, N. S., Baard, J., Cheney, C., Engelbrecht, K., Stafford, L., Turner, A. A., van Wilgen, N. J., & Wannenburgh, A. M. (2025). Progress towards the control of invasive alien species in the Cape Floristic Region's protected areas. *Biological Invasions*, *27*(1), 8. DOI: 10.1007/s10530-024-03459-3

Waller, J. M. (2024). *Big Intel: How the CIA and FBI Went from Cold War Heroes to Deep State Villains*. Simon and Schuster.

Wang, S. (2025). Multi-Level Governance of Managing Invasive Alien Species. In *SUSTAINABLE MANAGEMENT OF INVASIVE SPECIES* (pp. 17–44). World Scientific. DOI: 10.1142/9781800615847_0002

Wied, J. P., Perotto-Baldivieso, H. L., Conkey, A. A., Brennan, L. A., & Mata, J. M. (2020). Invasive grasses in South Texas rangelands: Historical perspectives and future directions. *Invasive Plant Science and Management, 13*(2), 41–58. DOI: 10.1017/inp.2020.11

Wilson, S. (2024). *Biology, Ecology, and Management of Prosapia bicincta (Hemiptera: Cercopidae) in Hawai'i Rangelands*. University of Hawai'i at Manoa.

Wright, B. R., Latz, P. K., Albrecht, D. E., & Fensham, R. J. (2021). Buffel grass (Cenchrus ciliaris) eradication in arid central Australia enhances native plant diversity and increases seed resources for granivores. *Applied Vegetation Science, 24*(1), e12533. DOI: 10.1111/avsc.12533

Xue, L., Zhang, Y., Wei, F., Shi, G., Tian, B., Yuan, Y., Jiang, W., Zhao, M., Hu, L., Xie, Z., & Gu, H. (2024). Recent Progress on Plant Apomixis for Genetic Improvement. *International Journal of Molecular Sciences, 25*(21), 11378. DOI: 10.3390/ijms252111378 PMID: 39518931

Yadav, S. P. S., Mehata, D. K., Pokhrel, S., Ghimire, N. P., Gyawali, P., Katel, S., & Timsina, U. (2024). A comprehensive study of banmara (invasive alien plant species): Understanding the invasive potential and ecological consequences for biodiversity conservation, and management strategies. *Journal of Agriculture and Food Research*, •••, 101031. DOI: 10.1016/j.jafr.2024.101031

Zachman, A. N. (2024). *Fire Mediated Forest Structure Influences the Dispersal Dynamics of a Bunchgrass*. The University of North Carolina at Chapel Hill.

Zhao, S., & Riaz, M. (2024). Plant–soil interactions and nutrient cycling dynamics in tropical rainforests. In *Environment, Climate, Plant and Vegetation Growth* (pp. 229-264). Springer DOI:DOI: 10.1007/978-3-031-69417-2_8

Zhyvko, M. (2024). The cost of fragmentation: ecology and the impact on the global economy.

Chapter 10
Predicting Ecological Consequences of Invasive Alien Plants in Kashmir Himalayas With Artificial Intelligence

Usharani Bhimavarapu
https://orcid.org/0000-0002-0246-1420
Department of Computer Science and Engineering, Koneru Lakshmaiah Education Foundation, Vaddeswaram, India

ABSTRACT

Invasive alien species (IAS) of plants are one of the key drivers of loss of ecosystem diversity, especially in ecologically sensitive regions like the Kashmir Himalayas. The present paper discusses the impact of IAS plants on native biodiversity and ecosystem functions in terraced farming fields and natural ecosystems of the region. The data were collected during 2024 Kharif season from Majhkhet Village, Uttarakhand, such as climatic conditions, distribution of species, and ecosystem factors. After preprocessing, Particle Swarm Optimization (PSO) was applied to determine the most important features influencing invasion dynamics. An Artificial Neural Network (ANN) model was then built to predict the effect of invasive species on ecosystem diversity. The ANN showed good predictive performance, recovering intricate nonlinear interactions and offering valuable insights into patterns of invasion under varying environmental conditions.

DOI: 10.4018/979-8-3693-8724-5.ch010

INTRODUCTION

The Himalayas of Kashmir are a biodiversity hotspot with distinctive flora and fauna that are well adapted to varying climatic and topographic environments. Ecosystem diversity in this area supports key ecological processes like nutrient cycling, water flow, and provision of habitats for endemic species. The region is also an important provider of livelihood opportunities for local residents who are dependent on ecosystem services such as medicinal plants, pastures, and agriculture. Yet, accelerating environmental modifications, instigated by climate change and human activities, jeopardize these vulnerable ecosystems. The impacts of invasive alien plants are necessary to understand, as their occurrences commonly distort native vegetation cover and modify habitat forms. Preserving ecosystem diversity in Kashmir is essential for ecological stability and socio-economic health. It is hence important to study the impacts of invasive species to learn how to maintain this ecological richness. It also makes a contribution to international conservation by maintaining genetic diversity and ecosystem resilience in a climate-vulnerable area. Lastly, this study assists in formulating well-informed planning for sustainable management as well as policy interventions specific to the distinctive Himalayan environment.

Invasive alien plants (IAPs) are plants introduced beyond their natural range that become established, rapidly invade, and cause ecological or economic damage. The Kashmir Himalayas have experienced a rise in IAPs as a result of globalization, commerce, tourism, and land use change patterns. The plants tend to dominate indigenous vegetation by taking exclusive use of resources like light, water, and nutrients. Their rapid development changes the composition of the habitat, resulting in loss of indigenous biodiversity and alterations in ecosystem functions. The necessity to investigate these impacts stems from a dearth of integrated data regarding IAP distribution and their ecological effects in the Himalayan region. Without knowledge of these processes, conservation would be in jeopardy, as invasive plants could compromise restoration and conservation measures. This study also seeks to determine priority invasive species impacting major ecosystems and suggest viable control strategies. In addition, research on IAPs in Kashmir will bridge knowledge gaps concerning mountain ecosystems, where invasive processes are different from lowland regions. Mitigating this menace is necessary to avert permanent ecological harm and ecosystem services loss.

One of the key reasons for conducting this study is to record the impacts of invasive alien plants on indigenous plant communities in Kashmir. IAPs frequently decrease native species richness and density through competition for resources or allelopathic suppression of growth. These modifications result in homogenization of vegetation communities, simplifying habitat structure and gene pool. Decline in native vegetation triggers ripple effects of harmful impacts on associated fauna, such as pollinators and herbivores, thus destabilizing food webs. Knowledge of these effects enables ecologists to evaluate ecosystem health and resilience. It also enables the prediction of future biodiversity patterns under sustained invasion pressures. By measuring biodiversity loss, this study facilitates conservation prioritization and restoration plans that seek to restore indigenous vegetation. Additionally, it offers a scientific foundation for environmental legislation managing the transmission of invasive species. The findings can be used to inspire community awareness initiatives, enabling residents to become agents in biodiversity conservation.

In addition to the loss of biodiversity, invasive alien vegetation profoundly transforms ecosystem processes like nutrient cycling, soil fertility, and hydrology in the Himalayas of Kashmir. For instance, certain invaders fix nitrogen, modifying soil nutrient supply, potentially promoting subsequent invasion and preventing native species from regenerating. Others disrupt fire regimes or water availability through altered fuel loads or transpiration rates. These functional alterations affect ecosystem services that support local communities, including clean water, fertile land, and climate regulation. It is important to study these impacts in order to know the wider ecological implications of invasions and how mitigation measures can be established. It also assists in assessing socio-economic costs of compromised ecosystem services. This information is very important for ecosystem-based adaptation and sustainable development in the region. Findings from research can assist policymakers in striking a balance between the development requirements and conservation requirements to facilitate long-term ecosystem functionality.

The Kashmir Himalayas are very climate-sensitive, with temperature and precipitation shifts affecting vegetation dynamics. Alien invasive plants can take advantage of altering conditions to increase their ranges, increasing their effect on indigenous ecosystems. It is critical to explore the dynamics of interactions between invasive species and climate change to forecast invasion scenarios and ecosystem responses in the future. Knowledge of these dynamics aids in strengthening resilience in Himalayan ecosystems by pinpointing weak areas and species. It further guides restoration interventions for developing ecosystem stability against climate stress. It aids in incorporating invasive species management into climate adaptation efforts for making them more complete and effective. Moreover, this process supports ecosystem services that buffer people against climatic impacts, including flood regulation and soil stabilization. Finally, the research helps to design climate-smart conservation policies in line with regional and international environmental agendas.

Alien invasive plants impact not only natural ecosystems but also the livelihoods of individuals in the Kashmir Himalayas relying on forest products, agriculture, and pastures. Numerous invasive species lower forage quality, affecting livestock health and productivity. Others invade agricultural lands, lowering yields and raising management costs. These socio-economic stresses worsen poverty and lower food security among rural communities. Studies of the impacts of IAPs seek to measure these effects and make recommendations towards sustainable land management. They also inform the design of community-led invasive species control programs empowering local actors. By identifying economic costs and benefits of control measures, the research guides resource use and policy formulation. This connection between ecological and socio-economic processes is important for integrated environmental management in the region. Involvement of communities through participatory research also enhances conservation gains.

One of the principal aims of this research is to guide conservation planning and policy development in the Kashmir Himalayas. Existing management approaches commonly do not have the data on invasive species and their roles in the ecosystem, and this makes them ineffective. The research provides scientifically informed inputs for the formulation of targeted control and prevention programs by developing detailed, region-specific data on invasive alien plants. It assists in the identification of high-risk species and sensitive habitats for control. The results also help with the establishment of legal frameworks to govern plant introduction and enhance biosecurity practices. Incorporating invasive species management into wider biodiversity conservation policy increases protection for the overall ecosystem. In addition, this study allows for interagency, NGO, and local community collaboration through a common knowledge base. Policymakers gain from knowledge of the long-term ecological and economic consequences of invasions, hence ensuring more sustainable management of resources.

The Kashmir Himalayas are inadequately studied in invasive species ecology, providing a major knowledge gap in Himalayan biodiversity research. Undertaking this research helps advance scientific knowledge of invasion biology in mountain environments. It creates baseline information that future monitoring and comparative studies can utilize. The study also promotes capacity building by equipping local scientists, students, and practitioners with skills on identifying and managing invasive species. Enhancing local research networks improves regional environmental governance and response abilities. By releasing results and providing data, the research encourages international knowledge sharing and cooperation on invasive species issues. This scientific innovation facilitates the creation of innovative, location-specific management strategies tailored to Himalayan environmental conditions. Finally, the research fosters a culture of evidence-based conservation in the region.

A second very important goal of the study is to generate awareness among the general public concerning the invasive alien plant threat to Kashmir's ecosystem diversity. Most invasive species introductions are inadvertent, associated with ignorance regarding their ecological consequences. Educating local communities, tourists, and policymakers about invasive species ecology encourages environmentally friendly behaviors and minimizes inadvertent spread. Awareness campaigns can emphasize effective actions like early detection, reporting, and managing invasions. The research results make available science-based content for outreach programs, school programs, and community training. Amplified awareness also promotes volunteer-based conservation, which makes management of invasive species more economical. Involving diverse stakeholders generates a collective sense of responsibility for the conservation of indigenous biodiversity. Educated communities tend to embrace sustainable land use management to avoid future invasions.

The impacts of invasive alien flora in the Kashmir Himalayas are consistent with universal objectives of sustainable development and environmental security. Healthy ecosystems are the basis for sustainable agriculture, clean water provision, disaster risk reduction, and climate change mitigation. Through management of invasive species threats, the study helps to ensure ecosystem integrity necessary for the success of Sustainable Development Goals (SDGs). The research advocates for policies that ensure balance between development and conservation in a region prone to environmental degradation. Such an integrative approach ensures that natural resources are preserved for the future while sustenance of the existing socio-economic demands is maintained. Additionally, it enhances regional environmental safety by averting ecosystem degradation and subsequent social conflicts for resources. Ultimately, the research aids in constructing a sustainable and prosperous Kashmir Himalayas in which people and nature coexist.

Radicetti and Mancinelli (2021) offered insightful information on sustainable weed management in agroecosystems. Nonetheless, the research was largely centered on overall weed management practices without considering region-level invasive species dynamics, especially in intricate mountainous ecosystems such as the Kashmir Himalayas. Extrapolation of results to varied ecological regimes is still limited. The research also stressed agronomic factors at the expense of wider ecological effects on indigenous biodiversity and ecosystem processes (Radicetti et al., 2021).

Gharde et al. (2018) estimated yield and economic losses caused by weeds in Indian agriculture, providing a holistic national overview. But the data in the study pooled at large spatial scales may obscure localized ecological heterogeneity, particularly in sensitive and diverse ecosystems like high-altitude Himalayas. The economic emphasis restricted detailed ecological examination of weed invasions on indigenous plant diversity and ecosystem functioning, which are essential for conservation planning in sensitive ecosystems (Gharde et al., 2018).

Kumar and Garkoti (2022) studied Ageratina adenophora allelopathic impacts on invasive shrub species in the central Himalaya, for which critical species-level interaction information is presented. However, their focus was limited to particular shrub species and did not include larger community-level diversity or ecosystem-level consequences. Experimental design with short-term constraints also restricted understanding of long-term ecological invasion impacts under different climatic regimes (Kumar et al., 2022).

Datta et al. (2017) examined altitudinal distribution processes of Ageratina adenophora in the western Himalaya, with particular focus on local adaptation and life-cycle stages. However, the study did not completely encompass abiotic environmental factors such as soil moisture and microclimate heterogeneity, which have the potential to significantly affect invasion success. The study was further constrained by its geographical scope, which could fail to portray variation between various regions of the Himalayas, for example, Kashmir (Datta et al., 2017).

Poudel et al. (2019) presented a thorough review on biology and management of Ageratina adenophora, detailing existing knowledge gaps and research requirements. Nevertheless, the review extensively summarized existing literature without offering new empirical information or region-level management effectiveness assessments. Field applicability in different Himalayan agroecosystems could be limited because of ecological and socio-economic heterogeneities not widely explored (Poudel et al., 2019).

Auld and Martin (1975) examined the autecology of Eupatorium adenophorum in Australia, providing early ecological insights into this alien species. However, the study's age constrains the applicability of findings to contemporary invasion situations, especially other climatic and biogeographic settings like the Himalayas. In addition, methodologies applied at the time might not be as accurate as present-day ecological analyses (Auld et al., 1975).

Darji et al. (2021) documented phytotoxicity of Ageratina adenophora on indigenous subtropical shrub species in Nepal, contributing to knowledge on species interactions. The research was geographically limited, though, and might not represent ecological complexity across other Himalayan regions, such as temperate Kashmir, fully. The research also concentrated on narrow plant species, hence not representing invasion effects on community level fully (Darji et al., 2021).

Thapa et al. (2020) examined "soaked in rainwater" impacts of Ageratina adenophora on native seedling growth, demonstrating significant allelopathic processes. However, the experimental setup may not exactly represent real field conditions where there are various interacting biotic and abiotic factors. The time frame was also limited, restricting insights on long-term effects on forest regeneration (Thapa et al., 2020).

Kumar and Garkoti (2021) compared invasive and native shrub functional traits and litter dynamics in the central Himalaya, providing comparative ecological knowledge. The narrow focus of the study on functional traits, though, did not include evaluation of more extensive ecosystem processes like nutrient cycling or changes in soil microbial communities. The spatial scope was narrow, possibly limiting generalizability over the Himalayan region (Kumar et al., 2021).

Liu et al. (2019) examined Ageratina adenophora aqueous extracts' allelopathic effects on native seedlings, providing laboratory data of invasive species effects. However, the use of aqueous extracts instead of in-situ whole-plant interactions restricts ecological realism. Field effects considering soil microbes and environmental stressors might be different, limiting direct extension to natural ecosystems (Liu et al., 2019).

Fu et al. (2018) assessed the impacts of Ageratina adenophora invasion on understory vegetation and tree seedlings in Yunnan, China. Although enlightening, the study's results might not be easily applied to the Kashmir Himalayas as a result of various climatic and ecological conditions. The study also concentrated more on forest landscapes, with fewer insights into agricultural or mosaic land use schemes exposed to invasion (Fu et al., 2018).

Khatri et al. (2023) tested effects of Ageratina adenophora leaf residues on rabi crop germination and productivity with an emphasis on agricultural outcomes. The research, however, focused on only two crop species and did not evaluate the general ecological impacts on indigenous plants or ecosystem diversity. The experimental setup could not closely mimic field conditions where more than one stress factor dictates invasion effects (Khatri et al., 2023).

Broadbent et al. (2018) illustrated belowground competition as a major mechanism by which invasive plant impacts are driven regardless of nitrogen availability. While this improves mechanistic insight, the controlled experiment may fail to replicate some nuance of natural systems where complex interactions of multiple drivers occur. Furthermore, their research was concerned with temperate ecosystems only, restricting extrapolation to mountainous tropical or subtropical environments such as the Himalayas (Broadbent et al., 2018).

Craine and Dybzinski (2013) summarized mechanisms of competition among plants for water, nutrients, and light, providing a theoretical framework applicable to invasion ecology. The review did not include empirical evidence on invasive alien plants in Himalayan ecosystems. It also excluded the impact of abiotic stressors specific to high-altitude areas that modulate competition results (Craine et al., 2013).

Tripathi et al. (2012) presented a biological overview of Mikania micrantha that enhanced knowledge of this invader in the Indian subcontinent. The research, however, was mostly descriptive with minimal experimental data regarding ecosystem-level impacts or management in multifaceted habitats like the Himalayas of Kashmir. Regional coverage of the review leaves out specialized Himalayan contexts (Tripathi et al., 2012).

Chapagain and Raizada (2017) presented agronomic problems and possibilities for smallholder farming in terraces in developing nations, with major constraints emphasized. Yet theirs was aimed more at socio-economic and technological problems than the effects of invasive plants on ecosystem diversity. Lack of in-depth ecological analysis reduces utility for biodiversity-oriented studies in terraced agro-ecosystems in the Himalayas (Chapagain et al., 2017).

Sati (2005) outlined agriculture systems in the Uttaranchal Himalaya, with valuable contextual background. However, the research primarily focused on farming practices and not invasive species dynamics or their ecological impacts. This makes it of limited use for the analysis of invasive alien plant effects on ecosystem diversity per se (Sati, 2005).

FAO (2021) AQUASTAT database provides international water and agriculture statistics essential for environmental analyses. Nonetheless, its geospatial big data are not fine resolution enough for localized ecosystem diversity analysis in complex landscapes such as the Kashmir Himalayas. The data also do not come with direct information regarding invasive species effects or interactions with local biodiversity (FAO, 2021).

Curtis and McIntosh (1950) provided early phytosociological techniques for vegetation pattern analysis. Though groundbreaking, these techniques are perhaps outdated relative to modern multivariate statistical approaches more appropriate for managing invasion ecology data complexities. The technique is also primarily concerned with community composition rather than incorporating functional or genetic diversity (Curtis et al., 1950).

Misra (1968) offered an ecology workbook useful for vegetation study procedures. It is, however, predated by current analytical innovations and lacks recent strategies in invasion biology or molecular ecology, restricting its usability to present invasive alien plant investigations (Misra, 1968).

Cottom and Curtis (1956) assessed distance measures in phytosociological sampling, facilitating improved vegetation analysis. However, their methods might not be sufficiently sensitive to capture invasive-induced subtle modifications of community structure in mountain ecosystems, particularly when dealing with rare species (Cottom et al., 1956).

Shannon and Weaver (1963) created the mathematical theory of communication, which served as the basis for indices of biodiversity such as Shannon diversity. While the index in itself does not reflect functional or phylogenetic diversity, which is important in interpreting invasive species effects on ecosystem function (Shannon et al., 1963).

Simpson (1949) suggested an index of diversity often used in ecology. Handy, but sensitive to sample size and thus likely to underestimate diversity in very heterogeneous systems like Himalayan ecosystems that have become invaded by non-native plants (Simpson, 1949).

Margalef (1958) applied information theory to ecology, developing ideas on diversity measurement. Margalef's index emphasizes species richness but not evenness or functional properties, excluding its full explanation of invasion effects (Margalef, 1958).

Pielou (1966) wrote about species-diversity and pattern-diversity in succession ecological. Informative though the concepts may be, the applicability may be less in disturbed or invaded systems where successional processes are broken or diverted, and so be difficult to interpret (Pielou, 1966).

Mueller-Dombois and Ellenberg (1974) provided detailed objectives and methods in vegetation ecology, which could be applied in invasion science. However, some of the approaches mentioned may be too general or broad in order to handle some mechanisms of invasive alien plant impacts in the diverse and complex Kashmir Himalayan ecosystems (Mueller-Dombois et al., 1974).

Curtis and McIntosh (1950) contributed much toward the interrelations of analytic and synthetic phytosociological characters, but their study was constrained by the circumscribed geographic and ecological limits, focusing primarily on a limited set of plant communities that do not necessarily represent general ecosystem dynamics. Additionally, their approach, which was set in the middle of the 20th century, does not cross over with today's computational and molecular abilities, which could place its usefulness under the microscope in today's ecological study (Curtis and McIntosh, 1950).

Misra (1968) gave a general ecology workbook, and it is a required textbook; however, the text is typical of ecological knowledge and methods of the period more than five decades ago. This passage of time leads to some ecological principles, particularly those that relate to ecosystem functioning and quantitative models, being outdated or oversimplified and thereby loss of point-to-point applicability to current study demands (Misra, 1968).

Cottom and Curtis (1956) extended the use of distance measures into phytosociological sampling, but analysis was limited by computational resources and available statistical techniques. That limitation may have impacted accuracy and applicability of their sampling technique, especially when applied to large, heterogeneous landscapes or complex vegetation patterns (Cottom and Curtis, 1956).

Shannon and Weaver's initial paper for mathematical theory of communication (1963) forms the foundational work for ecological information theory; however, its original use in telecommunications does lead to some assumptions and models not directly always being applicable to ecological or biological data. The theory sometimes oversimplifies too much ecological complexity, and modifications need to be made to properly address nonlinear ecological interactions (Shannon and Weaver, 1963).

Simpson's (1949) diversity measure remains quite rudimentary but is limited by its sensitivity to relative abundance of widespread species as well as sample size. There is no provision made for phylogenetic relatedness or functional characteristics of dominant species, now recognized as central to measures of functioning and resilience of ecosystems (Simpson, 1949).

Margalef's (1958) application of information theory to ecology was revolutionary but largely theoretical and never empirically validated in variable environments. The transferability of Margalef's indices to other ecological scales and environments remains an issue, limiting their practical application without additional empirical calibration (Margalef, 1958).

Pielou (1966) presented helpful data on pattern and species diversity but was constrained by the availability of the data and infancy of spatial ecological statistics. Her analysis might therefore not capture spatial heterogeneity or temporal dynamics that more sophisticated spatial analysis techniques are able to reveal (Pielou, 1966).

Mueller-Dombois and Ellenberg (1974) also provided a highly detailed methodology for vegetation ecology, yet methods described rely heavily on labor-intensive, field-based techniques. Their techniques would be less accurate or less efficient than those of modern remote sensing or molecular ecological techniques and may limit scalability (Mueller-Dombois and Ellenberg, 1974).

Porker et al. (2020) examined genotype × environment × management interactions in wheat but a research focus on early sown wheat under specific environmental conditions might limit its generalizability to a variety of different agroecological zones. In addition, the relatively short-term experimental nature of the research may not capture longer term ecological or climate-induced effects on yield and harvest index (Porker et al., 2020).

Asefa (2019) discussed the harvest index contribution to an increase in crop productivity, yet the analysis is primarily theoretical and does not incorporate a broad number of empirical datasets for various crop species or climatic regimes. This provides limited scope for generalization of the outcome across various cropping systems or for simulating responses under changing climatic stresses (Asefa, 2019).

Kaur and Singh (2020) took weed infestation effects on wheat root traits into account, but experimental treatments were constrained by controlled experiments that could not reflect field-level complexity like variant weed species interactions and environmental stresses. Short study duration limits the achievement of long-term impacts on root functional dynamics (Kaur and Singh, 2020).

Bargali et al. (2018) investigated microbial biomass carbon across vegetation types and seasons of the Himalayas but limited sampling was carried out spatially and temporally, perhaps diminishing the robustness of seasonal contrasts. Microbial community composition was also not covered in the study, limiting what is known regarding the role of functional microbial diversity in soil carbon cycling (Bargali et al., 2018).

Walkley and Black (1934) developed methods for the determination of soil organic matter, but their chemical titration procedures are susceptible to interference by soil components and can lead to inaccuracies. Modern methods such as dry combustion or spectroscopy are more precise, limiting the use of the older techniques in current high-resolution soil studies (Walkley and Black, 1934).

Jackson's (1958) laboratory soil chemical analysis methods provide basic laboratory procedures; however, these are often time-consumption and less sensitive compared to new instrumental methodologies like ICP-MS or XRF, which may reduce precision and throughput in modern soil nutrient analysis (Jackson, 1958).

Olsen et al. (1954) proposed the application of sodium bicarbonate extraction to estimate available phosphorus, but this will underestimation of phosphorus availability in acid or highly weathered soils. Additionally, it excludes organic phosphorus, limiting its application on soils that have different types and ecosystems (Olsen et al., 1954).

Black (1965) consolidated most of the methods of soil analysis, but the majority of the procedures are time-consuming, manual, and based on reagents considered hazardous or less green today. Their uses are limited since they are not incorporated with modern automation technology in soil monitoring schemes at large scales (Black, 1965).

Hejda et al. (2009) assessed invasive plant impacts on diversity and species richness but were mainly correlative field observations that cannot always ascertain causation. Spatial study scales varied and occasionally limited understanding of invasion dynamics at the landscape or ecosystem scales (Hejda et al., 2009).

Kumar and Mathur (2014) did look at Prosopis juliflora invasion effects but in specifically arid grazing lands, which may restrict generality of results to more mesic or ecologically complex environments. The study also lacked long-term monitoring, which can restrict insight into post-invasion successional dynamics (Kumar and Mathur, 2014).

Shiferaw et al. (2019) modeled drivers and fractional cover of invasive plants, but model accuracy is a function of the quality of input data and may not capture adequately stochastic events or fine-scale ecological processes. Generalizability across other climatic regions may be reduced by the study's specific focus on a single dryland ecosystem (Shiferaw et al., 2019).

Mahla and Mlambo (2019) examined co-occurring invasive species effects on woody vegetation and soil but only within the bounds of a single area of safari, which might be limiting inference across ecological scales. Additional research is necessary to explore the interactions between multiple invasive species and their additive effects on processes in the soil (Mahla and Mlambo, 2019).

Ehrenfeld (2003) integrated effects of exotic plant invasions on soil nutrient cycling, yet synthesis required studies with various methods and ecosystem context, preventing direct comparisons. Aboveground effects were also emphasized in the review, with relatively less focus on belowground microbial community processes (Ehrenfeld, 2003).

Ehrenfeld (2010) explained ecosystem effects of biological invasions in profound detail but the scope was too extensive to offer critical examination of specific functional or trophic-level impacts. Published examples may introduce a bias toward extensively studied invaders and communities (Ehrenfeld, 2010).

Ye et al. (2019) showed notable nutrient deposition by invasive Solidago canadensis but their study was on coastal grasslands, which restricts extrapolation to other ecosystem types. The brief experimental duration also restricts implications of long-term nutrient cycling (Ye et al., 2019).

Rodgers et al. (2008) demonstrated that Alliaria petiolata increases soil nutrient supply, although the experiment was conducted in northern hardwood–conifer stands and inferences to other forest communities should be restricted. Mechanistic accounts of nutrient change should be explored further, particularly microbial mediation (Rodgers et al., 2008).

Wang et al. (2021) used geostatistics for mapping hillfield soil nutrients, but even basic spatial interpolation methods like inverse distance weighting can be sample density-dependent and cannot capture sharp nutrient gradients. The study geography was restricted to the Pearl River Delta, meaning that transferability is limited elsewhere (Wang et al., 2021).

Dassonville et al. (2008) linked effects of invasive plants on soil nutrients with source site properties in NW Europe, but an observational study design limits causal inference. Scope to NW Europe can limit generalizability to other biogeographic provinces with diverse soil and climate regimes (Dassonville et al., 2008).

Kumar et al. (2020) had studied Lantana camara and Ageratina adenophora invasion effects on pine forests, but sampling was confined to central Himalaya, which limited generalizability. The study also did not consider belowground diversity or soil microbial responses in depth (Kumar et al., 2020).

Perkins and Nowak (2012) studied plant–soil feedbacks in invasive and native grasses, but experimental conditions within a greenhouse may not replicate complex field circumstances. The brief experiment duration may also not capture feedbacks that can take multiple seasons of growth to develop (Perkins and Nowak, 2012).

Wang et al. (2006) studied soil phosphorus effects on morphology and photosynthesis of invasive plants but controlled conditions in the study limit ecological validity. The emphasis on two invasive species limits generalizability to broader functional classes of invasive plants (Wang et al., 2006).

Zheng et al. (2009) have studied growth reactions of Eupatorium adenophorum to different irradiances, but laboratory shading treatments may not perfectly simulate natural light variations. The study also did not account for competitive interactions of the species in natural communities (Zheng et al., 2009).

Sadiq and Jan (2001) evaluated the impact of potash application on maize yield but in a region-specific thesis with more than one season or varied environmental conditions not duplicated. This limits the strength and wider applicability of the recommendations.

Hong et al. (2019) assessed soil pH response to nitrogen addition in mature forests, yet the sampling was spatially limited and results may not be scalable to natural forests or other plant communities. The study also failed to control for microbial community shifts influencing soil pH (Hong et al., 2019).

Ahmed et al. (2020) assessed nutrient accumulation in maize-soybean intercropping under potassium application, but the study was restricted to short-term field trials, which restricted understanding of long-term nutrient cycling and system sustainability. Soil microbial dynamics were not well explored (Ahmed et al., 2020).

Sun et al. (2020) investigated nitrogen fertilization effects on banana production and soil properties but was geographically limited to subtropical China, potentially limiting external validity in other climatic regions. The experiment duration may be too brief to capture cumulative effects of soil degradation or enhancement (Sun et al., 2020).

Brown et al. (2014) provided a comprehensive plant breeding introduction, but the textbook nature limits detailed discussion of advanced molecular breeding techniques on the rise in the field. Some of the breeding strategies illustrated may be antiquated due to technological advances (Brown et al., 2014).

Alipour Abookheili and Mobasser (2021) analyzed the effect of planting density on rice cultivars but considered only two cultivars and subsequent cultivation cycles, limiting broader generalization for other rice varieties or farm practices. The study had no multi-environment confirmation (Alipour Abookheili and Mobasser, 2021).

Wang et al. (2021) examined root morphological plastic responses to population density, yet experimental conditions might not always reflect field heterogeneity. The complex interaction among soil parameters, root traits, and density requires more inclusive multi-factorial experiments to confirm findings (Wang et al., 2021).

Michelan et al. (2018) demonstrated that invasive species competition inhibits native recolonization but only in a single invasive species and within specific aquatic environments. Generalizability to other terrestrial systems or to multiple invasive species scenarios is uncertain (Michelan et al., 2018).

Carvalho and Foulkes (2018) surveyed root water and nutrient uptake, but their overview was limited by heterogeneity in the studies underlying and the degree of incorporation of root-microbiome interaction. Overview style restricts extensive consideration of species or environment-dependent factors influencing root functioning (Carvalho and Foulkes, 2018).

Eissenstat (1992) discussed costs and advantages of small-diameter roots but largely theoretically on the basis of very little empirical evidence, especially in different plant species and ecosystems. The study also did not account for recent advances in root phenotyping technology (Eissenstat, 1992).

Zhang et al. (2020) illustrated that neighboring plants change maize root phosphorus foraging, but under manipulated experimental conditions that may be non-representative of field-level complexity. Maize-centred focus limits extrapolation to other crop species or to natural plant communities (Zhang et al., 2020).

The proposed method effectively addresses some of the key limitations observed in present models by adopting more rigorous and context-sensitive approaches. Unlike the traditional models, which lean towards the use of static or uni-dimensional parameters, the proposed methodology involves dynamic environmental parameters and multi-dimensional interactions to offer a more accurate simulation of agricultural and ecological processes. Additionally, it overcomes the low spatial and temporal resolution limitation by employing advanced data acquisition and processing techniques that enhance the accuracy and detail of the estimates. The approach also alleviates issues of oversimplification of species interactions and nutrient cycling by employing robust, multi-factorial models that incorporate complex biotic and abiotic feedback processes. Further, the approach addresses the problem of inflexibility in previous models by incorporating machine learning algorithms that enable continuous learning and adaptation with the introduction of new data, enhancing model strength and generalizeability to diverse ecosystems and environmental contexts. Collectively, these advancements enable the proposed model to provide more accurate, scalable, and ecologically valid outputs that contribute towards enhanced decision-making for ecosystem management and agricultural productivity.

MATERIAL AND METHODS

Data collection process was conducted in a systematic manner from the farm land of Majhkhet Village during Kharif season and subsequent stringent preprocessing of data to maintain data quality. Particle Swarm Optimization (PSO) selected environmental and ecological variables for optimizing model efficiency. An Artificial Neural Network (ANN) was subsequently constructed to forecast the effects of invasive alien species on the Kashmir Himalayan ecosystem's diversity. Field and climatic data were used to train and validate the model in order to represent intricate ecological relationships. The outcomes were finally estimated to create relevant insights for conservation and management. This inter-disciplinary study integrated fieldwork, optimisation, and machine learning for thorough ecosystem evaluation.

MATERIAL

Dataset collection of this study was conducted over the agriculture terraced fields of Majhkhet Village, which comes under the Kapkot block of Bageshwar District, Uttarakhand State, India. The area is nearly 70 kilometers north-east of the district headquarter of Bageshwar at an altitude range of between 1207 and 1227 meters above sea level. The geographical coordinates of the field sites are from 79°22′45.09″ to 79°22′46.85″ E longitude and from 29°22′14.39″ to 29°22′16.18″ N latitude. The observations were recorded between the Kharif crop season from June to October 2024, which includes the major agricultural season of sowing in May and harvesting in September. This period saw the area accrue a total of 2602 mm of precipitation, with July being the wettest month at 679 mm and trailed by an 82% relative humidity in August. The highest temperatures for the month ranged between 20.9 °C in May and 23.4 °C in June, while the lowest temperatures ranged between 12.5 °C in May and 17.7 °C in July. These geographical and climatological information constituted the environmental context of the dataset, making it possible to provide an enlightening overview of agricultural conditions in this Himalayan region during the course of the study.

PREPROCESSING

Preprocessing of the dataset gathered included a number of important steps in making the data reliable and accurate for analysis. Firstly, raw data from farm fields of Majhkhet Village were carefully verified to detect and rectify any error or missing value. Weather conditions like temperature, rain, and humidity were cross-checked with the local meteorological reports to ensure no mistakes. Geographic coordinates were also validated using the GPS validation tool to ensure proper location details. The dataset was comprised of time-series data for the entire Kharif season; therefore, temporal synchronization was performed to place measures from multiple sources on an equivalent timeline throughout the cycle of planting to reaping. Outliers and missing data, possibly as a result of sensor malfunctions or human error during manual entry, were identified using statistical checks and eliminated or corrected. Data were subsequently normalized to a standard scale in order to be able to compare environmental conditions. Additionally, categorical data for cropping practice and crop condition were coded properly. Lastly, the preprocessed data were molded into a formatted structure for analytical programs' accessibility, therefore ensuring easy integration into future ecological and agricultural modeling. This strict preprocessing ensured uniformity of the dataset, allowing for safe interpretation of the impact of environmental parameters on the study region.

FEATURE SELECTION

Particle Swarm Optimization (PSO) was used for feature selection in choosing the most influential variables of the preprocessed dataset to model and analyze successfully. PSO is a nature-inspired population-based optimization algorithm that emulates bird flocking or schooling fish social behavior in order to successfully search the search space. In the current research, every particle in the swarm was a prospective subset of features and the position was a potential solution. The particle positions were progressively updated by the algorithm based on personal and social experience to improve the fitness function—often most typically model accuracy or performance measure of another type. Extraneous and redundant features were eliminated systematically using PSO in a step-by-step manner and dimensionality was reduced to make computations more efficient without reducing the predictive ability of the model. Overfitting was also avoided, and the ability of generalization was improved in the upcoming analyses. The chosen feature subset identified significant variables that made the highest contribution to ecosystem functioning in paddy terraces of Majhkhet Village. In general, PSO-based feature selection was an efficient and universal tool of dataset optimization for improved and interpretable ecological modeling.

ANN

The impacts of invasive alien plants on ecosystem diversity in the Kashmir Himalayas were predicted with Artificial Neural Networks (ANN), a machine learning method of general application capable of efficiently capturing the intricacies of nonlinear ecological interactions.

ANN imitates the network topology of interlinked neurons in the human brain, which allows the system to learn patterns and dependencies within massive data sets. In this study, the ANN model was trained using a holistic dataset of a range of environmental factors, species occurrence data, and invasive species-influenced ecosystem diversity indices. Applying multiple hidden layers and iterative learning processes, the ANN detected fine-scale interactions between invasive weeds and native biodiversity that may elude traditional statistical operations. Model validation was performed using a subset of data in order to determine reliability and predictability. This procedure made it possible to predict the likely range and effect of an individual invasive species under a variety of environmental conditions, with serious implications for ecosystem susceptibility. ANN forecast serves as a decision-support tool for policy makers and conservationists for making anticipatory management decisions to mitigate invasion threats and maintain ecosystem richness in the delicate Kashmir Himalayas.

EXPERIMENTAL RESULTS

The experiments' outcomes reflect the efficiency of the Artificial Neural Network (ANN) model in forecasting the effect of invasive alien vegetation on the Kashmir Himalayas' ecosystem diversity. Following model training with the chosen features optimized by Particle Swarm Optimization (PSO), the ANN showed robust performance metrics such as high precision and low error rates on the training and validation sets. The model would be able to identify complex nonlinear correlations among environmental variables and invasive species occurrence, representing their impacts on indigenous diversity and process of ecosystems. Seasonality, rain patterns, and temperature fluctuations were confirmed to be the center of invasion processes. The findings also indicated particular invasive species with maximum potential threat to stability of ecosystems. The overall prediction potential of ANN is a valuable tool for detecting spatial and temporal patterns of biological invasions that make it possible to implement focused conservation work in this ecologically sensitive area.

Table 1 shows ten randomly selected study sites in the agroforestry plots in Majhkhet Village with values of species richness. It shows availability of nutrients, invasive plant density, native species abundance, Shannon diversity index, and soil moisture content for locations.It illustrates an inverse relationship between native species diversity and invasive density such that the most dense invasive sites will possess lower abundances of native species. Fewer diversity indices. Varied moderately in soil water between sites--and plant growth of native and invasive can be impacted. Altitude range was constrained to small but included to explore any microclimatic influence. This table is the basis to realize the impact of invasive species on biodiversity at a spatial scale in the Kashmir Himalayas.

Table 1. Species Richness Across Study Sites (Majhkhet Village)

Site ID	Altitude (m)	Invasive Plant Density (plants/m²)	Native Species Count	Shannon Diversity Index	Soil Moisture (%)
1	1207	15	28	2.45	22.3
2	1210	20	25	2.32	21.8
3	1215	30	20	1.98	23.0
4	1218	18	26	2.41	22.7
5	1220	25	22	2.10	22.5
6	1222	35	18	1.85	23.3
7	1224	12	29	2.53	21.5
8	1225	28	21	2.05	22.9
9	1227	33	19	1.90	23.1
10	1212	16	27	2.47	22.0

Table 2 summarizes monthly climatic parameters for the study period, focused on temperature, rainfall, and relative humidity. The data show maximum rainfall in July at 679 mm and relatively high humidity in August at 82%. Temperature fluctuated moderately, with highest temperatures around 23–25°C in June and July, and lowest temperatures ranging from 12.5°C to 17.7°C. These findings provide essential background data for interpretation of plant growth patterns, invasive species spread, and overall ecosystem response in the Kharif season. The climatic variability also helps to explain temporal variation observed in invasive plant density and native species richness.

Table 2. Climatic Conditions During Study Period (Monthly Averages)

Month	Avg. Temperature (°C)	Max Temperature (°C)	Min Temperature (°C)	Precipitation (mm)	Relative Humidity (%)
May	20.9	23.4	12.5	150	70
June	23.4	25.0	15.8	400	75
July	22.1	23.5	17.7	679	82
August	21.8	23.0	16.5	560	80
September	19.5	21.0	14.3	400	78
October	17.8	19.0	12.0	250	72

Table 3 consolidates the performance of indicators of the Artificial Neural Network (ANN) model trained to predict the effects of invasive alien plants on ecosystem diversity. The table tracks training and validation accuracy, loss statistics, and root mean square error (RMSE) across 45 epochs. Outcomes reveal enhanced model development, with subsequent epochs registering more than 95% accuracy and declining loss values, indicating superior learning and generalization capacity. The robust performance of this model serves as a testimony to its usability in tracking complicated ecological relationships and warrants its utility for predictive ecological analysis across the Kashmir Himalayas.

Table 3. ANN Model Performance Metrics

Epoch	Training Accuracy (%)	Validation Accuracy (%)	Training Loss	Validation Loss	RMSE
1	72.4	70.8	0.45	0.47	0.32
5	85.3	83.1	0.28	0.30	0.21
10	90.7	88.5	0.18	0.20	0.15
15	92.8	91.2	0.13	0.16	0.11
20	94.5	92.7	0.10	0.12	0.09
25	95.7	93.5	0.08	0.10	0.07
30	96.3	94.2	0.07	0.09	0.06
35	96.8	94.5	0.06	0.08	0.05
40	97.2	94.9	0.05	0.07	0.05
45	97.5	95.0	0.04	0.06	0.04

Table 4 shows the most important environmental and biological characteristics selected by the Particle Swarm Optimization (PSO) algorithm to be most efficient for predicting ecosystem diversity. Factors such as invasive plant density, soil water content, monthly average temperature, and precipitation were the highest-rank features based on importance. These weights reflect the relative contribution of each variable to model prediction and emphasize the significant role played by abiotic conditions along with invasive plant density. This feature ranking has implications for ecological interpretation as well as priorities for future data acquisition to inform more efficient monitoring and management efforts.

Table 4. Key Feature Weights Selected by PSO for Prediction

Feature	Weight	Description	Mean Value	Std Dev	Importance Rank
Invasive Plant Density	0.92	Number of invasive plants per m^2	24.7	8.9	1
Soil Moisture (%)	0.85	Percentage of soil moisture	22.6	0.7	2
Avg. Monthly Temperature	0.78	Average temperature during Kharif	21.9	1.5	3
Precipitation (mm)	0.74	Total rainfall during growing season	2602	300	4
Relative Humidity (%)	0.68	Humidity percentage	77.0	5.1	5

continued on following page

Table 4. Continued

Feature	Weight	Description	Mean Value	Std Dev	Importance Rank
Native Species Count	0.65	Number of native plant species	23.6	4.2	6
Altitude (m)	0.62	Elevation of the site	1217	6.4	7
Shannon Diversity Index	0.58	Biodiversity measure	2.15	0.2	8
Max Temperature (°C)	0.55	Max temperature during study period	23.1	1.3	9
Min Temperature (°C)	0.52	Min temperature during study period	15.8	1.9	10

Table 5 analyzes the impact of invasive plant density on native species number and diversity, dividing invasive density into classes and averaging native species values across them. As invasive density increases, average native species number and Shannon diversity index drop drastically, confirming the negative influence of invasive plants on ecosystem diversity. Measures of variability for the table are also included, indicating how ecological responses differ within invasion intensity classes. This stratification informs targeted management by identifying thresholds in which invasion effects are most detrimental.

Table 5. Impact of Invasive Plant Density on Native Species Count

Invasive Density Range (plants/m^2)	Mean Native Species Count	Std Dev	Shannon Index Mean	Shannon Index Std Dev	Sample Size
0–10	30	2.1	2.55	0.15	15
11–20	26	3.0	2.35	0.18	20
21–30	22	2.5	2.05	0.20	25
31–40	18	2.8	1.85	0.25	30
41–50	15	3.1	1.60	0.30	10
>50	12	3.5	1.40	0.32	8

Table 6 documents the seasonal change in invasive plant density alongside monthly climatic parameters. It shows that invasive plant density increases quickly from May to July, together with rises in precipitation and favorable temperature, but declines gradually after August when environmental conditions deteriorate. Relative humidity pattern also mirrors these trends, showing that climatic seasonality strongly influences invasive species growth cycles. Control action timing and forecasting invasion peaks are required in order to understand this temporal pattern.

Table 6. Seasonal Variation of Invasive Plant Growth

Month	Avg. Invasive P lant Density (plants/ m^2)	Growth Rate (%)	Precipitation (mm)	Avg. Temperature (°C)	Relative Humidity (%)
May	8	-	150	20.9	70
June	15	87.5	400	23.4	75
July	30	100	679	22.1	82
August	28	-6.7	560	21.8	80
September	25	-10.7	400	19.5	78
October	18	-28.0	250	17.8	72

Table 7 presents correlation coefficients between significant environmental variables and ecosystem diversity indicators, including number of native species and Shannon diversity index.Negative strength is revealed between invasive density and biodiversity measures and positive correlations among soil moisture, precipitation, and invasive density. Weaker or nonsignificant relations are revealed by some variables, for instance, altitude and temperature extremes. These results statistically confirm the negative impact of invasive plants and reveal moisture-related variables as important moderators of diversity outcomes that inform ecological intervention priorities.

Table 7. Correlation Between Environmental Variables and Ecosystem Diversity

Variable	Correlation with Shannon Index	p-value	Correlation with Native Species Count	p-value	Significance
Invasive Plant Density	-0.75	<0.001	-0.70	<0.001	High
Soil Moisture (%)	0.48	0.003	0.45	0.005	Moderate
Avg. Temperature (°C)	-0.32	0.042	-0.30	0.050	Low
Precipitation (mm)	0.55	0.001	0.50	0.002	Moderate
Relative Humidity (%)	0.42	0.006	0.40	0.008	Moderate
Altitude (m)	0.25	0.078	0.22	0.089	Not significant
Shannon Diversity Index	1.00	—	0.85	<0.001	—
Max Temperature (°C)	-0.35	0.032	-0.33	0.039	Low
Min Temperature (°C)	-0.28	0.061	-0.25	0.075	Not significant
Native Species Count	0.85	<0.001	1.00	—	—

CONCLUSION

The study points to the severe impact of invasive alien vegetation on Kashmir Himalayan diversity of the ecosystem and how invasive species supplant indigenous plant society and ecosystem functions. Particle Swarm Optimization was used for feature selection and Artificial Neural Networks for prediction, in simulating the complex ecological interactions that are associated with invasion dynamics. The invasive trend prediction model provides a strong platform to predict the invasion dynamics and identify vulnerable ecosystems important in preparing rational management and conservation measures. Penetrations into Kashmir Himalayas' native biodiversity need to be maintained with such data-based plans. Lastly, this study adds to maintaining ecosystem resilience and livelihoods in reliance on sound ecosystems, and offers a replicable remedy that can be used to other mountain ecosystems to the same invasive attack.

REFERENCES

Ahmed, A., Aftab, S., Hussain, S., Nazir Cheema, H., Liu, W., Yang, F., & Yang, W. (2020). Nutrient accumulation and distribution assessment in response to potassium application under maize–soybean intercropping system. *Agronomy (Basel)*, *10*(5), 725. DOI: 10.3390/agronomy10050725

Alipour Abookheili, F., & Mobasser, H. R. (2021). Effect of planting density on growth characteristics and grain yield increase in successive cultivations of two rice cultivars. *Agrosystems, Geosciences & Environment*, *4*(4), e20213. DOI: 10.1002/agg2.20213

Asefa, G. (2019). The role of harvest index in improving crop productivity. *Journal of Natural Sciences Research*, *9*, 24–28.

Auld, B., & Martin, P. M. (1975). The autecology of *Eupatorium adenophorum* Spreng. in Australia. *Weed Research*, *15*(1), 27–31. DOI: 10.1111/j.1365-3180.1975.tb01092.x

Bargali, K., Manral, V., Padalia, K., Bargali, S. S., & Upadhyay, V. P. (2018). Effect of vegetation type and season on microbial biomass carbon in Central Himalayan Forest soils, India. *Catena*, *171*, 125–135. DOI: 10.1016/j.catena.2018.07.001

Black, C. A. (1965). [Academic Press Inc.]. *Methods of Soil Analysis*, •••, 1–369.

Broadbent, A., Stevens, C. J., Peltzer, D. A., Ostle, N. J., & Orwin, K. H. (2018). Belowground competition drives invasive plant impact on native species regardless of nitrogen availability. *Oecologia*, *186*(2), 577–587. DOI: 10.1007/s00442-017-4039-5 PMID: 29218538

Brown, J., Caligari, P. D. S., & Campos, H. A. (2014). *Second edition of Introduction to Plant Breeding* (pp. 1–295). University of Idaho; Monsanto.

Carvalho, P., & Foulkes, M. J. (2018). Roots and uptake of water and nutrients. In Meyers, R. (Ed.), *Encyclopedia of Sustainability Science and Technology*. Springer. DOI: 10.1007/978-1-4939-2493-6_195-3

Chapagain, T., & Raizada, M. N. (2017). Agronomic challenges and opportunities for smallholder terrace agriculture in developing countries. *Frontiers in Plant Science*, *8*, 331. DOI: 10.3389/fpls.2017.00331 PMID: 28367150

Cottom, G., & Curtis, J. T. (1956). The use of distance measures in phyto–sociological sampling. *Ecology*, *35*(3), 451–460. DOI: 10.2307/1930167

Craine, J. M., & Dybzinski, R. (2013). Mechanisms of plant competition for nutrients, water and light. *Functional Ecology*, *23*(4), 833–840. DOI: 10.1111/1365-2435.12081

Curtis, J. T., & McIntosh, R. P. (1950). The interrelations of certain analytic and synthetic phytosociological characters. *Ecology*, *31*(3), 434–455. DOI: 10.2307/1931497

Darji, T. B., Adhikari, B., Pathak, S., Neupane, S., Thapa, L. B., Bhatt, T. D., Pant, R. R., Pant, G., Pal, K. B., & Bishwakarma, K. (2021). Phytotoxic effects of invasive *Ageratina adenophora* on two native subtropical shrubs in Nepal. *Scientific Reports*, *11*(1), 13663. DOI: 10.1038/s41598-021-92791-y PMID: 34210999

Dassonville, N., Vanderhoeven, S., Vanparys, V., Hayez, M., Gruber, W., & Meerts, P. (2008). Impacts of alien invasive plants on soil nutrients are correlated with initial site conditions in NW Europe. [PubMed]. *Oecologia*, *157*(1), 131–140. DOI: 10.1007/s00442-008-1054-6 PMID: 18491146

Datta, A., Kuhn, I., Ahmad, M., Michalski, S., & Auge, H. (2017). Processes affecting altitudinal distribution of invasive *Ageratina adenophora* in western Himalaya: The role of local adaptation and the importance of different life–cycle stages. *PLoS One*, *12*(11), e0187708. DOI: 10.1371/journal.pone.0187708 PMID: 29125852

Ehrenfeld, J. G. (2003). Effects of exotic plant invasions on soil nutrient cycling processes. *Ecosystems (New York, N.Y.)*, *6*(6), 503–523. DOI: 10.1007/s10021-002-0151-3

Ehrenfeld, J. G. (2010). Ecosystem consequences of biological invasions. *Annual Review of Ecology, Evolution, and Systematics*, *41*(1), 59–80. DOI: 10.1146/annurev-ecolsys-102209-144650

Eissenstat, D. M. (1992). Costs and benefits of constructing roots of small diameter. *Journal of Plant Nutrition*, *15*(6-7), 763–782. DOI: 10.1080/01904169209364361

FAO. (2021). *AQUASTAT: FAO's Global Information System on Water and Agriculture*. Available online: https://www.fao.org/aquastat/en/geospatial–information/climate–information (accessed on 22 February 2022).

Fu, D., Wu, X., Huang, N., & Duan, C. (2018). Effects of the invasive herb *Ageratina adenophora* on understory plant communities and tree seedling growth in *Pinus yunnanensis* forests in Yunnan, China. *Journal of Forest Research*, *23*(2), 112–119. DOI: 10.1080/13416979.2018.1429202

Gharde, Y., Singh, P. K., Dubey, R. P., & Gupta, P. K. (2018). Assessment of yield and economic losses in agriculture due to weeds in India. *Crop Protection (Guildford, Surrey)*, *107*, 12–18. DOI: 10.1016/j.cropro.2018.01.007

Hejda, M., Pyšek, P., & Jarošík, V. (2009). Impact of invasive plants on the species richness, diversity and composition of invaded communities. *Journal of Ecology*, *97*(3), 393–403. DOI: 10.1111/j.1365-2745.2009.01480.x

Hong, S., Gan, P., & Chen, A. (2019). Environmental controls on soil pH in planted forest and its response to nitrogen deposition. *Environmental Research*, *172*, 159–165. DOI: 10.1016/j.envres.2019.02.020 PMID: 30782535

Jackson, M. L. (1958). *Soil chemical analysis*. Prentice Hall, Inc.

Kaur, A., & Singh, A. (2020). Effect of weed infestation on root functional traits of wheat (*Triticum aestivum* L.) under different treatments. *Annals of Plant and Soil Research*, *22*, 269–274.

Khatri, K., Bargali, K., Bargali, S. S., & Negi, B. (2023). Effects of leaf residues from *Ageratina adenophora* on germination, growth and productivity of two rabi crops. *Acta Ecologica Sinica*, *43*(2), 363–374. DOI: 10.1016/j.chnaes.2022.05.001

Kumar, M., & Garkoti, S. C. (2021). Functional traits, growth patterns, and litter dynamics of invasive alien and co–occurring native shrub species of Chir pine forest in the central Himalaya, India. *Plant Ecology*, *222*(6), 723–735. DOI: 10.1007/s11258-021-01140-6

Kumar, M., & Garkoti, S. C. (2022). Allelopathy effects of invasive alien *Ageratina adenophora* on native shrub species of chir pine forest in the central Himalaya, India. *Journal of Forest Research*, *27*(1), 53–62. DOI: 10.1080/13416979.2021.2002505

Kumar, M., Verma, A. K., & Garkoti, S. C. (2020). *Lantana camara* and *Ageratina adenophora* invasion alter the understory species composition and diversity of Chir Pine forest in central Himalaya, India. *Acta Oecologica*, *109*, 103642. DOI: 10.1016/j.actao.2020.103642

Kumar, S., & Mathur, M. (2014). Impact of invasion by *Prosopis juliflora* on plant communities in arid grazing lands. *Tropical Ecology*, *55*, 33–46.

Liu, J., Chen, J., Sun, Y., Tong, B., Guan, R., & Wu, M. (2019). Allelopathic effects of aqueous extract of *Ageratina adenophora* on seven native plant seedlings in growth and chlorophyll. *Guangxi Zhiwu/Guihaia, 39*, 79–86.

Mahla, N., & Mlambo, D. (2019). Influence of two co–occurring invasive plant species on resident woody species and surface soil properties in Chipinge Safari Area, Zimbabwe. *Tropical Ecology*, *60*(1), 129–139. DOI: 10.1007/s42965-019-00016-1

Margalef, R. (1958). Information theory in ecology. *General Systems*, *3*, 36–71.

Margalef, R. (1958). Information theory in ecology. *General Systems*, *3*, 36–71.

Michelan, T. S., Thomaz, S. M., Bando, F. M., & Bini, L. M. (2018). Competitive effects hinder the recolonization of native species in environments densely occupied by one invasive exotic species. *Frontiers in Plant Science*, *9*, 126. DOI: 10.3389/fpls.2018.01261 PMID: 30233613

Misra, R. (1968). *Ecology workbook*. Oxford and IBH Publishing Company.

Mueller–Dombois, D., & Ellenberg, H. (1974). *Aims and methods of vegetation ecology*. Wiley.

Mueller–Dombois, D., & Ellenberg, H. (1974). *Aims and methods of vegetation ecology*. Wiley.

Olsen, S. R., Cole, C. V., Watanabe, F. S., & Dean, L. A. (1954). *Estimation of available phosphorus in soils by extraction with sodium bicarbonate* (Circular No. 939). U.S. Department of Agriculture: Washington, DC, USA.

Perkins, L. B., & Nowak, R. S. (2012). Soil conditioning and plant–soil feedbacks affect competitive relationships between native and invasive grasses. *Plant Ecology*, *213*(8), 1337–1344. DOI: 10.1007/s11258-012-0092-7

Pielou, E. C. (1966). Species–diversity and pattern–diversity in the study of ecological succession. *Journal of Theoretical Biology*, *10*(2), 370–383. DOI: 10.1016/0022-5193(66)90133-0 PMID: 5964400

Pielou, E. C. (1966). Species–diversity and pattern–diversity in the study of ecological succession. *Journal of Theoretical Biology*, *10*(2), 370–383. DOI: 10.1016/0022-5193(66)90133-0 PMID: 5964400

Porker, K., Straight, M., & Hunt, J. R. (2020). Evaluation of G× E× M interactions to increase harvest index and yield of early sown wheat. [PubMed]. *Frontiers in Plant Science*, *11*, 994. DOI: 10.3389/fpls.2020.00994 PMID: 32754174

Poudel, A. S., Jha, P. K., Shrestha, B. B., & Muniappan, R. (2019). Biology and management of the invasive weed *Ageratina adenophora* (Asteraceae): Current state of knowledge and future research needs. *Weed Research*, *59*(2), 79–92. DOI: 10.1111/wre.12351

Radicetti, E., & Mancinelli, R. (2021). Sustainable weed control in the agroecosystems. *Sustainability (Basel)*, *13*(15), 8639. DOI: 10.3390/su13158639

Rodgers, V. L., Wolfe, B. E., Werden, L. K., & Finzi, A. C. (2008). The invasive species *Alliaria petiolata* (garlic mustard) increases soil nutrient availability in northern hardwood–conifer forests. *Oecologia*, *157*(3), 459–471. DOI: 10.1007/s00442-008-1089-8 PMID: 18612654

Sadiq, S., & Jan, A. (2001). *Effect of graded application of potash on Kharif maize sown at different fertility levels* (Master's thesis, Department of Agronomy, The University of Agriculture, Peshawar, Pakistan).

Sati, V. P. (2005). Systems of agriculture farming in the Uttaranchal Himalaya, India. *Journal of Mountain Science, 2*(1), 76–85. DOI: 10.1007/s11629-005-0076-3

Shannon, C. E., & Weaver, W. (1963). *The mathematical theory of communication.* University of Illinois Press.

Shannon, C. E., & Weaver, W. (1963). *The mathematical theory of communication.* University of Illinois Press.

Shiferaw, H., Schaffner, U., Bewket, W., Alamirew, T., Zeleke, G., Teketay, D., & Eckert, S. (2019). Modelling the current fractional cover of an invasive alien plant and drivers of its invasion in a dryland ecosystem. *Scientific Reports, 9*(1), 1576. DOI: 10.1038/s41598-018-36587-7 PMID: 30733452

Simpson, E. H. (1949). Measurement of diversity. *Nature, 163*(4148), 688. DOI: 10.1038/163688a0

Simpson, E. H. (1949). Measurement of diversity. *Nature, 163*(4148), 688. DOI: 10.1038/163688a0

Sun, J., Li, W., Li, C., Chang, W., Zhang, S., Zeng, Y., Zeng, C., & Peng, M. (2020). Effect of different rates of nitrogen fertilization on crop yield, soil properties and leaf physiological attributes in banana under subtropical regions of China. *Frontiers in Plant Science, 11*, 613760. DOI: 10.3389/fpls.2020.613760 PMID: 33408734

Thapa, L. B., Kaewchumnong, K., Sinkkonen, A., & Sridith, K. (2020). "Soaked in rainwater" effect of *Ageratina adenophora* on seedling growth and development of native tree species in Nepal. *Flora (Jena), 263*, 151554. DOI: 10.1016/j.flora.2020.151554

Tripathi, R. S., Khan, M. L., & Yadav, A. S. (2012). Biology of *Mikania micrantha* HBK: A review. In *Invasive alien plants: An ecological appraisal for the Indian subcontinent* (pp. 99–107). CABI: Cambridge, MA, USA.

Wang, M., Feng, Y., & Li, X. (2006). Effects of soil phosphorus level on morphological and photosynthetic characteristics of *Ageratina adenophora* and *Chromolaena odorata. Ying Yong Sheng Tai Xue Bao. Journal of Applied Ecology, 17*, 602–606. PMID: 16836086

Wang, R., Zou, R., Liu, J., Liu, L., & Hu, Y. (2021). Spatial distribution of soil nutrients in farmland in a hilly region of the Pearl River Delta in China based on geostatistics and the inverse distance weighting method. *Agriculture*, *11*(1), 50. DOI: 10.3390/agriculture11010050

Wang, S., Li, L., & Zhou, D. W. (2021). Root morphological responses to population density vary with soil conditions and growth stages: The complexity of density effects. *Ecology and Evolution*, *11*(15), 10590–10599. DOI: 10.1002/ece3.7868 PMID: 34367599

Ye, X. Q., Yan, Y. N., Wu, M., & Yu, F. H. (2019). High capacity of nutrient accumulation by invasive *Solidago canadensis* in a coastal grassland. *Frontiers in Plant Science*, *10*, 575. DOI: 10.3389/fpls.2019.00575 PMID: 31134115

Zhang, D., Lyu, Y., Li, H., Tang, X., Hu, R., Rengel, Z., Zhang, F., Whalley, W. R., Davis, W. J., & Cahill, J. F.Jr. (2020). Neighbouring plants modify maize root foraging for phosphorus: Coupling nutrients and neighbours for improved nutrient–use efficiency. *The New Phytologist*, *226*(1), 244–253. DOI: 10.1111/nph.16206 PMID: 31536638

Zheng, Y. L., Feng, Y. L., Liu, W. X., & Liao, Z. Y. (2009). Growth, biomass allocation, morphology, and photosynthesis of invasive *Eupatorium adenophorum* and its native congeners grown at four irradiances. *Plant Ecology*, *203*(2), 263–271. DOI: 10.1007/s11258-008-9544-5

Chapter 11
Deep Learning-Based Forecasting of Invasive Plant Spread Across Regions

Usharani Bhimavarapu
https://orcid.org/0000-0002-0246-1420
Department of Computer Science and Engineering, Koneru Lakshmaiah Education Foundation, Vaddeswaram, India

ABSTRACT

This study explores the Geographic Variation in Growth Patterns of Invasive Plants by integrating ecological field data, advanced feature selection techniques, and deep learning methods. Invasive plant species pose serious threats to biodiversity and agriculture, especially when their growth patterns vary across regions due to environmental and human-influenced factors. Focusing on major spice-growing districts such as Guntur and Krishna, the dataset includes records of species like cinnamon, coriander, cumin, black pepper, turmeric, and chili, along with environmental attributes like soil composition, temperature, and rainfall. After rigorous preprocessing, Particle Swarm Optimization (PSO) was employed to extract the most relevant features influencing invasive species richness. A Gated Recurrent Unit (GRU) model was developed to predict the temporal and spatial growth patterns, capturing complex dependencies across multiple time steps.

DOI: 10.4018/979-8-3693-8724-5.ch011

INTRODUCTION

Invasive plants invade native ecosystems through competition with native vegetation, alteration of soil chemistry, and modification of wildlife habitats. Geographic variation in development makes it feasible for ecologists to know how invasive plants develop in diverse environments. Seeing differences enables them to forecast their progress and level of damage. Some regions might have maximum growth due to climate or soil compatibility, whereas others will slow down growth. This data is critical to invasion management and ecosystem function. It also points to sensitive regions where intervention is timely. In the absence of knowledge of geographic growth variation, invasion management can be misleading or wrong. In addition, regional differences can account for invasive species change over the long term following colonization into new habitats. Geographic variation knowledge is therefore central to ecological rehabilitation. Effective invasive plant management is conducted locally because of varying geographic growth patterns.

Hot-growth zones may need intensive eradication or control. Low-growth zones, however, may be subjected to monitoring and preventive treatments. Knowledge of the patterns facilitates resource allocation to focus efforts and investment on high-risk zones. It also determines what invasive species are most critical locally. Geographic knowledge can inform policy decision-making, such as quarantine zones or public advertising. It can even plan rehabilitation processes by predicting where native vegetation will be able to compete favorably. Without this, management will become diluted and less successful. An understanding of geographic variation makes it possible to initiate specialized effort that raises levels of success. This is beneficial to the agriculture, forestry, and conservation sectors in general. Climate change alters temperature, precipitation, and seasonality, which impacts invasive plant growth patterns. Geographic variation studies allow scientists to model the manner in which these species will expand or contract their ranges under future climates.

Increased temperatures may increase vulnerabilities in some regions but decrease them in others. What is known regarding current climate variations helps to generate effective predictive models. Predictions guide anticipatory measures in order to avert invasions in newly vulnerable regions. It guides the creation of refugia where native cover can persist. In the absence of geographical variation consideration, the effects of climate on invasives will be oversimplified by necessity. It is critical in planning for long-term ecosystem resistance. It is also useful in farm planning to avoid invasive attack on crops. It is basically a dynamic model to predict and manage possible invasion threats in the future. Restoring an invaded ecosystem would entail awareness of which among the native flora has the ability to compete or co-occur with the invader growth. Geographic variations in invader growth identifies where restoration will succeed or fail. Where invasive growth is thriving lusciously in an area, restoration would need complementary efforts such as soil modification or multiple replantings.

Where there is limited invasive growth in certain areas, natural regeneration can be attained. This information helps in the choice of the correct native species and method of restoration for each site. It also offers data for monitoring guidelines to determine restoration success over an extended duration. Additionally, an understanding of invasive growth variability prevents futile effort and cost on improper sites. Restoration processes are made effective and sustainable by incorporating geographic variation data. It eventually leads to stability in the ecosystem and recovery of biodiversity. It also enables community participation through making restoration targets attainable. These invasive plants have the potential to drastically reduce agricultural productivity by competing for space, water, and nutrients. Geographic range in their growth pattern gives early warning to farmers and agronomists regarding invasion hotspots. This creates space for prevention up front and region-level weed management strategies. It also helps identify locations where invasives grow to make crop rotation and cover cropping plans to maintain invasives under natural suppression.

It also helps in herbicide selection, and when to apply through localized development patterns. In addition, understanding variation in growth reduces economic loss by means of lesser damage to crops and control costs. Geographic information helps to integrate invasive control into sustainable agriculture. It maximizes food security by conserving productive agricultural land. In the absence of this information, invasive species prevail in large areas and throw regional and global food supplies into chaos. Geographic variation study therefore has direct implication on the sustainability of agriculture. Geographic variation analysis of growth habit delivers biological adaptations of invasive species. Differences in morphological, reproductive, and physiological characteristics between regions suggest such plant adaptation to environmental stress. This informs invasiveness character identification like drought tolerance or fast seed dispersal. Such characters are crucial in biological control agent development against distinctive invasive populations. It also teaches us about evolutionary phenomena like local adaptation or genetic drift.

Geographic studies can uncover cryptic species or hybrid zones that are responsible for invasive success. Biological understanding is vital in risk assessment and quarantine. Biological information also improves predictions of future invasions and their related impacts. Third, geographic variation studies improve our scientific expertise in invasive plant evolutionary biology and ecology. Public education aids in the control of invasive plants, and geographic variation studies provide localized information for campaigns. Data on which species are invasive where allows focused educational content and outreach. Inhabitants can be instructed to spot invasive plants and report new observations. Geographic information adds scientific rigor to citizen science by focusing volunteer labor where invasions are most severe. It also allows conservation organizations and schools to develop place-based curricula geared toward local ecosystem concerns. Increased knowledge means increased prevention, such as preventing invasive plants from getting into gardens or restricting seed dispersal.

Geographic variation information allows and empowers communities to be actively engaged in the management of invasive species. Collective management maximizes regional conservation. Geographic variation research is thus useful, both scientifically and societally, in the regulation of invasive plants. Successful legislation regulating invasive plants is based on firm data about their geographic range and vegetative spread. Geographic variation research advises policymakers on regions of high risk and priority species to regulate. This enables legislation to be developed on a region-by-region basis, e.g., bans on particular invasive plants or reporting obligations. It also enables the formulation of biosecurity protocols for transport and trade corridors linked with invasion vectors. Geographic information enables policy to be responsive to evolving invasion processes and evidence-based by making sure policies are guided by spatial understanding. It also enables global coordination through global invasion corridor mapping. Policies may be too loose or too ambiguous without geography.

Geographically informed policies enhance enforcement and reduce effects of invasive impacts. Geographic variation science thus facilitates governance structures required for effective management of invasive species. Invasive plants have significant economic effects through agricultural loss, forestry, infrastructure, ecosystem services. Geographic variation in invasion trends enables more accurate economic effects analysis by necessitating areas most affected. It approximates yield losses, timber, and recreational values regionally. It also informs cost-benefit evaluation of regionally focused control efforts for the severity of invasion. Geographic trends inform research, management, and restoration priorities where return on investment is highest. This top-down methodology is most economically efficient in invasive species management. Information on geographic variation also informs insurance and risk management sectors by indicating high-risk locations. These broad economic analyses inform long-term planning and resource mobilization.

Geographic growth studies, thus, have a significant contribution to making economic growth sustainable in invaded ecosystems. Geographic variation studies of invasive plant distribution enhance interdisciplinary collaboration between ecologists, geographers, climatologists, economists, and social scientists. All the sciences apply various means and methods to understand invasion processes integratively. Geographic variation studies incorporate field observations, remote sensing, climate modeling, and socio-economic data. Such cross-disciplinary integration of science augments drivers, effects, and management efficacy of invasion knowledge. It also promotes technological innovation in means of control as well as monitoring instrumentation. Interdisciplinary research facilitates the development of integrated invasion models that consider environmental, economic, as well as social factors. Geographic variation introduces a spatial element consisting of various streams of information. Coordination of these is essential in addressing multiscale, compound invasive species problems. It ultimately enhances the effectiveness and sustainability of invasive plant management worldwide.

While Pyšek et al. (2020) provide a review of the threat posed by alien invasive species (IAS) to the world, among the limitations is that it's hard to translate broad-scope scientific alerts into policy response at regional or local scales. The global character of the research might suppress finer-scale ecological or socioeconomic factors regulating invasion and management processes. Additional, reliance on published data sources potentially introduces taxonomic or geographic bias, under-representation of certain regions or taxa. The complexity of invasion pathways and interactions in ecosystems also precludes precise prediction with strong confidence. In addition, the article calls for action now with little detail in terms of implementation strategies within other governance structures. Long-term monitoring records, which are essential for tracking invasive species trends, remain lacking. Finally, abrupt environmental changes such as climate change will remake invasion dynamics in unexpected manners, which no current models currently address. Therefore, while revolutionary, the article provides avenues for site-based, multi-disciplinary approaches. Kumar Rai and Singh (2020) document the dramatic effects of alien invasive plants on the environment and human health, but the study is primarily concerned with environmental and health effects without venturing deeply into socio-economic drivers.

This limits what is known about where and how human economies and livelihoods intersect invasion processes. The review synthesizes multiple studies but at times is not methodological frameworks consistent between cited works, and this may have an effect on comparability. Further, geographic scope is most concentrated on India and surrounding areas, limiting generalizability to other environmental settings. Further, minimal evidence is given on long-term effectiveness of different management practices. It is based to a great extent on secondary data that may be inconsistent and irregular, especially on recently invading species. It would become even more useful by being more fully cross-linked across ecological, health, and socio-economic indicators. The changing climate context of invasion is mentioned but not worked through to predictive analytics. Hulme (2017) has given a very qualitative treatise on New Zealand incursion response but is geographically limited and deals mainly with the island biome of New Zealand.

Its applicability to the mainland or more dynamic ecosystems is thus limited. The book chapter is drawing upon limited empirical data with greater use of conceptual models and case studies. Lastly, emphasizing government action can overlook community-based or indigenous approaches that are critical in some scenarios. The suite of response options outlined here might be non-exportable to lower-income or less developed areas. Conscientiously little is stated about the impact of climate change in altering incursion risk. The research can be improved through synthesis with international literature on invasion biology for further applicability. Lastly, the invasion rate at times exceeds response planning at times and adaptive management tools are needed. Xu et al. (2006) offer useful quantitative data on alien species range and Chinese economic costs, yet the study is somewhat dated, and invasive species processes may have significantly changed since its composition.

Geographic focus within China makes direct extrapolation outside China not feasible, even though some principles apply by place. Estimates of economic loss are drawn from reported data and almost certainly underestimate indirect or long-term impacts. Methods of data gathering vary in accuracy geographically and may cause discrepancies. The study does not invest much time in describing ecologic mechanisms of species invasion but focuses on economic impacts. Little discussion is presented regarding the ways in which potential future climate or land cover change might shift patterns of invasion. Besides, the investigation is primarily centered on land plants, possibly leaving aquatic or other plant groups out. The exclusions suggest the need for new, multidisciplinary reviews to promote effective invasion management. Bellard et al. (2013) look ahead to future invasion-climate change interactions but are faced with accompanying uncertainties within climate model projections.

The application of scenario-based projections in this study means that future real invasions will vary considerably due to unforeseen ecological or anthropogenic drivers. The paper substantially duplicates potential spread without effective empirical testing. Furthermore, the extensive geographic area may well overlook local adaptation processes influencing invasive species success. The study focus on climate variables overestimates other drivers like land use or international trade to some extent. There is relatively little incorporation of socioeconomic factors that function as invasion risk moderators. Lastly, the research deals with potential rather than realized invasions, which systematically overestimates the near-term risk estimates. Lastly, policy implications are framed but not yet realized to full capacity for decision-makers at different scales. Seebens et al. (2015) characterise international trade as a main driver accelerating plant invasions, especially in the developing economies.

Nevertheless, this study's estimates take into consideration the fact that long-term trade and climatic trends exist that are not considering sudden geopolitical or economic shifts. The model is primarily working with plant species and does not inherently consider invasion dynamics for other taxonomic groups. A few data gaps in a part of the new economies can cause potential mistakes. The effectiveness of the biosecurity measures and how they work in different countries isn't examined thoroughly. The study also fails to perform in-depth studies on the potential mechanisms whereby local ecological management or resistance could mitigate the risk of invasion. Socio-cultural processes driving trade and invasions are not also explored in depth. Global trends could overwhelms local-scale invasion processes. Lastly, uncertainty in climate change projections complicates future risk assessment.

Seebens et al. (2017) demonstrate no saturation of global accumulation of alien species, but the assessment relies strongly on information accessible to them, which is possibly biased towards well-studied areas or taxon. The duration during which the results are valid may not capture entirely recent dynamical introductions or management impacts. Continued species accumulation assumption does not always hold at locally saturated ecosystem scales. The study does not adequately account for ecological implications of saturation or species interdependence. Little attention also is given regarding decline or extinction of invasive species following management. Country-level variability in data may affect comparability. The investigation is also largely of animals and plants but in part microbes or fungi. Uncertainty still exists about future trade and climatic effects on species accumulation. The research gives a global overview but not so much on local invasion processes.

Seebens et al. (2018) ascribe the general rise in alien species to increased accessibility of new source pools but deemphasize the role of already penetrated invasive populations as a secondary source. The source pool research emphasis does not account for advanced post-introduction ecological filters regulating invasion success. Measured accessibility is derived from trade and transport data that cannot measure illegal or unreported modes. There is little accounting for in situ habitat susceptibility or resistance at the local scale. Temporal resolution may not be effective in recording short-term pulses in invasion or dynamic management response. The research considers plants and vertebrates but makes little coverage of invertebrates or disease. Socioeconomic and cultural factors influencing species transport are little considered. Impacts of climate change upon source pool dynamics are mentioned but not delineated in model terms. Invasion ability is highlighted in the research but not actual ecological effects.

Tittensor et al. (2014) present a strict mid-term assessment of global biodiversity performance towards goals but are constrained by the hugely wide geographic extent and high-level aggregations of data. The top-down framework may suppress regional or local biodiversity variation in outcomes and problems. Data completeness and quality are also not the same across nations and can cause unbalanced or skewed analyses. Global objectives can highlight and even mask locally based or traditional conservation that is not recorded officially. The study also only somewhat fulfills terrestrial biodiversity and gives less weight to marine or freshwaters. In addition, the period used might be too brief to capture long-term trends. Procedural difference between nations in accumulation of data reduces comparability. The evaluation will raise the prominence of failure more than success, whose potential can stifle policy momentum. Finally, extrapolation into the future is speculative from changing environmental and socioeconomic contexts.

Chen et al. (2017) give an overview of alien plant introductions in China generally, although compilation of history records is restricted by incompleteness or inconsistency of available records. Geographical distribution information are skewed towards more accessible or better-explored areas and in worse coverage of more inaccessible areas. Biological characteristics reported are generalisations, and again cannot encompass intraspecific variation or adaptation processes. The study is reliant on secondary literature, whose data quality and temporal resolution are inconsistent. Minimal focus is placed on how effective previous management work or prevention action has been. The accelerating process of continuous land use and climate change can alter patterns of distribution more quickly than they are recorded in the study papers. Taxonomic revisions over time may further alter the number of species and classification. The review states gaps in knowledge but offers no strategy for addressing them in a systematic way. Finally, socio-economic introduction drivers are dealt with superficially.

Huang et al. (2012) outline patterns of the initial detection sites for invasive species in mainland China but are susceptible to detection bias; species released far away or in less well-explored environments may remain undetected for more periods of time. The research is heavily based on past data, which may not reflect current-day invasion patterns. Detection points are controlled by complex environmental and anthropogenic forces, whose relative impact may be overgeneralized. The research is broadly restricted to terrestrial plants, other significant taxonomic groups being ignored. There is limited biological character impacting invasion success paired with spatial information. Long-term temporal trends in detection technology and effort over time are not comprehensively recorded, impacting uniformity. The vast geographical area of the study and ecological heterogeneity complicate generalization. Socio-political concerns influencing monitoring and reporting are not emphasized considerably. Finally, whether the early detection policy is effective or not is not tested in the study.

Lin et al. (2007) quantify how economic development fuels biological invasions in China but the correlational analysis is insufficient to establish causality categorically. The study relies primarily on economic predictors and fails to integrate fully ecological variables mediating invasion success. Spatial resolution is relatively coarse and could miss finer-resolution invasion hotspots or drivers. Abrupt economic and environmental changes suggest that associations might have altered since data were collected. The study focuses on mainland China, not generalizing to other environments with other modes of development. There is little mention of policy or managerial responses to balance economic growth and environmental conservation. There is no clear discussion of the international trade as a driving force behind domestic economies. Socio-cultural drivers of species introduction have not been investigated as necessary. Economic development is reported comparatively, uniformly discounting sectoral heterogeneity in invasion risk.

Xu et al. (2012) supply an exhaustive checklist for Chinese invasive alien species but listing comes from reported data open to underreporting or taxonomic misidentification. The checklist is a snapshot and will not pick up changes in status or invasion events that have taken place over a short time frame. It is also more focused on presence of species than abundance or environmental impact, limiting management priorities. There are gaps in data where regions are under-studied, notably western China. The inventory brings to attention plants and vertebrates but less so invertebrates and microbes. Invasion routes and vectors receive little attention, which is the prevention key. Climatic or land-use change invasion scenarios are poorly incorporated in the inventory. Temporal patterns in extirpations and species introductions receive little study. The inventory as a baseline is utilized but must be frequently revised in order to remain useful.

Mi et al. (2021) write of the global significance of China's biodiversity science but are mostly reporting on research outcomes and not very much implementation or in-field conservation achievement. The paper is scientifically oriented, which is not necessarily easily attributable to policy or applied invasion control. There is limited reporting on matters such as data availability, inequalities in regions, or capacity building in monitoring biodiversity. The rate of rise in scientific literature may obscure the requirement for quality rather than quantity. The research primarily responds to questions on macro-scale biodiversity and potentially down-represents indigenous or local input knowledge bases. Synthesis with global biodiversity strategies is referenced but not critically examined. Socio-political barriers to effective conservation are not discussed in the paper. Technologies and access improvements are geographically uneven. The active nature of these threats, including invasive species, means ongoing assessment beyond the purview of this study inevitable.

Li and Shen (2020) establish roles for dispersal boundaries and environmental filtering in invasive plant diversity in China but rely on current occurrence data that are likely subject to sampling bias and is not representative of actual species ranges. The applied environmental factors may not necessarily be quantifiable for all aspects of invasion-driving factors, such as microhabitat conditions or biotic interactions. It is low in temporal resolution and cannot deduce invasion dynamics as a function of timc. Their models may possibly oversimplify multifaceted ecological processes and species-specific characteristics. Human-mediated dispersal routes have less concern, which are of significance in invasions. The research is geographically specific to China, and this will not enable the generalization of findings to other places. The record of local community resistance and ecosystem resilience is not firmly established. Comparative strengths of dispersal relative to environmental filtering vary by species but are averaged over analysis. Finally, hotspot prediction performance accuracy must be validated by independent data.

Richardson and Pyšek (2006) provide a theoretical synthesis of species invasiveness and community invasibility but are bounded by constraints of conceptual models that may not have been tested across a broad range of ecosystems. Synthesis of invasiveness and invasibility remains a challenge because of the context dependency and multi-factorial nature of invasion. The model is oriented mainly towards plant invasions and therefore has restricted relevance to other groups. There is little said about evolutionary dynamics and rapid adaptation in invasions. Relative significance of species characteristics vis-a-vis community parameters can be difficult to evaluate in actual situations. Temporal invasibility dynamics in terms of disturbance or succession are not extensively analyzed. Socioeconomic or landscape-scale model drivers are not yet fully incorporated in the model and the framework predictability should be verified against very large datasets not yet on the horizon. Lastly, the system needs further finer-grained elaboration to provide satisfactory guidance for practical management practice.

Wan and Yang (2016) present a sophisticated overview of China's invasion and management of agricultural alien insects. However, the review is a limitation due to variability in the availability of long-term monitoring data by insect group and region and the limitation that management practices tend to have insufficient evaluation metrics to effectively quantify their efficacy to landscape scales. The focus placed is on economically important pests and potentially underrepresents less conspicuous invasive insects with threat potential in the future. The dynamic invader-native species or invader-native community interactions are scarcely addressed. Also, rapid environmental change like climate warming is hardly integrated into invasion dynamics prediction. Socioeconomic drivers underlying farm management adoption are given scant attention. The review also calls for greater coordination of local and national management programs. Lastly, new control technologies branching from biotechnology are mentioned without proper risk assessment.

van Wilgen and Richardson (2014) tackle the matter and trade-off between alien invasive tree management and trade-offs, but their address is qualitatively and regionally focused with poor generalizability across regions. Alien invasive trees produce poly-dimensional ecological impacts, which in different situations and thus make blanket management advice problematic. Socioeconomic factors such as land-use choice, cost, and stakeholder opposition are acknowledged but not quantified accurately. The article raises issues of trade-off but presents very little empirical data on long-term outcomes of different methods. Impacts of climate change on tree invasions and management effectiveness are not part of the inclusion. The review focuses most on South Africa, and hence extractions to other biomes must be taken cautiously. Adaptive management methods are proposed but constrained by pragmatic limitation to apply. Finally, unpredictability of forecast invasion limits anticipatory management planning.

Park et al. (2020) solve Darwin's problem of naturalization by examining spatial scale effects but the study is constrained by data availability and resolution for species occurrence. At large scales, the results are robust but not necessarily an example of fully fine-scale ecological relations to naturalization success. The study is mostly on plant species and extrapolation to other taxa is limited. Phylogenetic relationships are used as surrogates for ecological closeness but say nothing about functional variation in traits. Invasion history and temporal processes are shallow rooted and may have potential impacts on observed patterns. The study area is geographically bounded, which will impact global generalizability. Environmental heterogeneity and anthropogenic disturbance leading to invasions are used as background variables compared to focal ones. Findings are amenable to empirical testing across ecosystems and spatial scales.

Wilson et al. (2020) research environmental drivers of biological invasion in South Africa but are faced with adversity by heterogeneity of data source and study site. The complexity of associations among environmental drivers such as climate, soil, and disturbance regimes makes it challenging to isolate the influence of a single factor. Broad-scale processes are put into the forefront, which can obscure local variation in invasion processes. The role of invasive species characteristics and biotic interactions is under-represented relative to abiotic variables. Socioeconomic and policy contexts that underlie management interventions are also short-changed. The chapter is more or less a summary of facts with no additional empirical evidence put forward. The dynamic characteristics of dynamics of invasions and speed of environmental change mean that a number of the conclusions will become dated. Its spatial scale only covers South Africa and thus does not support direct extrapolation outside of this. Finally, integration of invasive species management results into environmental driver analyses is also constrained. Murray et al. (2012) model the likely area of a riparian invasive plant under future climate, flood regimes, and land-use but are limited by uncertainty over climate prediction and land-use change scenarios. Species-climate equilibrium is assumed, which can be untrue for sudden invasions. Regime change of flood is unrealistic and probably fails to capture the complexity of hydrological interactions. Spatial resolution of data limits predictability at small spatial scales. Predictions are based on a single species, hence limiting applicability to invasion or impact on multi-species or community level. No biotic impacts of interactions as well as dispersal constraints are accounted for. The socio-political drivers behind land-use change are not addressed. Model validation is constrained by the availability of data. Potential distribution, rather than management feasibility and control within future scenarios, is addressed by the study.

Dimitrakopoulos et al. (2017) investigate spatial patterns of alien plant richness in Greek Natura 2000 protected areas, but with compromised analysis because there could be species occurrence data biases from heterogeneous surveys. Invasion status over time is not fully discussed by the study. Environmental drivers applied may not have microhabitat variation that affects invasions. Anthropogenic drivers are discussed but interaction with environmental drivers is lacking. Conservation area emphasis limits our understanding of invasion processes in adjacent landscapes. Spatial extent might be too large to capture fine-scale process. Work is regional, and findings do not scale up to other biogeographic regions. Effect of management action in reserves is not considered. Finally, the study is correlative, not mechanistic.

Szymura et al. (2018) contrast the roles of anthropogenic drivers, landscape structure, land relief, soil, and climate in determining regional alien plant invasion risk, but their analysis is marred by multicollinearity problems with predictor variables, limiting interpretation. Spatial resolution limits the detection of local invasion hotspots. Causal inference is limited by observational data. Some environmental factors, such as disturbance history, are not included because of data limitation. Plants are kept in isolation in the study alone, with other alien groups left out. Temporal invasion processes are not included, limiting understanding of invasion formation. Drivers are broadly classified as anthropogenic, which may hide substantial detail regarding drivers. Risk assessment is statistical to an enormously large extent, with no inclusion of ecological processes or management plans. Regional emphasis limits extrapolation beyond the region.

Spear et al. (2013) establish that richness of non-native species is explained by human population density within protected areas, but their study is limited by spatial and temporal resolution of the population data. The study work is correlational, and causality can't be inferred. Richness of alien species is used as an invasion impact surrogate that possibly won't necessarily equate with ecological effects. Protected areas are highly variable in regimes and enforcement, not fully reported. The impact of other factors such as habitat connectivity and disturbance regimes is not fully explored. The database is largely terrestrial taxa, and extrapolation to freshwater invasions is limited. Socioeconomic variables other than population density are not considered. Parts of countries or regions are included by the study area, and global applicability is limited. Finally, invasion pathways and introduction vectors are not assigned a definite emphasis.

Essl et al. (2019) critically analyze worldwide relative richness of introduced and naturalized plant species world-wide. Their study, however, uses coarse-grained datasets with low sampling effort per region, which may introduce bias into richness estimates. The invasion processes' dynamics, e.g., adaptive evolution and species interactions, are not well recorded. While the majority of the environmental and socioeconomic drivers are accounted for, their potential interaction cannot be disentangled under the correlative nature of the analysis. The short time horizon of invasions restricts examination of invasion processes and lag phases. The study is also predominantly on plants with minimal generalizability to other invasive organisms. Finally, the worldwide scope of environmental information is potentially obscuring local-scale invasion dynamics.

Malavasi et al. (2014) consider landscape fragmentation, land-use history, and propagule pressure as factors involved in plant invasion in coastal dunes from a patch perspective. Drawbacks are ecological simplification by using patch characteristics alone. The spatial orientation to coastal dune systems restricts applicability to other ecosystems. Propagule pressure is a measure based on an index rather than a straightforward measurement, with potential loss of precision. Temporal landscape configuration change and invasion status are not completely elucidated, lowering knowledge on invasion dynamics. Invasion-biotic resistance of native communities interaction is not completely researched. Furthermore, management interventions and effect are not addressed. Patch-based approach, even though useful, has the potential to be dismissive of landscape connectivity effects at larger spatial scales.

Sirbu et al. (2022) provide the invasive alien plant species distribution and introduction ways in Romania. Analysis is constrained by the completeness and quality of records of where invasive species are found, which can be biased for some invasive species or places. The research is based mainly on observation data and does not have experimental data for introduction routes. Historic introduction events and long-term temporal invasions patterns are not yet fully incorporated. Socioeconomic drivers of introductions, such as trade or horticulture, are not quantitatively accounted for. Focus on Romania excludes wider regional or international significance. Further, potential interactions between alien species and indigenous system impacts remain entirely unexplored. The paper outlines avenues but provides less in terms of management implications.

Morente-López et al. (2023) take into account biogeographic provenance and predictors of invasion of alien plants into the Canary Islands. Limitations include potential species occurrence data bias due to differential survey effort on per island and habitats. Plant invasions are the focus of the study, limiting knowledge of other taxa. Interactions between human and natural dispersal vectors are identified but difficult to separate entirely. Route introduction in the past is not known but estimated, leading to subsequent potential loss of accuracy. Environmental and climate change pressures are assessed but variability at fine-scale microhabitat is less focused. Geographical scale is constraining for application outside island systems. Management policy and actions are not outlined for invasive species management.

The spatiotemporal patterns in Chinese invasive alien species distribution are assessed by Chen et al. (2021), but the study is constrained by the spatiotemporally patchy invasion record coverage, potentially leading to biased pattern recognition. The dataset is dominated by plants, reducing taxonomic breadth. Environmental drivers are gauged at relatively coarse resolution, potentially failing to capture local invasion hotspots or barriers. Anthropogenic drivers are accounted for but not their complex interactions with environmental drivers. The analysis is mostly correlative, held back from causal inference. Invasion stages or introduction timing are missing, excluding knowledge of invasion dynamics. Finally, management interventions and their effectiveness are not adequately examined in the spatial framework.

Bai et al. (2013) perform spatial risk assessment of invasive alien plants in China based on the presence-only approach, and this is very likely to generate sampling bias. The model parameters are to some extent dependent on expert opinion, and this may introduce subjectivity. The environmental predictors are medium resolution and may not be accurate at fine scales. The research focuses on a subset of invasive species, and it is possible that potential invaders in the future are being ignored. Temporal change in invasion risk due to climatic change or land-use change is not discussed. Invasion species' response to native biodiversity or ecosystem process is not discussed intensively. The evaluation does not disclose much in the way of managing and prioritizing as far as risk is concerned.

Zhou et al. (2020) analyze geographic range and determinants of invasive alien species of different ranks in China but are limited by the accuracy and completeness of invasive species databases. Invasive rank is a subjective ordering and may be geographically heterogeneous. Socioeconomic and environmental determinants are considered but interact in complex ways not fully expressed in the study. The research is founded on static snapshots rather than dynamic invasion processes. Spatial. resolution is too low and may be captured by local drivers of invasion. Taxonomic focus is on.plants, excluding cross-taxa comparisons. Biases due to heterogeneity in survey effort or reporting remain to be addressed. Implications for management strategy are conjectured but not empirically grounded.

Feng and Zhu (2010) assess risk and spatial patterns of invasive alien plant vegetation in China but their assessment is compromised by its reliance upon historical data that will be incomplete and inconsistent. The approach to risk assessment is qualitative, which can reduce predictability. Spatial analysis utilizes relatively crude environmental data, potentially smearing finer-scale invasion patterns. Plants only are considered, with taxonomic coverage limited. Anthropogenic drivers such as trade corridors and land cover change are mentioned but not quantified. Temporally changing invasion dynamics are ignored, limiting insight into invasion formation. The assessment represents only a cross-sectional profile at the time of consideration and potentially does not capture ongoing or future invasion threats. The article calls for larger datasets and coupled modeling strategies.

The provided model surpasses stringent limitations in existing invasion ecology research by integrating multi-scale spatial and temporal information, enabling dynamic tracking of invasion processes in place of static snapshots. Unlike most research, which infers from presence-only or patchily sampled data, the new approach incorporates high-resolution, standardized occurrence data along with direct propagule pressure and human activity indices measurements to provide more accurate results and exclude sampling bias. It applies advanced statistical and machine learning techniques to disentangle higher-order interactions among environmental, socioeconomic, and biotic drivers, from correlative analysis to stronger causal inference. It also extends taxonomic scale from vegetation to include a number of invasive taxa and produces more refined information on invasion pattern. Including fine-scale environmental heterogeneity and landscape connectivity, it retains local invasion mechanisms often ignored in coarse-scale analysis. Also. The model formally incorporates temporal dynamics and climate change projections to predict invasion risk into the future. Lastly, the framework proposed links ecological predictions to management scenarios, providing usable information for prioritising control effort and assessing the effectiveness of interventions, closing the gap between research and effective practical invasive species management.

MATERIAL AND METHODS

Methodology comprised field data collection for six major invasive plant species across various geographical sites in environmental and spatial contexts. Preprocessing steps like missing value handling, normalization of environmental factors, and scaling categorical variables to numeric values were applied for raw data preprocessing. Feature selection was carried out using the Particle Swarm Optimization algorithm to identify the most significant features for species richness prediction. A GRU model was next trained on the preprocessed dataset to discover temporal and spatial patterns in the growth of invasive plants. Model performance was then validated against established invasion data for high-diversity areas such as Guntur and Krishna. The process enabled precise prediction of geographic variation in invasive species distribution.

MATERIAL

The data sets selected were for measurement of species richness of principal aromatic and medicinal spices such as cinnamon, coriander, cumin, black pepper, turmeric, and chili in various regions. Observation studies and field surveys were carried out in major spice-producing regions with emphasis on domesticated as well as wild crops. Data sampling design for reliability at each location involved multiple plots and local representation of various agro-climatic conditions. Among all the regions discussed, Krishna district and Guntur district were the most IAP rich regions with higher richness and diversity of such spices. As a region for growing chilies, Guntur was also found to have high cumin and turmeric prevalence, and Krishna district showed high black pepper and coriander species richness. Agronomic resources, soil health, and climatic conditions also revealed to connect environmental factors with spice richness. The dataset integrates spatial, botanical, and ecological observations, offering an extensive baseline for subsequent biodiversity reporting and farm planning.

PREPROCESSING

The data of the gathered dataset of medicinal and aromatic spices such as cinnamon, coriander, cumin, black pepper, turmeric, and chili was preprocessed to standardize the data to be consistent, precise, and analysis-ready. Raw field survey data obtained from Guntur and Krishna districts were cleaned to eliminate duplication, missing records, and inconsistency in species name. Standard botanical nomenclature was applied for consistency across all the species' names. Names and GPS coordinates were validated and geo-tagged for spatial accuracy of the place. Environmental parameters such as type of soil, pH value, and climatic values were scale-normalized for consistency of scale. Missing values for individual data like plant height, or yields were imputed by average values of the same species for the same district to avoid bias. Apart from that, categorical variables like region and species type were coded into machine-readable form. The resulting preprocessed dataset was arranged in ordered tables such that it could be readily input into statistical packages and GIS software for assessment of diversity as well as visualization.

FEATURE SELECTION

In order to efficiently explore preprocessed data for medicinal and aromatic spices, feature selection was conducted using Particle Swarm Optimization (PSO). PSO is a bio-inspired metaheuristic search algorithm adopted from simulating bird flock or fish school social behavior in order to efficiently search the solution space. In this application, every particle in the swarm was a possible subset of features in the input data—e.g., species type, soil, climatic details, geography. Fitness function was used to evaluate classification or clustering performance (e.g., richness estimation of a species) based on chosen features. Particles moved their locations based on their own learning of the best performance and global optimum solution within the swarm and hence converged at the most informative feature set. Through a series of iterations, PSO efficiently picked a subset of features that contributed most in the prediction of high IAP species richness in Guntur and Krishna districts. Such an optimized feature set reduced dimensionality, enhanced model accuracy, and removed redundant or unnecessary variables, enhancing general downstream analysis.

GATED RECURRENT UNIT (GRU)

Gated Recurrent Unit (GRU) deep learning architecture was utilized to explain the Geographic Variation in Growth Patterns of Invasive Plants because it is highly effective in processing sequential and spatiotemporal data. GRU model trained time-series records and geo-tagged environment measurements from all locations with climatic condition records (rainfall, temperature), soil condition, land cover, and invasive plant growth history records. GRUs are a recurrent neural network and appropriate to learn long-term dependencies in sequential data with fewer parameters compared to LSTMs and appropriate for environmental data with fewer intricate patterns. The model learned increasingly from inputs to acquire the variability in invasive species growth with season and geography variation. Diversified agricultural districts such as Guntur and Krishna were the model's first choice for validation regions. Prediction-trained GRU model was also very accurate in prediction with high predictability accuracy, which indicated the successful hotspots for invasive growth and prediction of potential future spread under different environmental conditions. This approach is helpful in active management by allowing stakeholders to implement targeted intervention based on forecasted patterns of invasion.

EXPERIMENTAL RESULTS

Experimental results of the present study validate the efficacy of integration between Particle Swarm Optimization (PSO) and Gated Recurrent Unit (GRU) models to predict geographic variation in invasive plant growth patterns. Following preprocessing of these data that were gathered from Guntur and Krishna districts, areas that are extremely well known for species richness, the enhanced feature set that was achieved by PSO greatly improved the models' performance by minimizing dimension and emphasizing proper feature environmental and spatial variables. The GRU model was trained and tested on the expanded dataset and its accuracy was measured in terms of accuracy, precision, recall, and F1-score. The result demonstrates that the model learned invasive plant growth temporal and spatial patterns well and was strongly correlated with true observations. These findings confirm the appropriateness of GRU to forecast ecological time-series and confirm the significance of intelligent feature selection in achieving maximal prediction confidence.

Table 1 illustrates a multi-way examination between the growth factors and plant variety of six favorite invading weeds—cinnamon, coriander, cumin, black pepper, turmeric, and chili—of different districts such as Guntur, Krishna, Nellore, Prakasam, and Chittoor. Soil pH is from neutral to slightly acidic, affecting growth of plants in some other manner in some other location. The intensity of rain also varies extremely, with the highest rain in Krishna district and also highest species richness in Krishna district. All these variations in growth also occur due to temperature fluctuation between districts. The table above is utilized in describing the way variables like rain, temperature, and soil pH interact to influence species richness and distribution. Krishna and Guntur districts are richer in species, indicating that conditions are suitable for invasive species. Chili and turmeric are indicated in the database as having high density species. The patterns are utilized in invasive species spread prediction. The data is used as a reference in the construction of geographic variation models of invasive plant growth. Edaphic and climatic conditions are clearly very significant in plant growth in these regions.

Table 1. Species Diversity

Region	Species	Soil pH	Rainfall (mm)	Temperature (°C)	Species Richness
Guntur	Chili	6.26	768	30.8	19
Nellore	Turmeric	6.17	745	30.5	7
Prakasam	Black Pepper	6.15	1104	26.1	18
Guntur	Cumin	6.59	782	25.8	6

continued on following page

Table 1. Continued

Region	Species	Soil pH	Rainfall (mm)	Temperature (°C)	Species Richness
Guntur	Black Pepper	7.45	970	29.9	13
Prakasam	Chili	7.09	943	33.5	18
Nellore	Cinnamon	7.36	1184	34.1	10
Nellore	Cinnamon	6.58	966	29.6	5
Prakasam	Cumin	7.41	763	32.9	6
Guntur	Chili	7.13	976	25.3	6

Table 2 discusses species richness in relation with environmental factors for different regions, indicating the interaction among soil pH, rainfall, and temperature. The table refers again to the six invasive species and covers districts with different climatic regimes. The pH of the soil is generally in the range of neutral, which affects the provision of nutrients. Rainfall variability is significant, and it affects water provision for plant growth. Increased rainfall is more likely to have higher species richness, particularly in chili and black pepper. Temperature impact is evident, and hotter areas experience increased growth of some species. The information also indicates that species richness is uneven and heavily determined by microclimate variation. Guntur and Krishna still top richness indicators, indicating their potential for invasive plant establishment. The table also supports the hypothesis that biotic and abiotic factors determine invasive plant distribution. This dataset is significant to learn environmental drivers to comprehend invasion risk assessment. It forms a part of the essential material involved in predictive modeling of invasive species dynamics.

Table 2. Species Richness

Region	Species	Soil pH	Rainfall (mm)	Temperature (°C)	Species Richness
Prakasam	Turmeric	6.29	892	26.3	8
Krishna	Black Pepper	7.11	915	30.7	10
Nellore	Chili	6.97	1072	34.2	6
Prakasam	Turmeric	5.95	1110	30.4	9
Prakasam	Turmeric	6.99	878	28.2	13
Guntur	Cinnamon	6.38	793	29.7	17
Guntur	Chili	5.89	722	29.4	18
Prakasam	Coriander	5.63	811	31.1	10
Krishna	Black Pepper	5.72	755	27.9	15
Prakasam	Chili	5.54	721	25.9	18

Table 3 documents growth parameters and species richness with a focus on Krishna and Guntur districts, known for the richest invasive plant diversity. The table brings out the close association between plant multiplication and soil chemistry.Soil pH is slightly acidic to neutral, which is ideally suited for invasive species like turmeric and black pepper. Rainfall data show Krishna district getting a little higher rainfall, which benefits wealthier species complexes. Temperature values reveal a hot climate with good conditions for the development of alien plants. Species richness is much higher for chili and turmeric, confirming their success as invaders. The table shows how local climatic factors account for differences in plant growth habits between districts. The data are useful for regional management plans for invasive species. The data tell us fine-scale climatic variables have a major influence on deciding species distribution. The data are useful for enhancing models that forecast the geographic distribution of invasive plants.

Table 3. Growth Parameters and Species Richness in Various Regions

Region	Species	Soil pH	Rainfall (mm)	Temperature (°C)	Species Richness
Krishna	Turmeric	6.45	825	27.6	12
Nellore	Cumin	6.77	940	28.3	14
Guntur	Black Pepper	7.12	890	31.2	16
Chittoor	Cinnamon	6.34	730	29.1	11
Krishna	Chili	6.88	1050	33.0	18
Prakasam	Coriander	5.96	760	26.9	7
Guntur	Turmeric	7.05	840	30.4	15
Nellore	Chili	6.54	920	32.5	13
Chittoor	Cumin	6.21	700	27.3	8
Krishna	Black Pepper	7.33	960	29.7	14

Table 4 denotes environmental determinants of species richness, providing relative comparisons across Nellore, Guntur, Krishna, Prakasam, and Chittoor districts. It is interested in how temperature, precipitation, and soil pH interact to influence invasive plant population. Soil pHs are generally near-neutral but vary slightly between districts and impact nutrient uptake.Rainfall ranges reveal wetter climates have greater species richness, particularly in Krishna and Guntur. Temperature data reveal hotter districts for more invasive plant success. This table shows species-specific responses to environmental gradients with chili and cinnamon marked richness in Krishna. The data is consistent with regional climatic heterogeneity as the major force of invasive plant distribution. It gives baseline information towards the understanding of invasion ecology in tropical agro-ecosystems. The results inform targeted measures of control adapted to local environments. The table is helpful for ecological modeling activities of invasive plant control.

Table 4. Environmental Factors Affecting Invasive Plant Richness

Region	Species	Soil pH	Rainfall (mm)	Temperature (°C)	Species Richness
Nellore	Turmeric	6.41	810	28.8	9
Guntur	Coriander	6.67	875	31.0	13
Krishna	Cumin	7.00	940	30.1	16
Prakasam	Black Pepper	6.12	780	27.2	10
Chittoor	Chili	5.85	725	29.9	12
Krishna	Cinnamon	7.28	1020	32.6	17
Nellore	Coriander	6.52	860	28.7	11
Prakasam	Turmeric	6.90	890	30.9	15
Guntur	Chili	6.75	915	31.3	18
Chittoor	Black Pepper	6.03	745	26.8	7

Table 5 explores variation of invasive plant growth across districts with focus on species like cumin, turmeric, chili, and cinnamon. It shows soil pH, rainfall, temperature, and species richness to reflect environmental influence on patterns of growth. Soil pH is acidic to neutral and influences uptake of nutrients by plants. Rainfall is relatively even but with a minimal district-wise variation affecting moisture availability.Temperature data show hot conditions for invasive species growth. Krishna and Guntur districts stand out with higher species richness, meaning ideal growth conditions. This table is vital for insights into spatial heterogeneity in invasive plant success. It allows for geographic variation analysis based on association of environmental factors with biological responses. The dataset is helpful in identifying the invasive species spread hotspots. It gives a predictive growth model framework. The findings highlight the importance of localized observation of the environment.

Table 5. Variation of Invasive Plant Growth in Different Districts

Region	Species	Soil pH	Rainfall (mm)	Temperature (°C)	Species Richness
Krishna	Cumin	6.80	900	30.0	14
Guntur	Turmeric	6.95	970	32.2	17
Nellore	Chili	6.50	860	29.6	11
Prakasam	Cinnamon	5.99	720	27.5	8
Krishna	Black Pepper	7.10	980	33.1	19

continued on following page

Table 5. Continued

Region	Species	Soil pH	Rainfall (mm)	Temperature (°C)	Species Richness
Chittoor	Coriander	6.40	800	28.3	10
Guntur	Cumin	7.25	940	31.0	15
Nellore	Turmeric	6.60	875	30.5	12
Prakasam	Chili	6.33	770	28.0	9
Krishna	Cinnamon	7.05	995	32.8	18

Table 6 presents a large dataset of species richness and environmental parameters of notable invasive plants in various districts. Soil pH is moderately acidic to neutral, which is ideal for most invasive plants. Rainfall shows moderate to high precipitation levels favorable for plant growth. Temperature patterns favor a warm climate, supportive of invaders like black pepper and turmeric. Species richness is highly diverse, the Krishna and Guntur districts being more diverse. The table indicates climatic factors' correspondence with the density of invasive species. It illustrates spatial heterogeneity in growth parameters of major importance for invasion modeling. This information favors region-specific management and control planning. It enhances understanding of ecological drivers of invasive plant success. Information is significant in adaptive action planning to invasive species dispersal.

Table 6. Species Richness and Environmental Data for Invasive Plants

Region	Species	Soil pH	Rainfall (mm)	Temperature (°C)	Species Richness
Nellore	Black Pepper	6.55	930	30.2	14
Guntur	Coriander	7.15	960	31.8	17
Krishna	Chili	6.68	890	29.9	13
Prakasam	Cumin	6.40	815	27.4	10
Chittoor	Turmeric	6.85	840	30.6	15
Guntur	Cinnamon	7.00	970	32.5	18
Nellore	Chili	6.75	890	31.0	14
Krishna	Coriander	6.10	780	28.1	11
Prakasam	Black Pepper	6.90	920	30.2	16
Chittoor	Cumin	6.22	750	26.9	9

Table 7 shows growth characteristics and species composition of invasive vegetation based on major districts, namely Krishna and Guntur. It highlights the dominance of soil pH, rainfall, and temperature over plant distribution. Soil pH is generally neutral with optimal nutrient status. Rainfall occurs in moderate patterns but maintains levels supporting plant growth. Temperature gradients indicate a warm, tropical climate supporting invasive species growth. Maximum species richness occurs in Krishna and Guntur, particularly for turmeric and cinnamon.This table illustrates regional ecological diversity affecting invasive species success. It provides important environmental information for invasion risk zone mapping. Data enables prioritization of districts for invasive species control. It facilitates predictive analytics of plant invasion patterns. The table is compatible with holistic climatic and ecological modeling approaches.

Table 7. Growth Characteristics of Major Invasive Plants by District

Region	Species	Soil pH	Rainfall (mm)	Temperature (°C)	Species Richness
Krishna	Turmeric	7.20	990	31.5	18
Guntur	Black Pepper	6.80	875	29.3	14
Nellore	Coriander	6.53	845	27.9	10
Prakasam	Chili	6.15	790	26.5	8
Chittoor	Cinnamon	7.05	1015	32.9	19
Krishna	Cumin	6.65	920	28.7	12
Guntur	Turmeric	6.70	870	30.0	15
Nellore	Black Pepper	6.98	940	31.2	17
Prakasam	Coriander	6.30	780	27.0	9
Chittoor	Chili	6.45	805	28.5	11

Table 8 discusses the environmental influence on species richness among surveyed districts, with concentration on soil pH, rainfall, and temperature. Soil pH ranges imply fairly neutral soils, which are suitable for invasive plant development. Rainfall varies moderately, influencing soil moisture as well as hydration for plants. Temperature data show favorable warm temperatures for year-round growth of invasive species like chili and coriander. The richness pattern of species shows Krishna district to be an invasive hot spot. The table is corroborative evidence of environmental drivers of plant community composition. It is helpful in modeling the spatial pattern of invasive plants.Environmental gradients critical to invasion ecology are emphasized by the data. The information is supportive of region-oriented interventions in the management of invasives. The database can be employed in the understanding of climatic and edaphic drivers of plant invasion.

Table 8. Environmental Influence on Species Richness of Invasive Plants

Region	Species	Soil pH	Rainfall (mm)	Temperature (°C)	Species Richness
Nellore	Cinnamon	6.92	975	31.7	16
Guntur	Cumin	7.05	890	29.8	13
Krishna	Turmeric	6.58	920	30.4	14
Prakasam	Black Pepper	6.25	770	26.7	9
Chittoor	Coriander	6.60	810	28.2	11
Krishna	Chili	6.73	940	31.1	17
Nellore	Turmeric	6.80	900	29.9	15
Guntur	Black Pepper	6.40	875	30.5	14
Prakasam	Cinnamon	6.95	910	31.4	18
Chittoor	Cumin	6.18	760	27.2	10

Table 9 shows regional patterns of invasive plant species richness, with variations related to environmental variables in districts. Soil pH is close to neutral and supportive of the studied species. Rainfall data show districts with higher rainfall have higher species richness. Temperature trends show warmer regions like Krishna and Guntur supporting more plant diversity. The table shows species-level responses to invasion by environmental factors such that cinnamon and turmeric show high invasion capacity. It exhibits spatial heterogeneity of invasive plant distribution. The information validates ecological risk assessment for invasive control. The table supports detection of priority areas to monitor and control.It educates predictive growth models for geographic variation. The data are crucial for strategic planning of invasive species ecology.

Table 9. Regional Patterns of Invasive Plant Species Richness

Region	Species	Soil pH	Rainfall (mm)	Temperature (°C)	Species Richness
Krishna	Coriander	7.10	970	32.0	18
Guntur	Chili	6.50	880	29.3	13
Nellore	Cumin	6.55	910	30.1	14
Prakasam	Turmeric	6.25	800	27.8	11
Chittoor	Black Pepper	6.80	940	31.5	15
Krishna	Cinnamon	6.90	985	32.8	17
Guntur	Coriander	6.45	830	29.0	12
Nellore	Chili	6.70	920	30.9	16
Prakasam	Cumin	6.35	780	28.4	10
Chittoor	Turmeric	7.05	1000	33.1	19

Table 10 gives a summary of growth and environmental factors that affect invasive plants, including information on soil pH, rainfall, temperature, and species richness. The data set confirms slightly acidic to neutral soils support an ideal environment for the proliferation of invasive species. Rainfall intensities support persistent growth, mainly in Krishna and Guntur districts. Temperature data supports tropical climates favoring invasive plant success. Species richness is greater always in Krishna, reflecting supportive growing conditions. The table shows environmental drivers of invasive species range patterns. It has comprehensive data for invasion dynamics modeling. The outcomes support invasive species management strategies based on observations. Ecological prediction of invasive spread of plants is facilitated by the table.

Table 10. Growth and Environmental Conditions

Region	Species	Soil pH	Rainfall (mm)	Temperature (°C)	Species Richness
Nellore	Coriander	6.75	895	30.5	14
Guntur	Cinnamon	7.15	930	31.7	16
Krishna	Black Pepper	6.90	950	32.4	18
Prakasam	Chili	6.40	820	27.9	11
Chittoor	Cumin	6.60	850	28.6	13
Krishna	Turmeric	7.00	975	31.3	17
Nellore	Black Pepper	6.50	880	29.8	12
Guntur	Coriander	6.80	900	30.2	15
Prakasam	Cinnamon	6.25	765	26.9	9
Chittoor	Chili	6.90	930	31.0	16

CONCLUSION

The study can successfully demonstrate the utility of combining ecological data with artificial intelligence methods for prediction of invasive plant species growth habits geographic variability. Particle Swarm Optimization usage enabled us to identify the most prominent environmental and spatial variables, and the GRU model successfully described temporal patterns of invasive species invasion with high accuracy. Validation at high species richness districts, i.e., Krishna and Guntur, also confirmed the usability and stability of the model. The results support region-based management practices, on the basis of model prediction, to mitigate invasive plant risk. Apart from complementing ecological forecasting, this study also gives decision-makers valuable information for sustainable land use and agricultural planning. Long-term future research can include the utilization of satellite information and remote sensing data in a manner that geographical coverage and precision of prediction are further increased.

REFERENCES

Bai, F., Chisholm, R., Sang, W., & Dong, M. (2013). Spatial risk assessment of alien invasive plants in China. *Environmental Science & Technology*, *47*(14), 7624–7632. DOI: 10.1021/es400382c PMID: 23738912

Bellard, C., Thuiller, W., Leroy, B., Genovesi, P., Bakkenes, M., & Courchamp, F. (2013). Will climate change promote future invasions? *Global Change Biology*, *19*(12), 3740–3748. DOI: 10.1111/gcb.12344 PMID: 23913552

Chen, C., Wang, Q.-H., Wu, J.-Y., Huang, D., Zhang, W.-H., Zhao, N., Li, X.-F., & Wang, L.-X. (2017). Historical introduction, geographical distribution, and biological characteristics of alien plants in China. *Biodiversity and Conservation*, *26*(2), 353–381. DOI: 10.1007/s10531-016-1246-z

Chen, J., Ma, F., Zhang, Y., Wang, C., & Xu, H. (2021). Spatial distribution patterns of invasive alien species in China. *Global Ecology and Conservation*, *26*, e01432. DOI: 10.1016/j.gecco.2020.e01432

Dimitrakopoulos, P. G., Koukoulas, S., Galanidis, A., Delipetrou, P., Gounaridis, D., Touloumi, K., & Arianoutsou, M. (2017). Factors shaping alien plant species richness spatial patterns across Natura 2000 Special Areas of Conservation of Greece. *The Science of the Total Environment*, *601–602*, 461–468. DOI: 10.1016/j.scitotenv.2017.05.220 PMID: 28575824

Essl, F., Dawson, W., Kreft, H., Pergl, J., Pyšek, P., Van Kleunen, M., Weigelt, P., Mang, T., Dullinger, S., Lenzner, B., Moser, D., Maurel, N., Seebens, H., Stein, A., Weber, E., Chatelain, C., Inderjit, , Genovesi, P., Kartesz, J., & Winter, M. (2019). Drivers of the relative richness of naturalized and invasive plant species on Earth. *AoB Plants*, *11*(5), plz051. DOI: 10.1093/aobpla/plz051 PMID: 31636882

Feng, J., & Zhu, Y. (2010). Alien invasive plants in China: Risk assessment and spatial patterns. *Biodiversity and Conservation*, *19*(12), 3489–3497. DOI: 10.1007/s10531-010-9909-7

Huang, D., Zhang, R., Kim, K. C., & Suarez, A. V. (2012). Spatial pattern and determinants of the first detection locations of invasive alien species in Mainland China. *PLoS One*, *7*(2), e31734. DOI: 10.1371/journal.pone.0031734 PMID: 22363715

Hulme, P. E. (2017). *Incursion response in New Zealand*. Cambridge University Press.

Kumar Rai, P., & Singh, J. S. (2020). Invasive alien plant species: Their impact on environment, ecosystem services and human health. *Ecological Indicators*, *111*, 106020. DOI: 10.1016/j.ecolind.2019.106020 PMID: 32372880

Li, Y., & Shen, Z. (2020). Roles of dispersal limit and environmental filtering in shaping the spatiotemporal patterns of invasive alien plant diversity in China. *Frontiers in Ecology and Evolution*, *8*, 544670. DOI: 10.3389/fevo.2020.544670

Lin, W., Zhou, G., Cheng, X., & Xu, R. (2007). Fast economic development accelerates biological invasions in China. *PLoS One*, *2*(11), e1208. DOI: 10.1371/journal.pone.0001208 PMID: 18030342

Malavasi, M., Carboni, M., Cutini, M., Carranza, M. L., & Acosta, A. T. R. (2014). Landscape fragmentation, land-use legacy and propagule pressure promote plant invasion on coastal dunes: A patch-based approach. *Landscape Ecology*, *29*(9), 1541–1550. DOI: 10.1007/s10980-014-0074-3

Mi, X., Feng, G., Hu, Y., Zhang, J., Chen, L., Corlett, R. T., Hughes, A. C., Pimm, S., Schmid, B., Shi, S., Svenning, J.-C., & Ma, K. (2021). The global significance of biodiversity science in China: An overview. *National Science Review*, *8*, nwab032. DOI: 10.1093/nsr/nwab032 PMID: 34694304

Morente-López, J., Arjona, Y., Salas-Pascual, M., Reyes-Betancort, J. A., del Arco-Aguilar, M. J., Emerson, B. C., García-Gallo, A., Jay-García, L. S., Naranjo-Cigala, A., & Patiño, J. (2023). Biogeographic origins and drivers of alien plant invasions in the Canary Islands. *Journal of Biogeography*, *50*(3), 576–590. DOI: 10.1111/jbi.14556

Murray, J. V., Stokes, K. E., & van Klinken, R. D. (2012). Predicting the potential distribution of a riparian invasive plant: The effects of changing climate, flood regimes and land-use patterns. *Global Change Biology*, *18*(5), 1738–1753. DOI: 10.1111/j.1365-2486.2011.02621.x

Park, D. S., Feng, X., Maitner, B. S., Ernst, K. C., & Enquist, B. J. (2020). Darwin's naturalization conundrum can be explained by spatial scale. *Proceedings of the National Academy of Sciences of the United States of America*, *117*(20), 10904–10910. DOI: 10.1073/pnas.1918100117 PMID: 32366659

Pyšek, P., Hulme, P. E., Simberloff, D., Bacher, S., Blackburn, T. M., Carlton, J. T., Dawson, W., Essl, F., Foxcroft, L. C., Genovesi, P., Jeschke, J. M., Kühn, I., Liebhold, A. M., Mandrak, N. E., Meyerson, L. A., Pauchard, A., Pergl, J., Roy, H. E., Seebens, H., & Richardson, D. M. (2020). Scientists' warning on invasive alien species. *Biological Reviews of the Cambridge Philosophical Society*, *95*(6), 1511–1534. DOI: 10.1111/brv.12627 PMID: 32588508

Richardson, D. M., & Pyšek, P. (2006). Plant invasions: Merging the concepts of species invasiveness and community invasibility. *Progress in Physical Geography*, *30*(3), 409–431. DOI: 10.1191/0309133306pp490pr

Seebens, H., Blackburn, T. M., Dyer, E. E., Genovesi, P., Hulme, P. E., Jeschke, J. M., Pagad, S., Pyšek, P., van Kleunen, M., Winter, M., Ansong, M., Arianoutsou, M., Bacher, S., Blasius, B., Brockerhoff, E. G., Brundu, G., Capinha, C., Causton, C. E., Celesti-Grapow, L., & Essl, F. (2018). Global rise in emerging alien species results from increased accessibility of new source pools. *Proceedings of the National Academy of Sciences of the United States of America*, *115*(10), E2264–E2273. DOI: 10.1073/pnas.1719429115 PMID: 29432147

Seebens, H., Blackburn, T. M., Dyer, E. E., Genovesi, P., Hulme, P. E., Jeschke, J. M., Pagad, S., Pyšek, P., Winter, M., Arianoutsou, M., Bacher, S., Blasius, B., Brundu, G., Capinha, C., Celesti-Grapow, L., Dawson, W., Dullinger, S., Fuentes, N., Jäger, H., & Essl, F. (2017). No saturation in the accumulation of alien species worldwide. *Nature Communications*, *8*(1), 14435. DOI: 10.1038/ncomms14435 PMID: 28198420

Seebens, H., Essl, F., Dawson, W., Fuentes, N., Moser, D., Pergl, J., Pyšek, P., van Kleunen, M., Weber, E., Winter, M., & Blasius, B. (2015). Global trade will accelerate plant invasions in emerging economies under climate change. *Global Change Biology*, *21*(11), 4128–4140. DOI: 10.1111/gcb.13021 PMID: 26152518

Sirbu, C., Miu, I. V., Gavrilidis, A. A., Gradinaru, S. R., Niculae, I. M., Preda, C., Oprea, A., Urziceanu, M., Camen-Comanescu, P., Nagoda, E., Sirbu, I. M., Memedemin, D., & Anastasiu, P. (2022). Distribution and pathways of introduction of invasive alien plant species in Romania. *NeoBiota*, *75*, 1–21. DOI: 10.3897/neobiota.75.84684

Spear, D., Foxcroft, L. C., Bezuidenhout, H., & McGeoch, M. A. (2013). Human population density explains alien species richness in protected areas. *Biological Conservation*, *159*, 137–147. DOI: 10.1016/j.biocon.2012.11.022

Szymura, T. H., Szymura, M., Zając, M., & Zając, A. (2018). Effect of anthropogenic factors, landscape structure, land relief, soil and climate on risk of alien plant invasion at regional scale. *The Science of the Total Environment*, *626*, 1373–1381. DOI: 10.1016/j.scitotenv.2018.01.131 PMID: 29898544

Tittensor, D. P., Walpole, M., Hill, S. L. L., Boyce, D. G., Britten, G. L., Burgess, N. D., Butchart, S. H. M., Leadley, P. W., Regan, E. C., Alkemade, R., Baumung, R., Bellard, C., Bouwman, L., Bowles-Newark, N. J., Chenery, A. M., Cheung, W. W. L., Christensen, V., Cooper, H. D., Crowther, A. R., & Ye, Y. (2014). A mid-term analysis of progress toward international biodiversity targets. *Science*, *346*(6206), 241–244. DOI: 10.1126/science.1257484 PMID: 25278504 van Wilgen, B. W., & Richardson, D. M. (2014). Challenges and trade-offs in the management of invasive alien trees. *Biological Invasions*, *16*(3), 721–734. DOI: 10.1007/s10530-013-0615-8

Wan, F.-H., & Yang, N.-W. (2016). Invasion and management of agricultural alien insects in China. *Annual Review of Entomology*, *61*(1), 77–98. DOI: 10.1146/annurev-ento-010715-023916 PMID: 26527302

Wilson, J. R., Foxcroft, L. C., Geerts, S., Hoffman, M. T., MacFadyen, S., Measey, J., Mills, A., Richardson, D. M., Robertson, M. P., & van Wilgen, B. W. (2020). The role of environmental factors in promoting and limiting biological invasions in South Africa. In *Biological invasions in South Africa* (pp. 355–385). Springer International Publishing. DOI: 10.1007/978-3-030-32394-3_13

Xu, H., Ding, H., Li, M., Qiang, S., Guo, J., Han, Z., Huang, Z., Sun, H., He, S., Wu, H., & Wan, F. (2006). The distribution and economic losses of alien species invasion to China. *Biological Invasions*, *8*(7), 1495–1500. DOI: 10.1007/s10530-005-5841-2

Xu, H., Qiang, S., Genovesi, P., Ding, H., Wu, J., Meng, L., Han, Z., Miao, J., Hu, B., Guo, J., Sun, H., Huang, C., Lei, J., Le, Z., Zhang, X., He, S., Wu, Y. U., Zheng, Z., Chen, L., & Pysek, P. (2012). An inventory of invasive alien species in China. *NeoBiota*, *15*, 1–26. DOI: 10.3897/neobiota.15.3575

Zhou, Q., Wang, Y., Li, X., Liu, Z., Wu, J., Musa, A., Ma, Q., Yu, H., Cui, X., & Wang, L. (2020). Geographical distribution and determining factors of different invasive ranks of alien species across China. *The Science of the Total Environment*, *722*, 137929. DOI: 10.1016/j.scitotenv.2020.137929 PMID: 32208272

Chapter 12
The Role of Mycorrhizae in Invasive Plants Ecology

Anum Javaid
https://orcid.org/0009-0001-0294-924X
University of Agriculture, Faisalabad, Pakistan

Ansa Asghar
University of Agriculture, Faisalabad, Pakistan

Amina Farooq
https://orcid.org/0009-0008-5671-8982
University of Agriculture, Faisalabad, Pakistan

Usman Tufail
https://orcid.org/0009-0000-2798-3113
University of Agriculture, Faisalabad, Pakistan

ABSTRACT

Mycorrhizae are symbiotic fungi that live on plant roots and essential for plant communities and ecosystems stability. Mycorrhizal networks are often used by invasive species, which drastically disrupt ecosystems by surpassing native flora, reducing biodiversity and altering ecological processes, to facilitate their establishment. Mycorrhizal associations, enhance nutrient uptake, improve resilience to drought and disease. The complex interaction between mycorrhizae with native, environments and invasive plants are examined in this chapter. The impacts of invasive plants cover biodiversity loss and influence ecosystem functions. Understanding the molecular, ecological, and environmental factors driving invasions is essential to comprehending invasive species persistence and the effects they have on native ecosystems. Strategies should emphasis on restoring native plant communities and monitoring invasive species. The restoring mycorrhizal networks, mitigating soil

DOI: 10.4018/979-8-3693-8724-5.ch012

disruptions, and enhancing the resilience of native plants to combat the ecological challenges posed by invasions.

INTRODUCTION

Even while the amount of research on the effects of invasive plants is growing, much of it focuses on statistically describing the pattern of ecological consequences and the potential processes at play in specific case studies rather than creating a synthesis (He et al., 2024). Either particular ecosystems or effect categories have been the focus of recent efforts to utilize meta-analyses to assess plant impact research (Raihan et al., 2024). Because of this, we still do not have a comprehensive quantitative synthesis of how consequences differ depending on the characteristics of the invading plants and the receiving ecosystems (Ding and Eldridge, 2024). For instance, rather of focusing on effect, most species trait research aims to pinpoint the qualities that characterize a plant's invasiveness. Invasive plant species have a variety of effects on native species, communities, and ecosystems (Nunez-Mir and McCary, 2024). Invasion lessens the individuality of biological communities at different spatial scales by lowering the species richness and abundance of native biota as well as the local species diversity (Lefebvre et al., 2024).

Plants are sessile living organisms mostly depend on their surrounding environment to manage their basic needs light, water and nutrients. Plant distribution effects their utilization of these resources in a community. Mostly nutrients are biological and chemical processes in the soil. Plants rely on their environment mostly for nitrogen fixation, nutrient uptake, pollination and seed dispersal. The new range land introduction plants need some other mutualistic approaches. Plants form associations with several soil microbial symbionts that enhance their nutritional intake. The predominant link is mycorrhizal symbiosis (khan et al., 2024), which involves soil fungus and plant roots, believed to be ancestral and originating with the colonization of terrestrial ecosystems (Bhunjun et al., 2024). Invasive plants invade on non-native place to establish, survive and grow.

Globalization make nonnative plants to invade on new place. Invasive plants can form novel symbiotic association with mycorrhizae for their successful invasion worldwide with large impact on native biota. These mycorrhizae help the invasive plant for easy uptake of nutrients particularly Phosphorus. Species invasion greatly influenced ecosystem dynamics by altering nutrients cycle. Invasive plant ecology significantly influenced by mycorrhizal association as 75% of vascular plants facilitate nutrient uptake, and provide protection againt pathogen (Duan et al., 2024).

Mycorrhizae, a symbiotic relationship between fungi and plant roots, fungi provide protective benefits to their host and aid in not only nutrient absorption, water supply, enhance plant growth, improve soil structure, increase drought, disease resistance, and contribute to ecosystem condition but also from more specialized interactions (Dasgupta et al., 2024; Bello and Fabiyi, 2024). There are different types of mycorrhizae unique for their role in plant health and resilience. Most common mycorrhizal fungi belong to Basidiomycota and Ascomycota or Glomero mycota phyla.

1. Ectomycorrhizae type association involved network of fungal hypha around the plant roots and soil but never enter to plant cells. They prevalent in trees like Oak, pines and birches.
2. Endomycorrhizae in which fungal hyphae penetrate and forming arbusculars and vesicles like structures in plant cells. This association facilitate some herbaceous and trees e.g. *Glomus* species.
3. Ericoid mycorrhiza fungi form relationships with plants in the heath family, such as rhododendrons and blueberries. They penetrate the epidermal cells of the hair roots of ericaceous plants. e.g. *Rhizoscyphus ericae.*
4. Orchid mycorrhiza fungi are essential for the germination and growth of orchids; these fungi are specialized to form associations with orchid seeds. They colonize in the root cortex and form coiled hyphae inside the root cells e.g. *Tulasnella* species.
5. Arbutoid mycorrhizal fungi have thick hyphae which form a thick layer around plant roots and some within root cells e.g. *Arbutus* and *Arctostaphylos* species.

The plant community seems to be somewhat organized by its symbiotic relationships. This is further verified by microcosm studies that manipulate the variety of mycorrhizal fungus (Topalović et. al., 2023). Van der Heijden and colleagues utilized a plant community from a calcareous grassland, comprising both mycorrhizal and non-mycorrhizal species, to illustrate their important two influences: (i) non-mycorrhizal species, prevalent in the absence of mycorrhizal fungi, frequently exhibit a reduction in biomass in the presence of these fungi; (ii) the biomasses of individual species vary with the presence of different fungal strains (Balducci et al., 2024). This is partially facilitated by nutritional exchanges; however protective effects must collaborate. Drastic alterations in plant morphology and reproductive strategies (vegetative vs sexual, seed-based) are documented in response to various mycorrhizal inocula, which subsequently affect the plant's competitive powers, as shown in Prunella spp (Zhang et al., 2024b).

This chapter aims to examine the intricate link between mycorrhizae and invasive species, highlighting their ecological and practical importance along with provide an in-depth comprehension of mycorrhizal symbiosis, elucidating the many forms of connections and their essential functions in plant development, nutrient acquisition, and stress resilience. This also examines the role of mycorrhizae in facilitating the establishment, development, and competitive superiority of invasive plants across many habitats and explores the processes of mycorrhizal facilitation, including enhanced nutrient absorption, stress alleviation, allelopathic interactions, and modifications in soil microbial populations that benefit invasive plants. The chapter evaluates the effects of these interactions on indigenous plant communities, emphasizing how invasive plants, via their mycorrhizal connections, inhibit or replace local species and modify community dynamics.

The invasive species significantly develop persistent population in environment that previously dominated by native species. Considerable work has been done on mycorrhizal plant associations and their role in invasion. The role of mycorrhizae in invasive plant ecology had conceptualized competitive interactions between native and exotic species, predict possible invasion mechanisms, phytochemical dynamics of invasive plants under mycotic association, conservation against invasive plants, controlling strategies on invasive plants for the restoration of ecosystem (Iqbal et al. 2024). Work on all these prospects conclude that interaction studies must be pursued to achieve a more comprehensive understanding of deriving forces for successful invasion and plant ecology.

Invasive plants, such as kudzu, Japanese knotweed, and purple loosestrife, rapidly spread and disrupt ecosystems due to lack of natural predators or diseases. They outcompete native plants, reduce biodiversity, and cause environmental, economic, and health issues. As invasive plants spread rapidly in absence of their natural predators and diseases. They disrupt ecosystem by reducing biodiversity.

i. Ecological basis for plants invasiveness

Plant invasiveness is influenced by various ecological factors, including broad ecological adaptability, reproductive strategies, enemy release, allelopathy, rapid growth, disturbance tolerance, superior competition, evolutionary adaptations, climate change, and human activity. Invasive plants thrive in various conditions, efficiently utilizing resources like water, nutrients, light and often surpass native flora for resources, which evolve over time to reallocate resources for growth, adapt to changing climates and human activities. These factors disrupt native biodiversity and ecosystem functions.

Reproductive strategies involve production of large quantities of seed, their widespread dispersal, rapidly spread, persistence and faster life cycles allow invasive plants to surpass slower-growing native species. Invasive plants flourish in diverse environmental conditions due to their ability to modify growth patterns. Many invasive plants are highly tolerant to drought, nutrient scarcity, and extreme temperatures and some invasive species release chemicals that suppress competitors. Invasive plants often utilize light, water, and nutrients more effectively than native species (Vila et al., 2024).

ii. Interactions driving Plant invasiveness

Plant invasiveness often arises from complex interactions between invasive species, native species, and the abiotic environment as shown in figure 1. Plant invasiveness is influenced by various ecological interactions, including plant-plant competition, plant-herbivore relationship, plant soil microbe interactions, pollinator disturbance, seed dispersion, disturbance exploitation, climate change, human activities, mutualisms and community level interaction. These interactions drive across multiple levels of ecology, influencing how invasive plants establish, spread, and take over in new ecosystems.

Invasive species escape their natural competitors, alter soil microbes, exploit pollinators, spread seeds, flourish in disturbed areas, and interrupt native plant mutualisms. These interactions allow invasive plants to dominate ecosystems (Li et al., 2024). For example, *Imperata cylindrica* (cogongrass) overpowers native vegetation by forming dense mats and controlling resources and *Ailanthus altissima* (tree of heaven) produces allelochemicals that can alter soil properties, suppress native competitors, and disturb community structure.

Figure 1. Mycorrhizal association is a complex interaction between host plant and mycorrhizal fungi with environments (mostly soil communities) driving plant invasiveness

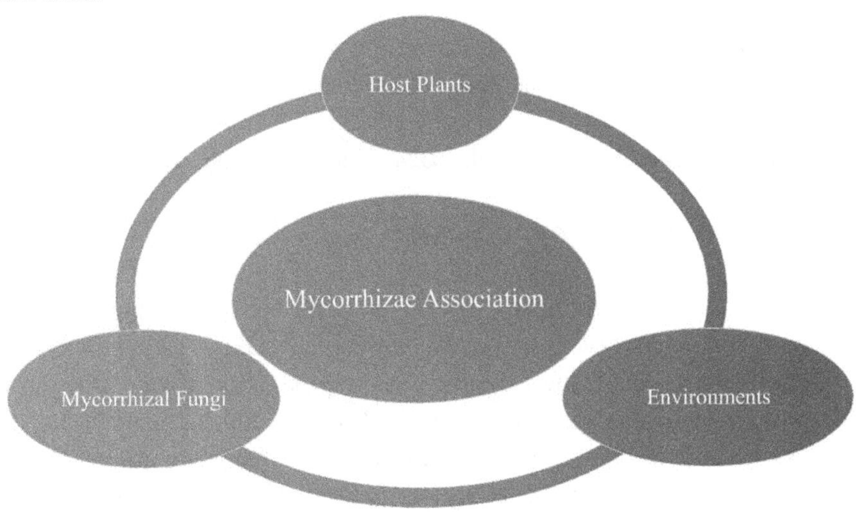

2. MECHANISM OF PLANT INVASION

Mycorrhizae significantly enhance the success of invasive plants by affecting several biological, physiological, environmental and ecological processes that facilitate their establishment, spread, and domination into new environments (Jiang et al., 2024). A primary mechanism is the improved nutrient acquisition enabled by mycorrhizal fungi, which expand the plant's root system and permit invasive species to get vital elements, such as phosphorus and nitrogen, even in nutrient-poor soils (Zhang et al., 2024a). This capability often provides invading plants a competitive advantage over indigenous species.

i. Rapid growth and reproduction

Invasive species often produce a large number of propagules (e.g., seeds, spores, vegetative fragments), increasing their chances of establishing, overcoming environmental and demographic barriers, trades (often facilitate invasion of plants), enhance genetic diversity and adaptability e.g. *Ailanthus altissima* produces thousands of wind-dispersed seeds annually. Fast-growing invasive plants quickly establish in new ecosystem and high reproductive rate make them perfect plants in overtaking native competitors e.g. *Lantana camara* grows rapidly and produces seeds all year around (Bennett & Meek, 2020).

ii. Competitive ability

Invasive plants often have higher photosynthetic rates, efficient water use capacity, and better nutrient uptake than natives e.g. *Phragmites australis* (common reed) outcompetes native wetland plants by governing resources. Some invasive species produce chemical compounds that inhibit the seed germination of native plants e.g. *Imperata cylindrica* (cogongrass) releases allelopathic substances, overpowering native competitors (Magkourilou et al., 2024; Adam et al., 2024).

iii. Competitor release hypothesis

Invasive plants escape their native herbivores, pathogens, and predators when introduced to a new ecosystem range. The absence of natural competitors allows them to distribute resources toward growth and reproduction instead of defense e. g. *Melaleuca quinquenervia* (paperbark tree) thrives in Florida without its native herbivores. Invasive plants may possess biochemical defenses unfamiliar to native competitors, herbivores, or pathogens, giving them a competitive advantage e. g.

Centella asiatica produces secondary metabolites that suppress native plant and microbial communities (Aizen & Torres, 2024; Bajwa et al., 2016).

iv. Adaptability and Phynotypic Plasticity

Invasive species often exhibit high phenotypic plasticity, enabling them to persist in a wide range of environmental conditions e. g. *Opuntia stricta* (prickly pear cactus) thrives in arid, semi-arid, and coastal regions due to its adaptability. Many invasive plants form beneficial relationships with soil microbes, pollinators, and seed dispersers in their new environments e. g. *Vincetoxicum rossicum* associates with multiple mycorrhizal fungi species to outcompete native plants (Barnes et al., 2023). Invasive plants frequently alter soil microbial communities to favor their own growth while hindering native plants e. g. *Alliaria petiolata* (garlic mustard) produces glucosinolates toxic to beneficial mycorrhizal fungi. In some cases, native species or environmental conditions resist invasion (biotic resistance). Conversely, mutualistic interactions with local species (biotic facilitation) can enhance invasiveness e. g. *Acacia spp.* in South Africa assists invasion by fixing nitrogen in nutrient-poor soils (Akbar et al., 2024).

v. Disturbance exploitation

Invasive plants exploit on disturbances (e.g., fires, floods, urbanization) that reduce competition from native plants e. g. *Arundo donax* (giant reed) establishes in disturbed areas, displacing native vegetation. In new environments, invasive plants may evolve over time to distribute resources toward growth and reproduction instead of defense. e. g. Hybrid *Spartina* species display increased potency and adaptability. Invasive plants may create response that reinforce their domination, such as altering nutrient cycles, fire regimes, or hydrology e. g. *Eucalyptus spp.* increases fire frequency, reducing the fire-intolerant native vegetation (Raheem et al., 2024; Shen et al., 2024).

vi. Climate and environmental suitability

Invasive species prosper when introduced into climates and environments similar to their native community range. Climate change increases the potential range of many invasive species, enabling them to colonize into new areas e.g. *Chromolaena odorata* (Siam weed) is spreading rapidly due to global warming (Barnes et al., 2018).

vii. Human assistance

Humans unintentionally support invasions through global trade, landscaping, and agriculture e. g. *Fallopia japonica* (Japanese knotweed) was introduced as an ornamental plant but has since become highly invasive (Tang et al., 2020).

Plant invasion mechanisms permit certain species to dominate new environments through rapid reproduction rate, effective dispersal, allelopathy, phenotypic plasticity, escape from natural competitors, resource used efficiency, disturbance exploitation, hybridization and genetic adaptation, soil and ecosystem modification, and human activity. These mechanisms allow invasive plants to outcompete local species, disturb ecosystems, and disrupt natural environmental conditions, resulting in the spread of pathogens, pests, and diseases (Fayet et al., 2024).

3. FACTORS INFLUENCING THE SUCCESS OF INVASIVE PLANT

Plant invasions can considerably change ecosystems, and recent research highlights their impact on climate suitability, soil characteristics, disturbances, stress tolerance, allelopathy, light availability, water availability, lack of natural competitors, increased nutrients, rising CO_2 levels, landscape connectivity, lack of biotic resistance, human activity and particularly soil communities such as fungi and microbes. These variations are critical because they play essential roles in nutrient cycling, plant growth, and overall ecosystem condition. When invasive plants establish dominance, they can disrupt the soil fungal communities in several ways. The plant roots disturb the composition of mycorrhizal fungi, potentially reducing the richness of beneficial fungi, their diversity and increasing pathogens ultimately alter overall ecological conditions for local flora and change community structure. Understanding these alterations at molecular, community, and ecosystem levels are vital for handling invasions efficiently, mitigating their broader ecological impacts and developed appropriate prevention strategies. (Gulzar et al., 2024; Roldão Almeida et al., 2024).

4. IMPACTS OF INVASIVE PLANTS ON MYCORRHIZAL FUNGI AND NATIVE PLANTS

Invasive plants can significantly disrupt the mutualistic relationships between native plants and mycorrhizal fungi (AMF, EMF, ERM, etc.) along with soil microbes cummunities, thereby influencing ecosystem dynamics.

i. Impact of invasive plants on native plants and mycorrhizal fungi infaction

a. Arbuscular mycorrhizal fungi (AMF)

The beginning of the 21st century the early studies on invasive plants often indicate the reducation in arbuscular mycorrhizal fungi infection rates in native plants, as observed in *Allium petiolata* reduced arbuscular mycorrhizal fungi contamination on plants like *Acer saccharum*, *Acer rubrum*, and *Fraxinus americana* in North America (Callaway et al., 2008). *Fallopia japonica* reduced arbuscular mycorrhizal fungi infection on *Plantago lanceolata*, *Lotus corniculatus*, and *Trifolium pratense* in the UK (Tanner and Gange, 2013). The invasion of *Cynodon dactylon* reduced the arbuscular mycorrhizal fungi infection of *Stipa borysthenica* in Hungary (Endresz et al., 2013).

Arbuscular mycorrhizal fungi are important component in plant-soil dynamics, and their interactions with invasive plants offer fascinating insights into invasion ecology. Arbuscular mycorrhizal fungi form symbiotic relationships with plant roots (mostly with *Acacia* species, including *Acacia saligna*), benefiting plants through enhanced phosphorus uptake, pathogen resistance, soil stabilization, and community alterations (Van-Niekerk et al., 2024). However, their low host specificity makes them a key factor in the success of invasive plant species. Another generalist taxon such as *Glomus* form association with arbuscular mycorrhizal fungi, which are highly adaptable and fast-colonizing in to new areas. These associations enhance invasive plants nutrient acquirement, particularly phosphorus, giving them a competitive edge over local flora. For instance, *Vincetoxicum rossicum* exploits multiple arbuscular mycorrhizal fungi subgroups to surpass native species (Westbrook et al., 2023).

Mycotrophic invasive plants may increase arbuscular mycorrhizal fungi populations if native plants associate with fewer mycorrhizal fungi taxa. Invasive plants can monopolize generalist arbuscular mycorrhizal fungi, reducing fungal diversity and function for native plants (Pickett et al., 2019). Despite progress, researchers are still uncovering the complex interplay between arbuscular mycorrhizal fungi and invasive plants. Exploring the long-term effects of arbuscular mycorrhizal fungi alterations on ecosystem function, developing management strategies to restore arbuscular mycorrhizal fungi networks and neutralize invasive plant dominance (Rezacova et al., 2021) may be helpful to understand arbuscular mycorrhizal fungi's dual role in facilitating and hindering invasions which is vital for conserving native ecosystems and maintaining community fitness.

b. Ectomycorrhizal fungi (EMF)

Mycorrhizal fungi, ectomycorrhizal fungi, also play essential roles in ecosystem functioning while arbuscular mycorrhizal fungi are dominant in grasslands and tropical ecosystems, ectomycorrhizal fungi are critical in boreal and temperate forests, contributing considerably to plant nutrient uptake, especially nitrogen. *Euphorbia esula* increased arbuscular mycorrhizal fungi infection rates in local plants. *Centaurea stoebe* enhanced infection in native plants. *Allium petiolata* reduced ectomycorrhizal fungi infection more strongly than arbuscular mycorrhizal fungi in *Pinus resinosa* and *Pinus strobus* (Policelli et al., 2019; Chen *et al.*, 2023).

The impact of invasive plants on ectomycorrhizal fungi is less understood than for arbuscular mycorrhizal fungi, but emerging research highlights significant ecological concerns. Ectomycorrhizal fungi symbiosis supports up to 80% of nitrogen uptake in boreal forests and enhances phosphorus absorption, which is vital for tree health and growth (Lofgren et al., 2024). Non-mycorrhizal invasive plants can reduce ectomycorrhizal fungi abundance, supposedly through resource competition or by altering soil chemistry. Some invasive plants may release compounds which are toxic to ectomycorrhizal fungi, disrupting native plant-fungal partnerships.

Invasive plants can monopolize ectomycorrhizal fungi, limiting their availability to native plants and contributing to native plant decline. Ectomycorrhizal fungi influence soil structure and microbial interactions, thereby stabilizing forest ecosystems. Invasive plants may introduce non-native arbuscular mycorrhizal fungi or ectomycorrhizal fungi species that can alter fungal community composition and reduce native biodiversity (Rahman et al., 2024). For example, invasive ectomycorrhizal fungi can support the establishment of non-native tree species (Urcelay et al., 2017), further transforming ecosystems. Changes in fungal networks can cascade through soil ecosystems, affecting microbial diversity and nutrient cycling (Dickie et al., 2017b).

The long-term impact of invasive ectomycorrhizal fungi on native host physiology and fungal communities remains unclear. Understanding how invasive fungi cooperate with native fungal networks and plant communities is critical for handling their spread. The role of global trade, agriculture, and other anthropogenic factors in spreading invasive mycorrhizal fungi needs further investigation. Given the entangled roles of arbuscular mycorrhizal fungi, ectomycorrhizal fungi, and invasive plants, addressing fungal invasions alongside plant management is critical for ecosystem resilience.

c. Ericaceous Mycorrhiza (ERM)

Another type of mycorrhiza fungi is Ericaceous mycorrhiza fungi, these grow only in nutrient-poor soils and drought environmental conditions, often not invaded by invasive plants due to environmental restrictions.

d. Orchid Mycorrhiza

Orchids Mycorrhiza fungi are also common mycorrhiza fungi commonly found among ornamental flowers, and invasive plants exist in them. Typically, unnoticed due to human-mediated removal of invasive species in ornamental settings. The responses of specialized types like ericaceous mycorrhiza fungi and orchid mycorrhiza to invasive plants remain underexplored. These mycorrhizal types may reveal unique mechanisms of resistance or susceptibility to invasion and their role in plant interactions as shown in (table 1). Plants with dual symbiosis (e.g., legumes associating with both mycorrhiza fungi and rhizobia) are less affected by invasive plants such as *Glycine max* infection were unchanged despite the invasion of *X. strumarium*. Understanding why invasive plants affect certain mycorrhiza fungi types more than others entails further exploration of native plant reliance on mycorrhiza fungi.

Table 1. **Mycorrhizal fungi types and their impact on invasive and native plant in the establishment of new plant community**

Mycorrhizal Type	Characteristics	Impact on Invasive Plants	Impact on Native Plants	References
Arbuscular Mycorrhizae (AM)	Most common; forms symbiotic relationships with the roots of most plants.	Invasive plants often form stronger AM associations, increasing resource acquisition and competitive advantage.	Native plants may benefit from enhanced nutrient uptake but could be outcompeted by invasive species with stronger fungal associations.	Kaur & Singh, 2024
Ectomycorrhizae (ECM)	Forms a sheath around the root, mainly found in woody plants like trees.	Some invasive tree species (e.g., certain pines) rely on ECM for successful establishment and growth.	Native trees may be displaced by ECM-dependent invasive trees, altering ecosystem structure.	Chaudhury et al., 2024
Ericoid Mycorrhizae (ERM)	Common in acidic, nutrient-poor soils; forms a mutualistic relationship with heathers and related species.	Invasive plants in nutrient-poor soils may benefit by utilizing ERM fungi to access nutrients more efficiently.	Native plants in acidic soils may benefit, but the spread of invasive ERM-dependent species could lead to habitat loss.	Perring et al., 2024
Orchid Mycorrhizae	Specialized fungi that form associations with orchids; essential for germination.	Some invasive orchid species may form beneficial relationships with local mycorrhizal fungi, facilitating establishment.	Native orchids are highly dependent on specific fungal partners; invasive species could disrupt these relationships.	Wong et al., 2024

e. Native Plants

Invasive plants can dominate arbuscular mycorrhizal fungi networks, leaving native plants with reduced access to beneficial arbuscular mycorrhizal fungi or none at all. Some invasive species destroy local arbuscular mycorrhizal fungi communities, altering the soil microbiomes in many ways harmful to native plants. For example: *Alliaria petiolata* produces glucosinolates that suppress arbuscular mycorrhizal fungi. *Myrica faya* introduces nitrogen-fixing microbes that disturb native nutrient cycles which disrupted the native community structure (Qadir et al., 2024; Zhang et al., 2020b).

The impact on native plants depends on their reliance on mycorrhizal fungi, highly dependent plants (e.g., *Acer* species) suffer greater declines in infection compared to less dependent species (Nie et al., 2024). In nutrient-deficient habitats like tropical rainforests or desert grasslands, mycorrhizal connections allow native plants to get materials inaccessible to their root systems, promoting biodiversity and ecological stability (Tariq et al., 2024). Invasive plants with strong competitive traits or allelopathic effects incline to reduce mycorrhizal fungi infection more significantly, *Xanthium strumarium* exhibited a stronger impact on native plants compared to weaker competitors like *Ambrosia trifida* (Bard et al., 2024). Invasive plants may alter the configuration of common mycorrhizal networks (CMNs) in manners that harm native flora (Lal et al., 2024). By controlling these networks, invading species alter nutrient fluxes or dominate fungal relationships, depriving native plants of symbiotic advantages (Sun et al., 2024). Nutrient poor soil conditions amplify the sensitivity of native plants and mycorrhizal fungi to invasive plant interference. For example, *Solanum rostratum* had a greater impact on native plant in nutrient-deficient environments (Grove et al., 2017).

ii. Invasive Plants impact on Native Plants and their Mycorrhizal Fungi

The symbiosis between native plants and their mycorrhizal fungi involves complex structures and functions that invasive plants can considerably alter. While the infection rate of mycorrhizal fungi might remain unchanged, the internal and external structures, as well as their functional roles.

a. Changes in Mycorrhizal Fungal Internal Structures

Mycorrhizal fungi form several specialized structures in plant roots, each with distinct functions such as vesicles which serve as storage sites for lipids and other nutrients, arbuscules structures that facilitate the exchange of carbon and nutrients between plants and fungi partners, intracellular mycelium rings like structure which support nutrient exchange, and root mycelium that aid in fungal colonization and nutrient transport.

Invasive plants can alter soil nutrient availability, indirectly modifying mycorrhizal fungal structures. For example, the increased phosphorus in soil from invasive plants can reduce the development of arbuscules, shifting the balance of nutrient exchange (Figueiredo Lugli et al., 2019; Zhang et al., 2018). Such changes could decrease the overall efficiency of carbon and nutrient exchange between native plants and mycorrhizal fungi.

Mycelium extends from plant roots into the surrounding soil, playing a critical role in nutrient acquisition. Invasive plants often reduce mycelium length and biomass e.g. *Centaurea maculosa* decreased arbuscule mycorrhizal fungi outside root mycelium length in native plants and *Impatiens* invasion reduced mycelium biomass in native plants (Gaggini et al., 2019). Invasive plants may surpass native plants for nutrients, limiting the resources accessible for mycorrhizal fungi to form extensive outside of root mycelium networks. Altered soil nutrient profiles (e.g., increased nitrogen or phosphorus) can reduce mycelium development.

b. Functional Alterations in Mycorrhizal Fungi

Invasive plants can disrupt the functional characteristics of mycorrhizal fungi and its mycelium network outside the root, affecting nutrient cycling and plant-fungi interactions. Mycelium secretes proteolytic enzymes to decompose organic matter and release nutrients. Reduced enzyme activity under invasive plant influence can limit nutrient availability for native plants (Zhang et al., 2020c).

Transporters within mycelium transfer soil nutrients to host plants and invasive plants can lower transport efficiency by altering fungal structure or soil chemistry (Genre et al., 2020). Some studies focus on mycelium length, while others emphasize biomass or enzymatic activity, complicating cross-study comparisons. Investigate functional attributes of mycelium (e.g., enzymatic activity, nutrient transport) alongside structural metrics may help to evaluate how invasive plants affect mycelium in diverse ecosystems with varying soil nutrient profiles. A more integrated approach that combines structural and ecological studies, developing standardized measures and exploring physiological characteristics are crucial for understanding and mitigating these impacts.

iii. Impact of Invasive Plants on Native Community Composition

Invasive plants also alter the composition of fungal communities, the extent of these effects depends on both the characteristics of the invasive and native plants as well as environmental conditions. For example, invasive plants can increase the abundance of specific arbuscules mycorrhizal fungi species in the rhizosphere, leading to changes in the community balance (Zhang et al., 2019). Han et al., (2024) found that invasive species *Xanthium strumarium* changed the arbuscules mycorrhizal fungi community composition of *Artemisia annua* but had no effect on *Setaria viridis*. *Impatiens glandulifera* invasion over one year did not affect arbuscules mycorrhizal fungi composition in native plants (Gaggini et al., 2018).

Factors such as the duration of invasion, nutrient availability, and the resilience of native arbuscules mycorrhizal fungi communities may play a role in this variability. Invasive plants, in conjunction with mycorrhizae, may modify soil microbial populations, often creating circumstances that are less conducive to native plants. These alterations may include variations in fungal diversity and abundance, favoring fungal species that preferentially interact with invaders (Wang et al., 2024). To analyze the impacts of invasive plants on arbuscules mycorrhizal fungi communities, researchers commonly use Denaturing Gradient Gel Electrophoresis (DGGE) to detects microbial diversity, Phospholipid Fatty Acid (PLFA) Analysis to identifies fungal biomarkers and High-Throughput Sequencing to provides detailed insight into arbuscules mycorrhizal fungi diversity and composition (Jiang et al., 2019).

iv. Impact on Native Plants and Mycorrhizal Fungi Networks

Native plants often form complex networks with multiple mycorrhizal fungi, even at the level of a single root apex. These networks are influenced by factors like the environment, plant dependency on fungi, and the interactions among fungal species. Understanding the effects of invasive plants on these networks is critical for sorting out the ecological impact of plant invasions (Sepp et al., 2019).

a. Structure of Mycorrhizal Fungi Networks

Different mycorrhizea fungi are linked within the network to form highly connective network with greater interaction and stability. These plants rely heavily on mycorrhizae for nutrient acquisition, while facultative plants can supply nutrients through other mechanisms (e.g., legume root nodules). Specialized plants tend to form simpler, more tightly connected networks. Environmental conditions like drought can reduce network complexity and stability, as shown by de Vries et al. (2018). Arbuscules mycorrhizal fungi networks show stronger connectivity with host plants and ectomycorrhizal fungi networks exhibit greater specificity to symbiotic partners (Yang et al., 2019).

b. Impact of Invasive Plants on Fungal Networks

Invasive plants alter the mycorrhizal fungi community by changing species composition and abundance. This likely affects the network's connectivity and modularity. Invasive plants may disrupt interactions between native fungi and plants, weakening network. Arbuscules mycorrhizal fungi network likely more impacted by invasive plants due to their widespread presence and sensitivity to environmental changes. Ectomycorrhizal fungi network may show stronger resistance due to higher specificity and adaptability. Although invasive plants have been shown to impact mycorrhizal fungi composition and abundance (Zhang et al., 2019), their effects on fungal network properties like connectivity and modularity remain underexplored.

v. Impact of Invasive Plants on Native Plants and Non-Mycorrhizal Fungi (NMF)

Invasive plants not only upset native plant-mycorrhizal fungi relationships but also affect non-mycorrhizal fungi communities which can significantly change soil condition, nutrient cycling, and plant resilience. Understanding the complex interactions between invasive plants, native plants, and non-mycorrhizal fungi is essential for assessing the broader ecological consequences. Types of non-mycorrhizal fungi that are disturbed due to invasive plants include pathogenic fungi which harm native plants by causing diseases, parasitic fungi exploit plant resources, often weakening the host species, neutral endophytes live within plant tissues without significantly benefiting or harming the plant and beneficial endophytes help plants survive in extreme environments, providing stress resistance or aiding nutrient uptake (Hill et al., 2019).

Invasive plants indirectly influence native plants by altering the balance of beneficial and harmful fungi in the soil. Invasive plants can alter fungal community dynamics, which indirectly affects mycorrhizal fungi infection rates, their structure and function, root mycelium length, and fungal community composition and their network.

vi. Impact of invasive plants on soil microbes

Plant-soil interactions represent a complex and vigorous relationship between plants and their soil environment, these can be abiotic (changes to physical and chemical properties of the soil), or biotic (microbial communities). Both were influence plants in such soil environment, which in turn affects plant growth and community dynamics. Microbes are recognized as 1) decomposers driving organic matter breakdown and nutrient release, 2) symbionts interaction between mycorrhizal fungi and nitrogen-fixing bacteria, 3) pathogens influencing plant community composition by selectively reducing certain species. In plant soil relationship microbes plays significant role in ecosystem functioning such as regulate carbon and nitrogen availability, enhance nutrient uptake, plant growth and survival in changing environment (Ullah et al., 2024) as shown in figure 2.

Soil pathogens have profound effects on natural ecosystems and agricultural systems, often amplifying the impact of invasive plants by disrupting native plant-pathogen dynamics (Agarwal et al., 2024). The pathogens brought over by invasive species have been shown not only devastate native plant populations but also create cascading effects on ecosystem structure, function, and resilience. Soil pathogens suppress dominant plants, influencing spatial and temporal plant diversity in natural systems. Some pathogens exhibit host specificity, targeting either specific genera or closely related plant species. Many pathogens, such as *Phytophthora* species, are introduced unintentionally but can have devastating effects due to the lack of host resistance in the invaded region. Small size and inconspicuous nature facilitate the spread of fungal pathogens faster than their host plants (Picket et al., 2018). Invasive pathogens have severely impacted North American forests, with examples including *Cryphonectria parasitica* (chestnut blight) (Rigling and Prospero, 2018) and *Phytophthora ramorum* (sudden oak death) (Grünwald et al., 2016) as shown in Table 2.

Loss of grounding tree species due to invasive pathogens can lead to ecosystem-wide changes, such as altered nutrient fluxes, water dynamics, and biodiversity. Invasive plants influence the abundance and diversity of native fungal pathogens in invaded sites, sometimes facilitating or inhibiting their growth through chemical like alkaloids that inhibit generalist pathogens, favoring host-specific pathogens, inhibit competitors, deter herbivores, or repel pathogens, giving them an advantage (Mohammed et al., 2022). If pathogens eliminate keystone species, the resulting changes can cascade through food webs and ecosystem functions, causing long-lasting damage. Invasive pathogens represent a significant ecological threat, underscoring the need for interdisciplinary research to manage their spread and mitigate their impacts (Zheng et al., 2017).

Figure 2. The plant associated with arbuscular mycorrhizal fungi improve soil conditions and nutrient cycling

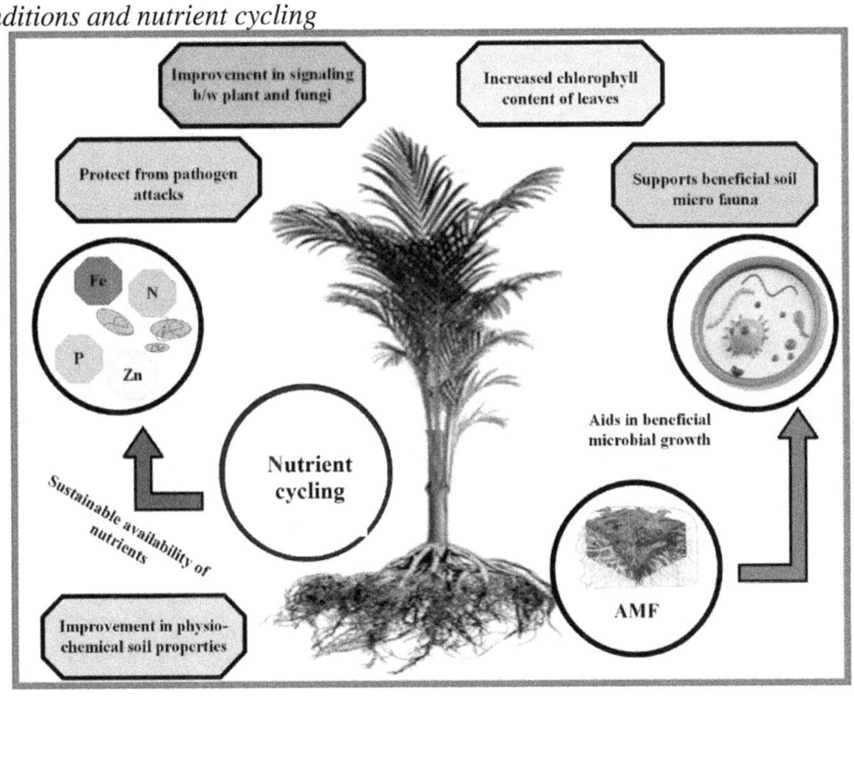

Table 2. Notable invasive pathogens spread through invasive plants and their symptoms

Fungal microbes carried by Invasive species	Diseases	Affected species	country	Symptoms	Reference
Phytophthora captiosa sp. nov.	Crown dieback of Eucalyptus	*Eucalyptus botryoides and E. saligna*	New Zealand	Leaf spots, petiole infection, twig and small branch lesions.	Dick et al.,2006
Phytophthora plurivora sp. nov.	Aetiology of tree and herbaceous species	*Alnus viridis* and *Juniperus communis*	Italy and Slovenia	Foliar necrosis, fruit rot, shoot blight and branch bleeding cankers and death of the crown.	Bregant et al., 2023
Phytophthora cinnamomi Rands.	Decline in *Eucalyptus marginata* forests	*Trachymene pilosa, Stylidium diuroides*	Australian	Thinning of canopy, root lesions, although above ground tissues remained symptomless.	Crone et al., 2013
Phytophthora infestans (Mont.) de Bary.	Irish potato famine and Late blight of potato and tomato	Potato, tomato	South America, Eastern United States	Foliage destroyed, potato tubers and Tomato Fruit infection and cell lesions.	Fry et al., 2013
Cryphonectria parasitica (Murrill) Barr.	American chestnut trees	*Castanea dentata, C. sativa, C. mollissima* and *C. crenata*	Northeastern USA forests	Perennial necrotic lesions (so-called cankers) on the bark of stems and branches of susceptible host trees, eventually leading to wilting of the plant.	Rigling and Prospero, 2018
Phytophthora ramorum sp. nov.	Sudden oak death and blight	Coast live oak and tanoak, *Syringa, Camellia* and many ornamental species	California, USA	Stem cankers, and leaf blight or stem dieback on ornamentals and understory forest species	Grünwald et al., 2008
Phytophthora kernoviae sp. nov.	Aerial bleeding lesion	*Fagus sylvatica, Rhododendron ponticum* and forest understorey species.	South-west England	Foliar necrosis and shoot dieback, and bleeding cankers on stems	Denman et al., 2009

continued on following page

Table 2. Continued

Fungal microbes carried by Invasive species	Diseases	Affected species	country	Symptoms	Reference
Phytophthora alni subsp. *uniformis*	Aetiology of root and stem	*Alnus* Mill. (green alder)	Italy	Shoot blight, progressive canopy dieback, bleeding cankers, epicormic shoots on trunk and root, and collar rot	Bregant et al., 2020
Hymenoscyphus fraxineus (T. Kowalski) Barak, Queloz, Hosoya	Leaf and twig dieback	Common ash (*Fraxinus excelsior* L.)	Poland and Denmark	Root and collar rot, bark lesions on stems and roots of mature declining ash trees	Orlikowski et al., 2011

5. MYCORRHIZAE GLOBAL PATTERN

As globalization advances, the introduction of invasive plants into non-native ecosystems has escalated, often resulting in their transformation into invasive species. These plants pose significant threats to biodiversity, the socio-economy, and human health. Among the various mechanisms enabling their successful invasion (Pathak *et al.*, 2019; Ji-Zhong et al., 2019), disrupting native species' symbiotic relationships, particularly with mycorrhizal fungi (MF), has emerged as a key strategy. Invasive plants often thrive due to the absence of predators, herbivores, pathogens, and more resources (e.g., carbon, nitrogen) previously selected to defense are now redirected resources toward growth and reproduction, enhancing competitiveness in the invaded regions (Zhao et al., 2020).

Invasive species exhibit high adaptability to environmental fluctuations, utilizing resources more efficiently than native plants. Invasive plants interfere with native plant-root mycorrhizal fungi symbiosis, affecting nutrient uptake and stress resilience (Tedersoo et al., 2020). Mycorrhizal fungi symbiosis enhances resistance to drought, heavy metal toxicity, and other environmental stressors (Delavaux & Bever, 2022; Kohler et al., 2017), and provides protective benefits against pathogens, improving plant resilience. Invasive plants can interfere with native plants' mycorrhizal relationships in several ways one of them is invasive plants may outcompete native species by fostering fungal communities more favorable to their growth.

Disruption of mycorrhizal fungi networks may increase the proportion of pathogenic fungi in native plants, exacerbating their decline. By disrupting mycorrhizal fungi networks, invasive plants can monopolize soil nutrients, further disadvantaging the native plants. Other invasive species alter soil microbiomes to suppress mycorrhizal fungi which are beneficial to native plants while enhancing fungal symbionts favorable to themselves (Dickie et al., 2017a). Continuous mycorrhizal fungi disruption leads to cumulative declines in native plant fitness, creating a reaction that promotes invasive species dominance than native plants (Rezacova et al., 2021). The ability of invasive plants to disrupt native plant-mycorrhizal fungi symbiosis is an important mechanism underpinning their ecological success in different environments. This disruption reduces native plants' nutrient uptake, stress resilience, and competitive ability, thereby facilitating invasive plants dominance.

6. STRESS ECOLOGY IN PLANT INVASION

The role of arbuscular mycorrhizal fungi in enhancing plant resistance to various abiotic and biotic stresses, such as drought, salinity, heavy metals, soil composition, extreme temperatures, and waterlogging. Mycorrhizal fungi symbiosis improve plant nutrient absorption, enabling plants to grow better under nutrient-deficient conditions as well enhanced uptake of water and essential nutrients like phosphorus (P) which is a primary mechanism to mediated stress resistance (Khan et al., 2024). Under salinity and drought, it helps adjust the osmotic potential by accumulating organic compounds (e.g., proline, glycine betaine, carbohydrates) and inorganic ions (e.g., K^+, Cl^-). Glomalin produced by arbuscular mycorrhizal fungi helps in soil structure stabilization, water retention, and heavy metal immobilization, aiding stress tolerance. Arbuscular mycorrhizal fungi colonization enhances antioxidant enzyme activities (e.g., SOD, POD), reducing oxidative damage caused by reactive oxygen species (ROS) under stresses. Arbuscular mycorrhizzal fungi influence the expression of plant genes involved in detoxification and stress tolerance, including aquaporin genes for water transport (Wang et al., 2017).

The reduction in mycorrhizal fungi diversity caused by various stresses leads to significant ecological changes in plant-fungi community composition results in a community with reduced diversity but a higher proportion of stress-tolerant species. the altering in the competitive dynamics within plant communities and the decline in the overall efficiency of nutrient cycling and soil condition, potentially impacting plant productivity and ecosystem stability (Branco et al., 2022).

i. Mycorrhizae associated invasive plant under abiotic stress

a. Drought Tolerance

Drought is a significant environmental stressor that extremely impacts plant productivity, limits water availability, affecting the plant's ability to maintain normal physiological processes, close their stomata while reduces transpiration, it also restricts the uptake of CO_2, directly reducing photosynthetic activity, limits the production of sugars, impacting the plant's energy supply, negatively affect plant growth and development which ultimately leads to lower agricultural yields and diminished productivity in affected crops (Orine et al., 2022). Mycorrhizal plants employ strategies to cope with water deficits involved drought mitigation and drought tolerance.

These strategies are facilitated by the symbiotic relationship between plants and arbuscular mycorrhizal fungi. The primarily focused of arbuscular mycorrhizal fungi on enhancing water absorption and availability arbuscular mycorrhizal fungi hyphae extend into the soil, increasing the effective surface area for water absorption and enhance apoplastic water flow. Studies demonstrate that arbuscular mycorrhizal fungi colonization, such as by *Rhizophagus intraradices*, enables plants like tomatoes to thrive under drought conditions by improving nutrient content and water use efficiency (Leventis et al., 2021).

The drought tolerance involves direct benefits that enhance the plant's intrinsic ability to tolerate drought stress. Arbuscular mycorrhizal fungi help regulate processes such as stomatal conductance, osmotic adjustments, and antioxidant production which improved nutrient acquisition under water stress, boosting resilience. The regulation of phytohormones like abscisic acid (ABA) and jasmonic acid (JA) to control water use efficiency and stomatal conductance, and accumulation of compatible solutes to maintain turgor pressure improved the drought tolerance in plants. The combined effects of drought mitigation and tolerance mechanisms significantly improve plant fitness under drought conditions, allowing them to maintain growth and productivity (Ouledali et al., 2019).

b. Waterlogged Tolerance

Flooding poses unique challenges for arbuscular mycorrhizal fungi due to their oxygen requirements, but evidence suggests that arbuscular mycorrhizal fungi can still play an important role in wetland ecosystems under certain conditions (Velez-Gavilan, 2024; Ramírez-Viga et al., 2018). Wetland ecosystems often show a higher diversity of arbuscular mycorrhizal fungi communities. Their diversity and composition in these environments are influenced by vegetation type, soil chemical properties, and the identity of arbuscular mycorrhizal fungi species (Solís-Rodríguez et al., 2020). Arbuscular mycorrhizal fungi enhance nutrient uptake under flooded conditions. For instance, in wetland rice (*Oryza sativa*), arbuscular mycorrhizal fungi transported significant amounts of phosphorus (P) via the mycorrhizal pathway (Bao et al., 2022).

In *Phragmites australis*, arbuscular mycorrhizal fungi helped the plant cope with drying-rewetting cycles, supporting phosphorus acquisition to improve plant growth under flood conditions (Usman et al., 2021). It also enhanced osmotic adjustment and reduction in arbuscular mycorrhizal fungi colonization in oxygen-deficient environments, yet some arbuscular mycorrhizal fungi adapt to wetland conditions. Their ability to support plants under medium-intensity flooding while enhancing nutrient uptake positions them as valuable allies in sustainable ecosystem management (Solís-Rodríguez et al., 2020).

c. Temperature Tolerance

Temperature is a critical environmental stressor that can noticeably impact plant growth and productivity. Arbuscular mycorrhizal fungi play a pivotal role in helping plants tolerate temperature extremes, including both heat and cold stresses. Arbuscular mycorrhizal fungi improve plant access to vital resources, ensuring better resilience under stress (Caradonia et al., 2019). Mycorrhizal plants maintain higher photosynthetic capacity, protect their photosynthetic apparatus from damage, enhance root development, improving water absorption and supporting photosynthesis under heat stress (Mathur and Jajoo, 2020). Arbuscular mycorrhizal fungi also helps plants to accumulate osmolytes to maintain cellular homeostasis in tolerant plants and mitigating oxidative damage caused by extremes conditions (Mathur et al., 2021). Mycorrhizal plants exhibit greater shoot and root dry weights compared to non-mycorrhizal plants at elevated temperatures.

Different arbuscular mycorrhizal fungi strains exhibit varying efficacy in alleviating cold stress such as *Glomus versiforme* is often more effective than *Rhizophagus irregularis* in mitigating cold stress in some cultivars. Their inoculation in barley (*Hordeum vulgare*) improved growth, photosynthesis, osmotic balance, and potassium uptake during cold stress. However, extremely low temperatures can hinder arbuscular mycorrhizal fungi growth and reduce hyphal network formation and fungal activity (Hajiboland et al., 2019). Arbuscular mycorrhizal fungi also rises the reduction in N_2O discharges by improving N uptake as a result, the soluble N in the soil decreases and it can negatively affect the denitrification process. This ability to enhance resource acquisition, protect against stress-induced damage, and contribute to climate resilience makes them a vital component of sustainable agricultural and ecological practices (Berruti et al., 2018).

d. Salinity Tolerance

Arbuscular mycorrhizal fungi are crucial allies in enhancing plant tolerance to salinity stress. Naturally occurring fungi in saline environments have demonstrated significant contributions to improving the growth and resilience of various plant species under such conditions (Soni et al., 2024). they enhance water absorption capacity and nutrient acquisition, which are critical for plants in saline conditions and improved uptake of essential nutrients like phosphorus (P), potassium (K), copper (Cu), iron (Fe), and zinc (Zn) has been reported. Arbuscular mycorrhizal fungi improve ionic balance by reducing the uptake of toxic ions like Na^+ and Cl^-, while increasing the K/Na ratio in plant tissues (Santander et al., 2020; Amanifar and Toghranegar 2020).

Arbuscular mycorrhizal fungi structures act as a selective filter, preventing excessive entry of harmful ions into plant roots, promote the accumulation of osmoregulators, such as proline, sugars, and glycinebetaine, which help plants maintain osmotic balance and increased concentrations of these compounds have been observed in wheat, maize, and other species under salinity stress (Li et al., 2020). Mycorrhizal plants exhibit better gas exchange, higher stomatal conductance, and photosynthetic efficiency under saline conditions (Parvin et al., 2020). The effectiveness of arbuscular mycorrhizal fungi in alleviating salinity stress varies with fungal species, plant species, and salinity levels. For example, *Diversispora versiformis* outperformed *Funneliformis mosseae* in mitigating salt stress in *Chrysanthemum morifolium* (Wang et al., 2018). It plays a pivotal role in enabling plants to flourish in saline environments. However, further research into species-specific responses and optimized inoculation strategies is necessary to maximize their efficacy.

e. Heavy Metal Tolerance

Arbuscular mycorrhizal fungi have demonstrated exceptional potential in mitigating heavy metal stress in plants, making them valuable tools for bioremediation and ecological restoration of mining and polluted sites. Over 80% of surveyed plants in mining sites are colonized by arbuscular mycorrhizal fungi, showing significant diversity even in heavily polluted soils (Chakraborty et al., 2024). These species in these environments are specifically adapted to tolerate heavy metals like Cu, Pb, Co, Cd, and Zn (Wang et al., 2021; Yin et al., 2021). Their hyphae bind heavy metals, reducing their bioavailability and toxicity to plants, express metal transporters gene, such as GintZnT1 in *Rhizophagus irregularis*, which regulate Zn, Cu, and Fe uptake and play a crucial role in heavy metal detoxification (Silva-Castro et al., 2022).

Arbuscular mycorrhizal fungi inoculation reduces the concentration of heavy metals in plant shoots and roots, while enhancing the uptake of essential elements like K, P, and Mg. They enhance plant survival, seed germination, and biomass (fresh and dry weights) in heavy metal-polluted soils. Mycorrhizal colonization improves nutrient availability, soil structure, and overall plant fitness under stress (Zhan et al., 2019). *Glomus* isolates reduced heavy metal concentrations in maize shoots and roots while increasing essential nutrient levels. Notably, the *Glomus* isolate Br1 from *Viola calaminaria* outperformed common strains in mitigating heavy metal toxicity. Soils contaminated with cadmium (Cd), arbuscular mycorrhizal fungi inoculation significantly reduced Cd uptake in maize, demonstrating arbuscular mycorrhizal fungi's potential in phytoremediation (Chen et al., 2023; Liu et al., 2022).

Figure 3. The inoculation of arbuscular mycorrhizal fungi on healthy plant under stress go through various processes to improve productivity.

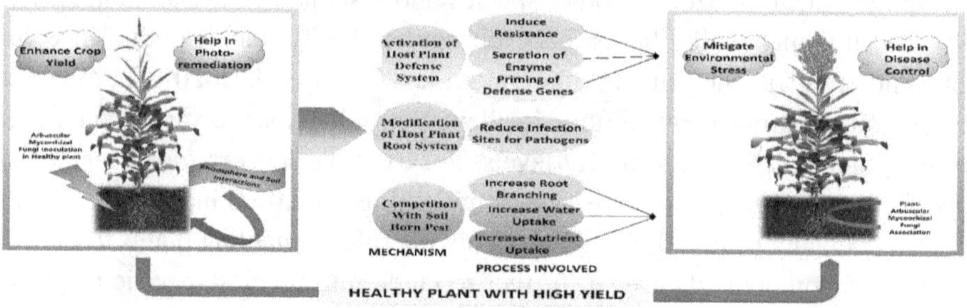

Arbuscular mycorrhizal fungi enhance the re-establishment of vegetation in mining sites, improving soil quality and structure for long-term ecosystem recovery. Co-inoculation strategies with arbuscular mycorrhizal fungi and beneficial micro-organisms may further enhance heavy metal remediation efficiency and enhance yield as shown in figure 3. Abiotic stresses can reduce arbuscular mycorrhizal fungi diversity and alter community composition, favoring stress-tolerant arbuscular mycorrhizal fungi species. They contribute significantly to plant stress tolerance, improving agricultural productivity under challenging environmental conditions by enhancing soil carbon storage, reducing greenhouse gas emissions, and improving plant resilience (Song et al., 2020; Conversa et al., 2019).

The detailed understanding of arbuscular mycorrhizal fungi's role in stress tolerance as shown in (table 3) emphasizes their potential applications in sustainable agriculture and environmental management.

Table 3. **Plants-arbuscular mycorrhiza fungi association responses under abiotic stresses**

Tolerance to different stresses	Invasive species	Fungus species association	Invasive species with mycorrhiza association responses under various abiotic stresses	References
Drought	*Zea mays* L.	*Rhizophagus intraradices*, strain BGCBJ09	Increased plant dry weight, uptake of P, N, K, and Mg in shoot, and water use efficiency	Quiroga et al., 2019 Hu et al., 2020
	Glycine max L.	*Septoglomus constrictum, Glomus spp., Glomus aggregatum*	Improved water content and P and N levels	Pavithra and Yapa, 2018
	Triticum aestivum L.	*Glomus mosseae*	Increased plant growth, osmotic potential, chlorophyll content and fluorescence, activities of antioxidant enzymes, ascorbic acid, enzymes of N and P metabolism, and contents of N, P, and K	Pons et al., 2020 Pal and Panday, 2017
	Poncirus trifoliate L.	*Funneliformis mosseae, Paraglomus occultum*	Increased plant growth and hyphal length, hyphal water absorption rate, and leaf water potential	Zhang et al., 2020b Zou et al., 2021
	Robinia pseudoacacia L.	*Funneliformis mosseae, Rhizophagus intraradices*	Increased dry biomass, WUE, and net photosynthetic rate	He et al., 2017
	Ipomoea batatas L. (Lam.)	*Glomus spp.*	Proline and soluble sugars adjust osmotic potential	Yooyongwech et al., 2017
	Pelargonium graveolens L. (Hérit.)	*Rhizophagus intraradices, Funneliformis mosseae*	Improved nutrient concentration, plant biomass, and essential oil content, and glomalin-related soil proteins (GRSP)	Amiri et al., 2015
	Glycine max L.	AMF	Enhanced leaf proline, photosynthesis, leaf area index, relative growth rate, fresh weight, and dry weight of seeds	Igiehon et al., 2021

continued on following page

Table 3. Continued

Tolerance to different stresses	Invasive species	Fungus species association	Invasive species with mycorrhiza association responses under various abiotic stresses	References
Heat	*Solanum lycopersicum* L.	*Rhizophagus irregularis*	Enhanced photosynthetic capacity, root hydraulic conductivity or aquaporin abundance and phosphorylation status	Calvo-Polanco et al., 2016
	Triticum aestivum L.	*R. irregularis, F. mosseae, F. geosporum*	Increased yield, seeds number, nutrient composition and allocation in roots	Cabral et al., 2016
	Zea mays L.	Rhizophagus intraradices, *Funneliformis mosseae, F. geosporum*	Increased leaf length, plant height, leaf number, chlorophyll a, photosynthetic rate, stomatal conductance, and transpiration rate	Mathur et al., 2018
	Asparagus officinalis L.	*Glomus intraradices*	Increase plant growth, enhanced biomass production, increase activity of antioxidative enzymes (SOD, ascorbate and peroxidase).	Yeasmin et al., 2019
Salinity	*Glycyrrhiza glabra* L.	*Funneliformis mosseae*	Improved secondary metabolites regulation in plant tissues, praline in shoot, growth parameters, high biomass as well as P, Zn, and Cu contents	Amanifar et al., 2019
	Aeluropus littoralis (Gouan) Parl.	*Claroideoglomus etunicatum*	Increased shoot and root dry mass, stomatal conductance, soluble sugars, free α-amino acids, and Na+ and K+ uptake	Dashtebani and Aliasgharzad, 2018
	Solanum lycopersicum L.	*Glomus intraradices*	Improved dry matter, ion uptake, growth parameters, and chlorophyll content	Khalloufi et al., 2017
	Cucumis sativus L.	*Glomus etunicatum, Glomus intraradices, Glomus mosseae*	Increased biomass, photosynthetic pigment synthesis, and enhanced antioxidant enzymes	Hashem et al., 2018
	Leymus chinensis (Trin.) Tzvelev.	*Glomus mosseae*	Increased colonization rate, seedling weight, water contents, and both P and N	Wang et al., 2022

continued on following page

Table 3. Continued

Tolerance to different stresses	Invasive species	Fungus species association	Invasive species with mycorrhiza association responses under various abiotic stresses	References
Heavy Metals	*Cajanus cajan* L.	*Rhizophagus irregularis*	Improved root biomass, nutrient status (P, N, Mg, Fe.), and proline biosynthesis	Garg and Singh, 2018
	Trigonella foenum-graecum L.	*Glomus monosporum, G. clarum, Gigaspora nigra, Acaulospora laevis*	Increased antioxidant enzymes activities and malondialdehyde content	Abdelhameed and Metwally, 2019
	Trigonella foenum-graecum L.	*Funneliformis mosseae, F. constrictum, Gigaspora margarita* and *Rhizophagus irregularis*	Improve shoot height, root length, total fresh and dry weights, increase total chlorophyll, growth attributes, proline, protein, free amino acids and sugars contents.	Abdelhameed, R. E., & Metwally, 2024
	Triticum aestivum L.	*Aspergillus terreus*	Improved plant growth, relative water content, root and shoot biomass, chlorophyll protein, and sugar contents, antioxidants and ion homeostasis.	Dilawar et al., 2024

ii. Mycorrhizae associated invasive plant under biotic stress

Arbuscular mycorrhizal fungi play a crucial role in improving biotic stresses in plants by promoting plant vigor and resilience against pathogens, parasitic plants, nematodes, and herbivores. Their complicated mechanisms of action make them valuable in sustainable agriculture and integrated pest management systems. Arbuscular mycorrhizal fungi colonization reduces damage caused by pathogenic fungi such as *Fusarium oxysporum* and *Alternaria solani*, as shown in crops like tomato, potato, and chickpea. In soybean, arbuscular mycorrhizal fungi significantly reduced the severity of charcoal root-rot disease. *Funneliformis mosseae* reduced nematode infections by altering root exudates, decreasing nematode penetration, reproduction, and morphometric parameters in tomato plants (Spagnoletti et al., 2021; Song et al., 2016).

Arbuscular mycorrhizal fungi enhances plant antioxidant enzyme production, aiding in pathogen defense and improve host nutrient status, alter root morphology, and create competitive environments in the rhizosphere. The synergistic effects with hormonal inducers (e.g., jasmonic acid, salicylic acid) further boost disease resistance (Kumar et al., 2024). Their effects on herbivore insects vary based on feeding specialization. For example, Arbuscular mycorrhizal fungi reduced the performance of chewing insects like *Arctia caja* but increased the success of sucking insects like *Mysus persicae*. Parasitism rates by predators like *Diglyphus isaea* on certain herbivores were influenced by Arbuscular mycorrhizal fungi species, showcasing species-specific interactions (Ng et al., 2023).

Arbuscular mycorrhizal fungi can inhibit the germination, attachment, and emergence of *Striga*, a significant parasitic plant in sub-Saharan Africa, while enhancing the growth and resilience of host plants like sugarcane. By reducing the soil *Striga* seed bank, arbuscular mycorrhizal fungi contribute to long-term management of parasitic plant infestations (Hassan et al., 2019). The combined applications of arbuscular mycorrhizal fungi and other beneficial organisms or treatments, such as hormonal sprays or co-inoculation with bacteria, amplify their protective effects against biotic stresses.

Arbuscular mycorrhizal fungi can be utilized to reduce reliance on chemical pesticides and fertilizers and ability to protect crops against multiple biotic stresses makes them integral to integrated pest management (IPM) strategies. They provide a natural and effective solution to combat biotic stresses in plants, enhancing plant resistance through different mechanisms which improved nutrition and microbial interactions. Their potential to reduce pathogen and pest damage while boosting plant growth highlights their importance in sustainable and resilient agricultural systems (Kumar et al., 2024; Kumar et al., 2017). The arbuscular mycorrhizal fungi overall response toward stresses, nutrient cycling, soil conditions and pathogen are shown in figure 4.

Figure 4. The overall plant-arbuscular mycorrhizal fungi interaction response toward stresses, soil conditions, nutrient cycling and soil pathogen attack on invasive plant establishment in new community structure.

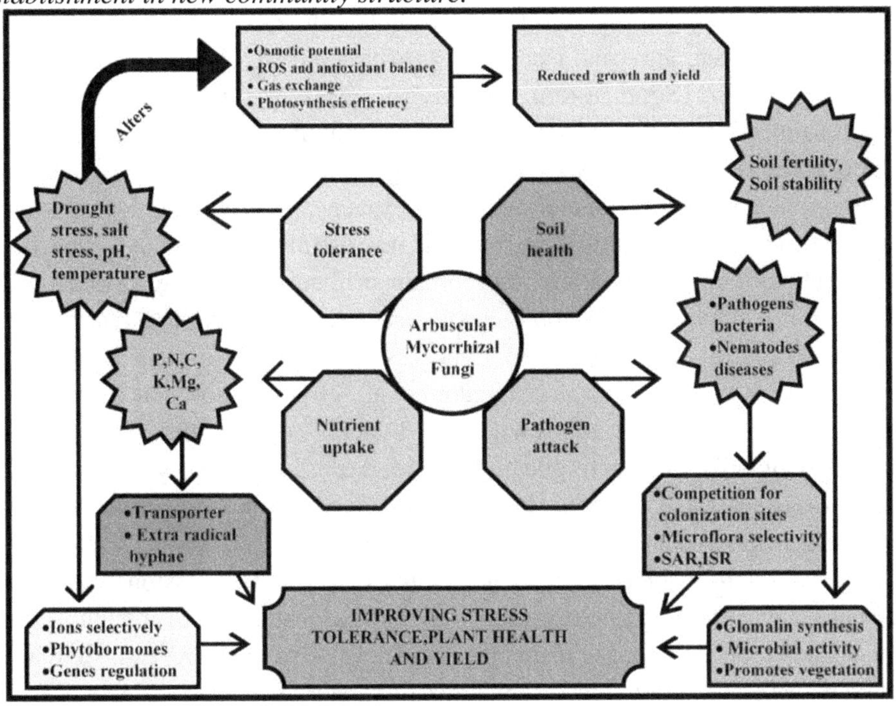

7. ECOLOGICAL RESTORATION

Ecosystems provide essential goods and services that support human life, including clean air, water, biodiversity, and fertile soil. These systems are generally self-regulating, capable of adapting to external disturbances. However, escalating natural and human-induced activities such as global warming, mining, overgrazing, pollution, and desertification have caused significant environmental problems, including ecological imbalance, degradation of ecosystem services, biodiversity loss, and decline in soil productivity (Husain et al., 2024; Grove et al., 2017a).

The invasion of non-native plant species poses a significant threat to the structural and functional integrity of native ecosystems. These invasive plants not only alter aboveground ecosystem (Wei et al., 2020) but also disrupt corresponding soil microbial communities (Du et al., 2024), including arbuscular mycorrhizal fungi, which play a vital role in ecosystem fitness and stability. Non-native plants often compete aggressively with native species, leading to biodiversity loss. Modify soil microbial communities, which can affect nutrient cycling, soil structure, and plant succession dynamics and impact wildlife, as demonstrated by pale swallow-wort's (*Vincetoxicum rossicum*) effects on monarch butterfly reproduction and arthropod communities (Casagrande and Dacey,2014; Milbrath et al., 2017). Chinese Privet (*Ligustrum sinense*) (Southeastern Alabama), Forms dense thickets, altering understory dynamics and associated soil microbial communities (Ivey and Leege, 2023).

Ecosystem restoration pursues to return degraded systems to their original state by using ecological principles and techniques by using pioneer plants species which can found under harsh conditions and pave the way for further ecological succession. Another is incorporating beneficial microorganism like arbuscular mycorrhizal fungi enhance nutrient availability, improve soil structure, and alleviate environmental stresses. Despite all this advancements, successful restoration is challenging due to abiotic stresses, and biotic stresses (Begum et al., 2019). Arbuscular mycorrhizal fungi form symbiotic associations with more than 80% of terrestrial plants and are distributed globally, even in fragile and degraded ecosystems. They are vital allies in the battle against ecosystem degradation, offering a natural and sustainable solution to restore ecosystem functionality. Their ability to support plant growth under stress, improve soil condition, and mediate ecological succession makes them indispensable in global restoration efforts. With continued research and innovation, arbuscular mycorrhizal fungi can display a transformative role in reversing environmental damage and promoting sustainable ecosystems (Perez-Castro et al., 2019).

Successful inoculation of degraded lands with arbuscular mycorrhizal fungi has improved plant growth and survival in challenging environments, such as in rocky desertified lands the arbuscular mycorrhizal fungi inoculation enhanced water retention and nutrient availability, and in mine sites it reduced heavy metal bioavailability and restored vegetative cover. They are used alongside pioneer plants and other soil amendments to create a synergistic effect, improving restoration outcomes (Wang, 2017; Wubs et al., 2016). To better understand and mitigate the impacts of invasive plants on mycorrhizal fungi and native ecosystems, future research should be address such as global environmental changes, integrated mechanisms (resource competition, allelopathy, signaling pathways, nutrient exchange), long-term and large-scale studies needed and develop targeted prevention or control measures based on specific stages of invasion, considering above mention mechanisms.

8. CHALLENGES AND FUTURE DIRECTIONS

The variety of mycorrhizal fungi, their sensitivity to host plants, and the differing environmental circumstances complicate predictions on the dynamics of these interactions across many habitats (Nizamani et al., 2024). Moreover, substantial information deficiencies exist about the variety, distribution, and functional functions of mycorrhizal fungi, which hinders the formulation of appropriate management methods. Numerous mycorrhizal species are inadequately comprehended, and their influence on either promoting or obstructing plant invasions constitutes a nascent area of research (Rajamuthuramalingam et al., 2024). A further difficulty pertains to the enduring impacts of invasive plants on soil microbial populations, referred to as soil microbial legacy effects.

Post-removal of invasive plants, the modified microbial communities they leave may persist in promoting their proliferation or inhibiting local species, so complicating ecosystem restoration efforts (Acero et al., 2023). The significance of mycorrhizal fungi in relation to global climate change remains ambiguous. Alterations in temperature, moisture, and soil conditions might modify mycorrhizal connections, thereby intensifying the proliferation of invasive plant species. The ambiguity about these dynamics poses a considerable obstacle to efficient management (Jiang et al., 2024). Future study should concentrate on the diversity, taxonomy, and functional characteristics of mycorrhizal fungi. Progress in high-throughput sequencing and metagenomic methodologies will elucidate the functions of diverse fungi in plant invasions. Moreover, there exists the possibility to formulate management measures that use mycorrhizal biology, such the introduction of native mycorrhizal fungus to inhibit invasive plants or the disruption of fungal networks linked with invasive plants to curtail their proliferation.

Incorporating this understanding into ecosystem restoration initiatives presents a potential opportunity, since the introduction of indigenous mycorrhizal fungus may facilitate the re-establishment of plant communities and bolster the resilience of native species. Moreover, understanding the responses of mycorrhizal fungi to climate change will be essential for forecasting future plant invasions and modifying management strategies appropriately. Localized case studies will be essential in formulating region-specific strategies for the control of invasive plants. Cooperation among ecologists, microbiologists, and land managers may provide customized strategies for various ecosystems, facilitating more efficient management of invasive species. By tackling these issues and investigating future avenues, we may improve our understanding of mycorrhizal dynamics and more effectively regulate plant invasions, therefore fostering biodiversity and ecological sustainability.

CONCLUSION

The function of mycorrhizal fungus in the ecology of invasive plants involves a complex and dynamic interplay of ecological interactions that may profoundly affect plant community structure and ecosystem functionality. Mycorrhizae, by their symbiotic associations with plants, provide essential advantages like nutrient uptake, stress resilience, and competitive superiority, which may be disproportionately exploited by invading species to establish and dominate novel ecosystems. These fungi often improve the resource-utilization efficiency of invasive plants, outcompeting native species and modifying soil microbial populations, thereby establishing feedback loops that further facilitate invasions. Additionally, climate change and globalization further intensify invasions by expanding suitable habitats and facilitating the movement of invasive species through trade and landscaping. Interactions between invasive species, native flora, and abiotic factors are complex. Invasive plants disrupt plant-soil microbe dynamics, and exploit disturbances, resulting in ecological imbalances. These interactions allow invasive species to establish persistent populations in ecosystems previously dominated by native plants. Understanding the factors influencing invasiveness is essential for managing invasions. Traits such as high reproductive rates, adaptability to diverse environments, and efficient resource utilization contribute to invasive success. Mycorrhizal fungi are critical in the ecology of invasions, significantly disrupt the mutualistic relationships between native plants, mycorrhizal fungi, and soil microbes, impacting ecosystem dynamics. Invasive plants alter fungal community structures and functions, reducing fungal biomass and enzymatic activity crucial for nutrient cycling. They also upset fungal networks, weakening connectivity and modularity.

Comprehending the processes via which mycorrhizal fungi affect both invasive and native plants, including nutrient cycling and soil microbial dynamics, is critical for conserving ecosystems. Developing integrated approaches to restore mycorrhizal networks, mitigate soil microbial disruptions, and enhance native plant resilience is essential for combating the ecological challenges posed by invasive species. The symbiotic relationships are enhancing their resilience to various abiotic and biotic stresses to improve plant water and nutrient absorption, promoting osmotic adjustments, accumulating protective organic compounds, and also ease oxidative damage by enhancing antioxidant enzyme activities. They also aid in plant resistance to biotic stresses and reduce damage from pathogens, herbivores and parasitic plants to improve the plant's nutrient status, altering root morphology, and increasing the production of defensive enzymes. In terms of ecological restoration, they play a vital role in rehabilitating degraded ecosystems and help restore soil condition, promote plant growth in harsh conditions, and improve nutrient cycling. Successful inoculation with fungus has improved plant survival in disturbed areas, and facilitating ecological succession.

Future study must concentrate on elucidating the intricacies of mycorrhizal connections, especially in the context of climate change, habitat loss, and alterations in soil microbial communities. However, effective management requires understanding the molecular, ecological, and environmental factors driving invasions. Developing strategies to understand the fungal networks, restoring native flora, and ecological restoration techniques may be formulated to inhibit invasive plants via biological and chemical methods, hence enhancing biodiversity and ecosystem resilience. Future research must highlight the association between invasive plants, mycorrhizal associations, and ecosystem dynamics to develop sustainable conservation approaches. This chapter underlines the necessity of a cohesive strategy that amalgamates ecological insight, technological progress, and practical management to address the ecological issues presented by invasive plants, while leveraging the capabilities of mycorrhizal fungi for sustainable ecosystem restoration. In conclusion, mycorrhizal fungus play a vital role in improving plant resilience to both abiotic and biotic stresses, enhancing plant yield, and supporting ecosystem stability. Their potential in sustainable agriculture, restoration ecology, and integrated pest management is vast, highlighting the importance of further research into their applications for improving soil condition and ecosystem services.

REFERENCE

Abdelhameed, R. E., & Metwally, R. A. (2019). Alleviation of cadmium stress by arbuscular mycorrhizal symbiosis. *International Journal of Phytoremediation, 21*(7), 663–671. DOI: 10.1080/15226514.2018.1556584 PMID: 30816051

Abdelhameed, R. E., & Metwally, R. A. (2024). The potential utilization of mycorrhizal fungi and glycine betaine to boost the fenugreek (*Trigonella foenum-graecum* L.) tolerance to chromium toxicity. *Journal of Soil Science and Plant Nutrition*, •••, 1–20. DOI: 10.1007/s42729-024-02131-y

Acero, V. P., Das, S., Rivellini, O., Purvis, E. M., Adewole, D. O., & Cullen, D. K. (2023). Emergent structural and functional properties of hippocampal multicellular aggregates. *Frontiers in Neuroscience, 17*, 1171115. DOI: 10.3389/fnins.2023.1171115 PMID: 37397454

Adam, A. M., Alshahrani, T. S., Alqarawi, A. A., Abd-ElGawad, A. M., Ahmed, E. M., & Sorrori, S. N. (2024). Arbuscular mycorrhiza fungi alleviate the allelopathic effects of invasive nicotiana glauca on the growth of *Acacia gerrardii* seedlings. *Forests, 15*(8), 1447. DOI: 10.3390/f15081447

Agarwal, R., Gupta, M., Sen, R., & Panchal, A., ES, N., & Raychoudhury, R. (. (2024). Investigation into how Odontotermes obesus maintains a predominantly Termitomyces monoculture in their fungus combs suggests a potential partnership with both fungi and bacteria. *Communications Biology, 7*(1), 1010. DOI: 10.1038/s42003-024-06708-2 PMID: 39154098

Aizen, M. A., & Torres, A. (2024). The Invasion Ecology of Mutualism. *Annual Review of Ecology, Evolution, and Systematics, 55*(1), 41–63. Advance online publication. DOI: 10.1146/annurev-ecolsys-102622-031210

Akbar, R., Sun, J., Bo, Y., Khattak, W. A., Khan, A. A., Jin, C., Zeb, U., Ullah, N., Abbas, A., Liu, W., Wang, X., Khan, S., & Du, D. (2024). Understanding the influence of secondary metabolites in plant invasion strategies: A comprehensive review. *Plants, 13*(22), 3162. DOI: 10.3390/plants13223162 PMID: 39599372

Amanifar, S., Khodabandeloo, M., Fard, E. M., Askari, M. S., & Ashrafi, M. (2019). Alleviation of salt stress and changes in glycyrrhizin accumulation by arbuscular mycorrhiza in liquorice (*Glycyrrhiza glabra*) grown under salinity stress. *Environmental and Experimental Botany, 160*, 25–34. DOI: 10.1016/j.envexpbot.2019.01.001

Amanifar, S., & Toghranegar, Z. (2020). The efficiency of arbuscular mycorrhiza for improving tolerance of Valeriana officinalis L. and enhancing valerenic acid accumulation under salinity stress. *Industrial Crops and Products*, *147*, 112234. DOI: 10.1016/j.indcrop.2020.112234

Amiri, R., Nikbakht, A., & Etemadi, N. (2015). Alleviation of drought stress on rose geranium [Pelargonium graveolens (L.) Herit.] in terms of antioxidant activity and secondary metabolites by mycorrhizal inoculation. *Scientia Horticulturae*, *197*, 373–380. DOI: 10.1016/j.scienta.2015.09.062

Bajwa, A. A., Chauhan, B. S., Farooq, M., Shabbir, A., & Adkins, S. W. (2016). What do we really know about alien plant invasion? A review of the invasion mechanism of one of the world's worst weeds. *Planta*, *244*(1), 39–57. DOI: 10.1007/s00425-016-2510-x PMID: 27056056

Balducci, M. G., Calevo, J., & Duffy, K. J. (2024). Orchid mycorrhizal communities associated with Orchis italica are shaped by ecological factors and geographical gradients. *Journal of Biogeography*. Advance online publication. DOI: 10.1111/jbi.15056

Bao, X., Zou, J., Zhang, B., Wu, L., Yang, T., & Huang, Q. (2022). Arbuscular mycorrhizal fungi and microbes interaction in rice mycorrhizosphere. *Agronomy (Basel)*, *12*(6), 1277. DOI: 10.3390/agronomy12061277

Bard, N. W., Cronk, Q. C., & Davies, T. J. (2024). Fungal endophytes can modulate plant invasion. *Biological Reviews of the Cambridge Philosophical Society*, *99*(5), 1652–1671. Advance online publication. DOI: 10.1111/brv.13085 PMID: 38629189

Barnes, C., Hilton, S., Nielsen, I. B., Ahlstrand, N. I., Hansen, A. J., Frøslev, T. G., & Bending, G. D. (2023). Above and belowground microbes are associated with variation of the leaf metabolome; But herbivory has nominal effects. DOI: 10.21203/rs.3.rs-3389986/v1

Barnes, C. J., van der Gast, C. J., McNamara, N. P., Rowe, R., & Bending, G. D. (2018). Extreme rainfall affects assembly of the root-associated fungal community. *The New Phytologist*, *220*(4), 1172–1184. DOI: 10.1111/nph.14990 PMID: 29350759

Begum, N., Qin, C., Ahanger, M. A., Raza, S., Khan, M. I., Ashraf, M., & Zhang, L. (2019). Role of arbuscular mycorrhizal fungi in plant growth regulation: Implications in abiotic stress tolerance. *Frontiers in Plant Science*, *10*, 1068. DOI: 10.3389/fpls.2019.01068 PMID: 31608075

Bello, T. T., & Fabiyi, O. A. (2024). Advancement in mycorrhizal fungi-based sustainable plant disease management. In *Mycorrhizal Symbiosis and Agroecosystem Restoration* (pp. 125–145). Springer., DOI: 10.1007/978-981-99-5030-0_6

Bennett, A. E., & Meek, H. C. (2020). The influence of arbuscular mycorrhizal fungi on plant reproduction. *Journal of Chemical Ecology*, *46*(8), 707–721. DOI: 10.1007/s10886-020-01192-4 PMID: 32583094

Berruti, A., Bianciotto, V., & Lumini, E. (2018). Seasonal variation in winter wheat field soil arbuscular mycorrhizal fungus communities after non-mycorrhizal crop cultivation. *Mycorrhiza*, *28*(5), 535–548. DOI: 10.1007/s00572-018-0845-9 PMID: 29931405

Bhunjun, C. S., Phukhamsakda, C., Hyde, K. D., McKenzie, E. H., Saxena, R. K., & Li, Q. (2024). Do all fungi have ancestors with endophytic lifestyles? *Fungal Diversity*, *125*(1), 73–98. DOI: 10.1007/s13225-023-00516-5

Branco, S., Schauster, A., Liao, H. L., & Ruytinx, J. (2022). Mechanisms of stress tolerance and their effects on the ecology and evolution of mycorrhizal fungi. *The New Phytologist*, *235*(6), 2158–2175. DOI: 10.1111/nph.18308 PMID: 35713988

Bregant, C., Rossetto, G., Meli, L., Sasso, N., Montecchio, L., Brglez, A., & Linaldeddu, B. T. (2023). Diversity of *Phytophthora* species involved in new diseases of mountain vegetation in Europe with the description of *Phytophthora pseudogregata* sp. nov. *Forests*, *14*(8), 1515. DOI: 10.3390/f14081515

Bregant, C., Sanna, G. P., Bottos, A., Maddau, L., Montecchio, L., & Linaldeddu, B. T. (2020). Diversity and pathogenicity of Phytophthora species associated with declining alder trees in Italy and description of Phytophthora alpina sp. nov. *Forests*, *11*(8), 848. DOI: 10.3390/f11080848

Cabral, C., Ravnskov, S., Tringovska, I., & Wollenweber, B. (2016). Arbuscular mycorrhizal fungi modify nutrient allocation and composition in wheat (Triticum aestivum L.) subjected to heat-stress. *Plant and Soil*, *408*(1-2), 385–399. DOI: 10.1007/s11104-016-2942-x

Callaway, R. M., Cipollini, D., Barto, K., Thelen, G. C., Hallett, S. G., Prati, D., & Klironomos, J. (2008). Novel weapons: Invasive plant suppresses fungal mutualists in America but not in its native Europe. *Ecology*, *89*(4), 1043–1055. DOI: 10.1890/07-0370.1 PMID: 18481529

Calvo-Polanco, M., Sánchez-Castro, I., Cantos, M., García, J. L., Azcón, R., Ruiz-Lozano, J. M., & Aroca, R. (2016). Effects of different arbuscular mycorrhizal fungal backgrounds and soils on olive plants growth and water relation properties under well-watered and drought conditions. *Plant, Cell & Environment, 39*(11), 2498–2514. DOI: 10.1111/pce.12807 PMID: 27448529

Caradonia, F., Francia, E., Morcia, C., Ghizzoni, R., Moulin, L., Terzi, V., & Ronga, D. (2019). Arbuscular mycorrhizal fungi and plant growth promoting rhizobacteria avoid processing tomato leaf damage during chilling stress. *Agronomy (Basel), 9*(6), 299. DOI: 10.3390/agronomy9060299

Casagrande, R. A., & Dacey, J. E. (2014). Monarch butterfly oviposition on swallow-worts (Vincetoxicum spp.). *Environmental Entomology, 36*(3), 631–636. DOI: 10.1603/0046-225X(2007)36[631:MBOOSV]2.0.CO;2 PMID: 17540075

Chakraborty, N., Das, A., Pal, S., Roy, S., Sil, S. K., Adak, M. K., & Hasanuzzaman, M. (2024). Exploring aluminum tolerance mechanisms in plants with reference to rice and arabidopsis: A comprehensive review of genetic, metabolic, and physiological adaptations in acidic soils. *Plants, 13*(13), 1760. DOI: 10.3390/plants13131760 PMID: 38999600

Chaudhury, R., Chakraborty, A., Rahaman, F., Sarkar, T., Dey, S., & Das, M. (2024). Mycorrhization in trees: Ecology, physiology, emerging technologies and beyond. *Plant Biology, 26*(2), 145–156. DOI: 10.1111/plb.13613 PMID: 38194349

Chen, J., Wang, L., Liang, X., Li, B., He, Y., & Zhan, F. (2023). An arbuscular mycorrhizal fungus differentially regulates root traits and cadmium uptake in two maize varieties. *Ecotoxicology and Environmental Safety, 264*, 115458. DOI: 10.1016/j.ecoenv.2023.115458 PMID: 37690173

Conversa, G., Miedico, O., Chiaravalle, A. E., & Elia, A. (2019). Heavy metal contents in green spears of asparagus (Asparagus officinalis L.) grown in Southern Italy: Variability among farms, genotypes and effect of soil mycorrhizal inoculation. *Scientia Horticulturae, 256*, 108559. DOI: 10.1016/j.scienta.2019.108559

Crone, M., McComb, J. A., O'Brien, P. A., & Hardy, G. E. S. J. (2013). Survival of *Phytophthora cinnamomi* as oospores, stromata, and thick-walled chlamydospores in roots of symptomatic and asymptomatic annual and herbaceous perennial plant species. *Fungal Biology, 117*(2), 112–123. DOI: 10.1016/j.funbio.2012.12.004 PMID: 23452949

Dasgupta, D., Dey, A., & Meena, M. C. (2024). *Soil Mycorrhizae and Their Industrial Applications. In industrial applications of soil microbes*. Bentham Science Publishers., DOI: 10.2174/9789815079753124030

Dashtebani, F., & Aliasgharzad, N. (2018). Germination, photosynthesis and growth of two halophytic grass *Puccinellia distans* and *Aeluropus littoralis* under salinity and their colonization with mycorrhizal species in their natural habitat in the Tabriz plain. *Journal of Plant Research (Iranian. Journal of Biology*, *30*(4), 817–833.

de Vries, F. T., Griffiths, R. I., Bailey, M., Craig, H., Girlanda, M., Gweon, H. S., & Bardgett, R. D. (2018). Soil bacterial networks are less stable under drought than fungal networks. *Nature Communications*, *9*(1), 3033. DOI: 10.1038/s41467-018-05516-7 PMID: 30072764

Delavaux, C. S., & Bever, J. D. (2022). Evidence for the evolution of native plant response to mycorrhizal fungi in post-agricultural grasslands. *Ecology and Evolution*, *12*(7), e9097. DOI: 10.1002/ece3.9097 PMID: 35845364

Denman, S., Kirk, S. A., Moralejo, E., & Webber, J. F. (2009). *Phytophthora ramorum* and *Phytophthora kernoviae* on naturally infected asymptomatic foliage. *Bulletin OEPP. EPPO Bulletin. European and Mediterranean Plant Protection Organisation*, *39*(1), 105–111. DOI: 10.1111/j.1365-2338.2009.02243.x

Dick, M. A., Dobbie, K., Cooke, D. E., & Brasier, C. M. (2006). *Phytophthora captiosa* sp. nov. and *P. fallax* sp. nov. causing crown dieback of *Eucalyptus* in New Zealand. *Mycological Research*, *110*(4), 393–404. DOI: 10.1016/j.mycres.2006.01.008 PMID: 16650734

Dickie, I. A., Bufford, J. L., Cobb, R. C., Desprez-Loustau, M. L., Grelet, G., Hulme, P. E., & Williams, N. M. (2017a). The emerging science of linked plant–fungal invasions. *The New Phytologist*, *215*(4), 1314–1332. DOI: 10.1111/nph.14657 PMID: 28649741

Dickie, I. A., Cooper, J. A., Bufford, J. L., Hulme, P. E., & Bates, S. T. (2017b). Loss of functional diversity and network modularity in introduced plant–fungal symbioses. *AoB Plants*, *9*(1), plw084. PMID: 28039116

Dilawar, N., Hamayun, M., Iqbal, A., Lee, B., Ali, S., Ahmad, A., & Hussain, A. (2024). Rhizofungus Aspergillus terreus Mitigates Heavy Metal Stress-Associated Damage in Triticum aestivum L. *Plants*, *13*(18), 2643. DOI: 10.3390/plants13182643 PMID: 39339618

Ding, J., & Eldridge, D. J. (2024). Woody encroachment: Social–ecological impacts and sustainable management. *Biological Reviews of the Cambridge Philosophical Society*, *99*(6), 1909–1926. Advance online publication. DOI: 10.1111/brv.13104 PMID: 38961449

Du, E., Jia, Y., Wu, C., Chen, X., & Zhang, F. (2024). Arbuscular mycorrhizal fungi and *Bacillus* promote *Flaveria bidentis* invasion success by inhibiting the growth of native species under different soil nutrient levels. *Plant and Soil*, *500*(1), 147–160. DOI: 10.1007/s11104-023-06052-0

Duan, S., Feng, G., Limpens, E., Bonfante, P., Xie, X., & Zhang, L. (2024). Cross-kingdom nutrient exchange in the plant–arbuscular mycorrhizal fungus–bacterium continuum. *Nature Reviews. Microbiology*, *22*(12), 1–18. DOI: 10.1038/s41579-024-01073-7 PMID: 39014094

Endresz, G., Somodi, I., & Kalapos, T. (2013). Arbuscular mycorrhizal colonisation of roots of grass species differing in invasiveness. *Community Ecology*, *14*(1), 67–76. DOI: 10.1556/ComEc.14.2013.1.8

Fayet, M., Long, M., Han, B., Belkorchia, A., Delbac, F., & Polonais, V. (2024). New insights into Microsporidia polar tube function and invasion mechanism. *The Journal of Eukaryotic Microbiology*, *13043*(5), e13043. Advance online publication. DOI: 10.1111/jeu.13043 PMID: 38973152

Figueiredo Lugli, L., Anderson, K. M., Aragao, L. E. O. C., Cordeiro, A. L., Cunha, H. F. V., Fuchslueger, L., & Meir, P. (2019). Multiple phosphorus acquisition strategies adopted by fine roots in low-fertility soils in Central Amazonia. *Plant and Soil*, *450*(1-2), 49–63. DOI: 10.1007/s11104-019-03963-9

Fry, W. E., McGrath, M. T., Seaman, A., Zitter, T. A., McLeod, A., Danies, G., & Smart, C. D. (2013). The 2009 late blight pandemic in the eastern United States–causes and results. *Plant Disease*, *97*(3), 296–306. DOI: 10.1094/PDIS-08-12-0791-FE PMID: 30722376

Gaggini, L., Rusterholz, H. P., & Baur, B. (2018). The invasive plant *Impatiens glandulifera* affects soil fungal diversity and the bacterial community in forests. *Applied Soil Ecology*, *124*, 335–343. DOI: 10.1016/j.apsoil.2017.11.021

Gaggini, L., Rusterholz, H. P., & Baur, B. (2019). The annual invasive plant *Impatiens glandulifera* reduces hyphal biomass of soil fungi in deciduous forests. *Fungal Ecology*, *39*, 242–249. DOI: 10.1016/j.funeco.2018.12.004

Garg, N., & Singh, S. (2018). Arbuscular mycorrhiza Rhizophagus irregularis and silicon modulate growth, proline biosynthesis and yield in Cajanus cajan L. Millsp. (pigeonpea) genotypes under cadmium and zinc stress. *Journal of Plant Growth Regulation*, *37*(1), 46–63. DOI: 10.1007/s00344-017-9708-4

Genre, A., Lanfranco, L., Perotto, S., & Bonfante, P. (2020). Unique and common traits in mycorrhizal symbioses. *Nature Reviews. Microbiology*, *18*(11), 649–660. DOI: 10.1038/s41579-020-0402-3 PMID: 32694620

Grove, S., Haubensak, K. A., Gehring, C., & Parker, I. M. (2017). Mycorrhizae, invasions, and the temporal dynamics of mutualism disruption. *Journal of Ecology*, *105*(6), 1496–1508. DOI: 10.1111/1365-2745.12853

Grünwald, N. J., Goss, E. M., & Press, C. M. (2008). *Phytophthora ramorum*: A pathogen with a remarkably wide host range causing sudden oak death on oaks and ramorum blight on woody ornamentals. *Molecular Plant Pathology*, *9*(6), 729–740. DOI: 10.1111/j.1364-3703.2008.00500.x PMID: 19019002

Grünwald, N. J., Larsen, M. M., Kamvar, Z. N., Reeser, P. W., Kanaskie, A., Laine, J., & Wiese, R. (2016). First report of the EU1 clonal lineage of Phytophthora ramorum on tanoak in an Oregon forest. *Plant Disease*, *100*(5), 1024–1024. DOI: 10.1094/PDIS-10-15-1169-PDN

Gulzar, R., Ahmad, R., Hassan, T., Rashid, I., & Khuroo, A. A. (2024). Environmental and anthropogenic drivers of invasive plant diversity and distribution in the Himalaya. *Ecological Informatics*, *81*, 102586. DOI: 10.1016/j.ecoinf.2024.102586

Hajiboland, R., Joudmand, A., Aliasgharzad, N., Tolrá, R., & Poschenrieder, C. (2019). Arbuscular mycorrhizal fungi alleviate low-temperature stress and increase freezing resistance as a substitute for acclimation treatment in barley. *Crop & Pasture Science*, *70*(3), 218–233. DOI: 10.1071/CP18385

Han, M., Zhang, H., Liu, M., Tang, J., Guo, X., Ren, W., & Kong, D. (2024). Increased dependence on nitrogen-fixation of a native legume in competition with an invasive plant. *Plant Diversity*, *46*(4), 510–518. Advance online publication. DOI: 10.1016/j.pld.2024.04.003 PMID: 39280977

Hashem, A., Alqarawi, A. A., Radhakrishnan, R., Al-Arjani, A. B. F., Aldehaish, H. A., & Egamberdieva, D., & Abd_Allah, E. F. (. (2018). Arbuscular mycorrhizal fungi regulate the oxidative system, hormones and ionic equilibrium to trigger salt stress tolerance in Cucumis sativus L. *Saudi Journal of Biological Sciences*, *25*(6), 110. DOI: 10.1016/j.sjbs.2018.03.009 PMID: 30174509

Hassan, M. M., Azrag, M. A., Rugheim, A. M., Abusin, R. M., Elnasikh, M. H., Modawi, H. I., & Babiker, A. G. E. (2019). Potential of *Trichoderma harzianum* as a biocontrol agent against *Striga hermonthica* in *Sorghum. International Journal of Current Microbiology and Applied Sciences*, *8*, 195–206. DOI: 10.20546/ijcmas.2019.803.027

He, F., Sheng, M., & Tang, M. (2017). Effects of Rhizophagus irregularis on photosynthesis and antioxidative enzymatic system in Robinia pseudoacacia L. under drought stress. *Frontiers in Plant Science*, *8*, 183. DOI: 10.3389/fpls.2017.00183 PMID: 28261240

He, Z.-Z., Stotz, G. C., Liu, X., Liu, J.-J., Wang, Y.-G., Yang, J., Li, L.-F., Zhang, W.-J., Nan, P., & Song, Z.-P. (2024). A global synthesis of the patterns of genetic diversity in endangered and invasive plants. *Biological Conservation*, *291*, 110473. DOI: 10.1016/j.biocon.2024.110473

Hill, P. W., Broughton, R., Bougoure, J., Havelange, W., Newsham, K. K., Grant, H., Murphy, D. V., Clode, P., Ramayah, S., Marsden, K. A., Quilliam, R. S., Roberts, P., Brown, C., Read, D. J., Deluca, T. H., Bardgett, R. D., Hopkins, D. W., & Jones, D. L. (2019). Angiosperm symbioses with non-mycorrhizal fungal partners enhance N acquisition from ancient organic matter in a warming maritime Antarctic. *Ecology Letters*, *22*(12), 2111–2119. DOI: 10.1111/ele.13399 PMID: 31621153

Hu, Y., Xie, W., & Chen, B. (2020). Arbuscular mycorrhiza improved drought tolerance of maize seedlings by altering photosystem II efficiency and the levels of key metabolites. *Chemical and Biological Technologies in Agriculture*, *7*(1), 1–14. DOI: 10.1186/s40538-020-00186-4

Husain, H. J., Wang, X., Pirasteh, S., Mafi-Gholami, D., Chouhan, B., Khan, M. L., & Gheisari, M. (2024). Review and assessment of the potential restoration of ecosystem services through the implementation of the biodiversity management plans for SDG-15 localization. *Heliyon*, *10*(8), e29877. Advance online publication. DOI: 10.1016/j.heliyon.2024.e29877 PMID: 38699718

Igiehon, N. O., Babalola, O. O., Cheseto, X., & Torto, B. (2021). Effects of rhizobia and arbuscular mycorrhizal fungi on yield, size distribution and fatty acid of soybean seeds grown under drought stress. *Microbiological Research*, *242*, 126640. DOI: 10.1016/j.micres.2020.126640 PMID: 33223380

Iqbal, B., Zhao, X., Khan, K. Y., Javed, Q., Nazar, M., Khan, I., & Du, D. (2024). Microplastics meet invasive plants: Unraveling the ecological hazards to agroecosystems. *The Science of the Total Environment*, *906*, 167756. DOI: 10.1016/j.scitotenv.2023.167756 PMID: 37832681

Ivey, M., & Leege, L. M. (2023). Life After Privet: Plant Community Dynamics in a Forested Wetland Following Removal of the Invasive Ligustrum sinense Lour. *Castanea*, *88*(1), 49–60. DOI: 10.2179/0008-7475.88.1.49

Ji-Zhong, W., Zhang, Z. X., & Chun-Jing, W. (2019). Effects of ecoregional vulnerability on habitat suitability of invasive alien plants: An assessment using 13 species on a global scale. *Environmental Earth Sciences*, *78*(5), 1–9.

Jiang, Y., Wang, M., Yan, X., Liu, M., & Guo, X. (2024). Exotic plants introduction changed soil nutrient cycle and symbiotic relationship with arbuscular mycorrhizal fungi in wetland ecological projects. *Frontiers in Plant Science*, *15*, 1410009. DOI: 10.3389/fpls.2024.1410009 PMID: 39049854

Jiang, Y. L., Chen, X. H., Miao, Q., & Qu, B. (2019). *Difference in fungal communities between in roots and in root-associated soil of nine orchids in Liaoning.*, DOI: 10.17521/cjpe.2019.0055

Kaur, M., & Singh, P. (2024). Unravelling the diversity, root colonization, and morphological features of arbuscular mycorrhizal fungi associated with invasive plant species of Sirmaur, Himachal Pradesh, India. *Asian Journal of Mycology*, *7*(2), 1–10.

Khalloufi, M., Martínez-Andújar, C., Lachaâl, M., Karray-Bouraoui, N., Pérez-Alfocea, F., & Albacete, A. (2017). The interaction between foliar GA3 application and arbuscular mycorrhizal fungi inoculation improves growth in salinized tomato (Solanum lycopersicum L.) plants by modifying the hormonal balance. *Journal of Plant Physiology*, *214*, 134–144. DOI: 10.1016/j.jplph.2017.04.012 PMID: 28482334

Khan, M., Khan, T., Tabassum, B., & Hashim, M. (2024). Microbiome-driven soil fertility: understanding symbiotic relationships. In *Progress in Soil Microbiome Research* (pp. 77–115). Springer. DOI: 10.1007/978-3-031-71487-0_4

Kohler, J., Roldán, A., Campoy, M., & Caravaca, F. (2017). Unraveling the role of hyphal networks from arbuscular mycorrhizal fungi in aggregate stabilization of semiarid soils with different textures and carbonate contents. *Plant and Soil*, *410*(1-2), 273–281. DOI: 10.1007/s11104-016-3001-3

Kumar, A., Dutta, J., Chandrawanshi, N. K., Ekka, A., & Sethi, S. K. (2024). *Arbuscular mycorrhizal fungi in alleviation of biotic stress tolerance in plants: A new direction in sustainable agriculture. In entrepreneurship with microorganisms.* Academic Press., DOI: 10.1016/B978-0-443-19049-0.00006-2

Kumar, M., Prasad, R., Kumar, V., Tuteja, N., & Varma, A. (2017). Mycorrhizal fungi under biotic and abiotic stress. *Mycorrhiza-Eco-Physiology, Secondary Metabolites. Nanomaterials (Basel, Switzerland)*, •••, 57–69.

Lal, M. K., Tiwari, R. K., Adavi, S. B., Kumar, A., Bamatov, I., Ivanova, E., Behera, L., Jena, R., & Kumar, R. (2024). Deciphering phenomics approaches for understanding plant–microbe interactions in nutrient absorption and utilization. *Plant Physiology Reports*, 1-17.

Lefebvre, S., Segar, J., & Staude, I. R. (2024). Non-natives are linked to higher plant diversity across spatial scales. *Journal of Biogeography*, *51*(7), 1290–1298. Advance online publication. DOI: 10.1111/jbi.14824

Leventis, G., Tsiknia, M., Feka, M., Ladikou, E. V., Papadakis, I. E., Chatzipavlidis, I., & Ehaliotis, C. (2021). Arbuscular mycorrhizal fungi enhance growth of tomato under normal and drought conditions, via different water regulation mechanisms. *Rhizosphere*, *19*, 100394. DOI: 10.1016/j.rhisph.2021.100394

Li, J., He, J. Z., Liu, M., Yan, Z. Q., Xu, X. L., & Kuzyakov, Y. (2024). Invasive plant competitivity is mediated by nitrogen use strategies and rhizosphere microbiome. *Soil Biology & Biochemistry*, *192*, 109361. DOI: 10.1016/j.soilbio.2024.109361

Liu, D., Zheng, K., Wang, Y., Zhang, Y., Lao, R., Qin, Z., Li, T., & Zhao, Z. (2022). Harnessing an arbuscular mycorrhizal fungus to improve the adaptability of a facultative metallophytic poplar (Populus yunnanensis) to cadmium stress: Physiological and molecular responses. *Journal of Hazardous Materials*, *424*, 127430. DOI: 10.1016/j.jhazmat.2021.127430 PMID: 34678563

Lofgren, L., Nguyen, N. H., Kennedy, P. G., Pérez-Pazos, E., Fletcher, J., Liao, H. L., Wang, H., Zhang, K., Ruytinx, J., Smith, A. H., Ke, Y.-H., Cotter, H. V. T., Engwall, E., Hameed, K. M., Vilgalys, R., & Branco, S. (2024). Suillus: An emerging model for the study of ectomycorrhizal ecology and evolution. *The New Phytologist*, *242*(4), 1448–1475. DOI: 10.1111/nph.19700 PMID: 38581203

Magkourilou, E., Bell, C., Daniell, T., & Field, K. (2024). The functionality of arbuscular mycorrhizal networks across scales of experimental complexity and ecological relevance. *Functional Ecology*. Advance online publication. DOI: 10.1111/1365-2435.14618

Mathur, S., Agnihotri, R., Sharma, M. P., Reddy, V. R., & Jajoo, A. (2021). Effect of high-temperature stress on plant physiological traits and mycorrhizal symbiosis in maize plants. *Journal of Fungi (Basel, Switzerland)*, *7*(10), 867. DOI: 10.3390/jof7100867 PMID: 34682289

Mathur, S., & Jajoo, A. (2020). Arbuscular mycorrhizal fungi protects maize plants from high temperature stress by regulating photosystem II heterogeneity. *Industrial Crops and Products*, *143*, 111934. DOI: 10.1016/j.indcrop.2019.111934

Mathur, S., Sharma, M. P., & Jajoo, A. (2018). Improved photosynthetic efficacy of maize (Zea mays) plants with arbuscular mycorrhizal fungi (AMF) under high temperature stress. *Journal of Photochemistry and Photobiology. B, Biology*, *180*, 149–154. DOI: 10.1016/j.jphotobiol.2018.02.002 PMID: 29425887

Milbrath, L. R., Davis, A. S., & Biazzo, J. (2017). Demography of invasive black and pale swallow-wort populations in New York. *Northeastern Naturalist*, *24*(1), 37–53. DOI: 10.1656/045.024.0104

Mohammed, A. E., Alotaibi, M. O., & Elobeid, M. (2022). Interactive influence of elevated CO2 and arbuscular mycorrhizal fungi on sucrose and coumarin metabolism in Ammi majus. *Plant Physiology and Biochemistry*, *185*, 45–54. DOI: 10.1016/j.plaphy.2022.05.029 PMID: 35660776

Ng, A., Wilson, B. A., & Frew, A. (2023). Belowground crop responses to root herbivory are associated with the community structure of native arbuscular mycorrhizal fungi. *Applied Soil Ecology*, *185*, 104797. DOI: 10.1016/j.apsoil.2022.104797

Nie, W., He, Q., Guo, H., Zhang, W., Ma, L., Li, J., & Wen, D. (2024). Arbuscular mycorrhizal fungi: Boosting crop resilience to environmental stresses. *Microorganisms*, *12*(12), 2448. DOI: 10.3390/microorganisms12122448 PMID: 39770651

Nizamani, M. M., Hughes, A. C., Qureshi, S., Zhang, Q., Tarafder, E., Das, D., Acharya, K., Wang, Y., & Zhang, Z.-G. (2024). Microbial biodiversity and plant functional trait interactions in multifunctional ecosystems. *Applied Soil Ecology*, *201*, 105515. DOI: 10.1016/j.apsoil.2024.105515

Nunez-Mir, G. C., & McCary, M. A. (2024). Invasive plants and their root traits are linked to the homogenization of soil microbial communities across the United States. *Proceedings of the National Academy of Sciences of the United States of America*, *121*(44), e2418632121. DOI: 10.1073/pnas.2418632121 PMID: 39446386

Orine, D., Defossez, E., Vergara, F., Uthe, H., van Dam, N. M., & Rasmann, S. (2022). Arbuscular mycorrhizal fungi prevent the negative effect of drought and modulate the growth-defence trade-off in tomato plants. *Journal of Sustainable Agriculture and Environment*, *1*(3), 177–190. DOI: 10.1002/sae2.12018

Orlikowski, L. B., Ptaszek, M., Rodziewicz, A., Nechwatal, J., Thinggaard, K., & Jung, T. (2011). *Phytophthora* root and collar rot of mature *Fraxinus excelsior* in forest stands in Poland and Denmark. *Forest Pathology*, *41*(6), 510–519. DOI: 10.1111/j.1439-0329.2011.00714.x

Ouledali, S., Ennajeh, M., Ferrandino, A., Khemira, H., Schubert, A., & Secchi, F. (2019). Influence of arbuscular mycorrhizal fungi inoculation on the control of stomata functioning by abscisic acid (ABA) in drought-stressed olive plants. *South African Journal of Botany*, *121*, 152–158. DOI: 10.1016/j.sajb.2018.10.024

Pal, A., & Pandey, S. (2017). Symbiosis of arbuscular mycorrhizal fungi and Pennisetum glaucum L. improves plant growth and glomalin-related soil protein in barren soil. *Int J Sci Invent Today*, *6*(6), 783–792.

Parvin, S., Van Geel, M., Yeasmin, T., Verbruggen, E., & Honnay, O. (2020). Effects of single and multiple species inocula of arbuscular mycorrhizal fungi on the salinity tolerance of a Bangladeshi rice (Oryza sativa L.) cultivar. *Mycorrhiza*, *30*(4), 431–444. DOI: 10.1007/s00572-020-00957-9 PMID: 32367433

Pathak, R., Negi, V. S., Rawal, R. S., & Bhatt, I. D. (2019). Alien plant invasion in the Indian Himalayan Region: State of knowledge and research priorities. *Biodiversity and Conservation*, *28*(12), 3073–3102. DOI: 10.1007/s10531-019-01829-1

Pérez Castro, S., Cleland, E. E., Wagner, R., Sawad, R. A., & Lipson, D. A. (2019). Soil microbial responses to drought and exotic plants shift carbon metabolism. *The ISME Journal*, *13*(7), 1776–1787. DOI: 10.1038/s41396-019-0389-9 PMID: 30872806

Perring, M., Rowe, E., Arrigoni, E., Bidartondo, M., Britton, A., Field, C., Gilliam, F., Grenier, M., Hayes, F., & Jones, L. (2024). Scoping future research for air pollution recovery indicators (APRI).(Workshop report). https://nora.nerc.ac.uk/id/eprint/537618

Pickett, B., Irvine, I. C., Bullock, E., Arogyaswamy, K., & Aronson, E. (2019). Legacy effects of invasive grass impact soil microbes and native shrub growth. *Invasive Plant Science and Management*, *12*(1), 22–35. DOI: 10.1017/inp.2018.32

Policelli, N., Bruns, T. D., Vilgalys, R., & Nuñez, M. A. (2019). Suilloid fungi as global drivers of pine invasions. *The New Phytologist*, *222*(2), 714–725. DOI: 10.1111/nph.15660 PMID: 30586169

Pons, C., Voß, A. C., Schweiger, R., & Müller, C. (2020). Effects of drought and mycorrhiza on wheat and aphid infestation. *Ecology and Evolution*, *10*(19), 10481–10491. DOI: 10.1002/ece3.6703 PMID: 33072274

Qadir, M., Iqbal, A., Hussain, A., Hussain, A., Shah, F., Yun, B.-W., & Mun, B.-G. (2024). Exploring plant-bacterial symbiosis for eco-friendly agriculture and enhanced resilience. *International Journal of Molecular Sciences*, *25*(22), 12198. DOI: 10.3390/ijms252212198 PMID: 39596264

Quiroga, G., Erice, G., Ding, L., Chaumont, F., Aroca, R., & Ruiz-Lozano, J. M. (2019). The arbuscular mycorrhizal symbiosis regulates aquaporins activity and improves root cell water permeability in maize plants subjected to water stress. *Plant, Cell & Environment*, *42*(7), 2274–2290. DOI: 10.1111/pce.13551 PMID: 30916398

Raheem, A., Yohanna, P., Li, G., Noh, N. J., Iqbal, B., Tang, J., Du, D., Alahmadi, T. A., Ansari, M. J., Zhan, A., & Son, Y. (2024). Unraveling the ecological threads: How invasive alien plants influence soil carbon dynamics. *Journal of Environmental Management*, *356*, 120556. DOI: 10.1016/j.jenvman.2024.120556 PMID: 38537457

Rahman, M. A., Das, A. K., Al Riyadh, Z., Suhag, M., & Rahman, M. M. (2024). Eucalyptus in Agriculture: Friend or Foe? Analyzing its impact on crop yields, soil dynamics, and farmers' perceptions in Bangladesh. *Agroforestry Systems*, *98*(8), 1–20. DOI: 10.1007/s10457-024-01077-5

Raihan, A., Ridwan, M., & Rahman, M. S. (2024). An exploration of the latest developments, obstacles, and potential future pathways for climate-smart agriculture. *Climate Smart Agriculture*, *100020*(2), 100020. Advance online publication. DOI: 10.1016/j.csag.2024.100020

Rajamuthuramalingam, T., da Silva, W., Zuverza-Mena, N., Dimkpa, C., & White, J. C. (2024). Nano-sized metal oxide fertilizers for sustainable agriculture: Balancing benefits, risks, and risk management strategies. *Nanoscale*, *16*(43), 19998–20026. Advance online publication. DOI: 10.1039/D4NR01354A PMID: 39417765

Rigling, D., & Prospero, S. (2018). Cryphonectria parasitica, the causal agent of chestnut blight: Invasion history, population biology and disease control. *Molecular Plant Pathology*, *19*(1), 7–20. DOI: 10.1111/mpp.12542 PMID: 28142223

Rigling, D., & Prospero, S. (2018). *Cryphonectria parasitica*, the causal agent of chestnut blight: Invasion history, population biology and disease control. *Molecular Plant Pathology*, *19*(1), 7–20. DOI: 10.1111/mpp.12542 PMID: 28142223

Roldão Almeida, M., Marchante, E., & Marchante, H. (2024). Knowing the invader: Increasing knowledge about an invasive plant to improve management. *Restoration Ecology*, *32*(5), 14175. DOI: 10.1111/rec.14175

Santander, C., Ruiz, A., García, S., Aroca, R., Cumming, J., & Cornejo, P. (2020). Efficiency of two arbuscular mycorrhizal fungal inocula to improve saline stress tolerance in lettuce plants by changes of antioxidant defense mechanisms. *Journal of the Science of Food and Agriculture*, *100*(4), 1577–1587. DOI: 10.1002/jsfa.10166 PMID: 31769028

Sepp, S. K., Davison, J., Jairus, T., Vasar, M., Moora, M., Zobel, M., & Öpik, M. (2019). Non-random association patterns in a plant–mycorrhizal fungal network reveal host–symbiont specificity. *Molecular Ecology*, *28*(2), 365–378. DOI: 10.1111/mec.14924 PMID: 30403423

Shen, K., He, Y., Xia, T., Guo, Y., Wu, B., Han, X., & Liu, Y. (2024). Arbuscular mycorrhizal fungi improve the competitive advantage of a native plant relative to a congeneric invasive plant in growth and nutrition. *Ecology and Evolution*, *14*(5), e11459. DOI: 10.1002/ece3.11459 PMID: 38774145

Silva-Castro, G. A., Cano, C., Moreno-Morillas, S., Bago, A., & García-Romera, I. (2022). Inoculation of indigenous arbuscular mycorrhizal fungi as a strategy for the recovery of long-term heavy metal-contaminated soils in a mine-spill area. *Journal of Fungi (Basel, Switzerland)*, *9*(1), 56. DOI: 10.3390/jof9010056 PMID: 36675877

Solís-Rodríguez, U. R. J., Ramos-Zapata, J. A., Hernández-Cuevas, L., Salinas-Peba, L., & Guadarrama, P. (2020). Arbuscular mycorrhizal fungi diversity and distribution in tropical low flooding forest in Mexico. *Mycological Progress*, *19*(3), 195–204. DOI: 10.1007/s11557-019-01550-x

Song, F., Li, J., Fan, X., Zhang, Q., Chang, W., Yang, F., & Geng, G. (2016). Transcriptome analysis of *Glomus mosseae/Medicago sativa* mycorrhiza on atrazine stress. *Scientific Reports*, *6*(1), 20245. DOI: 10.1038/srep20245 PMID: 26833403

Song, Z., Bi, Y., Zhang, J., Gong, Y., & Yang, H. (2020). Arbuscular mycorrhizal fungi promote the growth of plants in the mining associated clay. *Scientific Reports*, *10*(1), 2663. DOI: 10.1038/s41598-020-59447-9 PMID: 32060306

Soni, M., Sheetal, K., Renjith, P., & Subbulakshmi, V. Birbal, Nathawat, N., Panwar, N., & Dagar, J. C. (2024). Domestication of wild halophytes for profitable biosaline agriculture. In halophytes vis-à-vis saline agriculture: perspectives and opportunities for food security, (pp. 479-505). Springer.

Spagnoletti, F. N., Carmona, M., Balestrasse, K., Chiocchio, V., Giacometti, R., & Lavado, R. S. (2021). The arbuscular mycorrhizal fungus Rhizophagus intraradices reduces the root rot caused by Fusarium pseudograminearum in wheat. *Rhizosphere*, *19*, 100369. DOI: 10.1016/j.rhisph.2021.100369

Sun, K., Pan, Y.-T., Jiang, H.-J., Xu, J.-Y., Ma, C.-Y., Zhou, J., Liu, Y., Shabala, S., Zhang, W., & Dai, C.-C. (2024). Root endophyte-mediated alteration in plant H2O2 homeostasis regulates symbiosis outcome and reshapes the rhizosphere microbiota. *Journal of Experimental Botany*, *75*(10), 3153–3170. DOI: 10.1093/jxb/erae069 PMID: 38394357

Tang, J. Q., Guo, X. C., Lu, X. Y., Liu, M. C., Zhang, H. Y., Feng, Y. L., & Kong, D. L. (2020). A review on the effects of invasive plants on mycorrhizal fungi of native plants and their underlying mechanisms. *Acta Phytoecologica Sinica*, *44*(11), 1095–1112. DOI: 10.17521/cjpe.2020.0166

Tanner, R. A., & Gange, A. C. (2013). The impact of two non-native plant species on native flora performance: Potential implications for habitat restoration. *Plant Ecology*, *214*(3), 423–432. DOI: 10.1007/s11258-013-0179-9

Tariq, A., Graciano, C., Sardans, J., Zeng, F., Hughes, A. C., Ahmed, Z., Ullah, A., Ali, S., Gao, Y., & Peñuelas, J. (2024). Plant root mechanisms and their effects on carbon and nutrient accumulation in desert ecosystems under changes in land use and climate. *The New Phytologist*, *242*(3), 916–934. DOI: 10.1111/nph.19676 PMID: 38482544

Tedersoo, L., Bahram, M., & Zobel, M. (2020). How mycorrhizal associations drive plant population and community biology. *Science*, *367*(6480), eaba1223. DOI: 10.1126/science.aba1223 PMID: 32079744

Topalović, O., Koorem, K., & Wilschut, R. A. (2023). Drivers of host-microbiome interactions in the rhizosphere. *Frontiers in Plant Science*, *14*, 1294388. DOI: 10.3389/fpls.2023.1294388 PMID: 37828922

Ullah, A., Gao, D., & Wu, F. (2024). Common mycorrhizal network: The predominant socialist and capitalist responses of possible plant-plant and plant-microbe interactions for sustainable agriculture. *Frontiers in Microbiology*, *15*, 1183024. DOI: 10.3389/fmicb.2024.1183024 PMID: 38628862

Urcelay, C., Longo, S., Geml, J., Tecco, P. A., & Nouhra, E. (2017). Co-invasive exotic pines and their ectomycorrhizal symbionts show capabilities for wide distance and altitudinal range expansion. *Fungal Ecology*, *25*, 50–58. DOI: 10.1016/j.funeco.2016.11.002

Usman, M., Ho-Plágaro, T., Frank, H. E., Calvo-Polanco, M., Gaillard, I., Garcia, K., & Zimmermann, S. D. (2021). Mycorrhizal symbiosis for better adaptation of trees to abiotic stress caused by climate change in temperate and boreal forests. *Frontiers in Forests and Global Change*, *4*, 742392. DOI: 10.3389/ffgc.2021.742392

Van Niekerk, G., Young, E., & Fourie, H. (2024). *Veld*. Penguin random house South Africa. (Pg, 320). ISBN: 1779890222, 9781779890221

Velez-Gavilan, J. (2024). *Phragmites australis* (common reed). DOI: 10.1079/cabicompendium.40514

Vilà, M., Trillo, A., Castro-Díez, P., Gallardo, B., & Bacher, S. (2024). Field studies of the ecological impacts of invasive plants in Europe. DOI: 10.3897/neobiota.90.112368

Wang, F. (2017). Occurrence of arbuscular mycorrhizal fungi in mining-impacted sites and their contribution to ecological restoration: Mechanisms and applications. *Critical Reviews in Environmental Science and Technology*, *47*(20), 1901–1957. Advance online publication. DOI: 10.1080/10643389.2017.1400853

Wang, M., Lamit, L. J., Lilleskov, E. A., Basiliko, N., Moore, T. R., Bubier, J. L., Guo, G., Juutinen, S., & Larmola, T. (2024). Peatland fungal community responses to nutrient enrichment: A story beyond nitrogen. *Global Change Biology*, *30*(11), e17562. DOI: 10.1111/gcb.17562 PMID: 39492595

Wang, S., Augé, R. M., & Toler, H. D. (2017). Arbuscular mycorrhiza formation and its function under elevated atmospheric O3: A meta-analysis. *Environmental Pollution*, *226*, 104–117. DOI: 10.1016/j.envpol.2017.03.074 PMID: 28411495

Wang, X., Fang, L., Beiyuan, J., Cui, Y., Peng, Q., Zhu, S., & Zhang, X. (2021). Improvement of alfalfa resistance against Cd stress through rhizobia and arbuscular mycorrhiza fungi co-inoculation in Cd-contaminated soil. *Environmental Pollution*, *277*, 116758. DOI: 10.1016/j.envpol.2021.116758 PMID: 33652182

Wang, Y., Lin, J., Yang, F., Tao, S., Yan, X., Zhou, Z., & Zhang, Y. (2022). Arbuscular mycorrhizal fungi improve the growth and performance in the seedlings of Leymus chinensis under alkali and drought stresses. *PeerJ, 10*, e12890.2-1114.

Wang, Y., Wang, M., Li, Y., Wu, A., & Huang, J. (2018). Effects of arbuscular mycorrhizal fungi on growth and nitrogen uptake of Chrysanthemum morifolium under salt stress. *PLoS One*, *13*(4), e0196408. DOI: 10.1371/journal.pone.0196408 PMID: 29698448

Wei, X. P., Yu, J., Gao, T. G., Zhang, B. G., & Zhang, Z. (2020). First report and eradication of an invasive weed, *Solanum carolinense* L. *Zhiwu Jianyi*, *34*(3), 61–63.

Westbrook, A. S., Milbrath, L. R., Weinberg, J., & DiTommaso, A. (2023). Biology of Invasive Plants 3. *Vincetoxicum nigrum* (L.) Moench and *Vincetoxicum rossicum* (Kleopow) Barbarich. *Invasive Plant Science and Management*, *16*(1), 3–26. DOI: 10.1017/inp.2023.7

Wong, S., Kaur, J., Kumar, P., Karremans, A. P., & Sharma, J. (2024). Distinct orchid mycorrhizal fungal communities among co-occurring *Vanilla* species in Costa Rica: Root substrate and population-based segregation. *Mycorrhiza*, *34*(3), 1–22. DOI: 10.1007/s00572-024-01147-7 PMID: 38664239

Wubs, E. J., van der Putten, W. H., Bosch, M., & Bezemer, T. M. (2016). Soil inoculation steers restoration of terrestrial ecosystems. *Nature Plants*, *2*(8), 16107. DOI: 10.1038/nplants.2016.107 PMID: 27398907

Yeasmin, R., Bonser, S. P., Motoki, S., & Nishihara, E. (2019). Arbuscular mycorrhiza influences growth and nutrient uptake of asparagus (Asparagus officinalis L.) under heat stress. *HortScience*, *54*(5), 846–850. DOI: 10.21273/HORTSCI13587-18

Yin, Z., Zhang, Y., Hu, N., Shi, Y., Li, T., & Zhao, Z. (2021). Differential responses of 23 maize cultivar seedlings to an arbuscular mycorrhizal fungus when grown in a metal-polluted soil. *The Science of the Total Environment*, *789*, 148015. DOI: 10.1016/j.scitotenv.2021.148015 PMID: 34051499

Yooyongwech, S., Samphumphuang, T., Tisarum, R., Theerawitaya, C., & Cha-Um, S. (2017). Water-deficit tolerance in sweet potato [Ipomoea batatas (L.) Lam.] by foliar application of paclobutrazol: Role of soluble sugar and free proline. *Frontiers in Plant Science*, *8*, 1400. DOI: 10.3389/fpls.2017.01400 PMID: 28848596

Zhan, F., Li, B., Jiang, M., Li, T., He, Y., Li, Y., & Wang, Y. (2019). Effects of arbuscular mycorrhizal fungi on the growth and heavy metal accumulation of bermudagrass [Cynodon dactylon (L.) Pers.] grown in a lead–zinc mine wasteland. *International Journal of Phytoremediation*, *21*(9), 849–856. DOI: 10.1080/15226514.2019.1577353 PMID: 30994000

Zhang, C., van der Heijden, M. G., Dodds, B. K., Nguyen, T. B., Spooren, J., Valzano-Held, A., & Berendsen, R. L. (2024b). A tripartite bacterial-fungal-plant symbiosis in the mycorrhiza-shaped microbiome drives plant growth and mycorrhization. *Microbiome*, *12*(1), 13. DOI: 10.1186/s40168-023-01726-4 PMID: 38243337

Zhang, F., Li, Q., Yerger, E. H., Chen, X., Shi, Q., & Wan, F. (2018). AM fungi facilitate the competitive growth of two invasive plant species, *Ambrosia artemisiifolia* and *Bidens pilosa*. *Mycorrhiza*, *28*(8), 703–715. DOI: 10.1007/s00572-018-0866-4 PMID: 30220052

Zhang, H. Y., Goncalves, P., Copeland, E., Qi, S. S., Dai, Z. C., Li, G. L., & Thomas, T. (2020b). Invasion by the weed *Conyza canadensis* alters soil nutrient supply and shifts microbiota structure. *Soil Biology & Biochemistry*, *143*, 107739. DOI: 10.1016/j.soilbio.2020.107739

Zhang, P., Li, B., Wu, J., & Hu, S. (2019). Invasive plants differentially affect soil biota through litter and rhizosphere pathways: A meta-analysis. *Ecology Letters*, *22*(1), 200–210. DOI: 10.1111/ele.13181 PMID: 30460738

Zhang, Q., He, J., & Pan, Q. (2024a). The role of arbuscular mycorrhizal fungi in exotic plant invasion. [JSCET]. *Journal of Sustainable Civil Engineering & Technology, 3*(2), 43–51. DOI: 10.24191/jscet.v3i2.43-51

Zhang, Z., Liu, Y., Brunel, C., & van Kleunen, M. (2020c). Soil-microorganism-mediated invasional meltdown in plants. *Nature Ecology & Evolution, 4*(12), 1612–1621. DOI: 10.1038/s41559-020-01311-0 PMID: 33020599

Zhao, Y. Z., Liu, M. C., Feng, Y. L., Wang, D., Feng, W. W., Clay, K., & Kong, D. L. (2020). Release from below-and aboveground natural enemies contributes to invasion success of a temperate invader. *Plant and Soil, 452*(1-2), 19–28. DOI: 10.1007/s11104-020-04520-5

Zheng, Y., Hu, H. W., Guo, L. D., Anderson, I. C., & Powell, J. R. (2017). Dryland forest management alters fungal community composition and decouples assembly of root-and soil-associated fungal communities. *Soil Biology & Biochemistry, 109*, 14–22. DOI: 10.1016/j.soilbio.2017.01.024

Zou, Y. N., Wu, Q. S., & Kuča, K. (2021). Unravelling the role of arbuscular mycorrhizal fungi in mitigating the oxidative burst of plants under drought stress. *Plant Biology, 23*(S1), 50–57. DOI: 10.1111/plb.13161 PMID: 32745347

Compilation of References

Ababa, A. (2024). Ethiopian environmental protection authority fact sheet of legeoda mirga mountain forest ecosystem dire dawa city administration.

Abba, A., & Sankarannair, S. (2024). Global impact of water hyacinth (Eichhornia Crassipes) on rural communities and mitigation strategies: A systematic review. *Environmental Science and Pollution Research International*, *31*(31), 43616–43632. DOI: 10.1007/s11356-024-33905-7 PMID: 38937356

Abdelhameed, R. E., & Metwally, R. A. (2019). Alleviation of cadmium stress by arbuscular mycorrhizal symbiosis. *International Journal of Phytoremediation*, *21*(7), 663–671. DOI: 10.1080/15226514.2018.1556584 PMID: 30816051

Abdelhameed, R. E., & Metwally, R. A. (2024). The potential utilization of mycorrhizal fungi and glycine betaine to boost the fenugreek (*Trigonella foenum-graecum* L.) tolerance to chromium toxicity. *Journal of Soil Science and Plant Nutrition*, •••, 1–20. DOI: 10.1007/s42729-024-02131-y

Abella, S. R., Craig, D. J., Smith, S. D., & Newton, A. C. (2012). Identifying native vegetation for reducing exotic species during the restoration of desert ecosystems. *Restoration Ecology*, *20*(6), 781–787. DOI: 10.1111/j.1526-100X.2011.00848.x

Abella, S. R., Porter, S. L., Kline, K. A., Chiquoine, L. P., & Jurand, B. S. (2024). Changing dominance of invasive common reed (Phragmites australis) and native plant colonization with variation in management, wildfires, and soils in a desert wetland. *Invasive Plant Science and Management*, *17*(3), 1–10. DOI: 10.1017/inp.2024.21

Abensperg-Traun, M., Atkins, L., Hobbs, R., & Steven, D. (1998). Exotic plant invasion and understorey species richness: A comparison of two types of eucalypt woodland in agricultural Western Australia. *Pacific Conservation Biology*, *4*(1), 21–32. DOI: 10.1071/PC980021

Acero, V. P., Das, S., Rivellini, O., Purvis, E. M., Adewole, D. O., & Cullen, D. K. (2023). Emergent structural and functional properties of hippocampal multicellular aggregates. *Frontiers in Neuroscience*, *17*, 1171115. DOI: 10.3389/fnins.2023.1171115 PMID: 37397454

Acharya, S., McLaughlin, D., Kaplan, D., & Cohen, M. J. (2020). A proposed method for estimating interception from near-surface soil moisture response. *Hydrology and Earth System Sciences*, *24*(4), 1859–1870. DOI: 10.5194/hess-24-1859-2020

Adam, A. M., Alshahrani, T. S., Alqarawi, A. A., Abd-ElGawad, A. M., Ahmed, E. M., & Sorrori, S. N. (2024). Arbuscular mycorrhiza fungi alleviate the allelopathic effects of invasive nicotiana glauca on the growth of *Acacia gerrardii* seedlings. *Forests*, *15*(8), 1447. DOI: 10.3390/f15081447

Adedibu, P. A. (2023). Ecological problems of agriculture: impacts and sustainable solutions. *ScienceOpen preprints*.

Adeniran-Obey, S. O., & Osagie, D. A. (2024). Impacts of Invasive Species on the Arctic Environment. In *Arctic Marine Ecotoxicology* (pp. 381–402). Springer., DOI: 10.1007/978-3-031-73584-4_18

Adhikari BK, Barrington S, Martinez J, King S (2009) Effectiveness of three bulking agents for food waste composting. Waste management 29(1): 197-203

Afreen, T., Kumari, S., Bhadouria, R., Devi, R. S., Singh, S., & Tripathi, S. (2023). Plant Invasion and Soil Processes: A Mechanistic Understanding. In *Plant Invasions and Global Climate Change* (pp. 227–246). Springer. DOI: 10.1007/978-981-99-5910-5_10

Agarwal, R., Gupta, M., Sen, R., & Panchal, A., ES, N., & Raychoudhury, R. (. (2024). Investigation into how Odontotermes obesus maintains a predominantly Termitomyces monoculture in their fungus combs suggests a potential partnership with both fungi and bacteria. *Communications Biology*, *7*(1), 1010. DOI: 10.1038/s42003-024-06708-2 PMID: 39154098

Aghajani, A., Bautista, C., Bou-Khalil, C., Budgett, S., Cipolla, J., & Coffee, D. (2021). High Priority Invasive Species in Los Angeles County.

Ahmad, M., Bano, A., Zafar, M., Sultana, S., & Rashid, S. (2015). Interdependence of biodiversity, applied ethnobotany, and conservation in higher ecosystems of northern Pakistan under fast climatic changes. In *Climate change impacts on high-altitude ecosystems* (pp. 455–489). Springer. DOI: 10.1007/978-3-319-12859-7_18

Ahmed, A., Aftab, S., Hussain, S., Nazir Cheema, H., Liu, W., Yang, F., & Yang, W. (2020). Nutrient accumulation and distribution assessment in response to potassium application under maize–soybean intercropping system. *Agronomy (Basel), 10*(5), 725. DOI: 10.3390/agronomy10050725

Ahmed, D. A., Sousa, R., Bortolus, A., Aldemir, C., Angeli, N. F., Błońska, D., Briski, E., Britton, J. R., Cano-Barbacil, C., Clark-Ginsberg, A., Culic, I., Cuthbert, R. N., Dick, J., Dimarco, R. D., Essl, F., Everts, T., García-Berthou, E., Hauer, M., Kouba, A., & Haubrock, P. J. (2025). Parallels and discrepancies between non-native species introductions and human migration. *Biological Reviews of the Cambridge Philosophical Society, 100*(3), 1365–1395. DOI: 10.1111/brv.70004 PMID: 39980263

Aizen, M. A., & Torres, A. (2024). The Invasion Ecology of Mutualism. *Annual Review of Ecology, Evolution, and Systematics, 55*(1), 41–63. Advance online publication. DOI: 10.1146/annurev-ecolsys-102622-031210

Akbar, S., & Akbar, S. (2020) Justicia adhatoda L.(Acanthaceae) (Syns.: Adhatoda vasica Nees.; A. pubescens Moench.; A. zeylanica Medik.; Dianthera latifolia Salisb). Handbook of 200 Medicinal Plants: A Comprehensive Review of Their Traditional Medical Uses and Scientific Justifications: 1059-1066

Akbar, R., Sun, J., Bo, Y., Khattak, W. A., Khan, A. A., Jin, C., Zeb, U., Ullah, N., Abbas, A., Liu, W., Wang, X., Khan, S., & Du, D. (2024). Understanding the Influence of Secondary Metabolites in Plant Invasion Strategies: A Comprehensive Review. *Plants, 13*(22), 3162. DOI: 10.3390/plants13223162 PMID: 39599372

Akinola, R., Pereira, L. M., Mabhaudhi, T., De Bruin, F. M., & Rusch, L. (2020). A review of indigenous food crops in Africa and the implications for more sustainable and healthy food systems. *Sustainability, 12*(1), 1–15. DOI: 10.3390/su12083493 PMID: 33520291

Akintobi, T. H., Jacobs, T., & Sabbs, D. (2020). Community engagement of African Americans in the era of COVID-19: Considerations, challenges, implications, and recommendations for public health. *Preventing Chronic Disease*, •••, 2020. PMID: 32790605

Alexander, S., Aronson, J., Whaley, O., & Lamb, D. (2016). The relationship between ecological restoration and the ecosystem services concept. *Ecology and Society, 21*(1), art34. DOI: 10.5751/ES-08288-210134

Ali M, Hakeem KR, Ali M, Hakeem KR (2020) Phyto-pharmaceutical Potential and the Isolation of Novel Compounds from Adhatoda vasica L. Nees. Scientific Explorations of Adhatoda vasica: An Asian Health Remedy: 37-60

Alipour Abookheili, F., & Mobasser, H. R. (2021). Effect of planting density on growth characteristics and grain yield increase in successive cultivations of two rice cultivars. *Agrosystems, Geosciences & Environment*, *4*(4), e20213. DOI: 10.1002/agg2.20213

Allen, C. R., Fontaine, J. J., Pope, K. L., & Garmestani, A. S. (2011). Adaptive management for a turbulent future. *Journal of Environmental Management*, *92*(5), 1339–1345. DOI: 10.1016/j.jenvman.2010.11.019 PMID: 21168260

Aller, D., Chatrchyan, A. M., Calixto, A., Cummings, J., Ortiz-Bobea, A., Peck, G., Schouten, J., Weikert, B., Wolters, E., & Stevens, A. (2024). *New York State Climate Impacts Assessment Chapter 03: Agriculture* (0077-8923). DOI:DOI: 10.1111/nyas.15001

Allison, S. K. (2023). *Ecological restoration and environmental change: Renewing damaged ecosystems*. Routledge. Zavaleta, E. (2000). The economic value of controlling an invasive shrub. *Ambio*, *29*(8), 462–467. DOI: 10.1579/0044-7447-29.8.462

Alpert, P., & Maron, J. L. (2000). Carbon addition as a countermeasure against biological invasion by plants. *Biological Invasions*, *2*(1), 33–40. DOI: 10.1023/A:1010063611473

Al-Rekaby LS, Merza TK (2015) Effect of water stress and organic fertilizer (Azomin) on some growth parameters of Adhatoda vasica L.(Nees). Kufa Journal for Agricultural Sciences 7(2)

Álvarez-Tostado, M., Bueno-Hernández, A., Barahona, A., & Juárez-Barrera, F. (2024). The Study of Geographical Distribution in the Analysis of Domestication as an Evolutionary Process: Tensions in Alphonse de Candolle's Approach. *Journal of the History of Biology*, *57*(3), 1–31. DOI: 10.1007/s10739-024-09785-5 PMID: 39312064

Amanifar, S., Khodabandeloo, M., Fard, E. M., Askari, M. S., & Ashrafi, M. (2019). Alleviation of salt stress and changes in glycyrrhizin accumulation by arbuscular mycorrhiza in liquorice (*Glycyrrhiza glabra*) grown under salinity stress. *Environmental and Experimental Botany*, *160*, 25–34. DOI: 10.1016/j.envexpbot.2019.01.001

Amanifar, S., & Toghranegar, Z. (2020). The efficiency of arbuscular mycorrhiza for improving tolerance of Valeriana officinalis L. and enhancing valerenic acid accumulation under salinity stress. *Industrial Crops and Products*, *147*, 112234. DOI: 10.1016/j.indcrop.2020.112234

Amiri, R., Nikbakht, A., & Etemadi, N. (2015). Alleviation of drought stress on rose geranium [Pelargonium graveolens (L.) Herit.] in terms of antioxidant activity and secondary metabolites by mycorrhizal inoculation. *Scientia Horticulturae*, *197*, 373–380. DOI: 10.1016/j.scienta.2015.09.062

Anderson, K., & Gaston, K. J. (2013). Lightweight unmanned aerial vehicles will revolutionize spatial ecology. *Frontiers in Ecology and the Environment*, *11*(3), 138–146. DOI: 10.1890/120150

Anderson, T., & Mao, J. (2008). Herbicidal compositions and methods of controlling weed growth. Google Patents, Basit A, Shutian T, Khan A, Khan SM, Shahzad R, Khan A, Khan S, Khan M (2022) Anti-inflammatory and analgesic potential of leaf extract of Justicia adhatoda L.(Acanthaceae) in Carrageenan and Formalin-induced models by targeting oxidative stress. *Biomedicine and Pharmacotherapy*, *153*, 113322.

Arasumani, M., Bunyan, M., & Robin, V. (2021). Opportunities and challenges in using remote sensing for invasive tree species management, and in the identification of restoration sites in tropical montane grasslands. *Journal of Environmental Management*, *280*, 111759. DOI: 10.1016/j.jenvman.2020.111759 PMID: 33298397

Areces-Berazain, F. (2023). Pennisetum setaceum (fountain grass).

Aronson, J., Clewell, A. F., Blignaut, J. N., & Milton, S. J. (2006). Ecological restoration: A new frontier for nature conservation and economics. *Journal for Nature Conservation*, *14*(3-4), 135–139. DOI: 10.1016/j.jnc.2006.05.005

Asefa, G. (2019). The role of harvest index in improving crop productivity. *Journal of Natural Sciences Research*, *9*, 24–28.

Athavale, V. A., Selvakumar, P., Poorani, S., Anand, T., & Rathore, D. S. (2025). Designing Chatbots for Educational Leadership. In Musundire, A. (Ed.), *Chatbots in Educational Leadership and Management* (pp. 181–204). IGI Global Scientific Publishing., DOI: 10.4018/979-8-3693-8734-4.ch008

Auld, B., & Martin, P. M. (1975). The autecology of *Eupatorium adenophorum* Spreng. in Australia. *Weed Research*, *15*(1), 27–31. DOI: 10.1111/j.1365-3180.1975.tb01092.x

Aziz, T. (2021). Changes in land use and ecosystem services values in Pakistan, 1950–2050. *Environmental Development*, *37*, 100576. DOI: 10.1016/j.envdev.2020.100576

Babu, S., Niveditha, E., Sooraj Dev, K., & Shamil Haque, K. (2025). Forest Degradation: Loss of Biodiversity and Ecosystem Services. In Forest Degradation and Management: An Indian Perspective (pp. 125-137). Springer.

Bahri, T., Vasconcellos, M., Welch, D. J., & Johnson, J. (2021). Adaptive management of fisheries in response to climate change. *FAO Fisheries and Aquaculture Technical Paper*, (667), 2021.

Bai, F., Chisholm, R., Sang, W., & Dong, M. (2013). Spatial risk assessment of alien invasive plants in China. *Environmental Science & Technology*, *47*(14), 7624–7632. DOI: 10.1021/es400382c PMID: 23738912

Bajwa, A. A., Chauhan, B. S., Farooq, M., Shabbir, A., & Adkins, S. W. (2016). What do we really know about alien plant invasion? A review of the invasion mechanism of one of the world's worst weeds. *Planta*, *244*(1), 39–57. DOI: 10.1007/s00425-016-2510-x PMID: 27056056

Bakina, L. G., Chugunova, M., & Polyak, Y. M. (2021). Bioaugmentation: Possible scenarios due to application of bacterial preparations for remediation of oil-contaminated soil. *Environmental Science and Pollution Research International*, *28*(1), 1–12. PMID: 33009614

Bakker, J. D., & Wilson, S. D. (2004). Using ecological restoration to constrain biological invasion. *Journal of Applied Ecology*, *41*(6), 1058–1064. DOI: 10.1111/j.0021-8901.2004.00962.x

Balcha, L. D. (2022). Prosopis juliflora distribution, impacts, and control methods available in Ethiopia. *Int. J. Nat. Resour. Ecol. Manag*, *7*, 132–144.

Balcha, L. D. (2022). Prosopis juliflora Distribution, Impacts, and Control Methods Available in Ethiopia. *Int. J. Nat. Resour. Ecol. Manag*, *7*, 132–144.

Balch, J. K., Bradley, B. A., D'Antonio, C. M., & Gómez-Dans, J. (2013). Introduced annual grass increases regional fire activity across the arid western USA (1980–2009). *Global Change Biology*, *19*(1), 173–183. DOI: 10.1111/gcb.12046 PMID: 23504729

Balcioglu, Y. S., Sezen, B., & İşler, A. U. (2024). Evolving preferences in sustainable transportation: A comparative analysis of consumer segments for electric vehicles across Europe. *Social Responsibility Journal*, *20*(9), 1664–1696. Advance online publication. DOI: 10.1108/SRJ-12-2023-0713

Balducci, M. G., Calevo, J., & Duffy, K. J. (2024). Orchid mycorrhizal communities associated with Orchis italica are shaped by ecological factors and geographical gradients. *Journal of Biogeography*. Advance online publication. DOI: 10.1111/jbi.15056

Banerjee, A., Khuroo, A. A., Dehnen-Schmutz, K., Pant, V., Patwardhan, C., Bhowmick, A. R., & Mukherjee, A. (2021). An integrated policy framework and plan of action to prevent and control plant invasions in India. *Environmental Science & Policy*, *124*, 64–72. DOI: 10.1016/j.envsci.2021.06.003

Bao, X., Zou, J., Zhang, B., Wu, L., Yang, T., & Huang, Q. (2022). Arbuscular mycorrhizal fungi and microbes interaction in rice mycorrhizosphere. *Agronomy (Basel)*, *12*(6), 1277. DOI: 10.3390/agronomy12061277

Bard, N. W., Cronk, Q. C., & Davies, T. J. (2024). Fungal endophytes can modulate plant invasion. *Biological Reviews of the Cambridge Philosophical Society*, *99*(5), 1652–1671. Advance online publication. DOI: 10.1111/brv.13085 PMID: 38629189

Bargali, K., Manral, V., Padalia, K., Bargali, S. S., & Upadhyay, V. P. (2018). Effect of vegetation type and season on microbial biomass carbon in Central Himalayan Forest soils, India. *Catena*, *171*, 125–135. DOI: 10.1016/j.catena.2018.07.001

Barnes, C. J., van der Gast, C. J., McNamara, N. P., Rowe, R., & Bending, G. D. (2018). Extreme rainfall affects assembly of the root-associated fungal community. *The New Phytologist*, *220*(4), 1172–1184. DOI: 10.1111/nph.14990 PMID: 29350759

Barnes, C., Hilton, S., Nielsen, I. B., Ahlstrand, N. I., Hansen, A. J., Frøslev, T. G., & Bending, G. D. (2023). Above and belowground microbes are associated with variation of the leaf metabolome; But herbivory has nominal effects. DOI: 10.21203/rs.3.rs-3389986/v1

Barney, J. N., Tekiela, D. R., Barrios-Garcia, M. N., Dimarco, R. D., Hufbauer, R. A., Leipzig-Scott, P., Nuñez, M. A., Pauchard, A., Pyšek, P., Vítková, M., & Maxwell, B. D. (2015). Global Invader Impact Network (GIIN): Toward standardized evaluation of the ecological impacts of invasive plants. *Ecology and Evolution*, *5*(14), 2878–2889. DOI: 10.1002/ece3.1551 PMID: 26306173

Barros, A., Fuentes Lillo, E., Aschero, V., Pauchard, A., Alvarez, M. A., Wedegärtner, R., Clavel, J., Müllerová, J., Pergl, J., Zong, S., Vítková, M., Klinerová, T., Cavieres, L. A., Larson, C., Rew, L. J., Seipel, T., Meffre, C., Arellano, T., Essl, F., & Lembrechts, J. J. (2025). Beyond the Trail—Understanding Non-Native Plant Invasions in Mountain Ecosystems. *Global Ecology and Biogeography*, *34*(6), e70060. DOI: 10.1111/geb.70060

Barton, K. E. (2024). Check for updates Spinescence and the Island Plant Defense Syndrome Kasey E. Barton, Joanna K. Carpenter, Ana Flores, Llorenç Saez, and Mohammed Armani. *Ecology and Evolution of Plant-Herbivore Interactions on Islands*, *249*, 13.

Bashir, F., Altaf, M., Batool, S., Khan, M. S. H., Narejo, A. H., Iftakhar, A., Bashir, S. M., Hadi, A., & Ali, S. Diversity and Ecology of Birds from selected Sites of Islamabad, Pakistan.

Battisti, C., Poeta, G., & Fanelli, G. (2016). *An introduction to disturbance ecology*. Springer. DOI: 10.1007/978-3-319-32476-0

Bayle, G. (2019). Ecological and social impacts of eucalyptus tree plantation on the environment. *Journal of Biodiversity Conservation and Bioresource Management*, *5*(1), 93–104. DOI: 10.3329/jbcbm.v5i1.42189

Beaury, E. M., Fusco, E. J., Jackson, M. R., Laginhas, B. B., Moreoverlli, T. L., Allen, J. M., Pasquarella, V. J., & Bradley, B. A. (2020). Incorporating climate change into invasive species management: Insights from managers. *Biological Invasions*, *22*(2), 233–252. DOI: 10.1007/s10530-019-02087-6

Begum, N., Qin, C., Ahanger, M. A., Raza, S., Khan, M. I., Ashraf, M., & Zhang, L. (2019). Role of arbuscular mycorrhizal fungi in plant growth regulation: Implications in abiotic stress tolerance. *Frontiers in Plant Science*, *10*, 1068. DOI: 10.3389/fpls.2019.01068 PMID: 31608075

Bellard, C., Thuiller, W., Leroy, B., Genovesi, P., Bakkenes, M., & Courchamp, F. (2013). Will climate change promote future invasions? *Global Change Biology*, *19*(12), 3740–3748. DOI: 10.1111/gcb.12344 PMID: 23913552

Bell, J. K., Siciliano, S. D., & Lamb, E. G. (2020). A survey of invasive plants on grassland soil microbial communities and ecosystem services. *Scientific Data*, *7*(1), 1–11. DOI: 10.1038/s41597-020-0422-x PMID: 32152302

Bello, T. T., & Fabiyi, O. A. (2024). Advancement in mycorrhizal fungi-based sustainable plant disease management. In *Mycorrhizal Symbiosis and Agroecosystem Restoration* (pp. 125–145). Springer., DOI: 10.1007/978-981-99-5030-0_6

BenDor, T., Lester, T. W., Livengood, A., Yonavjak, L., & Davis, A. P. (2015). Estimating the Size and Impact of the Ecological Restoration Economy. *PLoS One*, *10*(6), e0128339. DOI: 10.1371/journal.pone.0128339 PMID: 26083034

Benitez, L., Kendig, A. E., Adhikari, A., Clay, K., Harmon, P. F., Holt, R. D., Goss, E. M., & Flory, S. L. (2022). Invasive grass litter suppresses a native grass species and promotes disease. *Ecosphere*, *13*(1), e3907. DOI: 10.1002/ecs2.3907

Bennett, A. E., & Meek, H. C. (2020). The influence of arbuscular mycorrhizal fungi on plant reproduction. *Journal of Chemical Ecology*, *46*(8), 707–721. DOI: 10.1007/s10886-020-01192-4 PMID: 32583094

Berio Fortini, L., Leopold, C. R., Perkins, K. S., Chadwick, O. A., Yelenik, S. G., Jacobi, J. D., Bishaw, K.II, & Gregg, M. (2021). Landscape-level effects of invasive plants and animals on water infiltration through Hawaiian tropical forests. *Biological Invasions, 23*(5), 1421–1437. DOI: 10.1007/s10530-021-02494-8

Berkes, F. (2004). Rethinking community-based conservation. *Conservation Biology, 18*(3), 621–630. DOI: 10.1111/j.1523-1739.2004.00077.x

Berkley, J., & Beratan, K. K. (2021). "Capturing practitioners'" how-to" knowledge in the form of recommendations for more effective planning of collaborative adaptive management projects. *Ecology and Society*, *26*(4), 2021. DOI: 10.5751/ES-12840-260424

Berruti, A., Bianciotto, V., & Lumini, E. (2018). Seasonal variation in winter wheat field soil arbuscular mycorrhizal fungus communities after non-mycorrhizal crop cultivation. *Mycorrhiza, 28*(5), 535–548. DOI: 10.1007/s00572-018-0845-9 PMID: 29931405

Berry, E., Choudhary, A. K., Mishra, G., Tandon, R., & Geeta, R. (2020). Justicia adhatoda reveals two morphotypes with possible functional significance. *Journal of Plant Research, 133*(6), 783–805. DOI: 10.1007/s10265-020-01224-w PMID: 32979146

Bertelsmeier, C. (2021). Globalization and the anthropogenic spread of invasive social insects. *Current Opinion in Insect Science, 46*, 16–23. DOI: 10.1016/j.cois.2021.01.006 PMID: 33545436

Bhaduri, D., Sihi, D., Bhowmik, A., & Verma, B. C. (2022). A review on effective soil health bio-indicators for ecosystem restoration and sustainability. *Frontiers in Environmental Science*, •••, 10. PMID: 36060788

Bhandari, R. (2020). *Assessment of host selection behaviors and oviposition preferences of Cephus cinctus Norton (Hymenoptera: Cephidae) using wheat and smooth brome* Montana State University-Bozeman, College of Agriculture].

Bhatt, A., Chen, X., Pompelli, M. F., Jamal, A., Mancinelli, R., & Radicetti, E. (2023). Characterization of invasiveness, thermotolerance and light requirement of nine invasive species in China. *Plants, 12*(5), 1192. DOI: 10.3390/plants12051192 PMID: 36904052

Bhattacharya A (2019) Effect of high-temperature stress on crop productivity. Effect of high temperature on crop productivity and metabolism of macro molecules

Bhunjun, C. S., Phukhamsakda, C., Hyde, K. D., McKenzie, E. H., Saxena, R. K., & Li, Q. (2024). Do all fungi have ancestors with endophytic lifestyles? *Fungal Diversity*, *125*(1), 73–98. DOI: 10.1007/s13225-023-00516-5

Biancolini, D., & Rondinini, C. (2025). Global Enhancers and Constraints of Alien Range Size in Mammals: The Roles of Species Attributes, Invasion History and Ecological Contexts. *Global Ecology and Biogeography*, *34*(7), e70081. DOI: 10.1111/geb.70081

Bickford, W. A., Goldberg, D., Zak, D., Snow, D. S., & Kowalski, K. (2022). Plant effects on and response to soil microbes in native and non-native *Phragmites australis*. *Ecological Applications*, *32*(4), 2565. DOI: 10.1002/eap.2565 PMID: 35138659

Bilgili, M., Tumse, S., & Nar, S. (2024). Comprehensive Overview on the Present State and Evolution of Global Warming, Climate Change, Greenhouse Gasses and Renewable Energy. *Arabian Journal for Science and Engineering*, *49*(11), 14503–14531. DOI: 10.1007/s13369-024-09390-y

Black, C. A. (1965). [Academic Press Inc.]. *Methods of Soil Analysis*, •••, 1–369.

Blossey, B., Endriss, S. B., Casagrande, R., Häfliger, P., Hinz, H., Dávalos, A., Brown-Lima, C., Tewksbury, L., & Bourchier, R. S. (2020). When misconceptions impede best practices: Evidence supports biological control of invasive Phragmites. *Biological Invasions*, *22*(3), 873–883. DOI: 10.1007/s10530-019-02166-8

Blumenthal, D. M., Jordan, N. R., & Svenson, E. L. (2003). Weed control as a rationale for restoration: The example of tallgrass prairie. *Conservation Ecology*, *7*(1), 6. DOI: 10.5751/ES-00480-070106

Bogale, G. A., & Tolossa, T. T. (2021). Climate change intensification impacts and challenges of invasive species and adaptation measures in Eastern Ethiopia. *Sustainable Environment*, *7*(1), 1875555. DOI: 10.1080/23311843.2021.1875555

Bolds, S. A., Lockaby, B. G., Ditchkoff, S. S., Smith, M. D., & VerCauteren, K. C. (2021). *Impacts of a large invasive mammal on water quality in riparian ecosystems* (0047-2425).

Bolfrey-Arku, G. E.-K. (2004). Management of the Noxious Weed Speargrass (Imperata Cylindrica (L.) Beauv.) in the Forest and Forest-Savanna Transition Agro-Ecological Zones of Ghana University of Cape Coast].

Bologo, N. (2023). *Impact of the invasive alien plant species clearing programme on socio-economic benefits and plant biodiversity along the northern Drakensberg*. Mpumalanga Province.

Bonanomi G, Incerti G, Abd El-Gawad AM, Sarker TC, Stinca A, Motti R, Cesarano G, Teobaldelli M, Saulino L, Cona F (2018) Windstorm disturbance triggers multiple species invasion in an urban Mediterranean forest. iForest-Biogeosciences and Forestry 11(1): 64

Borden, J. B., & Flory, S. L. (2021). Urban evolution of invasive species. *Frontiers in Ecology and the Environment, 19*(3), 184–191. DOI: 10.1002/fee.2295

Boutin, C., Elmegaard, N., & Kjær, C. (2004). Toxicity testing of fifteen non-crop plant species with six herbicides in a greenhouse experiment: Implications for risk assessment. *Ecotoxicology (London, England), 13*(4), 349–369. DOI: 10.1023/B:ECTX.0000033092.82507.f3 PMID: 15344515

Bowen, A. K., & Stevens, M. H. (2020). Temperature, topography, soil characteristics, and NDVI drive habitat preferences of a shade-tolerant invasive grass. *Ecology and Evolution, 10*(19), 10785–10797. DOI: 10.1002/ece3.6735 PMID: 33072296

Bowen, J. L., Spivak, A. C., Bernhard, A. E., & Fulweiler, R. W. (2024). Salt marsh nitrogen cycling: Where land meets sea. *Trends in Ecology & Evolution, 39*(1), 53–65. PMID: 37827901

Bradford, J. B., & Lauenroth, W. K. (2006). Controls over invasion of Bromus tectorum: The importance of climate, soil, disturbance and seed availability. *Journal of Vegetation Science, 17*(6), 693–704. DOI: 10.1111/j.1654-1103.2006.tb02493.x

Bradley, B. A., Beaury, E. M., Fusco, E. J., & Lopez, B. E. (2023). Invasive species policy must embrace a changing climate. *Bioscience, 73*(2), 124–133. DOI: 10.1093/biosci/biac097

Bradley, B., Curtis, C. A., Fusco, E., Abatzoglou, J., Balch, J., Dadashi, S., & Tuanmu, M. (2018). Cheatgrass (*Bromus tectorum*) distribution in the Intermountain Western United States and its relationship to fire frequency, seasonality, and ignitions. *Biological Invasions, 20*(6), 1493–1506. DOI: 10.1007/s10530-017-1641-8

Bradley, S. W., Kim, P. H., Klein, P. G., McMullen, J. S., & Wennberg, K. (2021). Policy for innovative entrepreneurship: Institutions, interventions, and societal challenges. *Strategic Entrepreneurship Journal, 15*(2), 167–184. DOI: 10.1002/sej.1395

Branco, S., Schauster, A., Liao, H. L., & Ruytinx, J. (2022). Mechanisms of stress tolerance and their effects on the ecology and evolution of mycorrhizal fungi. *The New Phytologist, 235*(6), 2158–2175. DOI: 10.1111/nph.18308 PMID: 35713988

Breed, M. F., Harrison, P. A., Blyth, C., Byrne, M., Gaget, V., Gellie, N. J., Groom, S. V. C., Hodgson, R., Mills, J. G., Prowse, T. A. A., Steane, D. A., & Mohr, J. J. (2019). The potential of genomics for restoring ecosystems and biodiversity. *Nature Reviews. Genetics*, *20*(10), 615–628. DOI: 10.1038/s41576-019-0152-0 PMID: 31300751

Bregant, C., Rossetto, G., Meli, L., Sasso, N., Montecchio, L., Brglez, A., & Linaldeddu, B. T. (2023). Diversity of *Phytophthora* species involved in new diseases of mountain vegetation in Europe with the description of *Phytophthora pseudogregata* sp. nov. *Forests*, *14*(8), 1515. DOI: 10.3390/f14081515

Bregant, C., Sanna, G. P., Bottos, A., Maddau, L., Montecchio, L., & Linaldeddu, B. T. (2020). Diversity and pathogenicity of Phytophthora species associated with declining alder trees in Italy and description of Phytophthora alpina sp. nov. *Forests*, *11*(8), 848. DOI: 10.3390/f11080848

Bremm, C., Carvalho, P. C., Fonseca, L., Amaral, G. A., Mezzalira, J. C., Perez, N. B., Nabinger, C., & Laca, E. A. (2016). Diet switching by mammalian herbivores in response to exotic grass invasion. *PLoS One*, *11*(2), e0150167. DOI: 10.1371/journal.pone.0150167 PMID: 26919613

Broadbent, A., Stevens, C. J., Peltzer, D. A., Ostle, N. J., & Orwin, K. H. (2018). Belowground competition drives invasive plant impact on native species regardless of nitrogen availability. *Oecologia*, *186*(2), 577–587. DOI: 10.1007/s00442-017-4039-5 PMID: 29218538

Brodie, J. F., Lieberman, S., Moehrenschlager, A., Redford, K. H., Rodríguez, J. P., Schwartz, M., Seddon, P. J., & Watson, J. E. (2021). Global policy for assisted colonization of species. *Science*, *372*(6541), 456–458. DOI: 10.1126/science.abg0532 PMID: 33926936

Brooks, M. L., D'Antonio, C. M., Richardson, D. M., Grace, J. B., Keeley, J. E., DiTomaso, J. M., Hobbs, R. J., Pellant, M., & Pyke, D. (2004). Effects of invasive alien plants on fire regimes. *Bioscience*, *54*(7), 677–688. DOI: 10.1641/0006-3568(2004)054[0677:EOIAPO]2.0.CO;2

Brown, J., Caligari, P. D. S., & Campos, H. A. (2014). *Second edition of Introduction to Plant Breeding* (pp. 1–295). University of Idaho; Monsanto.

Brundu, G., Pauchard, A., Pyšek, P., Pergl, J., Bindewald, A., Brunori, A., Canavan, S., Campagnaro, T., Celesti-Grapow, L., Dechoum, M. S., Dufour-Dror, J.-M., Essl, F., Flory, S. L., Genovesi, P., Guarino, F., Guangzhe, L., Hulme, P. E., Jäger, H., Kettle, C. J., & Richardson, D. M. (2020). Global guidelines for the sustainable use of non-native trees to prevent tree invasions and mitigate their negative impacts. [In.]. *NeoBiota*, *61*, 65–116. DOI: 10.3897/neobiota.61.58380

Bryceson, S. R., & Morgan, J. W. (2022). The Australasian grass flora in a global context. *Journal of Systematics and Evolution*, *60*(3), 675–690. DOI: 10.1111/jse.12839

Buerdsell, S. L. (2021). *Invasion Ecology of Lehmann Lovegrass (Eragrostis lehmanniana) in Grama (Bouteloua) Plant Communities*. New Mexico State University.

BUG. T. G. (2015). Pests and pest control. *Centre of excellence in marine biology, University of Karachi, Karachi, 126.*

Buisson, E., Archibald, S., Fidelis, A., & Suding, K. N. (2022). Ancient grasslands guide ambitious goals in grassland restoration. *Science*, *377*(6606), 594–598. DOI: 10.1126/science.abo4605 PMID: 35926035

Bunyard, B. A. (2024). *The Little Book of Fungi*. Princeton University Press.

Bushmann, S. L., Drummond, F. A., & Lage, C. (2025). A survey of wild bees (Hymenoptera: Apoidea) on Maine islands: Possible refugia for rare native and non-native species. *Northeastern Naturalist*, *32*(2), 278–302. DOI: 10.1656/045.032.0215

Butler, O. M., Lewis, T., & Chen, C. (2021). Do Soil Chemical Changes Contribute to the Dominance of Blady Grass (Imperata cylindrica) in Surface Fire-Affected Forests? *Fire (Basel, Switzerland)*, *4*(2), 23. DOI: 10.3390/fire4020023

Büyüktahtakın, I. E., & Haight, R. G. (2018). A review of operations research models in invasive species management: State of the art, challenges, and future directions. *Annals of Operations Research*, *271*(2), 357–403. DOI: 10.1007/s10479-017-2670-5

Bzdęga, K., Zarychta, A., & Urbisz, A. (2021). Geostatistical models with the use of hyperspectral data and seasonal variation–A new approach for evaluating the risk posed by invasive plants. *Ecological Informatics*, *61*, 101295.

Cabral, C., Ravnskov, S., Tringovska, I., & Wollenweber, B. (2016). Arbuscular mycorrhizal fungi modify nutrient allocation and composition in wheat (Triticum aestivum L.) subjected to heat-stress. *Plant and Soil*, *408*(1-2), 385–399. DOI: 10.1007/s11104-016-2942-x

Callaway, R. M., Cipollini, D., Barto, K., Thelen, G. C., Hallett, S. G., Prati, D., & Klironomos, J. (2008). Novel weapons: Invasive plant suppresses fungal mutualists in America but not in its native Europe. *Ecology*, *89*(4), 1043–1055. DOI: 10.1890/07-0370.1 PMID: 18481529

Callaway, R. M., Ridenour, W. M., Laboski, T., Weir, T., & Vivanco, J. M. (2005). Natural selection for resistance to the allelopathic effects of invasive plants. *Journal of Ecology*, *93*(3), 576–583. DOI: 10.1111/j.1365-2745.2005.00994.x

Calvo-Polanco, M., Sánchez-Castro, I., Cantos, M., García, J. L., Azcón, R., Ruiz-Lozano, J. M., & Aroca, R. (2016). Effects of different arbuscular mycorrhizal fungal backgrounds and soils on olive plants growth and water relation properties under well-watered and drought conditions. *Plant, Cell & Environment*, *39*(11), 2498–2514. DOI: 10.1111/pce.12807 PMID: 27448529

Camarretta, N., Harrison, P. A., Bailey, T., Potts, B., Lucieer, A., Davidson, N., & Hunt, M. (2020). Monitoring forest structure to guide adaptive management of forest restoration: A review of remote sensing approaches. [– Springer]. *New Forests*, *51*(4), 2020. DOI: 10.1007/s11056-019-09754-5

Caradonia, F., Francia, E., Morcia, C., Ghizzoni, R., Moulin, L., Terzi, V., & Ronga, D. (2019). Arbuscular mycorrhizal fungi and plant growth promoting rhizobacteria avoid processing tomato leaf damage during chilling stress. *Agronomy (Basel)*, *9*(6), 299. DOI: 10.3390/agronomy9060299

Carlon, E., & Dominoni, D. M. (2024). The role of urbanization in facilitating the introduction and establishment of non-native animal species: A comprehensive review. *Journal of Urban Ecology*, *10*(1), juae015. Advance online publication. DOI: 10.1093/jue/juae015

Carmes, A. A., Sühs, R. B., Silva, G. M., Fiaschi, P., & Dechoum, M. S. (2025). Factors explaining species richness of invasive plants in Neotropical forests. *Acta Botanica Brasílica*, *39*, e20240012. DOI: 10.1590/1677-941x-abb-2024-0012

Carol Adair, E., & Burke, I. C. (2010). Plant phenology and life span influence soil pool dynamics: Bromus tectorum invasion of perennial C 3–C 4 grass communities. *Plant and Soil*, *335*(1-2), 255–269. DOI: 10.1007/s11104-010-0413-3

Carpenter, A., Murray, T., Randall, J., Robison, R., & Team, G. I. S. (1999). Bromus tectorum. *The Nature Conservancy*.

Carroll, S. P. (2007). Natives adapting to invasive species: Ecology, genes, and the sustainability of conservation. *Ecological Research*, *22*(6), 892–901. DOI: 10.1007/s11284-007-0352-5

Carruthers, R. I., Jack Deloach, C., Herr, J. C., Anderson, G. L., & Knutson, A. E. (2008). Salt cedar areawide pest management in the western USA. In *Area-wide pest management: theory and implementation* (pp. 271–299). CABI. DOI: 10.1079/9781845933722.0271

Carvalho, P., & Foulkes, M. J. (2018). Roots and uptake of water and nutrients. In Meyers, R. (Ed.), *Encyclopedia of Sustainability Science and Technology*. Springer. DOI: 10.1007/978-1-4939-2493-6_195-3

Casagrande, R. A., & Dacey, J. E. (2014). Monarch butterfly oviposition on swallow-worts (Vincetoxicum spp.). *Environmental Entomology*, *36*(3), 631–636. DOI: 10.1603/0046-225X(2007)36[631:MBOOSV]2.0.CO;2 PMID: 17540075

Cavanaugh, T. W. (2019). Florida Science: The Science that Makes Florida Different.

Chadha, A., Florentine, S. K., Chauhan, B. S., Long, B., & Jayasundera, M. (2019). Influence of soil moisture regimes on growth, photosynthetic capacity, leaf biochemistry and reproductive capabilities of the invasive agronomic weed; Lactuca serriola. *PLoS One*, *14*(6), e0218191. DOI: 10.1371/journal.pone.0218191 PMID: 31251746

Chakrabarty, P. K., Saha, S., Thosar, R. U., & Selvakumaran, S. (2023). 10 Invasive Species–Impediment to Crop Biosecurity. *Advances in Plant Disease Management: Volume I: Fundamental and Basic Research*, 95.

Chakraborty, N., Das, A., Pal, S., Roy, S., Sil, S. K., Adak, M. K., & Hasanuzzaman, M. (2024). Exploring aluminum tolerance mechanisms in plants with reference to rice and arabidopsis: A comprehensive review of genetic, metabolic, and physiological adaptations in acidic soils. *Plants*, *13*(13), 1760. DOI: 10.3390/plants13131760 PMID: 38999600

Chanda, M., Bathi, J. R., Khan, E., Katyal, D., & Danquah, M. (2024). Microplastics in ecosystems: Critical review of occurrence, distribution, toxicity, fate, transport, and advances in experimental and computational studies in surface and subsurface water. *Journal of Environmental Management*, *370*, 122492. DOI: 10.1016/j.jenvman.2024.122492 PMID: 39307085

Chang, X., Wang, W., & Zhou, H. (2025). Nitrogen acquisition by invasive plants: Species preferential N uptake matching with soil N dynamics contribute to its fitness and domination. *Plants*, *14*(5), 748. DOI: 10.3390/plants14050748 PMID: 40094724

Chapagain, T., & Raizada, M. N. (2017). Agronomic challenges and opportunities for smallholder terrace agriculture in developing countries. *Frontiers in Plant Science*, *8*, 331. DOI: 10.3389/fpls.2017.00331 PMID: 28367150

Charles, K. M., & Stehlik, I. (2021). Assisted species migration and hybridization to conserve cold-adapted plants under climate change. [– Wiley Online Library]. *Conservation Biology*, *35*(2), 2021. DOI: 10.1111/cobi.13583 PMID: 32643822

Chaudhary Tharu, S., & Gyawali, S. (2025). Effect of adventure tourism in Nepal's environment and sustainability.

Chaudhury, R., Chakraborty, A., Rahaman, F., Sarkar, T., Dey, S., & Das, M. (2024). Mycorrhization in trees: Ecology, physiology, emerging technologies and beyond. *Plant Biology*, *26*(2), 145–156. DOI: 10.1111/plb.13613 PMID: 38194349

Chazdon, R. L. (2008). Beyond deforestation: Restoring forests and ecosystem services. *Science*, *320*(5882), 1458–1460. DOI: 10.1126/science.1155365 PMID: 18556551

Cheah, S., & Teh, C. (2020). Parameterization of the Farquhar-von Caemmerer-Berry C3 photosynthesis model for oil palm. *Photosynthetica*, *58*(3), 769–779. DOI: 10.32615/ps.2020.020

Chen, C., Wang, Q.-H., Wu, J.-Y., Huang, D., Zhang, W.-H., Zhao, N., Li, X.-F., & Wang, L.-X. (2017). Historical introduction, geographical distribution, and biological characteristics of alien plants in China. *Biodiversity and Conservation*, *26*(2), 353–381. DOI: 10.1007/s10531-016-1246-z

Chen, J., Ma, F., Zhang, Y., Wang, C., & Xu, H. (2021). Spatial distribution patterns of invasive alien species in China. *Global Ecology and Conservation*, *26*, e01432. DOI: 10.1016/j.gecco.2020.e01432

Chen, J., Wang, L., Liang, X., Li, B., He, Y., & Zhan, F. (2023). An arbuscular mycorrhizal fungus differentially regulates root traits and cadmium uptake in two maize varieties. *Ecotoxicology and Environmental Safety*, *264*, 115458. DOI: 10.1016/j.ecoenv.2023.115458 PMID: 37690173

Chen, P., Huang, Q., Zhuge, Y., Li, C., Zhu, P., & Hou, Y. (2021). The effects of plant–soil feedback on invasion resistance are soil context dependent. *Oecologia*, *197*(2), 257–269. DOI: 10.1007/s00442-021-05004-8 PMID: 34328557

Chen, S., Wang, Z., Zhang, M., Shi, X., Wang, L., An, W., Li, Z., Pan, F., & Yang, L. (2023). Practical evaluation of prelithiation strategies for next-generation lithium-ion batteries. *Carbon Energy*, *5*(8), e323. DOI: 10.1002/cey2.323

Chenxi Z (2016) Environmental Influences on Growth and Reproduction of Invasive Commelina benghalensis.

Cherie, D. A. (2022). Determination and Correlation of pH and Electrical Conductivity of Assosa Agricultural Research Center Sites Soil. *Asian Soil Research Journal, 6*(3), 6–10. DOI: 10.9734/asrj/2022/v6i330131

Chiariotti, A., Borghese, A., Boselli, C., & Barile, V. L. (2025). Water Buffalo's Adaptability to Different Environments and Farming Systems: A Review. *Animals (Basel), 15*(11), 1538. DOI: 10.3390/ani15111538 PMID: 40509004

Chibuike, G. U., & Obiora, S. C. (2014). Heavy metal polluted soils: Effect on plants and bioremediation methods. *Applied and Environmental Soil Science, 2014*(1), 752708. DOI: 10.1155/2014/752708

Chung, H. I., Choi, Y., Yoo, Y., Engler, R., Lee, K., & Jeon, S. W. (2022). Integrated spatial model based evaluation methodology for optimal invasive species management: Common ragweed in the Republic of Korea. *Environmental Research Letters, 17*(3), 034047. DOI: 10.1088/1748-9326/ac4dc7

Clarke, B., Otto, F., Stuart-Smith, R., & Harrington, L. (2022). Extreme weather impacts of climate change: An attribution perspective. *Environmental Research: Climate, 1*(1), 012001.

Clark, S. L. (2020). Using herbicides to restore native species and improve habitat on rangelands and wildlands. *Outlooks on Pest Management, 31*(2), 57–60. DOI: 10.1564/v31_apr_02

Cleverly, J. R., Smith, S. D., Sala, A., & Devitt, D. A. (2006). Invasive capacity of Tamarix ramosissima in a Mojave Desert floodplain: The role of drought. *Oecologia, 147*(2), 296–306. DOI: 10.1007/s00442-005-0284-3 PMID: 28307496

Clewell, A. F., & Aronson, J. (2006). Motivations for the Restoration of Ecosystems. *Conservation Biology, 20*(2), 420–428. DOI: 10.1111/j.1523-1739.2006.00340.x PMID: 16903103

Clewell, A. F., & Aronson, J. (2012). *Ecological restoration: Principles, values, and structure of an emerging profession*. Island Press.

Clewell, A. F., Rieger, J. P., & Munro, J. (2004). The Rights and Wrongs of Restoration. *Theoretical and Applied Genetics, 109*(4), 733–742. DOI: 10.1007/s12031-003-1004-8 PMID: 15146319

Clewley, G. D., Eschen, R., Shaw, R. H., & Wright, D. J. (2012). The effectiveness of classical biological control of invasive plants. *Journal of Applied Ecology, 49*(6), 1287–1295. DOI: 10.1111/j.1365-2664.2012.02209.x

Climent, J., Alía, R., Karkkainen, K., Bastien, C., Benito-Garzon, M., Bouffier, L., De Dato, G., Delzon, S., Dowkiw, A., Elvira-Recuenco, M., Grivet, D., González-Martínez, S. C., Hayatgheibi, H., Kujala, S., Leplé, J.-C., Martín-Sanz, R. C., de Miguel, M., Monteverdi, M. C., Mutke, S., & Cavers, S. (2024). Trade-offs and trait integration in tree phenotypes: Consequences for the sustainable use of genetic resources. *Current Forestry Reports*, *10*(3), 196–222. DOI: 10.1007/s40725-024-00217-5

Coakley, S., & Petti, C. (2021). Impacts of the invasive Impatiens glandulifera: Lessons learned from one of Europe's top invasive species. *Biology (Basel)*, *10*(7), 619. DOI: 10.3390/biology10070619 PMID: 34356476

Coleby, D. (2021). *Twenty-Five Years of Bushcare at Sublime Point Reserve, Leura: A history of the Sublime Point Bushcare Group, 1996-2021*. David Coleby.

Conversa, G., Miedico, O., Chiaravalle, A. E., & Elia, A. (2019). Heavy metal contents in green spears of asparagus (Asparagus officinalis L.) grown in Southern Italy: Variability among farms, genotypes and effect of soil mycorrhizal inoculation. *Scientia Horticulturae*, *256*, 108559. DOI: 10.1016/j.scienta.2019.108559

Corbin, J. D., & D'Antonio, C. M. (2004). Competition between native perennial and exotic annual grasses: Implications for an historical invasion. *Ecology*, *85*(5), 1273–1283. DOI: 10.1890/02-0744

Cosenza, B., & Carvalho, A. (2024). 19 Representing Amazonia: Perspectives from the Global North and the Global South. *Environmental Communication*, *31*, 383.

Costello, K. E., Lynch, S. A., McAllen, R., O'Riordan, R. M., & Culloty, S. C. (2022). Assessing the potential for invasive species introductions and secondary spread using vessel movements in maritime ports. *Marine Pollution Bulletin*, *177*, 113496. DOI: 10.1016/j.marpolbul.2022.113496 PMID: 35272109

Cottom, G., & Curtis, J. T. (1956). The use of distance measures in phyto–sociological sampling. *Ecology*, *35*(3), 451–460. DOI: 10.2307/1930167

Couch, C., Simbiano, F. J., Diawara, M., Loua, E., Mamy, L., Cheek, M., & Magassouba, S. (2025). Developing a threat risk register based on the IUCN threat hierarchy for five tropical Important Plant Areas in Guinea. *PeerJ*, *13*, e19629. DOI: 10.7717/peerj.19629 PMID: 40677745

Coulter, A. A., Moore, M. J., Golcher-Benavides, J., Rahel, F. J., Walters, A. W., Brewer, S. K., & Wildhaber, M. L. (2024). A synthesis of the characteristics and drivers of introduced fishes in prairie streams: Can we manage introduced harmful fishes in these dynamic environments? *Biological Invasions*, *26*(12), 1–23. DOI: 10.1007/s10530-024-03450-y

County, D. A., & District, N. (2025). Environmental Assessment Border Highway Connector Project.

Craine, J. M., & Dybzinski, R. (2013). Mechanisms of plant competition for nutrients, water and light. *Functional Ecology*, *23*(4), 833–840. DOI: 10.1111/1365-2435.12081

Crocker, E., Nelson, E., & Blossey, B. (2017). Soil conditioning effects of Phragmites australis on native wetland plant seedling survival. *Ecology and Evolution*, *7*(15), 5571–5579. DOI: 10.1002/ece3.3024 PMID: 28808539

Croft, L. (2024). *Natural and experimental admixture of invasive blowflies in New Zealand and Australia*. The University of Waikato.

Crone, M., McComb, J. A., O'Brien, P. A., & Hardy, G. E. S. J. (2013). Survival of *Phytophthora cinnamomi* as oospores, stromata, and thick-walled chlamydospores in roots of symptomatic and asymptomatic annual and herbaceous perennial plant species. *Fungal Biology*, *117*(2), 112–123. DOI: 10.1016/j.funbio.2012.12.004 PMID: 23452949

Cucu, M. A., Choudhary, R., Trkulja, V., Garg, S., & Matić, S. (2025). Utilizing Environmentally Friendly Techniques for the Sustainable Control of Plant Pathogens: A Review. *Agronomy (Basel)*, *15*(7), 1551. DOI: 10.3390/agronomy15071551

Cunard, C. (2016). *Plant species coexistence: the roles of soil biota and invasion history*. University of Georgia.

Curtis, J. T., & McIntosh, R. P. (1950). The interrelations of certain analytic and synthetic phytosociological characters. *Ecology*, *31*(3), 434–455. DOI: 10.2307/1931497

D'Antonio, C. M., & Chambers, J. C. (2006). Using ecological theory to manage or restore ecosystems affected by invasive plant species. *Ecological Applications*, *16*(4), 1031–1038. DOI: 10.1890/1051-0761(2006)016[1031:UETTMO]2.0.CO;2

D'Antonio, C. M., & Meyerson, L. A. (2002). Exotic plant species as problems and solutions in ecological restoration: A synthesis. *Restoration Ecology*, *10*(4), 703–713. DOI: 10.1046/j.1526-100X.2002.01051.x

D'Antonio, C. M., & Vitousek, P. M. (1992). Biological invasions by exotic grasses, the grass/fire cycle, and global change. *Annual Review of Ecology and Systematics*, *23*(1), 63–87. DOI: 10.1146/annurev.es.23.110192.000431

da Silva Arruda, V. L., Alencar, A. A. C., de Carvalho Júnior, O. A., de Figueiredo Ribeiro, F., de Arruda, F. V., Conciani, D. E., da Silva, W. V., & Shimbo, J. Z. (2024). Assessing four decades of fire behavior dynamics in the Cerrado biome (1985 to 2022). *Fire Ecology*, *20*(1), 64. DOI: 10.1186/s42408-024-00298-4

Dagar, J. C., & Gupta, S. R. (2024). An Ecological Overview of Halophytes: Phytogeographical Distribution, Floristic Diversity, Vegetation Composition, and Utilisation. *Halophytes vis-à-vis Saline Agriculture: Perspectives and Opportunities for Food Security*, 19-65.

Dai, Z., Xiong, X., Zhu, H., Xu, H., Leng, P., Li, J., Tang, C., & Xu, J. (2021). Association of biochar properties with changes in soil bacterial, fungal and fauna communities and nutrient cycling processes. *Biochar*, *3*(1), 1–15. DOI: 10.1007/s42773-021-00099-x

Dakhil, M. A., El-Keblawy, A., El-Sheikh, M. A., Halmy, M. W. A., Ksiksi, T., & Hassan, W. A. (2021). Global invasion risk assessment of Prosopis juliflora at biome level: Does soil matter? *Biology (Basel)*, *10*(3), 203. DOI: 10.3390/biology10030203 PMID: 33803081

Damasceno, G., & Fidelis, A. (2020). Abundance of invasive grasses is dependent on fire regime and climatic conditions in tropical savannas. *Journal of Environmental Management*, *271*, 111016. DOI: 10.1016/j.jenvman.2020.111016 PMID: 32778299

Daniels, J. (2024). *The effectiveness of anthropogenic in-stream barriers as a management strategy for non-native species*. University of Southampton.

Darji, T. B., Adhikari, B., Pathak, S., Neupane, S., Thapa, L. B., Bhatt, T. D., Pant, R. R., Pant, G., Pal, K. B., & Bishwakarma, K. (2021). Phytotoxic effects of invasive *Ageratina adenophora* on two native subtropical shrubs in Nepal. *Scientific Reports*, *11*(1), 13663. DOI: 10.1038/s41598-021-92791-y PMID: 34210999

Dasgupta, D., Dey, A., & Meena, M. C. (2024). *Soil Mycorrhizae and Their Industrial Applications. In industrial applications of soil microbes*. Bentham Science Publishers., DOI: 10.2174/9789815079753124030I

Dashtebani, F., & Aliasgharzad, N. (2018). Germination, photosynthesis and growth of two halophytic grass *Puccinellia distans* and *Aeluropus littoralis* under salinity and their colonization with mycorrhizal species in their natural habitat in the Tabriz plain. *Journal of Plant Research (Iranian. Journal of Biology*, *30*(4), 817–833.

Dassonville, N., Vanderhoeven, S., Vanparys, V., Hayez, M., Gruber, W., & Meerts, P. (2008). Impacts of alien invasive plants on soil nutrients are correlated with initial site conditions in NW Europe. [PubMed]. *Oecologia, 157*(1), 131–140. DOI: 10.1007/s00442-008-1054-6 PMID: 18491146

Datta, A., Kuhn, I., Ahmad, M., Michalski, S., & Auge, H. (2017). Processes affecting altitudinal distribution of invasive *Ageratina adenophora* in western Himalaya: The role of local adaptation and the importance of different life–cycle stages. *PLoS One, 12*(11), e0187708. DOI: 10.1371/journal.pone.0187708 PMID: 29125852

Datta, D., Chattopadhyay, R. N., & Guha, P. (2012). Community based mangrove management: A review on status and sustainability. *Journal of Environmental Management, 107*, 84–95. DOI: 10.1016/j.jenvman.2012.04.013 PMID: 22595074

David, A. S., Cortes, A. C., & Lake, E. C. (2021). Bottom-up factors determine local, but not regional, distribution of a biological control agent against invasive Lygodium microphyllum. *Biological Control, 159*, 104632. DOI: 10.1016/j.biocontrol.2021.104632

David, P., Thebault, E., Anneville, O., Duyck, P. F., Chapuis, E., & Loeuille, N. (2017). Impacts of invasive species on food webs: A review of empirical data. *Advances in Ecological Research, 56*, 1–60. DOI: 10.1016/bs.aecr.2016.10.001

Davies, K. W., & Sheley, R. L. (2011). Promoting native vegetation and diversity in exotic annual grass infestations. *Restoration Ecology, 19*(2), 159–165. DOI: 10.1111/j.1526-100X.2009.00548.x

De Alwis, C., Kalla, A., Pham, Q. V., & Kumar, P. (2021). Survey on 6G frontiers: Trends, applications, requirements, technologies and future research. *IEEE Open Journal of the Communications Society, 2*, 2021. DOI: 10.1109/OJCOMS.2021.3071496

de Carvalho-Souza, G. F., Kourantidou, M., Laiz, I., Nuñez, M. A., & González-Ortegón, E. (2024). How to deal with invasive species that have high economic value? *Biological Conservation, 292*, 110548. DOI: 10.1016/j.biocon.2024.110548

de Castro, A. F., Boanares, D., de Medeiros Sarmento, P. S., Caldeira, C. F., Michelan, T. S., Ramos, S. J., & Gastauer, M. (2024). Fertilization during mineland rehabilitation may shift competitive outcomes toward invasive species. *Restoration Ecology, 32*(4), e14108. DOI: 10.1111/rec.14108

de Oliveira MS, Cruz JN, de Aguiar EH (2022) Potential Use of Terpenoids in Weed Management. Terpenoids: Recent Advances in Extraction, Biochemistry and Biotechnology: 200

De Santis, V., Quadroni, S., Britton, R. J., Carosi, A., Gutmann Roberts, C., Lorenzoni, M., Crosa, G., & Zaccara, S. (2021). Biological and trophic consequences of genetic introgression between endemic and invasive *Barbus* fishes. *Biological Invasions, 23*(12), 3351–3368. DOI: 10.1007/s10530-021-02577-6 PMID: 34054333

de Vries, F. T., Griffiths, R. I., Bailey, M., Craig, H., Girlanda, M., Gweon, H. S., & Bardgett, R. D. (2018). Soil bacterial networks are less stable under drought than fungal networks. *Nature Communications, 9*(1), 3033. DOI: 10.1038/s41467-018-05516-7 PMID: 30072764

Debuse, V. J., & Lewis, T. (2014). Long-term repeated burning reduces Lantana camara regeneration in a dry eucalypt forest. *Biological Invasions, 16*(12), 2697–2711. DOI: 10.1007/s10530-014-0697-y

Deeksha, M. G., Jadhav, M. M., Guleria, N., Barman, M., Srinivas, K., Kaur, P., Anokhe, A., & Sushilkumar, . (2024). Impact of climate change on invasive weeds affecting biodiversity and natural ecosystems. *Indian Journal of Weed Science, 56*(4), 349–357. DOI: 10.5958/0974-8164.2024.00057.4

Delavaux, C. S., & Bever, J. D. (2022). Evidence for the evolution of native plant response to mycorrhizal fungi in post-agricultural grasslands. *Ecology and Evolution, 12*(7), e9097. DOI: 10.1002/ece3.9097 PMID: 35845364

Denman, S., Kirk, S. A., Moralejo, E., & Webber, J. F. (2009). *Phytophthora ramorum* and *Phytophthora kernoviae* on naturally infected asymptomatic foliage. *Bulletin OEPP. EPPO Bulletin. European and Mediterranean Plant Protection Organisation, 39*(1), 105–111. DOI: 10.1111/j.1365-2338.2009.02243.x

Desta, T. T., Teklemariam, H., & Mulugeta, T. (2023). Insights of smallholder farmers on the trade-offs of eucalyptus plantation. *Environmental Challenges, 10*, 100663. DOI: 10.1016/j.envc.2022.100663

Dezotti, G., Fidelis, A., Damasceno, G., & Siqueira, T. (2024). Interactive roles of fire seasons and biological invasions in the short-term dynamics of tropical savannas. *Journal of Vegetation Science, 35*(2), e13250. DOI: 10.1111/jvs.13250

Di Marco, S. (2019). *Integrating ecology and evolutionary theory: A game changer for biodiversity conservation?* Biodiversity Conservation Science.

Di Sacco, A., Hardwick, K. A., Blakesley, D., Brancalion, P. H. S., Breman, E., Cecilio Rebola, L., Chomba, S., Dixon, K., Elliott, S., Ruyonga, G., Shaw, K., Smith, P., Smith, R. J., & Antonelli, A. (2021). Ten golden rules for reforestation to optimize carbon sequestration, biodiversity recovery and livelihood benefits. [– Wiley Online Library]. *Global Change Biology*, *27*(7), 2021. DOI: 10.1111/gcb.15498 PMID: 33494123

Di, B., Firn, J., Buckley, Y. M., Lomas, K., Pausas, J. G., & Smith, A. L. (2022). Impact of roadside burning on genetic diversity in a high-biomass invasive grass. *Evolutionary Applications*, *15*(5), 790–803. DOI: 10.1111/eva.13369 PMID: 35603028

Dibben, A., & Raskin, B. (2025). *Silvohorticulture: A grower's guide to integrating trees into crops*. Chelsea Green Publishing.

Dickie, I. A., Bufford, J. L., Cobb, R. C., Desprez-Loustau, M. L., Grelet, G., Hulme, P. E., & Williams, N. M. (2017a). The emerging science of linked plant–fungal invasions. *The New Phytologist*, *215*(4), 1314–1332. DOI: 10.1111/nph.14657 PMID: 28649741

Dickie, I. A., Cooper, J. A., Bufford, J. L., Hulme, P. E., & Bates, S. T. (2017b). Loss of functional diversity and network modularity in introduced plant–fungal symbioses. *AoB Plants*, *9*(1), plw084. PMID: 28039116

Dick, M. A., Dobbie, K., Cooke, D. E., & Brasier, C. M. (2006). *Phytophthora captiosa* sp. nov. and *P. fallax* sp. nov. causing crown dieback of *Eucalyptus* in New Zealand. *Mycological Research*, *110*(4), 393–404. DOI: 10.1016/j.mycres.2006.01.008 PMID: 16650734

Diener, A., & Mudu, P. (2021). vegetation protects us from air pollution? A critical review on green spaces' mitigation abilities for air-borne particles from a public health perspective-with implications .... *The Science of the Total Environment*, *755*, 1–12. DOI: 10.1016/j.scitotenv.2021.148605

Dilawar, N., Hamayun, M., Iqbal, A., Lee, B., Ali, S., Ahmad, A., & Hussain, A. (2024). Rhizofungus Aspergillus terreus Mitigates Heavy Metal Stress-Associated Damage in Triticum aestivum L. *Plants*, *13*(18), 2643. DOI: 10.3390/plants13182643 PMID: 39339618

Dimitrakopoulos, P. G., Koukoulas, S., Galanidis, A., Delipetrou, P., Gounaridis, D., Touloumi, K., & Arianoutsou, M. (2017). Factors shaping alien plant species richness spatial patterns across Natura 2000 Special Areas of Conservation of Greece. *The Science of the Total Environment*, *601–602*, 461–468. DOI: 10.1016/j.scitotenv.2017.05.220 PMID: 28575824

Dindaroglu, T. (2024). Ecosystemic Alienation from the Perspective of Paraecology. *SilvaWorld*, *3*(1), 28–43. DOI: 10.61326/silvaworld.v3i1.250

Ding, J., & Eldridge, D. J. (2024). Woody encroachment: Social–ecological impacts and sustainable management. *Biological Reviews of the Cambridge Philosophical Society*, *99*(6), 1909–1926. Advance online publication. DOI: 10.1111/brv.13104 PMID: 38961449

DiTomaso, J. M., Kyser, G. B., & Pitcairn, M. J. (2013). Yellow star thistle management guide. California Invasive Plant Council. Retrieved from http://www.cal-ipc.org

DiTomaso, J. M. (2000). Invasive weeds in rangelands: Species, impacts, and management. *Weed Science*, *48*(2), 255–265. DOI: 10.1614/0043-1745(2000)048[0255:IWIRSI]2.0.CO;2

Divya, K. L., Mhatre, P. H., Venkatasalam, E. P., & Sudha, R. (2021). Crop simulation models as decision-supporting tools for sustainable potato production: A review. *Potato Research*, *64*(2), 149–164. DOI: 10.1007/s11540-020-09483-9

Dodd, A. P. (1940). *The biological campaign against prickly pear*. Commonwealth Prickly Pear Board.

Dong, L.-J., Ma, L.-N., & He, W.-M. (2021). Arbuscular mycorrhizal fungi help explain invasion success of Solidago canadensis. *Applied Soil Ecology*, *157*, 103763. DOI: 10.1016/j.apsoil.2020.103763

Dougherty, R. F., Quinn, L. D., Endres, A. B., Voigt, T. B., & Barney, J. N. (2014). Natural-history survey of the ornamental grass *Miscanthus sinensis* in the introduced range. *Invasive Plant Science and Management*, *7*(1), 113–120. DOI: 10.1614/IPSM-D-13-00037.1

Duan, S., Feng, G., Limpens, E., Bonfante, P., Xie, X., & Zhang, L. (2024). Cross-kingdom nutrient exchange in the plant–arbuscular mycorrhizal fungus–bacterium continuum. *Nature Reviews. Microbiology*, *22*(12), 1–18. DOI: 10.1038/s41579-024-01073-7 PMID: 39014094

Duca, M., Boian, I., & Domenco, R. (2022). The impact of droughts on sunflower production in the Republic of Moldova. *Notulae Botanicae Horti Agrobotanici Cluj-Napoca*, *50*(4), 1–16. DOI: 10.15835/nbha50413040

Du, E., Jia, Y., Wu, C., Chen, X., & Zhang, F. (2024). Arbuscular mycorrhizal fungi and *Bacillus* promote *Flaveria bidentis* invasion success by inhibiting the growth of native species under different soil nutrient levels. *Plant and Soil*, *500*(1), 147–160. DOI: 10.1007/s11104-023-06052-0

Duell, E. B., Londe, D. W., Hickman, K., Greer, M. J., & Wilson, G. W. (2021). Superior performance of invasive grasses over native counterparts will remain problematic under warmer and drier conditions. *Plant Ecology*, *222*(9), 993–1006. DOI: 10.1007/s11258-021-01156-y

Duque, J. F. M., Zapico, I., Bugosh, N., & Tejedor, M. (2021). A Somolinos quarry land stewardship history: From ancient and recent land degradation to sensitive geomorphic-ecological restoration and its monitoring. [– Elsevier]. *Ecological Engineering*, •••, 2021.

Durán, P., Tortella, G., Viscardi, S., Barra, P. J., Carrión, V. J., Mora, M. D. L. L., & Pozo, M. J. (2018). Microbial community composition in take-all suppressive soils. *Frontiers in Microbiology*, *9*, 2198. DOI: 10.3389/fmicb.2018.02198 PMID: 30283421

Dyer, A. R., Hardison, J. L., & Rice, K. J. (2012). Phenology constrains opportunistic growth response in Bromus tectorum L. *Plant Ecology*, *213*(1), 103–112. DOI: 10.1007/s11258-011-0010-4

Early, R., Bradley, B. A., Dukes, J. S., Lawler, J. J., Olden, J. D., Blumenthal, D. M., Gonzalez, P., Grosholz, E. D., Ibañez, I., Miller, L. P., Sorte, C. J. B., & Tatem, A. J. (2016). Global threats from invasive alien species in the twenty-first century and national response capacities. *Nature Communications*, *7*(1), 12485. DOI: 10.1038/ncomms12485 PMID: 27549569

EEA (European Environment Agency). (2017). Green Infrastructure and Urban Planning. EEA Report No 7/2017. https://www.eea.europa.eu/publications/green-infrastructure-and-urban-planning

Ehrenfeld, J. G. (2003). Effects of exotic plant invasions on soil nutrient cycling processes. *Ecosystems (New York, N.Y.)*, *6*(6), 503–523. DOI: 10.1007/s10021-002-0151-3

Ehrenfeld, J. G. (2010). Ecosystem consequences of biological invasions. *Annual Review of Ecology, Evolution, and Systematics*, *41*(1), 59–80. DOI: 10.1146/annurev-ecolsys-102209-144650

Eissenstat, D. M. (1992). Costs and benefits of constructing roots of small diameter. *Journal of Plant Nutrition*, *15*(6-7), 763–782. DOI: 10.1080/01904169209364361

El-Halwagi, M. M. (2025). *Sustainable design through process integration: fundamentals and applications to industrial pollution prevention, resource conservation, and profitability enhancement*. Elsevier.

Ellison, A. M., Felson, A. J., & Friess, D. A. (2020). Mangrove rehabilitation and restoration as experimental adaptive management. *Frontiers in Marine Science*, *7*, 2020. DOI: 10.3389/fmars.2020.00327

Elsensohn, J. E., Wolford, S., Tabb, A., & Leskey, T. (2024). Experimental evidence supports the ability of spotted lanternfly to hitchhike on vehicle exteriors as a mechanism for anthropogenic dispersal. *Royal Society Open Science*, *11*(7), 240493. DOI: 10.1098/rsos.240493 PMID: 39076365

Elsheikh, E. A., El-Keblawy, A., Mosa, K. A., Okoh, A. I., & Saadoun, I. (2021). Role of endophytes and rhizosphere microbes in promoting the invasion of exotic plants in arid and semi-arid areas: A review. *Sustainability (Basel)*, *13*(23), 13081. DOI: 10.3390/su132313081

Emery-Butcher, H. E., Beatty, S. J., & Robson, B. J. (2020). The impacts of invasive ecosystem engineers in freshwaters: A review. [– Wiley Online Library]. *Freshwater Biology*, *65*(5), 2020. DOI: 10.1111/fwb.13479

Endress, B. A., Averett, J. P., Naylor, B. J., Morris, L. R., & Taylor, R. V. (2020). Non-native species threaten the biotic integrity of the largest remnant Pacific Northwest Bunchgrass prairie in the United States. *Applied Vegetation Science*, *23*(1), 53–68. DOI: 10.1111/avsc.12464

Endresz, G., Somodi, I., & Kalapos, T. (2013). Arbuscular mycorrhizal colonisation of roots of grass species differing in invasiveness. *Community Ecology*, *14*(1), 67–76. DOI: 10.1556/ComEc.14.2013.1.8

Enright, W. D. (2000). The effect of terrestrial invasive alien plants on water scarcity in South Africa. *Physics and Chemistry of the Earth. Part B: Hydrology, Oceans and Atmosphere*, *25*(3), 237–242. DOI: 10.1016/S1464-1909(00)00010-1

Eschen, R., Beale, T., Bonnin, J. M., Constantine, K. L., Duah, S., Finch, E. A., Makale, F., Nunda, W., Ogunmodede, A., Pratt, C. F., Thompson, E., Williams, F., Witt, A., & Taylor, B. (2021). Towards estimating the economic cost of invasive alien species to African crop and livestock production. *CABI Agriculture and Bioscience*, *2*(1), 1–18. DOI: 10.1186/s43170-021-00052-9

Eskelinen, A., & Harrison, S. (2014). Exotic plant invasions under enhanced rainfall are constrained by soil nutrients and competition. *Ecology*, *95*(3), 682–692. DOI: 10.1890/13-0288.1 PMID: 24804452

Essl, F., Dawson, W., Kreft, H., Pergl, J., Pyšek, P., Van Kleunen, M., Weigelt, P., Mang, T., Dullinger, S., Lenzner, B., Moser, D., Maurel, N., Seebens, H., Stein, A., Weber, E., Chatelain, C., Inderjit, , Genovesi, P., Kartesz, J., & Winter, M. (2019). Drivers of the relative richness of naturalized and invasive plant species on Earth. *AoB Plants*, *11*(5), plz051. DOI: 10.1093/aobpla/plz051 PMID: 31636882

Eviner, V. T., Garbach, K., Baty, J. H., & Hoskinson, S. A. (2012). Measuring the effects of invasive plants on ecosystem services: Challenges and prospects. *Invasive Plant Science and Management*, *5*(1), 125–136. DOI: 10.1614/IPSM-D-11-00095.1

Ewel, J. (2012). 13. Invasibility: Lessons from South Florida. *Ecology of Biological Invasions of North America and Hawaii*, *58*, 214–230. DOI: 10.1007/978-1-4612-4988-7_13

Fahad, S., & Wang, J. (2020). Climate change, vulnerability, and its impacts in rural Pakistan: A review. *Environmental Science and Pollution Research International*, *27*(2), 1334–1338. DOI: 10.1007/s11356-019-06878-1 PMID: 31802341

Fahey, C., & Flory, S. L. (2022). Soil microbes alter competition between native and invasive plants. *Journal of Ecology*, *110*(1), 141–155. DOI: 10.1111/1365-2745.13807

Faiza, N., Weiguo, J., Aijun, Y., & Wenxing, S. (2017). Giant deforestation leads to drastic eco-environmental devastating effects since 2000; A case study of Pakistan. *JAPS: Journal of Animal & Plant Sciences*, *27*(4).

FAO. (2021). *AQUASTAT: FAO's Global Information System on Water and Agriculture*. Available online: https://www.fao.org/aquastat/en/geospatial–information/climate–information (accessed on 22 February 2022).

Faria, L., de Carvalho, B. M., Carneiro, L., Miiller, N. O. R., Pedroso, C. R., Occhi, T. V. T., Tonella, L. H., & Vitule, J. R. S. (2023). Invasive species policy in Brazil: A review and critical analysis. *Environmental Conservation*, *50*(1), 67–72. DOI: 10.1017/S0376892922000406

Fawzy, S., Osman, A. I., Doran, J., & Rooney, D. W. (2020). Strategies for mitigation of climate change: A review. [– Springer]. *Environmental Chemistry Letters*, *18*(6), 2020. DOI: 10.1007/s10311-020-01059-w

Fayet, M., Long, M., Han, B., Belkorchia, A., Delbac, F., & Polonais, V. (2024). New insights into Microsporidia polar tube function and invasion mechanism. *The Journal of Eukaryotic Microbiology*, *13043*(5), e13043. Advance online publication. DOI: 10.1111/jeu.13043 PMID: 38973152

Feagin, R. A., Lozada-Bernard, S. M., Ravens, T. M., Möller, I., Yeager, K. M., & Baird, A. H. (2009). Does vegetation prevent wave erosion of salt marsh edges? *Proceedings of the National Academy of Sciences of the United States of America*, *106*(25), 10109–10113. DOI: 10.1073/pnas.0901297106 PMID: 19509340

Fenesi, A., Saura-Mas, S., Blank, R. R., Kozma, A., Lózer, B. M., & Ruprecht, E. (2016). Enhanced fire-related traits may contribute to the invasiveness of downy brome (Bromus tectorum). *Invasive Plant Science and Management*, *9*(3), 182–194. DOI: 10.1614/IPSM-D-16-00006.1

Feng, J., & Zhu, Y. (2010). Alien invasive plants in China: Risk assessment and spatial patterns. *Biodiversity and Conservation*, *19*(12), 3489–3497. DOI: 10.1007/s10531-010-9909-7

Feng, Q., Wang, B., Chen, M., Wu, P., Lee, X., & Xing, Y. (2021). Invasive plants as potential sustainable feedstocks for biochar production and multiple applications: A review. [– Elsevier]. *Resources, Conservation and Recycling*, *164*, 2021. DOI: 10.1016/j.resconrec.2020.105204

Feng, Q., Zhao, W., Hu, X., Liu, Y., Daryanto, S., & Cherubini, F. (2020). Trading-off ecosystem services for better ecological restoration: A case study in the Loess Plateau of China. [– Elsevier]. *Journal of Cleaner Production*, *257*, 2020. DOI: 10.1016/j.jclepro.2020.120469

Figueiredo Lugli, L., Anderson, K. M., Aragao, L. E. O. C., Cordeiro, A. L., Cunha, H. F. V., Fuchslueger, L., & Meir, P. (2019). Multiple phosphorus acquisition strategies adopted by fine roots in low-fertility soils in Central Amazonia. *Plant and Soil*, *450*(1-2), 49–63. DOI: 10.1007/s11104-019-03963-9

Filazzola, A., Brown, C., Dettlaff, M. A., Batbaatar, A., Grenke, J., Bao, T., Peetoom Heida, I., & Cahill, J. F.Jr. (2020). The effects of livestock grazing on biodiversity are multi-trophic: A meta-analysis. *Ecology Letters*, *23*(8), 1298–1309. DOI: 10.1111/ele.13527 PMID: 32369874

Finch DM, Butler JL, Runyon JB, Fettig CJ, Kilkenny FF, Jose S, Frankel SJ, Cushman SA, Cobb RC, Dukes JS (2021) Effects of climate change on invasive species. Invasive species in forests and rangelands of the United States: a comprehensive science synthesis for the United States forest sector: 57-83

Finch, D. M., Butler, J. L., Runyon, J. B., Fettig, C. J., Kilkenny, F. F., Jose, S., Frankel, S. J., Cushman, S. A., Cobb, R. C., & Dukes, J. S. (2021). Effects of climate change on invasive species. *Invasive species in forests and rangelands of the United States: a comprehensive science synthesis for the United States forest sector*, 57-83.

Finley, D., Dovciak, M., & Dean, J. (2023). A data driven method for prioritizing invasive species to aid policy and management. *Biological Invasions*, *25*(7), 2293–2307. DOI: 10.1007/s10530-023-03041-3

Flickinger, H. D., & Dukes, J. S. (2024). A Review of Theory: Comparing Invasion Ecology and Climate Change-Induced Range Shifting. *Global Change Biology*, *30*(12), e17612. DOI: 10.1111/gcb.17612 PMID: 39676232

Florence, A. M., Higley, L. G., Drijber, R. A., Francis, C. A., & Lindquist, J. L. (2019). Cover crop mixture diversity, biomass productivity, weed suppression, and stability. *PLoS One*, *14*(3), e0206195. DOI: 10.1371/journal.pone.0206195 PMID: 30870424

Flores, R. I. (2019). *Comparison of eight remnant Tamaulipan biotic province plant communities in the Lower Rio Grande Valley using multivariate analyses*. The University of Texas Rio Grande Valley.

Flory, S. L., & Clay, K. (2010). Non-native grass invasion alters native plant composition in experimental communities. *Biological Invasions*, *12*(5), 1285–1294. DOI: 10.1007/s10530-009-9546-9

Forseth, I. N., & Innis, A. F. (2004). Kudzu (Pueraria montana): History, physiology, and ecology combine to make a major ecosystem threat. *Critical Reviews in Plant Sciences*, *23*(5), 401–413. DOI: 10.1080/07352680490505150

Fox, H., & Cundill, G. (2018). Towards increased community-engaged ecological restoration: A review of current practice and future directions. *Ecological Restoration*, *36*(3), 208–218. DOI: 10.3368/er.36.3.208

Frankham, R., Ballou, J. D., & Briscoe, D. A. (2010). *Introduction to conservation genetics*. Cambridge University Press. DOI: 10.1017/CBO9780511809002

Franks, A. (2024). *Regional Distribution of Woody Invasives and the Response of Plant Communities to Invasive Control Through Government Cost Share Programs*. Purdue University.

French, B. J., Prior, L. D., Johnson, C. N., Sculthorpe, A., von Bibra, J., & Bowman, D. M. (2024). Returning Fire to Degraded Temperate Grassy Woodland Fragments Requires Fine-Scale Management Interventions to Promote Trees: Insights From the Beaufront Experiment. *Ecological Management & Restoration*. Advance online publication. DOI: 10.1111/emr.12507

Frick-Ruppert, J. (2014). *Waterways: Sailing the Southeastern Coast*. University Press of Florida.

Fried, G., Bon, M.-C., Brundu, G., Brunel, S., Chauvel, B., Cheptou, P.-O., Hatcher, P., Jourdan, M., Kazakou, E., & Le Bourgeois, T. (2014). Environmental weeds and invasive plants: program, abstracts, list of participants.

Frouz, J. (2023). Plant-soil feedback across spatiotemporal scales from immediate effects to legacy. *Soil Biology & Biochemistry*, *177*, 108790.

Fry, W. E., McGrath, M. T., Seaman, A., Zitter, T. A., McLeod, A., Danies, G., & Smart, C. D. (2013). The 2009 late blight pandemic in the eastern United States–causes and results. *Plant Disease*, *97*(3), 296–306. DOI: 10.1094/PDIS-08-12-0791-FE PMID: 30722376

Fuchs, A., Davidson, I. C., Megonigal, J., Devaney, J. L., Simkanin, C., Noyce, G., Lu, M., & Cott, G. M. (2024). Stronger increase of methane emissions from coastal wetlands by non-native Spartina alterniflora than non-native Phragmites australis. PLANTS, PEOPLE, PLANET.

Fu, D., Wu, X., Huang, N., & Duan, C. (2018). Effects of the invasive herb *Ageratina adenophora* on understory plant communities and tree seedling growth in *Pinus yunnanensis* forests in Yunnan, China. *Journal of Forest Research*, *23*(2), 112–119. DOI: 10.1080/13416979.2018.1429202

Fuhlendorf, S. D., Engle, D. M., Kerby, J. A. Y., & Hamilton, R. (2009). Pyric herbivory: Rewilding landscapes through the recoupling of fire and grazing. *Conservation Biology*, *23*(3), 588–598. DOI: 10.1111/j.1523-1739.2008.01139.x PMID: 19183203

Funk, J. L. (2013). The physiology of invasive plants in low-resource environments. *Conservation Physiology*, *1*(1), cot026. DOI: 10.1093/conphys/cot026 PMID: 27293610

Funk, J. L., Cleland, E. E., Suding, K. N., & Zavaleta, E. S. (2008). Restoration through reassembly: Plant traits and invasion resistance. *Trends in Ecology & Evolution*, *23*(12), 695–703. DOI: 10.1016/j.tree.2008.07.013 PMID: 18951652

Funk, J. L., & McDaniel, S. (2010). Altering light availability to restore invaded forest: The predictive role of plant traits. *Restoration Ecology*, *18*(6), 865–872. DOI: 10.1111/j.1526-100X.2008.00515.x

Funk, J. L., Parker, I. M., Matzek, V., Flory, S. L., Aschehoug, E. T., D'Antonio, C. M., Dawson, W., Thomson, D. M., & Valliere, J. (2020). Keys to enhancing the value of invasion ecology research for management. *Biological Invasions*, *22*(8), 2431–2445. DOI: 10.1007/s10530-020-02267-9

Gaertner, M., Den Breeyen, A., Hui, C., & Richardson, D. M. (2009). Impacts of alien plant invasions on species richness in Mediterranean-type ecosystems: A meta-analysis. *Progress in Physical Geography, 33*(3), 319–338. DOI: 10.1177/0309133309341607

Gaffke, A. M., Dudley, T. L., Bean, D. W., Drus, G. M., Johnson, M. J., Knutson, A. E., Weaver, D. K., Sing, S. E., Orr, B. K., & Thompson10, D. C. (2022). Tamarix Biological Control in North America. *CLASSICAL BIOLOGICAL CONTROL*, 329.

Gaggini, L., Rusterholz, H. P., & Baur, B. (2018). The invasive plant *Impatiens glandulifera* affects soil fungal diversity and the bacterial community in forests. *Applied Soil Ecology, 124*, 335–343. DOI: 10.1016/j.apsoil.2017.11.021

Gaggini, L., Rusterholz, H. P., & Baur, B. (2019). The annual invasive plant *Impatiens glandulifera* reduces hyphal biomass of soil fungi in deciduous forests. *Fungal Ecology, 39*, 242–249. DOI: 10.1016/j.funeco.2018.12.004

Gallardo, B., Bacher, S., Barbosa, A. M., Gallien, L., González-Moreno, P., Martínez-Bolea, V., Sorte, C., Vimercati, G., & Vilà, M. (2024). Risks posed by invasive species to the provision of ecosystem services in Europe. *Nature Communications, 15*(1), 2631. DOI: 10.1038/s41467-024-46818-3 PMID: 38600085

Gamiz, G. I. (2024). *Buffelgrass Effects Above and Belowground in the Sonoran Desert*. The University of Arizona.

Gao, L., Huang, J., Chen, X., Chen, Y., & Liu, M. (2018). Contributions of natural climate changes and human activities to the trend of extreme precipitation. *Atmospheric Research, 205*, 60–69. DOI: 10.1016/j.atmosres.2018.02.006

Gao, L., Wang, R., Shen, G., Zhang, J., Meng, G., & Zhang, J. (2017). Effects of biochar on nutrients and the microbial community structure of tobacco-planting soils. *Journal of Soil Science and Plant Nutrition, 17*(4), 884–896. DOI: 10.4067/S0718-95162017000400004

Gao, W.-T., & Su, W.-H. (2024). Weed Management Methods for Herbaceous Field Crops: A Review. *Agronomy (Basel), 14*(3), 486. DOI: 10.3390/agronomy14030486

Gao, Y., Wang, X., Liu, Q., Zhang, J., Han, X., Yang, S.-I., & Lu, D. (2025). Converting larch monocultures to mixed-species forests: Ten-year impacts of canopy manipulation on understory regeneration. *Forest Ecology and Management, 585*, 122611. DOI: 10.1016/j.foreco.2025.122611

García-Díaz, P., Cassey, P., Norbury, G., Lambin, X., Montti, L., Pizarro, J. C., Powell, P. A., Burslem, D. F., Cava, M., Damasceno, G., Fasola, L., Fidelis, A., Huerta, M. F., Langdon, B., Linardaki, E., Moyano, J., Núñez, M. A., Pauchard, A., Phimister, E., & Tomasevic, J. A. (2021). Management policies for invasive alien species: Addressing the impacts rather than the species. *Bioscience*, *71*(2), 174–185. DOI: 10.1093/biosci/biaa139

García-Díaz, P., Montti, L., Powell, P. A., Phimister, E., Pizarro, J. C., Fasola, L., Langdon, B., Pauchard, A., Raffo, E., Bastías, J., Damasceno, G., Fidelis, A., Huerta, M. F., Linardaki, E., Moyano, J., Núñez, M. A., Ortiz, M. I., Rodríguez-Jorquera, I., Roesler, I., & Lambin, X. (2022). Identifying priorities, targets, and actions for the long-term social and ecological management of invasive non-native species. *Environmental Management*, *69*(1), 140–153. DOI: 10.1007/s00267-021-01541-3 PMID: 34586487

Garg, N., & Singh, S. (2018). Arbuscular mycorrhiza Rhizophagus irregularis and silicon modulate growth, proline biosynthesis and yield in Cajanus cajan L. Millsp. (pigeonpea) genotypes under cadmium and zinc stress. *Journal of Plant Growth Regulation*, *37*(1), 46–63. DOI: 10.1007/s00344-017-9708-4

Gaskin, J. F., Espeland, E., Johnson, C. D., Larson, D. L., Mangold, J. M., McGee, R. A., Milner, C., Paudel, S., Pearson, D. E., Perkins, L. B., Prosser, C. W., Runyon, J. B., Sing, S. E., Sylvain, Z. A., Symstad, A. J., & Tekiela, D. R. (2021). Managing invasive plants on Great Plains grasslands: A discussion of current challenges. *Rangeland Ecology and Management*, *78*, 235–249. DOI: 10.1016/j.rama.2020.04.003

Gathala, M. K., Laing, A. M., Tiwari, T. P., Timsina, J., Rola-Rubzen, F., Islam, S., Maharjan, S., Brown, P. R., Das, K. K., Pradhan, K., Chowdhury, A. K., Kumar, R., Datt, R., Anwar, M., Hossain, S., Kumar, U., Adhikari, S., Magar, D. B. T., Sapkota, B. K., & Gerard, B. (2021). Improving smallholder farmers' gross margins and labor-use efficiency across a range of cropping systems in the Eastern Gangetic Plains. *World Development*, *138*, 105266. DOI: 10.1016/j.worlddev.2020.105266

Genes, L., & Dirzo, R. (2022). Restoration of plant-animal interactions in terrestrial ecosystems. *Biological Conservation*, *265*, 109393. DOI: 10.1016/j.biocon.2021.109393

Genre, A., Lanfranco, L., Perotto, S., & Bonfante, P. (2020). Unique and common traits in mycorrhizal symbioses. *Nature Reviews. Microbiology*, *18*(11), 649–660. DOI: 10.1038/s41579-020-0402-3 PMID: 32694620

Georgiou, K., Angers, D., Champiny, R. E., Cotrufo, M. F., Craig, M. E., Doetterl, S., Grandy, A. S., Lavallee, J. M., Lin, Y., Lugato, E., Poeplau, C., Rocci, K. S., Schweizer, S. A., Six, J., & Wieder, W. R. (2025). Soil carbon saturation: What do we really know? *Global Change Biology*, *31*(5), e70197. DOI: 10.1111/gcb.70197 PMID: 40345163

Gerber, E., Krebs, C., Murrell, C., Moretti, M., Rocklin, R., & Schaffner, U. (2008). Exotic invasive knotweeds (Fallopia spp.) negatively affect native plant and invertebrate assemblages in European riparian habitats. *Biological Conservation*, *141*(3), 646–654. DOI: 10.1016/j.biocon.2007.12.009

Gerwing, T. G., Hawkes, V. C., Gann, G. D., & Murphy, S. D. (2022). Restoration, reclamation, and rehabilitation: On the need for, and positing a definition of, ecological reclamation. *Restoration Ecology*, *30*(7), e13461. DOI: 10.1111/rec.13461

Ghaffar, A., Rahman, M. H. U., Ahmed, S., Haider, G., Ahmad, I., Khan, M. A., Afzaal, M., Ahmed, S., Fahad, S., & Hussain, J. (2022). Adaptations in cropping system and pattern for sustainable crops production under climate change scenarios. In *Improvement of plant production in the era of climate change* (pp. 1–34). CRC Press. DOI: 10.1201/9781003286417-1

Gharde, Y., Singh, P. K., Dubey, R. P., & Gupta, P. K. (2018). Assessment of yield and economic losses in agriculture due to weeds in India. *Crop Protection (Guildford, Surrey)*, *107*, 12–18. DOI: 10.1016/j.cropro.2018.01.007

Ghasemi, M., Charrahy, Z., & González-García, A. (2023). Mapping cultural ecosystem services provision: An integrated model of recreation and ecotourism opportunities. *Land Use Policy*, *131*, 106732. DOI: 10.1016/j.landusepol.2023.106732

Gibbons, S. M., Lekberg, Y., Mummey, D. L., Sangwan, N., Ramsey, P. W., & Gilbert, J. A. (2017). Invasive plants rapidly reshape soil properties in a grassland ecosystem. *mSystems*, *2*(2), 00178–00116. DOI: 10.1128/mSystems.00178-16 PMID: 28289729

Gibson, D. J., & Newman, J. A. (Eds.). (2019). *Grasslands and climate change*. Cambridge University Press. DOI: 10.1017/9781108163941

Gilani, S. A., Fujii, Y., Kikuchi, A., Shinwari, Z. K., & Watanabe, K. N. (2011). Ecological consequences, genetic and chemical variations in fragmented populations of a medicinal plant, Justicia adhatoda and implications for its conservation. *Pakistan Journal of Botany*, *43*, 29–37.

Gilliom, R. J., Barbash, J. E., Crawford, C. G., Hamilton, P. A., Martin, J. D., Nakagaki, N., Nowell, L. H., Scott, J. C., Stackelberg, P. E., Thelin, G. P., & Wolock, D. M. (2006). The quality of our nation's waters—Pesticides in the nation's streams and ground water, 1992–2001. U.S. Geological Survey Circular 1291. Retrieved from https://pubs.usgs.gov/circ/2005/1291/

Gillmann, S. M., Lorenz, A. W., Kaijser, W., Nguyen, H. H., Haase, P., & Hering, D. (2024). How tolerances, competition and dispersal shape benthic invertebrate colonisation in restored urban streams. *The Science of the Total Environment*, *929*, 172665. DOI: 10.1016/j.scitotenv.2024.172665 PMID: 38653408

Gioria, M., & Osborne, B. A. (2014). Resource competition in plant invasions: Emerging patterns and research needs. *Frontiers in Plant Science*, *5*, 501. DOI: 10.3389/fpls.2014.00501 PMID: 25324851

Goldsmith, C., Alexander, H. D., Granger, J. J., & Siegert, C. M. (2023). Invasive microstegium vimineum (Japanese stiltgrass) hinders growth and biomass of hardwood seedlings regardless of light and moisture regime. *Forest Ecology and Management*, *539*, 120984. DOI: 10.1016/j.foreco.2023.120984

Gómez-Aparicio, L., & Canham, C. D. (2008). Neighborhood models of the effects of invasive tree species on ecosystem processes. *Ecological Monographs*, *78*(1), 69–86. DOI: 10.1890/06-2036.1

Gonçalves, R. V. S., Cardoso, J. C. F., Oliveira, P. E., & Oliveira, D. C. (2021). Changes in the Cerrado vegetation structure: Insights from more than three decades of ecological succession. *Web Ecology*, *21*(1), 55–64. DOI: 10.5194/we-21-55-2021

Gonzales, E. K. (2008). *The effects of herbivory, competition, and disturbance on island meadows* (Doctoral dissertation, University of British Columbia).

González-García, A., Palomo, I., González, J. A., García-Díez, V., García-Llorente, M., & Montes, C. (2022). Biodiversity and ecosystem services mapping: Can it reconcile urban and protected area planning? *The Science of the Total Environment*, *803*, 150048. DOI: 10.1016/j.scitotenv.2021.150048 PMID: 34500265

Green, S. J., & Grosholz, E. D. (2021). Functional eradication as a framework for invasive species control. *Frontiers in Ecology and the Environment*, *19*(1), 1–10. DOI: 10.1002/fee.2277

Grenz, J., & Clements, D. R. (2023). Water, wind, and fire: extreme climate events enhance the spread of invasive plants in sensitive North American ecosystems. In *Plant invasions and global climate change* (pp. 113–137). Springer. DOI: 10.1007/978-981-99-5910-5_6

Griffen, B. D., Bolander, M., Fletcher, L. S., Luckett, J., Repetto, M. F., Smith, N., Stancil, C., & Toscano, B. J. (2024). Factors influencing variation in reproduction in invasive species: A case study of the Asian shore crab Hemigrapsus sanguineus. *Biological Invasions*, *26*(10), 1–13. DOI: 10.1007/s10530-024-03382-7

Grove, S., Haubensak, K. A., Gehring, C., & Parker, I. M. (2017). Mycorrhizae, invasions, and the temporal dynamics of mutualism disruption. *Journal of Ecology*, *105*(6), 1496–1508. DOI: 10.1111/1365-2745.12853

Grünwald, N. J., Goss, E. M., & Press, C. M. (2008). *Phytophthora ramorum*: A pathogen with a remarkably wide host range causing sudden oak death on oaks and ramorum blight on woody ornamentals. *Molecular Plant Pathology*, *9*(6), 729–740. DOI: 10.1111/j.1364-3703.2008.00500.x PMID: 19019002

Grünwald, N. J., Larsen, M. M., Kamvar, Z. N., Reeser, P. W., Kanaskie, A., Laine, J., & Wiese, R. (2016). First report of the EU1 clonal lineage of Phytophthora ramorum on tanoak in an Oregon forest. *Plant Disease*, *100*(5), 1024–1024. DOI: 10.1094/PDIS-10-15-1169-PDN

Gulzar, R., Ahmad, R., Hassan, T., Rashid, I., & Khuroo, A. A. (2024). Environmental and anthropogenic drivers of invasive plant diversity and distribution in the Himalaya. *Ecological Informatics*, *81*, 102586. DOI: 10.1016/j.ecoinf.2024.102586

Guo, Q., Brockway, D. G., Larson, D. L., Wang, D., & Ren, H. (2018). Improving ecological restoration to curb biotic invasion—A practical guide. *Invasive Plant Science and Management*, *11*(4), 163–174. DOI: 10.1017/inp.2018.29

Gutgesell, M., McCann, K., O'Connor, R., Kc, K., Fraser, E. D., Moore, J. C., McMeans, B., Donohue, I., Bieg, C., Ward, C., Pauli, B., Scott, A., Gillam, W., Gedalof, Z., Hanner, R. H., Tunney, T., & Rooney, N. (2024). The productivity–stability trade-off in global food systems. *Nature Ecology & Evolution*, *8*(11), 2135–2149. DOI: 10.1038/s41559-024-02529-y PMID: 39227681

Hager, H. A., Rupert, R., Quinn, L. D., & Newman, J. A. (2015). Escaped *Miscanthus sacchariflorus* reduces the richness and diversity of vegetation and the soil seed bank. *Biological Invasions*, *17*(6), 1833–1844. DOI: 10.1007/s10530-014-0839-2

Haggblade, S., Smale, M., Kergna, A., Theriault, V., & Assima, A. (2017). Causes and consequences of increasing herbicide use in Mali. *European Journal of Development Research*, *29*(3), 648–674. DOI: 10.1057/s41287-017-0087-2

Hajiboland, R., Joudmand, A., Aliasgharzad, N., Tolrá, R., & Poschenrieder, C. (2019). Arbuscular mycorrhizal fungi alleviate low-temperature stress and increase freezing resistance as a substitute for acclimation treatment in barley. *Crop & Pasture Science*, *70*(3), 218–233. DOI: 10.1071/CP18385

Hallstein, E., & Iseman, T. (2021). *Nature-based solutions in agriculture: Project design for securing investment*. Food & Agriculture Org.

Halmy, M. W. A. (2024). Importance of Geospatial Data for Monitoring and Managing Invasive Species for Sustainable Development: A Semantic Review. *Annals of Library and Information Studies*, *71*(4), 417–434.

Hamilton, J., & Miller, J. M. (2016). Adaptive introgression as a resource for management and genetic conservation in a changing climate. *Conservation Biology*, *30*(1), 256–268. DOI: 10.1111/cobi.12574 PMID: 26096581

Han, M., Zhang, H., Liu, M., Tang, J., Guo, X., Ren, W., & Kong, D. (2024). Increased dependence on nitrogen-fixation of a native legume in competition with an invasive plant. *Plant Diversity*, *46*(4), 510–518. Advance online publication. DOI: 10.1016/j.pld.2024.04.003 PMID: 39280977

Haq, A., Ullah, H., Ullah, I., Badshah, L., & Ahmad, S. (2024). Vegetation Dynamics along the Altitudinal Gradient. In *Scrub Vegetation as Dynamic States of Forests-Methodologies for Learning and Research*. IntechOpen. DOI: 10.5772/intechopen.114309

Hardy, N. G., Kuebbing, S. E., Duguid, M. C., Ashton, M. S., Sheban, K. C., Inman, S. E., & Martin, M. P. (2025). Non-native invasive plants in tropical dry forests: A global review of presence, impacts, and management. *Restoration Ecology*, *33*(1), e14288. DOI: 10.1111/rec.14288

Harms, N. E., Knight, I. A., Pratt, P. D., Reddy, A. M., Mukherjee, A., Gong, P., Coetzee, J., Raghu, S., & Diaz, R. (2021). Climate mismatch between introduced biological control agents and their invasive host plants: Improving biological control of tropical weeds in temperate regions. *Insects*, *12*(6), 549. DOI: 10.3390/insects12060549 PMID: 34204761

Harris, J. A. (2003). Measurements of the soil microbial community for estimating the success of restoration. *European Journal of Soil Science*, *54*(4), 801–808. DOI: 10.1046/j.1351-0754.2003.0559.x

Harrison, G. R., Jones, L. C., Ellsworth, L. M., Strand, E. K., & Prather, T. S. (2024). Cheatgrass alters flammability of native perennial grasses in laboratory combustion experiments. *Fire Ecology*, *20*(1), 103. DOI: 10.1186/s42408-024-00338-z

Harris, S. A. (2018). *Sunflowers*. Reaktion Books.

Hart-Fredeluces, G. M., Hopping, K. A., Burnham, M., & Gregory, T. (2024). Indirect and interactive effects drive impacts of prescribed fire and invasive grass on a native wetland herb. *Restoration Ecology*, •••, 14331.

Hartzell, S., Bartlett, M. S., & Porporato, A. (2017). The role of plant water storage and hydraulic strategies in relation to soil moisture availability. *Plant and Soil*, *419*(1-2), 503–521. DOI: 10.1007/s11104-017-3341-7

Harvey, J. A., & Fortuna, T. M. (2012). Chemical and structural effects of invasive plants on herbivore–parasitoid/predator interactions in native communities. *Entomologia Experimentalis et Applicata*, *144*(1), 14–26. DOI: 10.1111/j.1570-7458.2012.01252.x

Hashem, A., Alqarawi, A. A., Radhakrishnan, R., Al-Arjani, A. B. F., Aldehaish, H. A., & Egamberdieva, D., & Abd_Allah, E. F. (. (2018). Arbuscular mycorrhizal fungi regulate the oxidative system, hormones and ionic equilibrium to trigger salt stress tolerance in Cucumis sativus L. *Saudi Journal of Biological Sciences*, *25*(6), 110. DOI: 10.1016/j.sjbs.2018.03.009 PMID: 30174509

Hassan, M. M., Azrag, M. A., Rugheim, A. M., Abusin, R. M., Elnasikh, M. H., Modawi, H. I., & Babiker, A. G. E. (2019). Potential of *Trichoderma harzianum* as a biocontrol agent against *Striga hermonthica* in *Sorghum. International Journal of Current Microbiology and Applied Sciences*, *8*, 195–206. DOI: 10.20546/ijcmas.2019.803.027

Hauber, D. P., Saltonstall, K., White, D. A., & Hood, C. S. (2011). Genetic variation in the common reed, Phragmites australis, in the Mississippi River Delta marshes: Evidence for multiple introductions. *Estuaries and Coasts*, *34*(4), 851–862. DOI: 10.1007/s12237-011-9391-9

Hayes, M. A., Jesse, A., Hawke, B., Baldock, J., Tabet, B., Lockington, D., & Lovelock, C. E. (2017). Dynamics of sediment carbon stocks across intertidal wetland habitats of Moreton Bay, Australia. *Global Change Biology*, *23*(10), 4222–4234. DOI: 10.1111/gcb.13722 PMID: 28407457

Hazelton, E. L., Mozdzer, T. J., Burdick, D. M., Kettenring, K. M., & Whigham, D. F. (2014). Phragmites australis management in the United States: 40 years of methods and outcomes. *AoB Plants*, *6*, plu001. DOI: 10.1093/aobpla/plu001 PMID: 24790122

Hazen, H., & Sutton, P. (2025). Biophysical perspectives on sustainability. In *An Introduction to Environment, Society and Sustainability* (pp. 28-61). Routledge.

He, F., Sheng, M., & Tang, M. (2017). Effects of Rhizophagus irregularis on photosynthesis and antioxidative enzymatic system in Robinia pseudoacacia L. under drought stress. *Frontiers in Plant Science*, *8*, 183. DOI: 10.3389/fpls.2017.00183 PMID: 28261240

Hejda, M., Pyšek, P., & Jarošík, V. (2009). Impact of invasive plants on the species richness, diversity and composition of invaded communities. *Journal of Ecology*, *97*(3), 393–403. DOI: 10.1111/j.1365-2745.2009.01480.x

Heller, N. E., & Zavaleta, E. S. (2009). Biodiversity management in the face of climate change: A review of 22 years of recommendations. *Biological Conservation*, *142*(1), 14–32. DOI: 10.1016/j.biocon.2008.10.006

Hellmers, H., & Warrington, I. (2018). Temperature and plant productivity. In *Handbook of Agricultural Productivity* (pp. 11–22). CRC Press.

Hessenauer, P., Fijarczyk, A., Martin, H., Prunier, J., Charron, G., Chapuis, J., Bernier, L., Tanguay, P., Hamelin, R., & Landry, C. (2020). Hybridization and introgression drive genome evolution of Dutch elm disease pathogens. *Nature Ecology & Evolution*, *4*(4), 626–638. DOI: 10.1038/s41559-020-1133-6 PMID: 32123324

He, Z.-Z., Stotz, G. C., Liu, X., Liu, J.-J., Wang, Y.-G., Yang, J., Li, L.-F., Zhang, W.-J., Nan, P., & Song, Z.-P. (2024). A global synthesis of the patterns of genetic diversity in endangered and invasive plants. *Biological Conservation*, *291*, 110473. DOI: 10.1016/j.biocon.2024.110473

Hilger-Tisdale, A. N. (2025). *Effects of Plant Community Composition.* Nutrition, and Human Disturbance on the Behavior and Distribution of Large Herbivores University of Idaho.

Hill, P. W., Broughton, R., Bougoure, J., Havelange, W., Newsham, K. K., Grant, H., Murphy, D. V., Clode, P., Ramayah, S., Marsden, K. A., Quilliam, R. S., Roberts, P., Brown, C., Read, D. J., Deluca, T. H., Bardgett, R. D., Hopkins, D. W., & Jones, D. L. (2019). Angiosperm symbioses with non-mycorrhizal fungal partners enhance N acquisition from ancient organic matter in a warming maritime Antarctic. *Ecology Letters*, *22*(12), 2111–2119. DOI: 10.1111/ele.13399 PMID: 31621153

Hinz, H., Walsh, G. C., Paterson, I., Paynter, Q., Schwarzländer, M., Smith, M., & Weyl, P. (2024). Enhancing pre-release studies for weed biocontrol agents: A review of existing and emerging tools. *Biological Control*, *198*, 105607. DOI: 10.1016/j.biocontrol.2024.105607

Hobbs, R. J., & Harris, J. A. (2001). Restoration ecology: Repairing the Earth's ecosystems in the new millennium. *Restoration Ecology*, *9*(2), 1–9. DOI: 10.1046/j.1526-100x.2001.009002239.x

Hobbs, R. J., & Humphries, S. E. (1995). An integrated approach to the ecology and management of plant invasions. *Conservation Biology*, *9*(4), 761–770. DOI: 10.1046/j.1523-1739.1995.09040761.x

Hobbs, R. J., & Norton, D. A. (2004). Ecological filters, thresholds, and gradients in resistance to ecosystem reassembly after invasion. *Ecology*, *85*(12), 3391–3397. DOI: 10.1890/04-0242

Hoff, H. K., Littleford-Colquhoun, B. L., Kartzinel, R. Y., Anderson, H. M., Geremia, C., McGarvey, L. M., Segal, C., & Kartzinel, T. R. (2025). The apportionment of dietary diversity in wildlife. *Proceedings of the National Academy of Sciences of the United States of America*, *122*(29), e2502691122. DOI: 10.1073/pnas.2502691122 PMID: 40658848

Hoffmann, A. A., & Sgrò, C. M. (2011). Climate change and evolutionary adaptation. *Nature*, *470*(7335), 479–485. DOI: 10.1038/nature09670 PMID: 21350480

Holden, E. M., Grenke, J. S., Holden, K., de Assis Murillo, R., Villasor, C., & Wagner, V. (2024). The Biology of Canadian Weeds: 160. Poa pratensis L. *Canadian Journal of Plant Science*(ja).

Holland, O. C., Stark, J. S., Wilson, K. A., Baker, C. M., Shaw, J. D., & Helmstedt, K. J. (2025). Invasive species pose a threat to native species' abundances in an East Antarctic coastal marine ecosystems. *Journal of Applied Ecology*, 1365-2664.70062. DOI: 10.1111/1365-2664.70062

Holling, C. S. (1978). *Adaptive environmental assessment and management*. Wiley-Interscience.

Holl, K. D., & Aide, T. M. (2011). When and where to actively restore ecosystems? *Forest Ecology and Management*, *261*(10), 1558–1563. DOI: 10.1016/j.foreco.2010.07.004

Hong, S., Gan, P., & Chen, A. (2019). Environmental controls on soil pH in planted forest and its response to nitrogen deposition. *Environmental Research*, *172*, 159–165. DOI: 10.1016/j.envres.2019.02.020 PMID: 30782535

Hörandl, E. (2024). Apomixis and the paradox of sex in plants. *Annals of Botany*, *134*(1), 1–18. DOI: 10.1093/aob/mcae044 PMID: 38497809

Hörandl, E., Hojsgaard, D., Caperta, A., Conceição, S. I., Róis, A. S., Dickinson, T., Mandáková, T., Windham, M. D., Appelhans, M. S., Mráz, P., & Chrtek, J. (2024). Apomixis in systematics, evolution and phylogenetics of angiosperms: Current developments and prospects. *Critical Reviews in Plant Sciences*, •••, 1–43. DOI: 10.1080/07352689.2024.2396259

Hossain, M. A., Akhter, S., Sohrab, M. H., Afroz, F., Begum, M. N., & Rony, S. R. (2023). Applying an optimization technique for the extraction of antioxidant components from Justicia adhatoda leaves. *Engineering and Science*, *24*(2), 913. DOI: 10.30919/es913

Hossain, R., Islam, M. S., Dey, D., & Islam, M. T. (2022). Vinca rosea (Madagascar Periwinkle) and Adhatoda vesica (Malabar Nut). In *Herbs, Shrubs, and Trees of Potential Medicinal Benefits* (pp. 301–338). CRC Press. DOI: 10.1201/9781003205067-16

Hou, Q.-Q., Chen, B.-M., Peng, S.-L., & Chen, L.-Y. (2014). Effects of extreme temperature on seedling establishment of nonnative invasive plants. *Biological Invasions*, *16*(10), 2049–2061. DOI: 10.1007/s10530-014-0647-8

Howell, A., Winkler, D. E., Phillips, M. L., McNellis, B., & Reed, S. C. (2020). Experimental warming changes phenology and shortens growing season of the dominant invasive plant Bromus tectorum (cheatgrass). *Frontiers in Plant Science*, *11*, 570001. DOI: 10.3389/fpls.2020.570001 PMID: 33178240

Huang, C., & Asner, G. P. (2009). Applications of remote sensing to alien invasive plant studies. *Sensors (Basel)*, *9*(6), 4869–4889. DOI: 10.3390/s90604869 PMID: 22408558

Huang, C., Li, N., Zhang, Z., Liu, Y., Chen, X., & Wang, F. (2020). Assessment of the economic cascading effect on future climate change in China: Evidence from agricultural direct damage. *Journal of Cleaner Production*, *276*, 123951. DOI: 10.1016/j.jclepro.2020.123951

Huang, D., Zhang, R., Kim, K. C., & Suarez, A. V. (2012). Spatial pattern and determinants of the first detection locations of invasive alien species in Mainland China. *PLoS One*, *7*(2), e31734. DOI: 10.1371/journal.pone.0031734 PMID: 22363715

Huebner, C. D., & Jones, T. (2022). Invasive plants field & reference guide: an ecological perspective of plant invaders of forests & woodlands. *US Department of Agriculture, Forest Service, Northern Research Station and Eastern Region State & Private Forestry. R9-TP-001-22.*

Hui, Y., Atkinson, J. F., Zhu, Z., Schlea, D., & Redder, T. (2021). Invasive dreissenid mussel effects on phosphorus dynamics in Lake Ontario: Insights from integrated hydrodynamic–ecological modeling. *Canadian Journal of Fisheries and Aquatic Sciences*, *78*(1), 14–26. DOI: 10.1139/cjfas-2020-0164

Hulme, P. E. (2017). *Incursion response in New Zealand*. Cambridge University Press.

Hulme, P. E. (2021). Unwelcome exchange: International trade as a direct and indirect driver of biological invasions worldwide. *One Earth*, *4*(5), 666–679. DOI: 10.1016/j.oneear.2021.04.015

Hulvey, K. B., Standish, R. J., Hallett, L. M., Starzomski, B. M., Murphy, S. D., Nelson, C. R., ... & Suding, K. N. (2013). Incorporating novel ecosystems into management frameworks. *Novel ecosystems: intervening in the new ecological world order*, 157-171.

Hunt, V., Fant, J., Steger, L., Hartzog, P. E., Lonsdorf, E., Jacobi, S. K., & Larkin, D. (2017). PhragNet: Crowdsourcing to investigate ecology and management of invasive Phragmites australis in North America. *Wetlands Ecology and Management*, *25*(5), 607–618. DOI: 10.1007/s11273-017-9539-x

Husain, H. J., Wang, X., Pirasteh, S., Mafi-Gholami, D., Chouhan, B., Khan, M. L., & Gheisari, M. (2024). Review and assessment of the potential restoration of ecosystem services through the implementation of the biodiversity management plans for SDG-15 localization. *Heliyon*, *10*(8), e29877. Advance online publication. DOI: 10.1016/j.heliyon.2024.e29877 PMID: 38699718

Hussain, A., & Shah, S. U. H. (2020). Economic and Environmental Costs and Benefits of Paper Mulberry: A Case Study of Islamabad-Pakistan. *Journal of Health Research and Society*, *2*, 5–12. DOI: 10.34002/jhrs.v2i0.20

Hu, Y., Jiang, S., Li, J., Gao, J., Gao, Y., & Gu, S. (2009). Effects of light intensity and quality on plant growth and development. [Nat Sci Ed]. *J Inn Mongolia Agri Univ*, *30*, 296.

Hu, Y., Xie, W., & Chen, B. (2020). Arbuscular mycorrhiza improved drought tolerance of maize seedlings by altering photosystem II efficiency and the levels of key metabolites. *Chemical and Biological Technologies in Agriculture*, *7*(1), 1–14. DOI: 10.1186/s40538-020-00186-4

Iacarella, J. C., Mankiewicz, P. S., & Ricciardi, A. (2015). Negative competitive effects of invasive plants change with time since invasion. *Ecosphere*, *6*(7), 1–14. DOI: 10.1890/ES15-00147.1

Ighalo, J. O., Ohoro, C. R., Ojukwu, V. E., Oniye, M., Shaikh, W. A., Biswas, J. K., Seth, C. S., Mohan, G. B. M., Chandran, S. A., & Rangabhashiyam, S. (2025). Biochar for ameliorating soil fertility and microbial diversity: From production to action of the black gold. *iScience*, *28*(1), 111524. DOI: 10.1016/j.isci.2024.111524 PMID: 39807171

Igiehon, N. O., Babalola, O. O., Cheseto, X., & Torto, B. (2021). Effects of rhizobia and arbuscular mycorrhizal fungi on yield, size distribution and fatty acid of soybean seeds grown under drought stress. *Microbiological Research*, *242*, 126640. DOI: 10.1016/j.micres.2020.126640 PMID: 33223380

Iglesias-Carrasco, M., Torres, J., Cruz-Dubon, A., Candolin, U., Wong, B. B., & Velo-Antón, G. (2025). Global impacts of exotic eucalypt plantations on wildlife. *Biological Reviews of the Cambridge Philosophical Society*, *100*(4), 1734–1753. DOI: 10.1111/brv.70022 PMID: 40159998

Ilyas, M., Shah, S., Lai, Y.-W., Sher, J., Bai, T., Zaman, F., Bibi, F., Koul, M., Wani, S. H., Majrashi, A., Alharby, H. F., Hakeem, K. R., Wang, Y.-J., & Rather, S. A. (2022). Leaf functional traits of invasive grasses conferring high-cadmium adaptation over natives. *Frontiers in Plant Science*, *13*, 869072. DOI: 10.3389/fpls.2022.869072 PMID: 35720536

Inácio, M., Gomes, E., Bogdzevič, K., Kalinauskas, M., Zhao, W., & Pereira, P. (2022). Mapping and assessing coastal recreation cultural ecosystem services supply, flow, and demand in Lithuania. *Journal of Environmental Management*, *323*, 116175. DOI: 10.1016/j.jenvman.2022.116175 PMID: 36088764

Inderjit, & Cahill JF (2015) Linkages of plant–soil feedbacks and underlying invasion mechanisms. AoB plants 7: plv022

Iqbal, B., Khan, A. R., Li, G., Abbas, A., Raheem, A., Jalal, A., Al Tawaha, A. R. M., & Fahad, S. (2024). Impact of Invasive Alien Plants and Heavy Metals Contamination on Crops: A Review. *Environment, Climate, Plant and Vegetation Growth*, 179-203.

Iqbal, B., Zhao, X., Khan, K. Y., Javed, Q., Nazar, M., Khan, I., & Du, D. (2024). Microplastics meet invasive plants: Unraveling the ecological hazards to agroecosystems. *The Science of the Total Environment*, *906*, 167756. DOI: 10.1016/j.scitotenv.2023.167756 PMID: 37832681

Isha, K. P., Kumar, P., & Singh, A. N. (2025). An Overview of Justicia adhatoda: A Medicinal Plant but Native Invader in India. *Conservation*, *5*(1), 2. DOI: 10.3390/conservation5010002

Ishihara, K. L., Honda, M. D., Bageel, A., & Borthakur, D. (2018). *Leucaena leucocephala: a leguminous tree suitable for eroded habitats of Hawaiian Islands. In ravine lands: greening for livelihood and environmental security*. Springer.

Ivey, M., & Leege, L. M. (2023). Life After Privet: Plant Community Dynamics in a Forested Wetland Following Removal of the Invasive Ligustrum sinense Lour. *Castanea*, *88*(1), 49–60. DOI: 10.2179/0008-7475.88.1.49

Izhar, S. K., Rizvi, S. F., Siddiqui, S., & Afaq, U. (2023). Invasive weeds and their management in India. *Journal of Survey in Fisheries Sciences*, *10*(2S), 2571–2596.

Jackson, M. L. (1958). *Soil chemical analysis*. Prentice Hall, Inc.

James, J. J., Gornish, E. S., DiTomaso, J. M., Davy, J., Doran, M. P., Becchetti, T., Lile, D., Brownsey, P., & Laca, E. A. (2015). Managing medusahead (Taeniatherum caput-medusae) on rangeland: A meta-analysis of control effects and assessment of stakeholder needs. *Rangeland Ecology and Management*, *68*(3), 215–223. DOI: 10.1016/j.rama.2015.03.006

Jamil, M. D., Waheed, M., Akhtar, S., Bangash, N., Chaudhari, S. K., Majeed, M., Hussain, M., Ali, K., & Jones, D. A. (2022). Invasive plants diversity, ecological status, and distribution pattern in relation to edaphic factors in different habitat types of district Mandi Bahauddin, Punjab, Pakistan. *Sustainability (Basel)*, *14*(20), 13312. DOI: 10.3390/su142013312

Jamin, J., Diehl, D., Meyer, M., David, J., Schaumann, G. E., & Buchmann, C. (2022). Physico-chemical soil properties affected by invasive plants in Southwest Germany (Rhineland-Palatinate)—A case study. *Soil Systems*, *6*(2), 35. DOI: 10.3390/soilsystems6040093

Jamwal, M., Puri, S., Sharma, N., Prakash, S., & Pundir, A. (2023). Altitudinal variation in phytochemical, physicochemical, and morphological aspects of Justicia adhatoda L. plant growing wildly in Western Himalayas. *Journal of Applied Biology and Biotechnology*, *11*(3), 85–96.

Jeffries, D. L., Lawson-Handley, L., Lamatsch, D., Olsén, K. H., Sayer, C. D., & Hänfling, B. (2024). Towards the conservation of the crucian carp in Europe: Prolific hybridization but no evidence for introgression between native and non-native taxa. *Molecular Ecology*, *33*(19), 17515. DOI: 10.1111/mec.17515 PMID: 39212263

Jehangir, S., Khan, S. M., Ejaz, U., Zahid, N., Rashid, N., Noshad, Q., Din, Z. U., & Shoukat, A. (2024). Alien flora of Pakistan: Taxonomic composition, invasion status, geographic origin, introduction pathways, and ecological patterns. *Biological Invasions*, *26*(8), 1–17. DOI: 10.1007/s10530-024-03311-8

Jenkins, J. (2024). Using Annual Forages to Replace Declining Cool-Season Grass Pasture Availability.

Jiang, Y. L., Chen, X. H., Miao, Q., & Qu, B. (2019). *Difference in fungal communities between in roots and in root-associated soil of nine orchids in Liaoning*., DOI: 10.17521/cjpe.2019.0055

Jiang, Y., Wang, M., Yan, X., Liu, M., & Guo, X. (2024). Exotic plants introduction changed soil nutrient cycle and symbiotic relationship with arbuscular mycorrhizal fungi in wetland ecological projects. *Frontiers in Plant Science*, *15*, 1410009. DOI: 10.3389/fpls.2024.1410009 PMID: 39049854

Jian-Hua, H., Sheng, Q., Kang-Ning, D., & Yuan-Xing, G. (2010). Wind-dispersed traits of cypselas in ten Asteraceae alien invasive species. *Acta Phytoecologica Sinica*, *34*(8), 957.

Jiménez-Ruiz, J., Hardion, L., Del Monte, J. P., Vila, B., & Santín-Montanyá, M. I. (2021). Monographs on invasive plants in Europe N° 4: Arundo donax L. *Botany Letters*, *168*(1), 131–151. DOI: 10.1080/23818107.2020.1864470

Ji-Zhong, W., Zhang, Z. X., & Chun-Jing, W. (2019). Effects of ecoregional vulnerability on habitat suitability of invasive alien plants: An assessment using 13 species on a global scale. *Environmental Earth Sciences*, *78*(5), 1–9.

Jogesh, T., Stanley, M. C., & Berenbaum, M. R. (2014). Evolution of tolerance in an invasive weed after reassociation with its specialist herbivore. *Journal of Evolutionary Biology*, *27*(11), 2334–2346. DOI: 10.1111/jeb.12469 PMID: 25417738

Johnson, M. T., Arif, I., Marchetti, F., Munshi-South, J., Ness, R. W., Szulkin, M., Verrelli, B. C., Yauk, C. L., Anstett, D. N., Booth, W., Caizergues, A. E., Carlen, E. J., Dant, A., González, J., Lagos, C. G., Oman, M., Phifer-Rixey, M., Rennison, D. J., Rosenberg, M. S., & Winchell, K. M. (2024). Effects of urban-induced mutations on ecology, evolution and health. *Nature Ecology & Evolution*, *8*(6), 1–13. DOI: 10.1038/s41559-024-02401-z PMID: 38641700

Jones, A. D., Rastogi, D., Vahmani, P., Stansfield, A. M., Reed, K. A., Thurber, T., Ullrich, P. A., & Rice, J. S. (2023). Continental United States climate projections based on thermodynamic modification of historical weather. *Scientific Data*, *10*(1), 664. DOI: 10.1038/s41597-023-02485-5 PMID: 37770463

Kachigunda, B., Coupland, G. T., Mengersen, K., Perera, D. I., Van Der Merwe, J., & McKirdy, S. (2023). A systems approach to biosecurity pays off: A case study investigating introduction pathways and interceptions of non-indigenous species at a biosecurity border. *Biological Invasions*, *25*(9), 2849–2871. DOI: 10.1007/s10530-023-03077-5

Kaiser-Bunbury, C. N., Mougal, J., Whittington, A. E., Valentin, T., Gabriel, R., Olesen, J. M., & Blüthgen, N. (2017). Ecosystem restoration strengthens pollination network resilience and function. *Nature*, *542*(7640), 223–227. DOI: 10.1038/nature21071 PMID: 28135718

Kaiyang, X. U., Wanhui, Y. E., Guomin, L. I., & Jing, L. (2005). Phenotypic plasticity in response to light intensity in the invasive species {\sl Alternanthera philoxeroides}. Wuhan zhi wu xue yan jiu= Wuhan Botanical Research, 23(6), 560-563.

Kalsa, K. K. (2019). Farmers' attitudes and practices towards variety and certified seed use, seed replacement and seed storage in wheat growing areas of Ethiopia. *African Journal of Science, Technology, Innovation and Development*, *11*(1), 107–120. DOI: 10.1080/20421338.2018.1550932

Kao, Y. C., Adlerstein, S., & Rutherford, E. (2014). The relative impacts of nutrient loads and invasive species on a Great Lakes food web: An Ecopath with Ecosim analysis. *Journal of Great Lakes Research*, *40*, 35–52. DOI: 10.1016/j.jglr.2014.01.010

Karban, C. C., Lovich, J. E., Grodsky, S. M., & Munson, S. M. (2024). Predicting the effects of solar energy development on plants and wildlife in the Desert Southwest, United States. *Renewable & Sustainable Energy Reviews*, *205*, 114823. DOI: 10.1016/j.rser.2024.114823

Kato-Noguchi, H., & Kurniadie, D. (2021). Allelopathy of Lantana camara as an invasive plant. *Plants*, *10*(5), 1028. DOI: 10.3390/plants10051028 PMID: 34065417

Kaufman, S. R., & Kaufman, W. (2023). *Invasive plants: guide to identification and the impacts and control of common North American species*. Rowman & Littlefield.

Kaur, A., & Singh, A. (2020). Effect of weed infestation on root functional traits of wheat (*Triticum aestivum* L.) under different treatments. *Annals of Plant and Soil Research*, *22*, 269–274.

Kaur, M., Kaur, J., Sood, A., Katnoria, J. K., Nagpal, A. K., Bhatia, A., Verma, P., Kumar, V., Singh, I., & Bhatti, S. S. (2025). Potentially toxic metal contamination in semi-arid agricultural soils: Sources, risk analysis, and spatial distribution. *Environmental Monitoring and Assessment*, *197*(7), 783. DOI: 10.1007/s10661-025-14237-6 PMID: 40542220

Kaur, M., & Singh, P. (2024). Unravelling the diversity, root colonization, and morphological features of arbuscular mycorrhizal fungi associated with invasive plant species of Sirmaur, Himachal Pradesh, India. *Asian Journal of Mycology*, *7*(2), 1–10.

Kaur, S., Kaur, R., & Chauhan, B. S. (2018). Understanding crop-weed-fertilizer-water interactions and their implications for weed management in agricultural systems. *Crop Protection (Guildford, Surrey)*, *103*, 65–72. DOI: 10.1016/j.cropro.2017.09.011

Kaushik, Y., & Arora, P. (2024). Investigating the sustainable energy generation potential of an invasive weed: Lantana camara. *Environmental Science and Pollution Research International*, *31*(54), 62493–62509. DOI: 10.1007/s11356-024-35322-2 PMID: 39453581

Kay, D., Kelly, J., Schaeffer, T., Tuong, T., Leru, P., Addison, W., & Neamtu, R. (2020). Pollen Allergies in Romania.

Kazancoglu I, & Sagnak M, (2021). "Circular economy and the policy: A framework for improving the corporate environmental management in supply chains." Journal of Strategy and the Environment, 2021 – Wiley Online Library

Kearney, S. G., Watson, J. E., Reside, A. E., Fisher, D. O., Maron, M., Doherty, T. S., Legge, S. M., Woinarski, J. C., Garnett, S. T., Wintle, B. A., Ritchie, E. G., Driscoll, D. A., Lindenmayer, D., Adams, V. M., Ward, M. S., & Carwardine, J. (2023). Threat-abatement framework confirms habitat retention and invasive species management are critical to conserve Australia's threatened species. *Biological Conservation*, *277*, 109833. DOI: 10.1016/j.biocon.2022.109833

Keeley, J. E. (2006). Fire management impacts on invasive plants in the Western United States. *Conservation Biology*, *20*(2), 375–384. DOI: 10.1111/j.1523-1739.2006.00339.x PMID: 16903098

Keet, J. H., Ellis, A. G., Hui, C., Novoa, A., & Le Roux, J. J. (2021). Impacts of invasive Australian acacias on soil bacterial community composition, microbial enzymatic activities, and nutrient availability in fynbos soils. *Microbial Ecology*, *81*(2), 257–269. DOI: 10.1007/s00248-021-01683-1 PMID: 33515051

Kehoe, R., Sanders, D., Cruse, D., Silk, M., Gaston, K. J., Bridle, J. R., & van Veen, F. (2020). Longer photoperiods through range shifts and artificial light lead to a destabilizing increase in host–parasitoid interaction strength. *Journal of Animal Ecology*, *89*(11), 2508–2516. DOI: 10.1111/1365-2656.13328 PMID: 32858779

Kenis, M., Benelli, G., Biondi, A., Calatayud, P.-A., Day, R., Desneux, N., Harrison, R. D., Kriticos, D., Rwomushana, I., & Van den Berg, J. (2022). Invasiveness, biology, ecology, and management of the fall armyworm, Spodoptera frugiperda.

Kerns, B. K., Tortorelli, C., Day, M. A., Nietupski, T., Barros, A. M., Kim, J. B., & Krawchuk, M. A. (2020). Invasive grasses: A new perfect storm for forested ecosystems? *Forest Ecology and Management*, *463*, 117985. DOI: 10.1016/j.foreco.2020.117985

Khair, S. M., Mushtaq, S., Reardon-Smith, K., & Ostini, J. (2019). Diverse drivers of unsustainable groundwater extraction behaviour operate in an unregulated water scarce region. *Journal of Environmental Management*, *236*, 340–350. DOI: 10.1016/j.jenvman.2018.12.077 PMID: 30738305

Khalid, N., Naveed, S., & Khan, I. U. (2023). 47. Allelopathic effect of Tamarix aphylla on the emergence and growth parameters of Triticum aestivum and Zea mays. [PAB]. *Pure and Applied Biology*, *12*(1), 540–554. DOI: 10.19045/bspab.2023.120056

Khalloufi, M., Martínez-Andújar, C., Lachaâl, M., Karray-Bouraoui, N., Pérez-Alfocea, F., & Albacete, A. (2017). The interaction between foliar GA3 application and arbuscular mycorrhizal fungi inoculation improves growth in salinized tomato (Solanum lycopersicum L.) plants by modifying the hormonal balance. *Journal of Plant Physiology*, *214*, 134–144. DOI: 10.1016/j.jplph.2017.04.012 PMID: 28482334

Khan, I., Ahmad, B., Azam, S., Hassan, F., Aziz, A., Rehman, N., Ullah, F., & Liaqat, Z. (2018). Pharmacological activities of Justicia adhatoda. *Pakistan Journal of Pharmaceutical Sciences*, *31*(2). PMID: 29618423

Khan, I., Ibrahim, M., Khan, M. I., Hashim, S., Afzal, S., & Nawab, K. (2023). Management of Wild Onion (Asphodelus tenuifolius Cav.) in Chickpea Crop at District Karak. *Pakistan Journal of Weed Science Research*, *29*(3).

Khan, M. A., Gul, B., & Khan, H. (2015). Impact of climatic change on flora of high altitudes in Pakistan. In *Climate Change Impacts on High-Altitude Ecosystems* (pp. 361–381). Springer. DOI: 10.1007/978-3-319-12859-7_14

Khan, M. F., & Panikkar, P. (2009). Assessment of impacts of invasive fishes on the food web structure and ecosystem properties of a tropical reservoir in India. *Ecological Modelling*, *220*(18), 2281–2290. DOI: 10.1016/j.ecolmodel.2009.05.020

Khan, M., Khan, T., Tabassum, B., & Hashim, M. (2024). Microbiome-driven soil fertility: understanding symbiotic relationships. In *Progress in Soil Microbiome Research* (pp. 77–115). Springer. DOI: 10.1007/978-3-031-71487-0_4

Khan, N., & Fahad, S. (2020, January 12). Economic Review of Parthenium Hysterophorus L. Plant in the World. *Plant World*.

Khan, N., Fahad, S., Faisal, S., Akbar, A., & Naushad, M. (2020). Socio-economic and medicinal review of Eucalyptus tree in the world. *Available at SSRN* 3644215. DOI: 10.2139/ssrn.3644215

Khan, N., Ullah, R., Okla, M. K., Abdel-Maksoud, M. A., Saleh, I. A., Abu-Harirah, H. A., AlRamadneh, T. N., & AbdElgawad, H. (2023). Spatial distribution of the four invasive plants and their impact on natural communities' dynamics across the arid and semi-arid environments in northwest Pakistan. *Frontiers in Plant Science*, *14*, 1207222. DOI: 10.3389/fpls.2023.1207222 PMID: 37692447

Khatri, K., Bargali, K., Bargali, S. S., & Negi, B. (2023). Effects of leaf residues from *Ageratina adenophora* on germination, growth and productivity of two rabi crops. *Acta Ecologica Sinica*, *43*(2), 363–374. DOI: 10.1016/j.chnaes.2022.05.001

Khatsuriya, P. A., Makwana, V. M., & Dodia, P. P. (2024). Biodiversity Resilience Amidst Climate Shifts: A Global Perspective. In *Global Challenges for the Environment and Climate Change* (pp. 75-99). IGI Global. DOI: 10.4018/979-8-3693-2845-3.ch005

Khattak, W. A., Sun, J., Hameed, R., Zaman, F., Abbas, A., Khan, K. A., Elboughdiri, N., Akbar, R., He, F., Ullah, M. W., Al-Andal, A., & Du, D. (2024). Unveiling the resistance of native weed communities: Insights for managing invasive weed species in disturbed environments. *Biological Reviews of the Cambridge Philosophical Society*, *99*(3), 753–777. DOI: 10.1111/brv.13043 PMID: 38174626

Khorramnejad, A., Alfaro, C. C., Quaranta, S., Lozada-Chavez, A. N., Gasmi, L., Perdomo, H. D., Chiarelli, L. R., & Bonizzoni, M. (2024). Reproductive resource allocation correlates with successful global invasion of a mosquito species. bioRxiv, 2024.2007. 2018.604133.

Khuroo, A. A., Ahmad, R., Hamid, M., Rather, Z. A., Malik, A. H., & Rashid, I. (2021). An annotated inventory of invasive alien flora of India. *Invasive alien species: observations and issues from around the world, 2*, 16-37.

Kihara, J., Bolo, P., Kinyua, M., Nyawira, S. S., & Sommer, R. (2020). Soil health and ecosystem services: Lessons from sub-Sahara Africa (SSA). *Geoderma*, *361*, 114094. DOI: 10.1016/j.geoderma.2020.114342

Kim, G., Newman, G., & Jiang, B. (2020). Urban regeneration: Community engagement process for vacant land in declining cities. [– Elsevier]. *Cities (London, England)*, *102*, 2020. DOI: 10.1016/j.cities.2020.102730 PMID: 32831449

Knapp, L. S. P., Coyle, D. R., Dey, D. C., Fraser, J. S., Hutchinson, T., Jenkins, M. A., Kern, C. C., Knapp, B. O., Maddox, D., & Pinchot, C. (2023). Invasive plant management in eastern North American Forests: A systematic review. *Forest Ecology and Management*, *550*, 121517. DOI: 10.1016/j.foreco.2023.121517

Kohler, J., Roldán, A., Campoy, M., & Caravaca, F. (2017). Unraveling the role of hyphal networks from arbuscular mycorrhizal fungi in aggregate stabilization of semiarid soils with different textures and carbonate contents. *Plant and Soil*, *410*(1-2), 273–281. DOI: 10.1007/s11104-016-3001-3

Kotta, J. (2023). *Biological Invasions in a Changing World: Introduction to the Special Issue* (Vol. 15). MDPI.

Kourtev, P. S., Ehrenfeld, J. G., & Häggblom, M. (2002). Exotic plant species alter the microbial community structure and function in the soil. *Ecology*, *83*(11), 3152–3166. DOI: 10.1890/0012-9658(2002)083[3152:EPSATM]2.0.CO;2

Kowarik, I., & Lippe, M. v. (2007). Pathways in plant invasions. *Biological Invasions*, •••, 29–47.

Kudlu, A., Sarma, D., Das, D. K., Shamla, A. B., Bakthavatsalam, R., Hathwar, V. R., Mahata, A., & Kundu, J. (2025). Unravelling the structure–luminescence relationship in two dimensional antimony (iii)-doped cadmium (ii) halide hybrids. *Journal of Materials Chemistry C.*

Kumar P, Singh AN (2025) An Overview of Justicia adhatoda: A Medicinal Plant but Native Invader in India. Conservation (2673-7159) 5(1)

Kumar, A., Dutta, J., Chandrawanshi, N. K., Ekka, A., & Sethi, S. K. (2024). *Arbuscular mycorrhizal fungi in alleviation of biotic stress tolerance in plants: A new direction in sustainable agriculture. In entrepreneurship with microorganisms.* Academic Press., DOI: 10.1016/B978-0-443-19049-0.00006-2

Kumar, M., & Garkoti, S. C. (2021). Functional traits, growth patterns, and litter dynamics of invasive alien and co–occurring native shrub species of Chir pine forest in the central Himalaya, India. *Plant Ecology*, *222*(6), 723–735. DOI: 10.1007/s11258-021-01140-6

Kumar, M., & Garkoti, S. C. (2022). Allelopathy effects of invasive alien *Ageratina adenophora* on native shrub species of chir pine forest in the central Himalaya, India. *Journal of Forest Research*, *27*(1), 53–62. DOI: 10.1080/13416979.2021.2002505

Kumar, M., Prasad, R., Kumar, V., Tuteja, N., & Varma, A. (2017). Mycorrhizal fungi under biotic and abiotic stress. *Mycorrhiza-Eco-Physiology, Secondary Metabolites. Nanomaterials (Basel, Switzerland)*, •••, 57–69.

Kumar, M., Verma, A. K., & Garkoti, S. C. (2020). *Lantana camara* and *Ageratina adenophora* invasion alter the understory species composition and diversity of Chir Pine forest in central Himalaya, India. *Acta Oecologica*, *109*, 103642. DOI: 10.1016/j.actao.2020.103642

Kumar, N., & Aggarwal, N. K. (2024). A review on valorization, management, and applications of the hazardous weed Parthenium hysterophorus. *Systems Microbiology and Biomanufacturing*, *4*(2), 607–619. DOI: 10.1007/s43393-023-00226-8

Kumar, R., Guleria, N., Deeksha, M. G., Kumari, N., Kumar, R., Jha, A. K., Parmar, N., Ganguly, P., de Aguiar Andrade, E. H., Ferreira, O. O., de Oliveira, M. S., & Chandini, . (2024). From an Invasive Weed to an Insecticidal Agent: Exploring the Potential of Lantana camara in Insect Management Strategies—A Review. *International Journal of Molecular Sciences*, *25*(23), 12788. DOI: 10.3390/ijms252312788 PMID: 39684506

Kumar, R., Salaria, N., Kumari, I., & Hajam, Y. A. (2025). *Plant Tissue Culture Technology: An Innovative Tool for the Propagation, Management, and Conservation of Himalayan Biodiversity*. CRC Press. DOI: 10.1201/9781779643445

Kumar, S., & Mathur, M. (2014). Impact of invasion by *Prosopis juliflora* on plant communities in arid grazing lands. *Tropical Ecology*, *55*, 33–46.

Kushwaha, R. K., Rodrigues, V., Kumar, V., Patel, H., Raina, M., & Kumar, D. (2020). Soil microbes-medicinal plants interactions: ecological diversity and future prospect. In *Plant microbe symbiosis* (pp. 263–286). Springer. DOI: 10.1007/978-3-030-36248-5_14

Labadi, S., Giliberto, F., Rosetti, I., & Shetabi, L. (2021). Heritage and the sustainable development goals: Policy guidance for heritage and development actors. *International Journal of Heritage Studies*, **•••**, 2021.

Lafaie, L., Poenou, G., Hanon, O., López-Jiménez, L., Nieto, J. A., Lorenzo, A., Porras, J. A., Lumbierres, M., Bertoletti, L., & Monreal, M. (2024). Anticoagulation and venous thromboembolism in patients aged 90 years and older: Data from the RIETE registry. *Journal of the American Geriatrics Society*, *72*(1), 113–125. DOI: 10.1111/jgs.18626 PMID: 37814983

Lakoba, V. T., & Barney, J. N. (2020). Home climate and habitat drive ecotypic stress response differences in an invasive grass. *AoB Plants*, *12*(6), plaa062. Advance online publication. DOI: 10.1093/aobpla/plaa062 PMID: 33408848

Lal, M. K., Tiwari, R. K., Adavi, S. B., Kumar, A., Bamatov, I., Ivanova, E., Behera, L., Jena, R., & Kumar, R. (2024). Deciphering phenomics approaches for understanding plant–microbe interactions in nutrient absorption and utilization. *Plant Physiology Reports*, 1-17.

Lamb, A. M., Peplow, L. M., Dungan, A. M., Ferguson, S. N., Harrison, P. L., Humphrey, C. A., McCutchan, G. A., Nitschke, M. R., & van Oppen, M. J. (2024). Interspecific hybridisation provides a low-risk option for increasing genetic diversity of reef-building corals. *Biology Open*, *13*(9), bio060482. DOI: 10.1242/bio.060482 PMID: 39207257

Lamb, D. (2010). Ecological restoration. In *Regreening the bare hills: tropical forest restoration in the Asia-Pacific region* (pp. 325–358). Springer.

Lambertini, C. (2019). Why are tall-statured energy grasses of polyploid species complexes potentially invasive? A review of their genetic variation patterns and evolutionary plasticity. *Biological Invasions*, *21*(10), 3019–3041. DOI: 10.1007/s10530-019-02053-2

Lambert, M., & Donihue, C. (2020). Urban biodiversity management using evolutionary tools. *Nature Ecology & Evolution*, *4*(7), 903–910. DOI: 10.1038/s41559-020-1193-7 PMID: 32393868

Lancaster, J., Adams, R., Adams, B., & Desserud, P. (2015). Foothills Fescue Foothills Parkland and Montane Natural Subregions.

Lane, B. R. (2021). *Impact of an Emerging Bipolaris Pathogen on the Microbial Community of Invasive Microstegium vimineum*. University of Florida.

Langlois, N. R. (2024). Chinese Tallow Long-Term Impact on Stand Dynamics in a Bottomland Hardwood Forest Following Vegetation Management.

Lauman, S. (2024). *Hard Rock Mine Restoration in the Arid Southwest*. The University of Arizona.

Lavergne, S., & Molofsky, J. (2007). Increased genetic variation and evolutionary potential drive the success of an invasive grass. *Proceedings of the National Academy of Sciences of the United States of America*, *104*(10), 3883–3888. DOI: 10.1073/pnas.0607324104 PMID: 17360447

Lázaro-Lobo A, Alonso Á, Fernández R D, (2023). "Impacts of plant invasions on ecosystem functionality: A perspective for ecosystem health and ecosystem services." Plant Invasions and Ecosystem Health, 241-262.

Le Maitre, D. C. (2004). Predicting invasive species impacts on hydrological processes: The consequences of plant physiology for landscape processes. *Weed Technology*, *18*(4), 1408–1410. DOI: 10.1614/0890-037X(2004)018[1408:PISIOH]2.0.CO;2

Le Maitre, D. C., Blignaut, J. N., Clulow, A., Dzikiti, S., Everson, C. S., Görgens, A. H., & Gush, M. B. (2020). Impacts of plant invasions on terrestrial water flows in South Africa. In *Biological Invasions in South Africa* (pp. 431–457). Springer International Publishing. DOI: 10.1007/978-3-030-32394-3_15

Le Maitre, D. C., Gaertner, M., Marchante, E., Ens, E. J., Holmes, P. M., Pauchard, A., O'Farrell, P. J., Rogers, A. M., Blanchard, R., Blignaut, J., & Richardson, D. M. (2011). Impacts of invasive Australian acacias: Implications for management and restoration. *Diversity & Distributions*, *17*(6), 1015–1029. DOI: 10.1111/j.1472-4642.2011.00816.x

Lee, S. W., Lee, W. C., Lee, S. H., & Kim, S. O. (2020). Changes of soil properties through the remediation processes and techniques for the restoration of remediated soils. *Economic and Environmental Geology*, *53*(1), 1–12. DOI: 10.9719/EEG.2024.57.1.1

Lefebvre, S., Segar, J., & Staude, I. R. (2024). Non-natives are linked to higher plant diversity across spatial scales. *Journal of Biogeography*, *51*(7), 1290–1298. Advance online publication. DOI: 10.1111/jbi.14824

Leger, A. M., Ball, K. R., Rathke, S. J., & Blankinship, J. C. (2022). Mulch more so than compost improves soil health to reestablish vegetation in a semiarid rangeland. *Restoration Ecology*, *30*(2), e13546. DOI: 10.1111/rec.13698

Lemma, F., & Hailu, H. (2022). *IMPACTS OF Lantana camaraON ELEPHANT HABITAT USE AND FEED AVAILABILITY IN BABILE ELEPHANT SANCTUARY*. EASTERN ETHIOPIA Haramaya University.

Lenz, T. I., & Facelli, J. M. (2003). Shade facilitates an invasive stem succulent in a chenopod shrubland in South Australia. *Austral Ecology*, *28*(5), 480–490. DOI: 10.1046/j.1442-9993.2003.01304.x

Leventis, G., Tsiknia, M., Feka, M., Ladikou, E. V., Papadakis, I. E., Chatzipavlidis, I., & Ehaliotis, C. (2021). Arbuscular mycorrhizal fungi enhance growth of tomato under normal and drought conditions, via different water regulation mechanisms. *Rhizosphere*, *19*, 100394. DOI: 10.1016/j.rhisph.2021.100394

Lewis, R. R.III. (2005). Ecological engineering for successful management and restoration of mangrove forests. *Ecological Engineering*, *24*(4), 403–418. DOI: 10.1016/j.ecoleng.2004.10.003

Lhoumeau, S., Elias, R. B., Seidel, D., Gabriel, R., & Borges, P. A. (2025). Landscape to microhabitat: Uncovering the multiscale complexity of native and exotic forests on Terceira Island (Azores, Portugal). *PLoS One*, *20*(6), e0326304. DOI: 10.1371/journal.pone.0326304 PMID: 40522958

Li S, Wang P, Yuan W, Zushang S, Bullard SH (2018) Compositions and methods to selectively control invasive species. Google Patents, Lira Junior MA, Fracetto GGM, Araújo ASF, Fracetto FJC, Nóbrega RSA, Silva K, Galdino AC (2017) Rhizobial diversity for tropical pulses and forage and tree legumes in Brazil. Microbes for Legume Improvement: 135-151

Liebman, M., & Gallandt, E. R. (1997). Many little hammers: Ecological management of crop-weed interactions. *Ecological Applications*, *7*(1), 43–56. DOI: 10.1890/1051-0761(1997)007[0043:MLHEMO]2.0.CO;2

Lieurance, D., Canavan, S., Behringer, D. C., Kendig, A. E., Minteer, C. R., Reisinger, L. S., Romagosa, C. M., Flory, S. L., Lockwood, J. L., Anderson, P. J., Baker, S. M., Bojko, J., Bowers, K. E., Canavan, K., Carruthers, K., Daniel, W. M., Gordon, D. R., Hill, J. E., Howeth, J. G., & Wanamaker, C. (2023). Identifying invasive species threats, pathways, and impacts to improve biosecurity. *Ecosphere*, *14*(12), e4711. DOI: 10.1002/ecs2.4711

Li, J., He, J. Z., Liu, M., Yan, Z. Q., Xu, X. L., & Kuzyakov, Y. (2024). Invasive plant competitivity is mediated by nitrogen use strategies and rhizosphere microbiome. *Soil Biology & Biochemistry*, *192*, 109361. DOI: 10.1016/j.soilbio.2024.109361

Lin X, He J, Chen Z, Lyu Z, Dai B, Yu F, (2023). "Diffbir: Towards blind image restoration with generative diffusion prior." arXiv preprint arXiv.

Lindenmayer, D. B., & Likens, G. E. (2009). Adaptive monitoring: A new paradigm for long-term research and monitoring. *Trends in Ecology & Evolution*, *24*(9), 482–486. DOI: 10.1016/j.tree.2009.03.005 PMID: 19409648

Linder, H. P., Lehmann, C. E., Archibald, S., Osborne, C. P., & Richardson, D. M. (2018). Global grass (P oaceae) success underpinned by traits facilitating colonization, persistence and habitat transformation. *Biological Reviews of the Cambridge Philosophical Society*, *93*(2), 1125–1144. DOI: 10.1111/brv.12388 PMID: 29230921

Lin, J., Zhang, B., Zou, J., Luo, Z., Yang, H., Zhou, P., Chen, X., & Zhou, W. (2023). Induction of tetraploids in Paper Mulberry (Broussonetia papyrifera (L.) L'Hér. ex Vent.) by colchicine. *BMC Plant Biology*, *23*(1), 574. DOI: 10.1186/s12870-023-04487-2 PMID: 37978431

Lin, M., Chen, Y., Cheng, L., Zheng, Y., Wang, W., Sardans, J., Song, Z., Guggenberger, G., Zou, Y., Ding, X., Tariq, A., Zeng, F., Fahad Alrefaei, A., & Peñuelas, J. (2023). Response of topsoil Fe-bound organic carbon pool and microbial community to Spartina alterniflora invasion in coastal wetlands. *Catena, 220*, 106695. DOI: 10.1016/j.catena.2023.107414

Lin, W., Zhou, G., Cheng, X., & Xu, R. (2007). Fast economic development accelerates biological invasions in China. *PLoS One*, *2*(11), e1208. DOI: 10.1371/journal.pone.0001208 PMID: 18030342

Li, S., Jiang, Y., Xu, J., Wang, D., Ding, Z., Zhu, T., Chen, B., Yang, Y., Wei, M., Guo, R., Hou, Y., Chen, Y., Sun, C., Wei, K., Qaid, S. M. H., Lu, H., Tan, H., Di, D., Chen, J., & Yuan, M. (2024). High-efficiency and thermally stable FACsPbI3 perovskite photovoltaics. *Nature*, *635*(8037), 1–3. DOI: 10.1038/s41586-024-08103-7 PMID: 39348872

Liu, J., Chen, J., Sun, Y., Tong, B., Guan, R., & Wu, M. (2019). Allelopathic effects of aqueous extract of *Ageratina adenophora* on seven native plant seedlings in growth and chlorophyll. *Guangxi Zhiwu/Guihaia, 39*, 79–86.

Liu, D., Zheng, K., Wang, Y., Zhang, Y., Lao, R., Qin, Z., Li, T., & Zhao, Z. (2022). Harnessing an arbuscular mycorrhizal fungus to improve the adaptability of a facultative metallophytic poplar (Populus yunnanensis) to cadmium stress: Physiological and molecular responses. *Journal of Hazardous Materials*, *424*, 127430. DOI: 10.1016/j.jhazmat.2021.127430 PMID: 34678563

Liu, J., Duan, X., Li, G., Cai, Z., Wei, S., Song, Q., & Zheng, Z. (2023). Changes in Bacterial Communities and Their Effects on Soil Carbon Storage in Spartina alterniflora Invasion Areas, Coastal Wetland Bare Flats, and Sueada salsa .... *International Journal of Environmental Research and Public Health*, *20*(11), 5769. DOI: 10.3390/ijerph20054308 PMID: 36901319

Liu, M., Chen, X., Zhang, Y., Zhang, X., & Chen, Y. (2025). Nonlinear relationship between diversity of rare and common species and ecosystem multifunctionality in alpine meadows along an altitude gradient. *Journal of Environmental Management*, *376*, 124374. DOI: 10.1016/j.jenvman.2025.124374 PMID: 39914212

Liu, Y., Xu, F., Ding, L., Zhang, G., Bai, B., Han, Y., Xiao, L., Song, Y., Li, Y., Wan, S., & Li, G. (2023). Microplastics reduce nitrogen uptake in peanut plants by damaging root cells and impairing soil nitrogen cycling. *Journal of Hazardous Materials*, *443*, 130173. DOI: 10.1016/j.jhazmat.2022.130384 PMID: 36444071

Li, X., Zhang, F., Hui, E. C., & Lang, W. (2020). Collaborative workshop and community participation: A new approach to urban regeneration in China. [– Elsevier]. *Cities (London, England)*, *102*, 2020. DOI: 10.1016/j.cities.2020.102743

Li, Y., & Shen, Z. (2020). Roles of dispersal limit and environmental filtering in shaping the spatiotemporal patterns of invasive alien plant diversity in China. *Frontiers in Ecology and Evolution*, *8*, 544670. DOI: 10.3389/fevo.2020.544670

Li, Z., Shen, Q., Fan, W., Dong, S., Wang, Z., Xu, Y., Ma, T., & Cao, Y. (2024). … Solutions vs. Human-Induced Approaches for Alpine Grassland Ecosystem:"Climate-Help" Overwhelms "Human Act" to Promote Ecological Restoration in the Three …. *Remote Sensing (Basel)*, *16*(7), 2024. DOI: 10.3390/rs16071156

Lockwood, J. L., Hoopes, M. F., & Marchetti, M. P. (2013). *Invasion ecology*. John Wiley & Sons.

Lofgren, L., Nguyen, N. H., Kennedy, P. G., Pérez-Pazos, E., Fletcher, J., Liao, H. L., Wang, H., Zhang, K., Ruytinx, J., Smith, A. H., Ke, Y.-H., Cotter, H. V. T., Engwall, E., Hameed, K. M., Vilgalys, R., & Branco, S. (2024). Suillus: An emerging model for the study of ectomycorrhizal ecology and evolution. *The New Phytologist*, *242*(4), 1448–1475. DOI: 10.1111/nph.19700 PMID: 38581203

Lokare, P. D., & Pandya, J. B. (2024). *Entrepreneurial Botany*. Book Saga Publications. DOI: 10.60148/978-8197603839

Lombardini, L., & Rossi, L. (2019). Ecophysiology of plants in dry environments. In *Dryland ecohydrology* (pp. 71–100). Springer. DOI: 10.1007/978-3-030-23269-6_4

Lone, S., Yadav, A., Sharma, A., Tafazul, M., Badkhane, Y., & Raghuwanshi, D. (2013). A review on Adhatoda vasica Nees-An important and high demanded medicinal plant. *Indo American Journal of Pharm Research*, *3*, 2600.

Longley, P. A., Goodchild, M. F., Maguire, D. J., & Rhind, D. W. (2015). *Geographic Information Systems and Science*. Wiley.

Louda, S. M., Kendall, D., Connor, J., & Simberloff, D. (2003). Ecological effects of an insect introduced for the biological control of weeds. *Science*, *277*(5329), 1088–1090. DOI: 10.1126/science.277.5329.1088

Lowry, C. J., Blocklove, K. M., Eckert, K. E., & Koirala, N. (2024). If you can't beat them, join them: Using invasive plants to understand rhizodeposits as drivers of plant-soil feedbacks. *Frontiers in Environmental Science*, •••, 12.

Lucero, J. E., Faist, A. M., Lortie, C. J., & Callaway, R. M. (2022). Risk of facilitated invasion depends upon invader identity, not environmental severity, along an aridity gradient. *Frontiers in Ecology and Evolution*, *10*, 886690. DOI: 10.3389/fevo.2022.886690

Lugo, A. E. (2020). Effects of extreme disturbance events: From ecesis to social–ecological–technological systems. *Ecosystems (New York, N.Y.)*, *23*(8), 1726–1747. DOI: 10.1007/s10021-020-00491-x

Lu, Y., Kronzucker, H. J., Yu, M., Shabala, S., & Shi, W. (2024). Nitrogen-loss and carbon-footprint reduction by plant-rhizosphere exudates. *Trends in Plant Science*, *29*(1), 53–65. DOI: 10.1016/j.tplants.2023.09.007 PMID: 37802692

MacDonald, G. E. (2004). Cogongrass (Imperata cylindrica)—Biology, ecology, and management. *Critical Reviews in Plant Sciences*, *23*(5), 367–380. DOI: 10.1080/07352680490505114

Macedo, M. I., Ferreira, F. A., Dabić, M., & Ferreira, N. C. (2024). Structuring and analyzing initiatives that facilitate organizational transformation processes: A sociotechnical approach. *Technological Forecasting and Social Change*, *209*, 123739. DOI: 10.1016/j.techfore.2024.123739

Magableh, G. M. (2021). "Supply chains and the COVID-19 pandemic: A comprehensive framework." European Management Review, 2021 – Wiley Online Library Miller-Rushing A J, Athearn N, Blackford T, (2021). "COVID-19 pandemic impacts on conservation research, management, and public engagement in US national parks. [– Elsevier]. *Biological Conservation*, •••, 2021.

Magkourilou, E., Bell, C., Daniell, T., & Field, K. (2024). The functionality of arbuscular mycorrhizal networks across scales of experimental complexity and ecological relevance. *Functional Ecology*. Advance online publication. DOI: 10.1111/1365-2435.14618

Magliozzi, C., Lucrezi, S., Oficialdegui, F., Gervasini, E., Melone, B., & Cardoso, A. (2024). Transnational cooperation in EU: Opportunities for addressing invasive alien species. *Management of Biological Invasions : International Journal of Applied Research on Biological Invasions*, *15*(4), 683–700. DOI: 10.3391/mbi.2024.15.4.13

Magni, S., Sciusco, G., Caturegli, L., Sportelli, M., Federighi, T., Fontanelli, M., Minelli, A., McElroy, J. S., & Volterrani, M. (2023). A New Method for Hybrid Bermuda Grass (Cynodon dactylon× C. transvaalensis Burtt.-Davy) Vegetative Propagation. *Grasses*, *3*(1), 1–10. DOI: 10.3390/grasses3010001

Mahla, N., & Mlambo, D. (2019). Influence of two co-occurring invasive plant species on resident woody species and surface soil properties in Chipinge Safari Area, Zimbabwe. *Tropical Ecology, 60*(1), 129–139. DOI: 10.1007/s42965-019-00016-1

Mahmood, Q., Bilal, M., & Jan, S. (2014). Herbicides, pesticides, and plant tolerance: an overview. *Emerging technologies and management of crop stress tolerance*, 423-448.

Mahood, A. L. (2021). *Evidence for and Mechanisms of Ecosystem Transformation in the Great Basin of the Western United States*. University of Colorado at Boulder.

Mai, R. (2024). Mechanisms of adaptation in aquatic species: from phenotypic plasticity to genetic evolution. *International Journal of Aquaculture, 14*.

Mairaing, W., Jotisankasa, A., Leksungnoen, N., Hossain, M., Ngernsaengsaruay, C., Rangsiwanichpong, P., Pilumwong, J., Pramusandi, S., Semmad, S., & Ahmmed, A. N. F. (2024). A Biomechanical Study of Potential Plants for Erosion Control and Slope Stabilization of Highland in Thailand. *Sustainability (Basel), 16*(4), 2445. DOI: 10.3390/su16156374

Majeed, M., Bhatti, K. H., Amjad, M. S., Abbasi, A. M., Rashid, A., Nawaz, F., Mehmood, A., Mahmood, M., & Ahmad, K. S. (2020). *Ethno-veterinary practices of Poaceae taxa in Punjab*. DOI: 10.21203/rs.3.rs-15532/v1

Makwana, V. M., Patel, F. P., & Dodia, P. P. (2024a). Urbanization's Ecological Toll: Exploring Its Role in Biodiversity Decline. In *Impact of Societal Development and Infrastructure on Biodiversity Decline* (pp. 245-270). IGI Global. DOI: 10.4018/979-8-3693-6950-0.ch014

Makwana, V. M., Thakkar, K., & Dodia, P. P. (2024b). Exploring the Synergy of Soil-Animal Interactions: A Blueprint for Biodiversity Preservation. In *Ecological Aspects of Soil and Land Preservation* (pp. 30-50). IGI Global. DOI: 10.4018/979-8-3693-3374-7.ch002

Malavasi, M., Carboni, M., Cutini, M., Carranza, M. L., & Acosta, A. T. R. (2014). Landscape fragmentation, land-use legacy and propagule pressure promote plant invasion on coastal dunes: A patch-based approach. *Landscape Ecology, 29*(9), 1541–1550. DOI: 10.1007/s10980-014-0074-3

Mamolos, A. P., Nikolaidou, A. E., Pavlatou-Ve, A. K., Kostopoulou, S. K., & Kalburtji, K. L. (2011). Ecological threats and agricultural opportunities of the aquatic cane-like grass Phragmites australis in wetlands. *Genetics, Biofuels and Local Farming Systems*, 251-275.

Manfreda, S., McCabe, M. F., Miller, P. E., Lucas, R., Pajuelo Madrigal, V., Mallinis, G., Ben Dor, E., Helman, D., Estes, L., Ciraolo, G., Müllerová, J., Tauro, F., De Lima, M. I., De Lima, J. L. M. P., Maltese, A., Frances, F., Caylor, K., Kohv, M., Perks, M., & Toth, B. (2018). On the use of unmanned aerial systems for environmental monitoring. *Remote Sensing (Basel)*, *10*(4), 641. DOI: 10.3390/rs10040641

Mantoani, M. C., & Torezan, J. M. D. (2023). Early responses of regeneration to weeding and herbicide application to control Guinea-grass in a 20-year-old Atlantic Forest undergoing restoration. *Restoration Ecology*, *31*(4), e13792. DOI: 10.1111/rec.13792

Manu, M., & Li, J. (2025). Urban pavement crevices as microhabitats: Soil properties and spontaneous vegetation in roadside environments. *International Journal of Sustainable Development and World Ecology*, •••, 1–13. DOI: 10.1080/13504509.2025.2531435

Manzo, E., & Tomasello, S. (2025). Recent Long-Distance-Dispersal Explains the Range Disjunction of the Old-Word Cockleburs (Xanthium strumarium). *Journal of Biogeography*, *52*(5), e15104. DOI: 10.1111/jbi.15104

Manzoor, R., Zafar, M., Yaqoob, T., Ahmad, M., Ramadan, M. F., Althobaiti, A. T., Demirpolat, A., Çobanoğlu, D. N., Sultana, S., Makhkamov, T., Mamarakhimov, O., Yuldashev, A., Khakimova, D., Nizomova, M., Ochilov, U., & Majeed, S. (2023). Micromorphological sculptural diversity in foliar epidermis and trichomes features among invasive species. *Microscopy and Microanalysis*, *29*(4), 1531–1555. DOI: 10.1093/micmic/ozad063 PMID: 37488823

Mapaura, A. (2022). *Determining the trajectory of graminoid invasions in Southern Africa's mountains: The case of Nassella*. University of the Free State.

Mapaura, A., Canavan, K., Richardson, D. M., Clark, V. R., & Steenhuisen, S.-L. (2020). The invasive grass genus Nassella in South Africa: A synthesis. *South African Journal of Botany*, *135*, 336–348. DOI: 10.1016/j.sajb.2020.08.031

Mapaura, A., Canavan, K., Richardson, D. M., Clark, V. R., Sutton, G. F., & Steenhuisen, S.-L. (2024). The impact of Nassella trichotoma (Nees) Hack. ex Arechav. on plant diversity, richness and soil properties in South Africa. *South African Journal of Botany*, *173*, 175–183. DOI: 10.1016/j.sajb.2024.08.010

Margalef, R. (1958). Information theory in ecology. *General Systems*, *3*, 36–71.

Maron, J. L., Vilà, M., Bommarco, R., Elmendorf, S., & Beardsley, P. (2004). Rapid evolution of an invasive plant. *Ecological Monographs*, *74*(2), 261–280. DOI: 10.1890/03-4027

Martin, D. M., Mazzotta, M., & Bousquin, J. (2018). Combining ecosystem services assessment with structured decision making to support ecological restoration planning. *Environmental Management*, *62*(3), 608–618. DOI: 10.1007/s00267-018-1038-1 PMID: 29637276

Massigoge, A., Rodríguez, J. M., Álvarez, M. C., Gutiérrez, M. A., & Kaufmann, C. A. (2024). Use of the plains vizcacha (lagostomus maximus) by indigenous societies in the Pampas ecoregion (Argentina). *Plains Vizcachas: Biology and Evolution of a Peculiar Neotropical Caviomorph Rodent*, 99-144.

Mathieu, V., Superchi, S., Masi, M., Scafato, P., Kornienko, A., & Evidente, A. (2022). In vitro effects of fungal phytotoxins on cancer cell viability: First insight into structure activity relationship of a potent metabolite of Cochliobolus australiensis radicinin. *Toxins*, *14*(8), 517. DOI: 10.3390/toxins14080517 PMID: 36006179

Mathur, S., Agnihotri, R., Sharma, M. P., Reddy, V. R., & Jajoo, A. (2021). Effect of high-temperature stress on plant physiological traits and mycorrhizal symbiosis in maize plants. *Journal of Fungi (Basel, Switzerland)*, *7*(10), 867. DOI: 10.3390/jof7100867 PMID: 34682289

Mathur, S., & Jajoo, A. (2020). Arbuscular mycorrhizal fungi protects maize plants from high temperature stress by regulating photosystem II heterogeneity. *Industrial Crops and Products*, *143*, 111934. DOI: 10.1016/j.indcrop.2019.111934

Mathur, S., Sharma, M. P., & Jajoo, A. (2018). Improved photosynthetic efficacy of maize (Zea mays) plants with arbuscular mycorrhizal fungi (AMF) under high temperature stress. *Journal of Photochemistry and Photobiology. B, Biology*, *180*, 149–154. DOI: 10.1016/j.jphotobiol.2018.02.002 PMID: 29425887

Matloob, H., & Hameed, M. (2024). Osmoregulation facilitated by leaf structural and functional traits in Malabar nut (Justicia Adhatoda L.) along elevation gradient. *Pakistan Journal of Botany*, *56*(4), 1419–1429. DOI: 10.30848/PJB2024-4(2)

Maust, A. (2025). Quantifying the Effects of Wildfire Reintroduction on Native Pollinators in Ponderosa Pine Forests of the Pacific Northwest University of Washington].

McCary, M. A., Mores, R., Farfan, M. A., & Wise, D. H. (2016). Invasive plants have different effects on trophic structure of green and brown food webs in terrestrial ecosystems: A meta-analysis. *Ecology Letters*, *19*(3), 328–335. DOI: 10.1111/ele.12562 PMID: 26757702

McKinney, J., & Cleland, E. E. (2014). Root inputs influence soil water holding capacity and differentially influence the growth of native versus exotic annual species in an arid ecosystem. *Restoration Ecology*, *22*(6), 766–773. DOI: 10.1111/rec.12139

McKone, M. J., & Hernández, D. L. (2021). Community-level assisted migration for climate-appropriate prairie restoration. *Restoration Ecology, 29*(7), e13416. DOI: 10.1111/rec.13416

Mealor, B. A., & Hild, A. L. (2006). Potential selection in native grass populations by exotic invasion. *Molecular Ecology, 15*(8), 2291–2300. DOI: 10.1111/j.1365-294X.2006.02931.x PMID: 16780441

Mee, M. T., & Wang, H. H. (2012). Engineering ecosystems and synthetic ecologies. *Molecular BioSystems, 8*(10), 2470–2483. DOI: 10.1039/c2mb25133g PMID: 22722235

Meena, B., & Sharma, K. (2024). Biological Control of Weeds. *International Journal of Environment, Agriculture and Biotechnology, 9*(4), 091–100. DOI: 10.22161/ijeab.94.13

Meinard, Y., Dereniowska, M., Philippot, V., Glatron, S., Maris, V., & Georges, J.-Y. (2022). A heuristic for innovative invasive species management actions and strategies. *Ecology and Society, 27*(4), art24. DOI: 10.5751/ES-13615-270424

Mekonnen, Z. A., Riley, W. J., Berner, L. T., Bouskill, N. J., Torn, M. S., Iwahana, G., Breen, A. L., Myers-Smith, I. H., Criado, M. G., Liu, Y., Euskirchen, E. S., Goetz, S. J., Mack, M. C., & Grant, R. F. (2021). Arctic tundra shrubification: A review of mechanisms and impacts on ecosystem carbon balance. *Environmental Research Letters, 16*(4), 1–13. DOI: 10.1088/1748-9326/abf28b PMID: 33897808

Mendes, S. B., Olesen, J. M., Nogales, M., Marrero, P., Romero, J., Nieves, C., & Heleno, R. (2025). Urbanization of seed dispersal networks. *Conservation Biology*, •••, 70097. DOI: 10.1111/cobi.70097 PMID: 40605764

Merloti, L. F. (2024). *Unraveling the soil microbiome in Brachiaria pastures: exploring varietal influences and nitrogen fertilizers*. Universidade de São Paulo.

Meyer, S. E., Callaham, M. A., Stewart, J. E., & Warren, S. D. (2021). Invasive species response to natural and anthropogenic disturbance. *Invasive species in forests and rangelands of the United States: A comprehensive science synthesis for the United States forest sector*, 85-110.

Meyerson L A, Pauchard A, Brundu G, Carlton J T, (2022). "Moving toward global strategies for managing invasive alien species." Global Plant Invasions, 1-14.

Meyerson, L. A., Pauchard, A., Brundu, G., Carlton, J. T., Hierro, J. L., Kueffer, C., & Packer, J. G. (2022). Moving toward global strategies for managing invasive alien species. In *Global plant invasions* (pp. 331–360). Springer International Publishing. DOI: 10.1007/978-3-030-89684-3_16

Micael, J., & Gíslason, S. (2024). Tides of change: Monitoring NIS in Icelandic harbours. *Journal of Coastal Conservation, 28*(5), 72. DOI: 10.1007/s11852-024-01072-z

Michelan, T. S., Thomaz, S. M., Bando, F. M., & Bini, L. M. (2018). Competitive effects hinder the recolonization of native species in environments densely occupied by one invasive exotic species. *Frontiers in Plant Science, 9*, 126. DOI: 10.3389/fpls.2018.01261 PMID: 30233613

Milanović, M., Knapp, S., Pyšek, P., & Kühn, I. (2020). Linking traits of invasive plants with ecosystem services and disservices. *Ecosystem Services, 45*, 101194. DOI: 10.1016/j.ecoser.2020.101072

Milbrath, L. R., Davis, A. S., & Biazzo, J. (2017). Demography of invasive black and pale swallow-wort populations in New York. *Northeastern Naturalist, 24*(1), 37–53. DOI: 10.1656/045.024.0104

Milián-García, Y., Pyne, C., Chen, A., Lindsay, K., & Hanner, R. H. (2025). Uncovering the hidden within shipping containers: Molecular biosurveillance confirms a pathway for introducing multiple regulated and invasive species. *Biological Invasions, 27*(3), 91. DOI: 10.1007/s10530-025-03549-w

Miller, J. H., Manning, S. T., & Enloe, S. F. (2010). A management guide for invasive plants in southern forests. USDA Forest Service. Retrieved from https://www.srs.fs.usda.gov/pubs/gtr/gtr_srs131.pdf

Miller, B. P., Sinclair, E. A., Menz, M. H., Elliott, C. P., Bunn, E., Commander, L. E., Dalziell, E., David, E., Davis, B., Erickson, T. E., Golos, P. J., Krauss, S. L., Lewandrowski, W., Mayence, C. E., Merino-Martín, L., Merritt, D. J., Nevill, P. G., Phillips, R. D., Ritchie, A. L., & Stevens, J. C. (2017). A framework for the practical science necessary to restore sustainable, resilient, and biodiverse ecosystems. *Restoration Ecology, 25*(4), 605–617. DOI: 10.1111/rec.12475

Miniat, C. F., Fraterrigo, J. M., Brantley, S. T., Callaham, M. A., Cordell, S., Dukes, J. S., Giardina, C. P., Jose, S., & Lovett, G. (2021). Impacts of invasive species on forest and grassland ecosystem processes in the United States. *Invasive species in forests and rangelands of the United States: A comprehensive science synthesis for the United States forest sector*, 41-55.

Mircea, D. M., Calone, R., Estrelles, E., Soriano, P., Sestras, R. E., Boscaiu, M., Sestras, A. F., & Vicente, O. (2023). Responses of different invasive and non-invasive ornamental plants to water stress during seed germination and vegetative growth. *Scientific Reports, 13*(1), 13281. DOI: 10.1038/s41598-023-40517-7 PMID: 37587264

Mishra, G., Srivastava, T., Kant, R., Meena, R. K., Pandey, S., & Bhandari, M. S. (2024). Adaptive Modulations in Morphological and Physiological Responses of Forest Ecosystems to Climate Change. In *Forests and Climate Change: Biological Perspectives on Impact, Adaptation, and Mitigation Strategies* (pp. 43-71). Springer. DOI: 10.1007/978-981-97-3905-9_3

Mishra, A., Rajak, P., Raghubanshi, A. S., & Singh, H. (2025). Understanding competition between two invasive woody plants of India under an altered rainfall regime. *Plant Ecology*, *226*(1), 27–45. DOI: 10.1007/s11258-024-01470-1

Misra, R. (1968). *Ecology workbook*. Oxford and IBH Publishing Company.

Mitsch, W. J., & Jørgensen, S. E. (2004). *Ecological engineering and ecosystem restoration*. Wiley.

Mi, X., Feng, G., Hu, Y., Zhang, J., Chen, L., Corlett, R. T., Hughes, A. C., Pimm, S., Schmid, B., Shi, S., Svenning, J.-C., & Ma, K. (2021). The global significance of biodiversity science in China: An overview. *National Science Review*, *8*, nwab032. DOI: 10.1093/nsr/nwab032 PMID: 34694304

Mohammeda, A., & Bahmudb, S. (2023). The Status, Ecological and Biodiversity Impacts of Invasive Plant Species in Ethiopia: A review Paper. *GSJ, 11*(11).

Mohammed, A. E., Alotaibi, M. O., & Elobeid, M. (2022). Interactive influence of elevated CO2 and arbuscular mycorrhizal fungi on sucrose and coumarin metabolism in Ammi majus. *Plant Physiology and Biochemistry*, *185*, 45–54. DOI: 10.1016/j.plaphy.2022.05.029 PMID: 35660776

Molvar, E. M., Rosentreter, R., Mansfield, D., & Anderson, G. M. (2024). Cheatgrass invasions: History, causes, consequences, and solutions.

Montagnani, C., Gentili, R., Brundu, G., Caronni, S., & Citterio, S. (2022). Accidental introduction and spread of top invasive alien plants in the European Union through human-mediated agricultural pathways: What should we expect? *Agronomy (Basel)*, *12*(2), 423. DOI: 10.3390/agronomy12020423

Mooney, H. A., & Cleland, E. E. (2001). The evolutionary impact of invasive species. *Proceedings of the National Academy of Sciences of the United States of America*, *98*(10), 5446–5451. DOI: 10.1073/pnas.091093398 PMID: 11344292

Moore, E., D'Amico, V., & Trammell, T. L. E. (2023). Plant community dynamics following non-native shrub removal depend on invasion intensity and forest site characteristics. *Ecosphere*, *14*(1), e02323. DOI: 10.1002/ecs2.4351

Morales-Rodríguez, C., Vannini, A., Scanu, B., González-Moreno, P., Turco, S., Drais, M. I., Brandano, A., Varo Martínez, M. Á., Mazzaglia, A., Deidda, A., Cidre, A., Gomes-Laranjo, J., Liberati, D., & Ruiz-Gómez, F. J. (2025). Challenges to Mediterranean Fagaceae ecosystems affected by Phytophthora cinnamomi and climate change: Integrated pest management perspectives. *Current Forestry Reports*, *11*(1), 9. DOI: 10.1007/s40725-024-00237-1

Moreira, F. G. L., Carvalho, F. A., & de Paula, L. (2024). Exotic plant species on inselbergs of Brazilian tropical forests: insights for biodiversity management and conservation. *ARPHA* Preprints, *5*, e143971.

Moreno-Mateos, D., Barbier, E. B., Jones, P. C., Jones, H. P., Aronson, J., López-López, J. A., McCrackin, M. L., Meli, P., Montoya, D., & Rey Benayas, J. M. (2017). Anthropogenic ecosystem disturbance and the recovery debt. *Nature Communications*, *8*(1), 14163. DOI: 10.1038/ncomms14163 PMID: 28106039

Morente-López, J., Arjona, Y., Salas-Pascual, M., Reyes-Betancort, J. A., del Arco-Aguilar, M. J., Emerson, B. C., García-Gallo, A., Jay-García, L. S., Naranjo-Cigala, A., & Patiño, J. (2023). Biogeographic origins and drivers of alien plant invasions in the Canary Islands. *Journal of Biogeography*, *50*(3), 576–590. DOI: 10.1111/jbi.14556

Morin, L., Reid, A. M., Sims-Chilton, N. M., Buckley, Y. M., Dhileepan, K., Hastwell, G. T., Nordblom, T. L., & Raghu, S. (2009). Review of approaches to evaluate the effectiveness of weed biological control agents. *Biological Control*, *51*(1), 1–15. DOI: 10.1016/j.biocontrol.2009.05.017

Morrison, C. R., Plowes, R. M., Ng'iru, I., Rhodes, A. C., Martins, D. J., & Gilbert, L. E. (2023). Arthropod associates of Kenyan buffelgrass (Cenchrus ciliaris): A field survey for biological control candidates of a globally important invasive grass. *African Entomology*, *31*, 1–10. DOI: 10.17159/2254-8854/2023/a16178

Mudereri, B. T., Abdel-Rahman, E. M., Dube, T., Landmann, T., Khan, Z., Kimathi, E., Owino, R., & Niassy, S. (2020). Multi-source spatial data-based invasion risk modeling of Striga (Striga asiatica) in Zimbabwe. *GIScience & Remote Sensing*, *57*(1), 1–18. DOI: 10.1080/15481603.2020.1744250

Mudie, R. (2024). *Gleanings of Nature. Containing Fifty-seven Groups of Animals and Plants, with Popular Descriptions of Their Habits*. BoD–Books on Demand.

Mueller–Dombois, D., & Ellenberg, H. (1974). *Aims and methods of vegetation ecology*. Wiley.

Müller-Kiefer, J., & Tomasello, S. (2022). Invasive or threatened? The decline of the old-world Xanthium strumarium (Asteraceae) in Italy based on herbarium records. *Flora Mediterranea*, *32*, 17–24.

Müller, L. M., & Bahn, M. (2022). Drought legacies and ecosystem responses to subsequent drought. *Global Change Biology*, *28*(1), 211–223. DOI: 10.1111/gcb.16270 PMID: 35607942

Mungi, N. A., Galetti, M., Ratnam, J., Chala, D., Sanín, M. J., & Svenning, J.-C. (2025). Alien plants and novel ecosystems in the Greater Tropics. Nature Reviews Biodiversity, 1-17.

Murphy, B. A., & Hall, J. A. (2024). How a strong measurement validity review can go astray: A look at and recommendations for future measurement-focused reviews. *Clinical Psychology Review*, *114*, 102506. DOI: 10.1016/j.cpr.2024.102506 PMID: 39615954

Murray, J. V., Stokes, K. E., & van Klinken, R. D. (2012). Predicting the potential distribution of a riparian invasive plant: The effects of changing climate, flood regimes and land-use patterns. *Global Change Biology*, *18*(5), 1738–1753. DOI: 10.1111/j.1365-2486.2011.02621.x

Naeem, Z., Ahmad, M., Hameed, M., Bibi, S., Rani, W., Ratnasekera, D., Soufan, W., Adnan, M., & El Sabagh, A. (2024). Adaptability potential of *Justicia Adhatoda L.* To xeric environments through root morphophysiological and architectural modulations. *Applied Ecology and Environmental Research*, *22*(4), 2977–2998. DOI: 10.15666/aeer/2204_29772998

Nagel, M., Pence, V., Ballesteros, D., Lambardi, M., Popova, E., & Panis, B. (2024). Plant cryopreservation: Principles, applications, and challenges of banking plant diversity at ultralow temperatures. *Annual Review of Plant Biology*, *75*(1), 75. DOI: 10.1146/annurev-arplant-070623-103551 PMID: 38211950

Nagler, P. L., Sridhar, B. M., Olsson, A. D., van Leeuwen, W. J., & Glenn, E. P. (2018). Hyperspectral remote sensing tools for quantifying plant litter and invasive species in arid ecosystems. In *Advanced Applications in Remote Sensing of Agricultural Crops and Natural Vegetation* (pp. 97–129). CRC Press. DOI: 10.1201/9780429431166-6

Nagy, R. C., Fusco, E. J., Balch, J. K., Finn, J. T., Mahood, A., Allen, J. M., & Bradley, B. A. (2021). A synthesis of the effects of cheatgrass invasion on US Great Basin carbon storage. *Journal of Applied Ecology*, *58*(10), 2415–2426. DOI: 10.1111/1365-2664.13770

Nahla Zakariya, P. Invasive plants: Unveiling the invasion process, ecological effects and management approaches. *The Western Ghats*, 101.

Nasir, M., Ramash, R., Fatima, H., Ashraf, S., Munir, I., Asghar, S., Adnan, M., Masood, A., & Chaudhari, S. K. (2024). Phytochemical Characterization and Assessment of Crude Extracts from Justicia adhatoda for Phytotoxic and Cytotoxic Activity. *Scientifica, 2024*(1), 1374346. DOI: 10.1155/2024/1374346 PMID: 39359907

Natale, K., & Kiviat, E. (2025). BEST MANAGEMENT PRACTICES FOR PRIORITY INVASIVE PLANTS IN THE LOWER HUDSON VALLEY.

Navarro, A., Young, M., Allan, B., Carnell, P., Macreadie, P., & Ierodiaconou, D. (2020). The application of Unmanned Aerial Vehicles (UAVs) to estimate above-ground biomass of mangrove ecosystems. [– Elsevier]. *Remote Sensing of Environment, 242*, 2020. DOI: 10.1016/j.rse.2020.111747

Nawaz, A., Ayub, M. A., Nadeem, F., & Al-Sabahi, J. N. (2016). Lantana (Lantana camara): A medicinal plant having high therapeutic potentials–A comprehensive review. *International Journal of Chemical and Biochemical Science, 10*, 52–59.

Nawaz, H., Ishaq, M. W., Qadeer, S., Mahmood, M. H., Adeel, M., Anwar, I., Mazahr, M., Faisal, M., Arshad, M. A., Umar, M., & Zaheer, M. (2025). Assessing the Threat: Parthenium Adverse Effects on Biodiversity, Human Communities, and Environmental Integrity. *Sch Bull, 11*(3), 21–41. DOI: 10.36348/sb.2025.v11i03.001

Nechols, J. R., Hough, A. R., Margolies, D. C., Ruberson, J. R., McCornack, B. P., Sandercock, B. K., & Murray, L. (2020). Effect of temperature on plant resistance to arthropod pests. *Environmental Entomology, 49*(3), 537–545. DOI: 10.1093/ee/nvaa033 PMID: 32280953

Negi, G. C., Sharma, S., Vishvakarma, S. C., Samant, S. S., Maikhuri, R. K., Prasad, R. C., & Palni, L. M. (2019). Ecology and use of Lantana camara in India. *Botanical Review, 85*(2), 109–130. DOI: 10.1007/s12229-019-09209-8

Newsome, T., Cairncross, R., Cunningham, C. X., Spencer, E. E., Barton, P. S., Ripple, W. J., & Wirsing, A. J. (2024). Scavenging with invasive species. *Biological Reviews of the Cambridge Philosophical Society, 99*(2), 562–581. DOI: 10.1111/brv.13035 PMID: 38148253

Ng, A., Wilson, B. A., & Frew, A. (2023). Belowground crop responses to root herbivory are associated with the community structure of native arbuscular mycorrhizal fungi. *Applied Soil Ecology, 185*, 104797. DOI: 10.1016/j.apsoil.2022.104797

Nguyen, M. A., Ortega, A. E., Nguyen, K. Q., Kimball, S., Goulden, M. L., & Funk, J. L. (2016). Evolutionary responses of invasive grass species to variation in precipitation and soil nitrogen. *Journal of Ecology*, *104*(4), 979–986. DOI: 10.1111/1365-2745.12582

Nie, W., He, Q., Guo, H., Zhang, W., Ma, L., Li, J., & Wen, D. (2024). Arbuscular mycorrhizal fungi: Boosting crop resilience to environmental stresses. *Microorganisms*, *12*(12), 2448. DOI: 10.3390/microorganisms12122448 PMID: 39770651

Nikolakopoulou, M., Argerich, A., Bernal, S., Gacia, E., Ribot, M., Martí, E., Sorolla, A., & Sabater, F. (2020). Effect of three emergent macrophyte species on nutrient retention in aquatic environments under excess nutrient loading. *Environmental Science & Technology*, *54*(19), 11753–11762. DOI: 10.1021/acs.est.0c03216 PMID: 33175518

Ni, M., Luo, H., Xu, H., Chu, C., & Fang, S. (2023). High temperature can improve the performance of invasive plants by facilitating root growth. *American Journal of Botany*, *110*(10), e16227. DOI: 10.1002/ajb2.16227 PMID: 37561668

Ning, Z. H., Chen, C., Zhang, S. Y., & Wang, A. D. (2023). Lateral Hydrological Connectivity Driven by Tidal Flooding Regulates Range-Expansion of Invasive Spartina alterniflora in Tidal Channel-Salt Marsh Systems. *Journal of Coastal Research*, *39*(3), 539–548.

Ning, Z., Chen, C., Xie, T., Zhu, Z., Wang, Q., Cui, B., & Bai, J. (2021). Can the native faunal communities be restored from removal of invasive plants in coastal ecosystems? A global meta-analysis. [– Wiley Online Library]. *Global Change Biology*, *27*(19), 2021. DOI: 10.1111/gcb.15765 PMID: 34170600

Nizamani, M. M., Hughes, A. C., Qureshi, S., Zhang, Q., Tarafder, E., Das, D., Acharya, K., Wang, Y., & Zhang, Z.-G. (2024). Microbial biodiversity and plant functional trait interactions in multifunctional ecosystems. *Applied Soil Ecology*, *201*, 105515. DOI: 10.1016/j.apsoil.2024.105515

Norris, R. F., Caswell-Chen, E. P., & Kogan, M. (2003). *Concepts in integrated pest management*. Prentice Hall.

Ntiri, E. S., Calatayud, P.-A., Van Den Berg, J., Schulthess, F., & Le Ru, B. P. (2016). Influence of temperature on intra-and interspecific resource utilization within a community of lepidopteran maize stemborers. *PLoS One*, *11*(2), e0148735. DOI: 10.1371/journal.pone.0148735 PMID: 26859748

Nuñez, M. A., Chiuffo, M. C., Torres, A., Paul, T., Dimarco, R. D., Raal, P., Policelli, N., Moyano, J., García, R. A., Van Wilgen, B. W., Pauchard, A., & Richardson, D. M. (2017). Ecology and management of invasive Pinaceae around the world: Progress and challenges. *Biological Invasions, 19*(11), 3099–3120. DOI: 10.1007/s10530-017-1483-4

Nunez-Mir, G. C., & McCary, M. A. (2024). Invasive plants and their root traits are linked to the homogenization of soil microbial communities across the United States. *Proceedings of the National Academy of Sciences of the United States of America, 121*(44), e2418632121. DOI: 10.1073/pnas.2418632121 PMID: 39446386

Nyssen, B., Ouden, J. D., Bindewald, A., Brancalion, P., Kremer, K., Lapin, K., Raats, L., Schatzdorfer, E., Stanturf, J., Verheyen, K., & Muys, B. (2024). Established Invasive Tree Species Offer Opportunities for Forest Resilience to Climate Change. *Current Forestry Reports, 10*(6), 1–31. DOI: 10.1007/s40725-024-00232-6

Oancea, S. (2025). Occurrence, Pharmacological Properties, Toxic Effects, and Possibilities of Using Berries from Selected Invasive Plants. *Antioxidants, 14*(4), 399. DOI: 10.3390/antiox14040399 PMID: 40298647

Oduor, A. M. (2013). Evolutionary responses of native plant species to invasive plants: A review. *The New Phytologist, 200*(4), 986–992. DOI: 10.1111/nph.12429 PMID: 24712050

Olden, J. D., Chen, K., García-Berthou, E., King, A. J., South, J., & Vitule, J. R. (2022). Invasive species in streams and rivers. In *Encyclopedia of Inland waters* (Vol. 2). DOI: 10.1016/B978-0-12-819166-8.00083-9

Olsen, S. R., Cole, C. V., Watanabe, F. S., & Dean, L. A. (1954). *Estimation of available phosphorus in soils by extraction with sodium bicarbonate* (Circular No. 939). U.S. Department of Agriculture: Washington, DC, USA.

Olugbenga, O. O., & Chongs, M. E. (2024). Epidemiology and Genetic Relationship of Toxigenic Fungi in Maize Grains From North Central Nigerian States. *Environmental Health Insights, 18*, 11786302241249858. DOI: 10.1177/11786302241249858 PMID: 38721401

Olupot, W. (2022). The potential of natural succession to restore degraded areas of a Ugandan rainforest dominated by the exotic paper mulberry Broussonetia papyrifera (L.) L'Hér. ex Vent. *Forest Ecology and Management, 504*, 119861. DOI: 10.1016/j.foreco.2021.119861

Orine, D., Defossez, E., Vergara, F., Uthe, H., van Dam, N. M., & Rasmann, S. (2022). Arbuscular mycorrhizal fungi prevent the negative effect of drought and modulate the growth-defence trade-off in tomato plants. *Journal of Sustainable Agriculture and Environment, 1*(3), 177–190. DOI: 10.1002/sae2.12018

Orlikowski, L. B., Ptaszek, M., Rodziewicz, A., Nechwatal, J., Thinggaard, K., & Jung, T. (2011). *Phytophthora* root and collar rot of mature *Fraxinus excelsior* in forest stands in Poland and Denmark. *Forest Pathology, 41*(6), 510–519. DOI: 10.1111/j.1439-0329.2011.00714.x

Osman, K. T. (2014). *Soil degradation, conservation and remediation* (Vol. 820). Springer. DOI: 10.1007/978-94-007-7590-9

Ouledali, S., Ennajeh, M., Ferrandino, A., Khemira, H., Schubert, A., & Secchi, F. (2019). Influence of arbuscular mycorrhizal fungi inoculation on the control of stomata functioning by abscisic acid (ABA) in drought-stressed olive plants. *South African Journal of Botany, 121*, 152–158. DOI: 10.1016/j.sajb.2018.10.024

Overholt, W. A., & Franck, A. R. (2017). The invasive legacy of forage grass introductions into Florida. *Natural Areas Journal, 37*(2), 254–264. DOI: 10.3375/043.037.0214

Owen, M. D., Beckie, H. J., Leeson, J. Y., Norsworthy, J. K., & Steckel, L. E. (2015). Integrated pest management and weed management in the United States and Canada. *Pest Management Science, 71*(3), 357–376. DOI: 10.1002/ps.3928 PMID: 25346235

Pal, A., & Pandey, S. (2017). Symbiosis of arbuscular mycorrhizal fungi and Pennisetum glaucum L. improves plant growth and glomalin-related soil protein in barren soil. *Int J Sci Invent Today, 6*(6), 783–792.

Palik, B. J., Clark, P. W., D'Amato, A. W., Swanston, C., & Nagel, L. (2022). Operationalizing forest-assisted migration in the context of climate change adaptation: Examples from the eastern USA. [– Wiley Online Library]. *Ecosphere, 13*(10), 2022. DOI: 10.1002/ecs2.4260

Palma, E., Mabey, A. L., Vesk, P. A., & Catford, J. A. (2021). Characterising invasive species. In *Routledge Handbook of Biosecurity and Invasive Species* (pp. 15–39). Routledge. DOI: 10.4324/9781351131599-3

Palmer, M. A., Ambrose, R. F., & Poff, N. L. (1997). Ecological theory and community restoration ecology. *Restoration Ecology, 5*(4), 291–300. DOI: 10.1046/j.1526-100X.1997.00543.x

Park, D. S., Feng, X., Maitner, B. S., Ernst, K. C., & Enquist, B. J. (2020). Darwin's naturalization conundrum can be explained by spatial scale. *Proceedings of the National Academy of Sciences of the United States of America, 117*(20), 10904–10910. DOI: 10.1073/pnas.1918100117 PMID: 32366659

Parmar, D., Verma, S., Sharma, D., & Singh, E. (2024). Semiochemical based integrated livestock pest control. *Tropical Animal Health and Production, 56*(2), 49. DOI: 10.1007/s11250-024-03890-7 PMID: 38236343

Parshad, R. D., Quansah, E., Black, K., & Beauregard, M. (2016). Biological control via "ecological" damping: An approach that attenuates non-target effects. *Mathematical Biosciences, 273*, 23–44. DOI: 10.1016/j.mbs.2015.12.010 PMID: 26767801

Parvin, S., Van Geel, M., Yeasmin, T., Verbruggen, E., & Honnay, O. (2020). Effects of single and multiple species inocula of arbuscular mycorrhizal fungi on the salinity tolerance of a Bangladeshi rice (Oryza sativa L.) cultivar. *Mycorrhiza, 30*(4), 431–444. DOI: 10.1007/s00572-020-00957-9 PMID: 32367433

Pathak, R., Negi, V. S., Rawal, R. S., & Bhatt, I. D. (2019). Alien plant invasion in the Indian Himalayan Region: State of knowledge and research priorities. *Biodiversity and Conservation, 28*(12), 3073–3102. DOI: 10.1007/s10531-019-01829-1

Paudyal, K., Baral, H., Putzel, L., Bhandari, S., & Keenan, R. J. (2017). Change in land use and ecosystem services delivery from community-based forest landscape restoration in the Phewa Lake watershed, Nepal. *International Forestry Review, 19*(4), 88–101. DOI: 10.1505/146554817822330524

Pavlova, A., Beheregaray, L. B., Coleman, R., Gilligan, D., Harrisson, K. A., Ingram, B. A., Kearns, J., Lamb, A. M., Lintermans, M., Lyon, J., Nguyen, T. T. T., Sasaki, M., Tonkin, Z., Yen, J. D. L., & Sunnucks, P. (2017). Severe consequences of habitat fragmentation on genetic diversity of an endangered Australian freshwater fish: A call for assisted gene flow. *Evolutionary Applications, 10*(6), 531–550. DOI: 10.1111/eva.12484 PMID: 28616062

Pearson, D., & Ortega, Y. (2009). Managing invasive plants in natural areas: moving beyond weed control. *In: Kingely, Rudolph V., ed. Weeds: Management, Economic Impacts and Biology. Hauppauge, NY: Nova Science Publishers, Inc. p. 1-21.*, 1-21.

Pecl, G. T., Stuart-Smith, J., Walsh, P., Bray, D. J., Kusetic, M., Burgess, M., Frusher, S. D., Gledhill, D. C., George, O., Jackson, G., Keane, J., Martin, V. Y., Nursey-Bray, M., Pender, A., Robinson, L. M., Rowling, K., Sheaves, M., & Moltschaniwskyj, N. (2019). Redmap Australia: Challenges and successes with a large-scale citizen science-based approach to ecological monitoring and community engagement on climate change. *Frontiers in Marine Science, 6*, 349. DOI: 10.3389/fmars.2019.00349

Peng, W., Ochilov, A., Mirzaev, K., Sharafetdinovich, T. J., Nizamov, A., Ergashev, E., Ergashev, T. K., Artikova, S., Babakulov, B., & Ergashev, T. (2024). Operational management science of sustainable brick manufacturing using fluoride-enriched marble slurry and parthenium weeds. *Fluoride*, *57*(8), 1–9.

Pereira, G. W., Valente, D. S. M., de Queiroz, D. M., Santos, N. T., & Fernandes-Filho, E. I. (2022). Soil mapping for precision agriculture using support vector machines combined with inverse distance weighting. *Precision Agriculture*, *23*(4), 1189–1204. DOI: 10.1007/s11119-022-09880-9

Pérez Castro, S., Cleland, E. E., Wagner, R., Sawad, R. A., & Lipson, D. A. (2019). Soil microbial responses to drought and exotic plants shift carbon metabolism. *The ISME Journal*, *13*(7), 1776–1787. DOI: 10.1038/s41396-019-0389-9 PMID: 30872806

Perkins, L. B., & Nowak, R. S. (2012). Soil conditioning and plant–soil feedbacks affect competitive relationships between native and invasive grasses. *Plant Ecology*, *213*(8), 1337–1344. DOI: 10.1007/s11258-012-0092-7

Perring, M., Rowe, E., Arrigoni, E., Bidartondo, M., Britton, A., Field, C., Gilliam, F., Grenier, M., Hayes, F., & Jones, L. (2024). Scoping future research for air pollution recovery indicators (APRI).(Workshop report). https://nora.nerc.ac.uk/id/eprint/537618

Phang, L. Y., Mingyuan, L., Mohammadi, M., Tee, C. S., Yuswan, M. H., Cheng, W.-H., & Lai, K.-S. (2024). Phytoremediation as a viable ecological and socioeconomic management strategy. *Environmental Science and Pollution Research International*, *31*(1), 253–265. DOI: 10.1007/s11356-024-34585-z PMID: 39103580

Pickett, B., Irvine, I. C., Bullock, E., Arogyaswamy, K., & Aronson, E. (2019). Legacy effects of invasive grass impact soil microbes and native shrub growth. *Invasive Plant Science and Management*, *12*(1), 22–35. DOI: 10.1017/inp.2018.32

Pielou, E. C. (1966). Species–diversity and pattern–diversity in the study of ecological succession. *Journal of Theoretical Biology*, *10*(2), 370–383. DOI: 10.1016/0022-5193(66)90133-0 PMID: 5964400

Pik, D., Lucero, J. E., Lortie, C. J., & Braun, J. (2020). Light intensity and seed density differentially affect the establishment, survival, and biomass of an exotic invader and three species of native competitors. *Community Ecology*, *21*(3), 259–272. DOI: 10.1007/s42974-020-00027-2

Pili, A. N., Tingley, R., Sy, E. Y., Diesmos, M. L. L., & Diesmos, A. C. (2020). Niche shifts and environmental non-equilibrium undermine the usefulness of ecological niche models for invasion risk assessments. *Scientific Reports*, *10*(1), 1–12. DOI: 10.1038/s41598-020-64568-2 PMID: 32409706

Ping, Q., Xu, S., He, X., & Sun, S. (2025). Nitrogen addition alters litter chemical traits to regulate decomposition: A meta-analysis. *The Science of the Total Environment*, *983*, 179705. DOI: 10.1016/j.scitotenv.2025.179705 PMID: 40403540

Policelli, N., Bruns, T. D., Vilgalys, R., & Nuñez, M. A. (2019). Suilloid fungi as global drivers of pine invasions. *The New Phytologist*, *222*(2), 714–725. DOI: 10.1111/nph.15660 PMID: 30586169

Pons, C., Voß, A. C., Schweiger, R., & Müller, C. (2020). Effects of drought and mycorrhiza on wheat and aphid infestation. *Ecology and Evolution*, *10*(19), 10481–10491. DOI: 10.1002/ece3.6703 PMID: 33072274

Ponti, L., Gutierrez, A. P., de Campos, M. R., Desneux, N., Biondi, A., & Neteler, M. (2021). Biological invasion risk assessment of Tuta absoluta: Mechanistic versus correlative methods. *Biological Invasions*, *23*(1), 1–15. DOI: 10.1007/s10530-021-02613-5 PMID: 33488272

Porker, K., Straight, M., & Hunt, J. R. (2020). Evaluation of G× E× M interactions to increase harvest index and yield of early sown wheat. [PubMed]. *Frontiers in Plant Science*, *11*, 994. DOI: 10.3389/fpls.2020.00994 PMID: 32754174

Poudel, A. S., Jha, P. K., Shrestha, B. B., & Muniappan, R. (2019). Biology and management of the invasive weed *Ageratina adenophora* (Asteraceae): Current state of knowledge and future research needs. *Weed Research*, *59*(2), 79–92. DOI: 10.1111/wre.12351

Pouyat R V, Page-Dumroese D S, Patel-Weynand T, (2020). "Forest and rangeland soils of the United States under changing conditions: a comprehensive science synthesis."

Powles, S. B. (2008). Evolved glyphosate-resistant weeds around the world: Lessons to be learnt. *Pest Management Science*, *64*(4), 360–365. DOI: 10.1002/ps.1525 PMID: 18273881

Pradeep, E., Mukherjee, S. K., Selvakumar, P., Koppaka, A. K., Chandel, P. S., & Vats, C. (2025). Empowering Social Media Users With Ethical AI. In Chakraborty, S. (Ed.), *Ethical AI Solutions for Addressing Social Media Influence and Hate Speech* (pp. 91–118). IGI Global Scientific Publishing., DOI: 10.4018/979-8-3693-9904-0.ch005

Pratt, P. D., Moran, P. M., Pitcairn, M. J., Reddy, A., & O'Brien, J. (2021). Biological control of invasive plants in California's Delta: Past, present, and future. *Journal of Aquatic Plant Management*, *59*(s), 55–66. DOI: 10.57257/japm-20215955

Prentis, P. J., Wilson, J. R., Dormontt, E. E., Richardson, D. M., & Lowe, A. J. (2008). Adaptive evolution in invasive species. *Trends in Plant Science*, *13*(6), 288–294. DOI: 10.1016/j.tplants.2008.03.004 PMID: 18467157

Preston, J., Debney, A., Gamble, C., Hardy, M., Underwood, G., Garbutt, A., Harley, J., Baker, R., Dunk, R., & Grigg, M. (2025). Seascape connectivity: evidence, knowledge gaps and implications for temperate coastal ecosystem restoration practice and policy. npj Ocean Sustainability, 4(1), 33.

Pretorius, I., Schou, W. C., Richardson, B., Ross, S. D., Withers, T. M., Schmale, D. G.III, & Strand, T. M. (2023). In the wind: Invasive species travel along predictable atmospheric pathways. *Ecological Applications*, *33*(3), e2806. DOI: 10.1002/eap.2806 PMID: 36660794

Prevéy, J. S., & Seastedt, T. R. (2015). Increased winter precipitation benefits the native plant pathogen Ustilago bullata that infects an invasive grass. *Biological Invasions*, *17*(10), 3041–3047. DOI: 10.1007/s10530-015-0934-z

Prober, S. M., Byrne, M., McLean, E. H., Steane, D. A., Potts, B. M., Vaillancourt, R. E., & Stock, W. D. (2015). Climate-adjusted provenancing: A strategy for climate-resilient ecological restoration. *Frontiers in Ecology and Evolution*, *3*, 65. DOI: 10.3389/fevo.2015.00065

Pujolar, J., Breitburg, D., Lee, J., Decker, M. B., & Jaspers, C. (2023). Hybridization and adaptive introgression in a marine invasive species in native habitats. *iScience*, *26*(12), 108430. DOI: 10.1016/j.isci.2023.108430 PMID: 38077133

Puspitaloka, D., Kim, Y. S., Purnomo, H., & Fulé, P. Z. (2020). Defining ecological restoration of peatlands in Central Kalimantan, Indonesia. *Restoration Ecology*, *28*(2), 435–446. DOI: 10.1111/rec.13097

Pyšek, P., Hulme, P. E., Simberloff, D., Bacher, S., Blackburn, T. M., Carlton, J. T., Dawson, W., Essl, F., Foxcroft, L. C., Genovesi, P., Jeschke, J. M., Kühn, I., Liebhold, A. M., Mandrak, N. E., Meyerson, L. A., Pauchard, A., Pergl, J., Roy, H. E., Seebens, H., & Richardson, D. M. (2020). Scientists' warning on invasive alien species. *Biological Reviews of the Cambridge Philosophical Society*, *95*(6), 1511–1534. DOI: 10.1111/brv.12627 PMID: 32588508

Pyšek, P., & Richardson, D. M. (2010). Invasive species, environmental change and management, and health. *Annual Review of Environment and Resources*, *35*(1), 25–55. DOI: 10.1146/annurev-environ-033009-095548

Pywell, R. F., Bullock, J. M., Hopkins, A., Walker, K. J., Sparks, T. H., Burke, M. J., & Peel, S. (2002). Restoration of species-rich grassland on arable land: Assessing the limiting processes using a multi-site experiment. *Journal of Applied Ecology*, *39*(2), 294–309. DOI: 10.1046/j.1365-2664.2002.00718.x

Qadir, M., Iqbal, A., Hussain, A., Hussain, A., Shah, F., Yun, B.-W., & Mun, B.-G. (2024). Exploring plant-bacterial symbiosis for eco-friendly agriculture and enhanced resilience. *International Journal of Molecular Sciences*, *25*(22), 12198. DOI: 10.3390/ijms252212198 PMID: 39596264

Qazi, A. W., Saqib, Z., & Zaman-ul-Haq, M. (2023). Diversity and distribution of endemic flora in Pakistan. *Proceedings of the Pakistan Academy of Sciences: B. Life and Environmental Sciences, 60*(2), 165-179.

Quarles, W. (2007). Global warming means more pests. *The IPM Practitioner*, *29*(9/10), 1–8.

Quilodrán, C., Tsoupas, A., & Currat, M. (2020). The spatial signature of introgression after a biological invasion with hybridization. *Frontiers in Ecology and Evolution*, *8*, 8. DOI: 10.3389/fevo.2020.569620

Quinn, L. D., Stewart, J. R., Yamada, T., Toma, Y., Saito, M., Shimoda, K., & Fernández, F. G. (2012). Environmental tolerances of *Miscanthus sinensis* in invasive and native populations. *BioEnergy Research*, *5*(1), 139–148. DOI: 10.1007/s12155-011-9163-1

Quiroga, G., Erice, G., Ding, L., Chaumont, F., Aroca, R., & Ruiz-Lozano, J. M. (2019). The arbuscular mycorrhizal symbiosis regulates aquaporins activity and improves root cell water permeability in maize plants subjected to water stress. *Plant, Cell & Environment*, *42*(7), 2274–2290. DOI: 10.1111/pce.13551 PMID: 30916398

Qureshi, H., & Anwar, T. (2024). Role of allelopathy in invasion of invasive species. *Allelopathy Journal*, *62*(2), 115–138. DOI: 10.26651/allelo.j/2024-62-2-1490

Rädecker, N., Pogoreutz, C., Gegner, H. M., Cárdenas, A., Roth, F., Bougoure, J., Guagliardo, P., Wild, C., Pernice, M., Raina, J.-B., Meibom, A., & Voolstra, C. R. (2021). Heat stress destabilizes symbiotic nutrient cycling in corals. *Proceedings of the National Academy of Sciences of the United States of America*, *118*(10), 1–8. DOI: 10.1073/pnas.2022653118 PMID: 33500354

Radicetti, E., & Mancinelli, R. (2021). Sustainable weed control in the agroecosystems. *Sustainability (Basel), 13*(15), 8639. DOI: 10.3390/su13158639

Rafaqat, W., Sanchez, P., Botnen, D., & Fernandez-Anez, N. (2025). Analysing historical events and current management strategies of wildfires in Norway. *Scientific Reports, 15*(1), 24905. DOI: 10.1038/s41598-025-08760-2 PMID: 40640312

Raghurama, M., & Sankaran, M. (2021). Restoring tropical forest–grassland mosaics invaded by woody exotics. *Restoration Ecology, 29*(8), e13491. DOI: 10.1111/rec.13491

Raheem, A., Yohanna, P., Li, G., Noh, N. J., Iqbal, B., Tang, J., Du, D., Alahmadi, T. A., Ansari, M. J., Zhan, A., & Son, Y. (2024). Unraveling the ecological threads: How invasive alien plants influence soil carbon dynamics. *Journal of Environmental Management, 356*, 120556. DOI: 10.1016/j.jenvman.2024.120556 PMID: 38537457

Rahman, M. A., Das, A. K., Al Riyadh, Z., Suhag, M., & Rahman, M. M. (2024). Eucalyptus in Agriculture: Friend or Foe? Analyzing its impact on crop yields, soil dynamics, and farmers' perceptions in Bangladesh. *Agroforestry Systems, 98*(8), 1–20. DOI: 10.1007/s10457-024-01077-5

Rahman, S., Nishi, K., Muslim, M., & Rahman, M. (2021). Effects of organic manures on the growth and yield of malabar nut (Justicia adhatoda L.). *Dhaka University Journal of Biological Sciences, 30*(1), 79–86. DOI: 10.3329/dujbs.v30i1.51811

Rai R K, Shrestha L, Joshi S, Clements D R, (2022). "Biotic and economic impacts of plant invasions." Global Plant Invasions, 147-164.

Raihan, A., & Bari, A. M. (2024). Energy-economy-environment nexus in China: The role of renewable energies toward carbon neutrality. *Innovation and Green Development, 3*(3), 100139. DOI: 10.1016/j.igd.2024.100139

Raihan, A., Ridwan, M., & Rahman, M. S. (2024). An exploration of the latest developments, obstacles, and potential future pathways for climate-smart agriculture. *Climate Smart Agriculture, 100020*(2), 100020. Advance online publication. DOI: 10.1016/j.csag.2024.100020

Rai, P. K. (2015). What makes the plant invasion possible? Paradigm of invasion mechanisms, theories and attributes. *Environmental Skeptics and Critics, 4*(2), 36.

Rai, P. K. (2022). Environmental degradation by invasive alien plants in the anthropocene: Challenges and prospects for sustainable restoration. *Anthropocene Science, 2*(1), 1–13.

Rai, P. K., & Singh, J. S. (2020). Invasive alien plant species: Their impact on environment, ecosystem services and human health. *Ecological Indicators*, *117*, 106613.

Rajamuthuramalingam, T., da Silva, W., Zuverza-Mena, N., Dimkpa, C., & White, J. C. (2024). Nano-sized metal oxide fertilizers for sustainable agriculture: Balancing benefits, risks, and risk management strategies. *Nanoscale*, *16*(43), 19998–20026. Advance online publication. DOI: 10.1039/D4NR01354A PMID: 39417765

Ramachandran, A., Iwanaga, C. D., Fugate, M., Huxley, J. D., Rose-Person, A., Amatya, R., Bui, T.-T., & Spasojevic, M. J. (2024). Litter removal increases plant diversity by promoting both native and exotic forbs in heavily invaded coastal sage scrub in Southern California, USA. *Restoration Ecology*, ●●●, 14340.

Ranabhat, G. (2025). Vegetation Diversity in Southwestern Illinois Sand Prairies Southern Illinois University at Edwardsville].

Ranjitha, P. K., Selvakumar, P., Anute, N., Manjunath, T., Sureshkrishna, G., & Pougajendy, S. (2025). Transportation Challenges In Islands Tourism. In Hassan, V. (Ed.), *Navigating Mass Tourism to Island Destinations: Preservation and Cultural Heritage Challenges* (pp. 287–312). IGI Global Scientific Publishing., DOI: 10.4018/979-8-3693-9107-5.ch011

Rashid, M., Abbas, S. H., & Rehman, A. (2014). The status of highly alien invasive plants in Pakistan and their impact on the ecosystem: A review. *Innov J Agric Sci*, *2*, 1–4.

Ratcliffe, H., Kendig, A., Vacek, S., Carlson, D., Ahlering, M., & Dee, L. E. (2024). Extreme precipitation promotes invasion in managed grasslands. *Ecology*, *105*(1), e4190. DOI: 10.1002/ecy.4190 PMID: 37877294

Ratcliff, J. J., Hall, A. J., Porto, E., Saville, B. R., Lewis, R. J., Allen, J. W., Frankel, M., Wright, D. W., Barrow, D. L., & Pradilla, G. (2023). Early Minimally Invasive Removal of Intracerebral Hemorrhage (ENRICH): Study protocol for a multi-centered two-arm randomized adaptive trial. *Frontiers in Neurology*, *14*, 1126958. DOI: 10.3389/fneur.2023.1126958 PMID: 37006503

Ravali, K., Sarathchandra, G., Kumary, S. U., Sujatha, P. L., Kannan, T. A., & Parthiban, M. (2024, July 18). An in silico Evaluation of Anti-diabetic Potential of Vasicine, A Quinazoline Alkaloid of Justicia adhatoda Linnaeus. *Indian Journal of Animal Research*, *1*(Of), 4. DOI: 10.18805/IJAR.B-5278

Ravhuhali, K. E., Mudau, H. S., Moyo, B., Hawu, O., & Msiza, N. H. (2021). Prosopis species—An invasive species and a potential source of browse for livestock in semi-arid areas of South Africa. *Sustainability (Basel), 13*(13), 7369. DOI: 10.3390/su13137369

Ravichandran, K. R., & Thangavelu, M. (2017). Role and influence of soil microbial communities on plant invasion. *Ecological Questions, 27*, 9–23. DOI: 10.12775/EQ.2017.024

Reaser, J. K., Burgiel, S. W., Kirkey, J., Brantley, K. A., Veatch, S. D., & Burgos-Rodríguez, J. (2020). The early detection of and rapid response (EDRR) to invasive species: A conceptual framework and federal capacities assessment. *Biological Invasions, 22*(1), 1–19. DOI: 10.1007/s10530-019-02156-w

Reaser, J. K., Simpson, A., Guala, G. F., Morisette, J. T., & Fuller, P. (2020). Envisioning a national invasive species information framework. *Biological Invasions, 22*(1), 21–36. DOI: 10.1007/s10530-019-02141-3

Reed, M. S. (2008). Stakeholder participation for environmental management: A literature review. *Biological Conservation, 141*(10), 2417–2431. DOI: 10.1016/j.biocon.2008.07.014

Reid, A. J., Carlson, A. K., Creed, I. F., Eliason, E. J., Gell, P. A., Johnson, P. T., Kidd, K. A., MacCormack, T. J., Olden, J. D., Ormerod, S. J., Smol, J. P., Taylor, W. W., Tockner, K., Vermaire, J. C., Dudgeon, D., & Cooke, S. J. (2019). Emerging threats and persistent conservation challenges for freshwater biodiversity. *Biological Reviews of the Cambridge Philosophical Society, 94*(3), 849–873. DOI: 10.1111/brv.12480 PMID: 30467930

Reis, B. P., Kövendi-Jakó, A., Szitár, K., Török, K., & Halassy, M. (2021). Long-term effect of mowing on the restoration of Pannonian sand grassland to replace invasive black locust plantation. *Restoration Ecology, 29*(S1), e13152. DOI: 10.1111/rec.13152

Reiss, B. F. (2025). *The Social Life of Plants: Botanical Representations in Moche Art*. The University of New Mexico.

Ren, G., Du, Y., Yang, B., Wang, J., Cui, M., Dai, Z., Adomako, M. O., Rutherford, S., & Du, D. (2023). Influence of precipitation dynamics on plant invasions: Response of alligator weed (Alternanthera philoxeroides) and co-occurring native species to varying water availability across plant communities. *Biological Invasions, 25*(2), 519–532. DOI: 10.1007/s10530-022-02931-2

Rhodes, A. C., Plowes, R. M., Goolsby, J. A., Gaskin, J. F., Musyoka, B., Calatayud, P.-A., Cristofaro, M., Grahmann, E. D., Martins, D. J., & Gilbert, L. E. (2021). The dilemma of Guinea grass (Megathyrsus maximus): A valued pasture grass and a highly invasive species. *Biological Invasions, 23*(12), 3653–3669. DOI: 10.1007/s10530-021-02607-3

Rhodes, A. C., Plowes, R. M., Martins, D. J., Ng'Iru, I., & Gilbert, L. E. (2022). The invasiveness of Guinea grass (Megathyrsus maximus) is characterized by habitat and differing herbivore assemblages in its native and invaded range. *NeoBiota, 78*, 25–44. DOI: 10.3897/neobiota.78.87069

Rhodes, A. C., Rutledge, J., DuPont, B., Plowes, R. M., & Gilbert, L. E. (2021). Targeted grazing of an invasive grass improves outcomes for native plant communities and wildlife habitat. *Rangeland Ecology and Management, 75*, 41–50. DOI: 10.1016/j.rama.2020.11.007

Riar, M. K., Carley, D. S., Zhang, C., Schroeder-Moreno, M. S., Jordan, D. L., Webster, T. M., & Rufty, T. W. (2016). Environmental influences on growth and reproduction of invasive Commelina benghalensis. *International Journal of Agronomy, 2016*(1), 5679249.

Ribeiro, P. C., Menendez, E., da Silva, D. L., Bonieck, D., Ramírez-Bahena, M. H., Resende-Stoianoff, M. A., Peix, A., Velázquez, E., Mateos, P. F., & Scotti, M. R. (2017). Invasion of the Brazilian campo rupestre by the exotic grass Melinis minutiflora is driven by the high soil N availability and changes in the N cycle. *The Science of the Total Environment, 577*, 202–211. DOI: 10.1016/j.scitotenv.2016.10.162 PMID: 27823824

Richardson, D. M., & Pyšek, P. (2006). Plant invasions: Merging the concepts of species invasiveness and community invasibility. *Progress in Physical Geography, 30*(3), 409–431. DOI: 10.1191/0309133306pp490pr

Richardson, D. M., & Rejmánek, M. (2011). Trees and shrubs as invasive alien species–a global review. *Diversity & Distributions, 17*(5), 788–809. DOI: 10.1111/j.1472-4642.2011.00782.x

Rigling, D., & Prospero, S. (2018). Cryphonectria parasitica, the causal agent of chestnut blight: Invasion history, population biology and disease control. *Molecular Plant Pathology, 19*(1), 7–20. DOI: 10.1111/mpp.12542 PMID: 28142223

Robichaud, C. D., Basso, J. V., & Rooney, R. C. (2022). Control of invasive Phragmites australis (European common reed) alters macroinvertebrate communities. *Restoration Ecology, 30*(4), e13548. DOI: 10.1111/rec.13548

Rodgers, V. L., Wolfe, B. E., Werden, L. K., & Finzi, A. C. (2008). The invasive species *Alliaria petiolata* (garlic mustard) increases soil nutrient availability in northern hardwood–conifer forests. *Oecologia, 157*(3), 459–471. DOI: 10.1007/s00442-008-1089-8 PMID: 18612654

Rohela, G. K., Saini, P., Syam, S., Roy, P., Gonal, B., Aziz, D., Ayoub, O. B., & Maheswari, M. (2025). Mulberry Trees: A Sustainable Solution for Urban Forestry and Improved Air Quality. In *Sustainable Urban Environment and Waste Management: Theory and Practice* (pp. 223-246). Springer. DOI: 10.1007/978-981-96-1140-9_12

Roldão Almeida, M., Marchante, E., & Marchante, H. (2024). Knowing the invader: Increasing knowledge about an invasive plant to improve management. *Restoration Ecology, 32*(5), 14175. DOI: 10.1111/rec.14175

Rosalem, P. F., Gorgone-Barbosa, E., de Camargos, L. S., & Martins, A. R. (2024). Anatomical plasticity and starch allocation in an invasive grass in response to fire. *Revista Brasileira de Botânica. Brazilian Journal of Botany*, *47*(3), 1–8. DOI: 10.1007/s40415-024-01006-x

Rose, R. A., Byler, D., Eastman, J. R., Fleishman, E., Geller, G., Goetz, S., Guild, L., Hamilton, H., Hansen, M., Headley, R., Hewson, J., Horning, N., Kaplin, B. A., Laporte, N., Leidner, A., Leimgruber, P., Morisette, J., Musinsky, J., Pintea, L., & Wilson, C. (2015). Ten ways remote sensing can contribute to conservation. *Conservation Biology*, *29*(2), 350–359. DOI: 10.1111/cobi.12397 PMID: 25319024

Rossiter, N. A., Setterfield, S. A., Douglas, M. M., & Hutley, L. B. (2003). Testing the grass-fire cycle: Alien grass invasion in the tropical savannas of northern Australia. *Diversity & Distributions*, *9*(3), 169–176. DOI: 10.1046/j.1472-4642.2003.00020.x

Rout, M. E., Chrzanowski, T. H., Smith, W. K., & Gough, L. (2013). Ecological impacts of the invasive grass Sorghum halepense on native tallgrass prairie. *Biological Invasions*, *15*(2), 327–339. DOI: 10.1007/s10530-012-0289-7

Roychoudhury, N. (2025). Forest Invasive Species and Their Management. In Textbook of Forest Science (pp. 315-331). Springer. DOI: 10.1007/978-981-97-8289-5_14

Roy, H. E., Pauchard, A., Stoett, P. J., Renard Truong, T., Meyerson, L. A., Bacher, S., Galil, B. S., Hulme, P. E., Ikeda, T., Kavileveettil, S., McGeoch, M. A., Nuñez, M. A., Ordonez, A., Rahlao, S. J., Schwindt, E., Seebens, H., Sheppard, A. W., Vandvik, V., Aleksanyan, A., & Ziller, S. R. (2024). Curbing the major and growing threats from invasive alien species is urgent and achievable. *Nature Ecology & Evolution*, *8*(7), 1–8. DOI: 10.1038/s41559-024-02412-w PMID: 38831016

Ruiz-Jaen, M. C., & Aide, T. M. (2005). Restoration success: How is it being measured? *Restoration Ecology*, *13*(3), 569–577. DOI: 10.1111/j.1526-100X.2005.00072.x

Russo, A., Esperon-Rodriguez, M., St-Denis, A., & Tjoelker, M. G. (2025). Native vs. Non-Native Plants: Public Preferences, Ecosystem Services, and Conservation Strategies for Climate-Resilient Urban Green Spaces. *Land (Basel)*, *14*(5), 954. DOI: 10.3390/land14050954

Ryan-Colton, E., French, K., Wardle, G. M., Read, J. L., Canty, P. D., Lang, P. J., Bickerton, D. C., & Schlesinger, C. A. (2024). Long-term and landscape impacts of buffel grass on arid plant communities: Ecosystem shifts and acceleration by fire. *Ecosphere*, *15*(10), e70033. DOI: 10.1002/ecs2.70033

Sabbagh, E., Lakzayi, M., Keshtehgar, A., & Rigi, K. (2014). The effect of salt stress on respiration, PSII function, chlorophyll, carbohydrate and nitrogen content in crop plants. *Intl J Farm & Alli Sci*, *3*(9), 988–993.

Sadiq, S., & Jan, A. (2001). *Effect of graded application of potash on Kharif maize sown at different fertility levels* (Master's thesis, Department of Agronomy, The University of Agriculture, Peshawar, Pakistan).

Sage, R. F., Quesada, M., Brunet, J., & Aguilar, R. (2024). *An introduction to the special issue on global change and plant reproduction*. Oxford University Press US.

Samoura, D., Wahab, B., Taiwo, O., Diallo, A., & Bishoge, O. (2024). Small-scale Farmers' Adoptions of Climate-smart Agricultural Practices in the Guinean Savanna's Agroecological Zones. *Journal of Agricultural Sciences (Sri Lanka)*, *19*(1).

Samways, M. J., Gaigher, R., Deacon, C., & Pryke, J. S. (2025). Heterogeneous conservation corridors of remnant vegetation protect biodiversity in South African timber mosaics. *Biodiversity and Conservation*, *34*(2), 357–379. DOI: 10.1007/s10531-024-02993-9

Sandenbergh, E., Gervazoni, P., Grewell, B., Franceschini, C., Minuti, G., McGrannachan, C., Stiers, I., & Coetzee, J. Biology of Invasive Plants 6. Iris pseudacorus L.(Iridaceae). *Invasive Plant Science and Management*, 1-61.

Sanders, S. K., van Kleunen, M., Allan, E., & Thakur, M. P. (2025). Effects of extreme drought on the invasion dynamics of non-native plants. *Trends in Plant Science*, *30*(3), 291–300. DOI: 10.1016/j.tplants.2024.10.009 PMID: 39523141

Santander, C., Ruiz, A., García, S., Aroca, R., Cumming, J., & Cornejo, P. (2020). Efficiency of two arbuscular mycorrhizal fungal inocula to improve saline stress tolerance in lettuce plants by changes of antioxidant defense mechanisms. *Journal of the Science of Food and Agriculture, 100*(4), 1577–1587. DOI: 10.1002/jsfa.10166 PMID: 31769028

Santosa, E. B. (2025). Blessing in disguise of the Invasive Alien Species (IAS): Environmental problem into local-based advantages potentials. Media Konservasi, 30(2).

Sarada, K., & Rao, G. (2008). Antimicrobial activity of essential oils of Adhatoda vasica (Basaka) fresh and shade dried leaves and roots from Anantapur District, Andhra Pradesh. *Indian J Microb World, 10*, 86–90.

Saravanan S, Karthi S (2015) distinction in morphological and biochemical changes in valuable medicinal herb Justicia Adhatoda under elevated CO2.

Sarker, S., & Leta, O. T. (2025). Review of Watershed Hydrology and Mathematical Models. *Eng, 6*(6), 129. DOI: 10.3390/eng6060129

Sati, V. P. (2005). Systems of agriculture farming in the Uttaranchal Himalaya, India. *Journal of Mountain Science, 2*(1), 76–85. DOI: 10.1007/s11629-005-0076-3

Saura-Mas, S., & Lloret, F. (2005). Wind effects on dispersal patterns of the invasive alien Cortaderia selloana in Mediterranean wetlands. *Acta Oecologica, 27*(2), 129–133. DOI: 10.1016/j.actao.2004.12.001

Scotton, M., & Andreatta, D. (2021). Anti-erosion rehabilitation: Effects of revegetation method and site traits on introduced and native plant cover and richness. *The Science of the Total Environment, 793*, 148714. DOI: 10.1016/j.scitotenv.2021.145915 PMID: 33647653

Seastedt, T. R., Hobbs, R. J., & Suding, K. N. (2008). Management of novel ecosystems: Are novel approaches required? *Frontiers in Ecology and the Environment, 6*(10), 547–553. DOI: 10.1890/070046

Seastedt, T. R., & Pyšek, P. (2011). Mechanisms of plant invasions of North American and European grasslands. *Annual Review of Ecology, Evolution, and Systematics, 42*(1), 133–153. DOI: 10.1146/annurev-ecolsys-102710-145057

Sedibana, L., Kganyago, M., Jaca, T., & Yessoufou, K. (2025). First Spectral Reflectance Dataset of a Highly Invasive and Aggressive Alien Plant Species, Equisetum hyemale (snakegrass), in South Africa. Towards Early Detection and Effective Eradication. *Data in Brief, 60*, 111582. DOI: 10.1016/j.dib.2025.111582 PMID: 40534718

Seebens, H., Meyerson, L. A., Rahlao, S. J., Lenzner, B., Tricarico, E., Aleksanyan, A., Courchamp, F., Keskin, E., Saeedi, H., & Tawake, A. (2023). Trends and status of alien and invasive alien species.

Seebens, H., Blackburn, T. M., Dyer, E. E., Genovesi, P., Hulme, P. E., Jeschke, J. M., Pagad, S., Pyšek, P., van Kleunen, M., Winter, M., Ansong, M., Arianoutsou, M., Bacher, S., Blasius, B., Brockerhoff, E. G., Brundu, G., Capinha, C., Causton, C. E., Celesti-Grapow, L., & Essl, F. (2018). Global rise in emerging alien species results from increased accessibility of new source pools. *Proceedings of the National Academy of Sciences of the United States of America*, *115*(10), E2264–E2273. DOI: 10.1073/pnas.1719429115 PMID: 29432147

Seebens, H., Blackburn, T. M., Dyer, E. E., Genovesi, P., Hulme, P. E., Jeschke, J. M., Pagad, S., Pyšek, P., Winter, M., Arianoutsou, M., Bacher, S., Blasius, B., Brundu, G., Capinha, C., Celesti-Grapow, L., Dawson, W., Dullinger, S., Fuentes, N., Jäger, H., & Essl, F. (2017). No saturation in the accumulation of alien species worldwide. *Nature Communications*, *8*(1), 14435. DOI: 10.1038/ncomms14435 PMID: 28198420

Seebens, H., Essl, F., Dawson, W., Fuentes, N., Moser, D., Pergl, J., Pyšek, P., van Kleunen, M., Weber, E., Winter, M., & Blasius, B. (2015). Global trade will accelerate plant invasions in emerging economies under climate change. *Global Change Biology*, *21*(11), 4128–4140. DOI: 10.1111/gcb.13021 PMID: 26152518

Segrestin, J., Lisner, A., Götzenberger, L., Hájek, T., Janíková, E., Jílková, V., Konečná, M., Švancárová, T., & Lepš, J. (2025). Biodiversity loss disrupts seasonal carbon dynamics in a species-rich temperate grassland. *Ecology*, *106*(5), e70091. DOI: 10.1002/ecy.70091 PMID: 40342156

Sepp, S. K., Davison, J., Jairus, T., Vasar, M., Moora, M., Zobel, M., & Öpik, M. (2019). Non-random association patterns in a plant–mycorrhizal fungal network reveal host–symbiont specificity. *Molecular Ecology*, *28*(2), 365–378. DOI: 10.1111/mec.14924 PMID: 30403423

Setterfield, S. A., Rossiter-Rachor, N. A., Douglas, M. M., Wainger, L., Petty, A. M., Barrow, P., Shepherd, I. J., & Ferdinands, K. B. (2013). Adding fuel to the fire: The impacts of non-native grass invasion on fire management at a regional scale. *PLoS One*, *8*(5), e59144. DOI: 10.1371/journal.pone.0059144 PMID: 23690917

Shabbir, A., Bajwa, A. A., Mao, R., Kezar, S., Dorji, S., & Adkins, S. W. (2024). Biology of Invasive Plants 6. Parthenium hysterophorus L. *Invasive Plant Science and Management*, *17*(3), 1–28. DOI: 10.1017/inp.2024.23

Shabbir, A., Zalucki, M. P., Dhileepan, K., Khan, N., & Adkins, S. W. (2023). The current and potential distribution of parthenium weed and its biological control agent in Pakistan. *Plants*, *12*(6), 1381. DOI: 10.3390/plants12061381 PMID: 36987069

Shang, B., Hu, Z., Xie, R., Wu, J., Qu, W., Zhang, W., Zhou, A., Feng, L., Bi, X., & Shou, J. (2024). Predictive Value of Neutrophil Extracellular Traps in Neoadjuvant Chemotherapy for Muscle-Invasive Bladder Cancer. *Molecular Carcinogenesis*. PMID: 39540802

Shannon, C. E., & Weaver, W. (1963). *The mathematical theory of communication*. University of Illinois Press.

Sharma, S. K., & Kumar, K. (2025). Forest Fire Effects on Structure, Composition, and Functioning of Ecosystem. In Forest Fire and Climate Change: Insights into Science (pp. 79-99). Springer.

Sharma, G. P., Raghubanshi, A. S., & Singh, J. S. (2005). Lantana invasion: An overview. *Weed Biology and Management*, *5*(4), 157–167. DOI: 10.1111/j.1445-6664.2005.00178.x

Sharma, J., Singh, R., Garai, S., Rahaman, S. M., Khatun, M., Ranjan, A., Mishra, S. N., & Tiwari, S. (2022). Climate change and dispersion dynamics of the invasive plant species Chromolaena odorata and Lantana camara in parts of the central and eastern India. *Ecological Informatics*, *72*, 101824. DOI: 10.1016/j.ecoinf.2022.101824

Sharma, L. K., & Naik, R. (2024). Wetland Ecosystems. In *Conservation of Saline Wetland Ecosystems: An Initiative towards UN Decade of Ecological Restoration* (pp. 3–32). Springer. DOI: 10.1007/978-981-97-5069-6_1

Sharma, S. K., & Verma, C. M. (2016). Performance of *Cenchrus ciliaris* Linn. strains in an arid rangeland of western Rajasthan. *Annals of Arid Zone*, *55*(1), 1–9.

Sheley, R. L., James, J. J., Smith, B. S., & Vasquez, E. A. (2010). Applying ecologically-based invasive plant management. *Rangeland Ecology and Management*, *63*(6), 605–613. DOI: 10.2111/REM-D-09-00187.1

Sheley, R. L., Mangold, J. M., & Anderson, J. L. (2006). Potential for successional theory to guide restoration of invasive plant-dominated rangeland. *Ecological Monographs*, *81*(2), 181–198. DOI: 10.1890/0012-9615(2006)076[0365:PFSTTG]2.0.CO;2

Shende, A. D., Tekade, S. A., Deshmukh, A. A., Tembhurkar, S. P., & Selvakumar, P. (2025). Water Quality Monitoring and Control With AI. In Ruiz-Vanoye, J., & Díaz-Parra, O. (Eds.), *Smart Water Technology for Sustainable Management in Modern Cities* (pp. 51–80). IGI Global Scientific Publishing., DOI: 10.4018/979-8-3693-8074-1.ch003

Shen, K., He, Y., Xia, T., Guo, Y., Wu, B., Han, X., & Liu, Y. (2024). Arbuscular mycorrhizal fungi improve the competitive advantage of a native plant relative to a congeneric invasive plant in growth and nutrition. *Ecology and Evolution*, *14*(5), e11459. DOI: 10.1002/ece3.11459 PMID: 38774145

Shen, Y., Fang, Y., Oelmann, Y., Vancov, T., Chen, H., Du, Z., Chang, S. X., & Cai, Y. (2025). Livestock excretion alleviates soil microbial carbon and phosphorus limitations relative to trampling and defoliation in an alpine meadow. *Agriculture, Ecosystems & Environment*, *389*, 109687. DOI: 10.1016/j.agee.2025.109687

Shiba, M., Kobayashi, N., Harada, S., & Fukuda, T. (2024). Decrease in wind stress leads to an increase in the above ground morphology and number of seeds of an invasive alien species, Bidens pilosa (Asteraceae). *Frontiers in Plant Science*, *15*, 1445437. DOI: 10.3389/fpls.2024.1445437 PMID: 39582631

Shiferaw, H., Schaffner, U., Bewket, W., Alamirew, T., Zeleke, G., Teketay, D., & Eckert, S. (2019). Modelling the current fractional cover of an invasive alien plant and drivers of its invasion in a dryland ecosystem. *Scientific Reports*, *9*(1), 1576. DOI: 10.1038/s41598-018-36587-7 PMID: 30733452

Shimoda, S., & Wagai, R. (2020). Ecosystem dynamics after abandonment of rice paddy fields: Does alien plant invasion enhance carbon storage? *Ecosystems (New York, N.Y.)*, *23*(5), 931–944. DOI: 10.1007/s10021-019-00426-1

Shin, Y. J., Midgley, G. F., Archer, E. R., Arneth, A., Barnes, D. K., Chan, L., Hashimoto, S., Hoegh-Guldberg, O., Insarov, G., Leadley, P., Levin, L. A., Ngo, H. T., Pandit, R., Pires, A. P. F., Pörtner, H.-O., Rogers, A. D., Scholes, R. J., Settele, J., & Smith, P. (2022). Actions to halt biodiversity loss generally benefit the climate. *Global Change Biology*, *28*(9), 2846–2874. DOI: 10.1111/gcb.16109 PMID: 35098619

Shivanna, K. (2009). Pollination biology, breeding system and reproductive success of Adhatoda vasica, an important medicinal plant. *Current Science*, •••, 408–412.

Shugart, H., & White, P. (2024). *The Little Book of Trees*. Princeton University Press.

Siles, G., Alcántara, J. M., Rey, P. J., & Bastida, J. M. (2010). Defining a target map of native species assemblages for restoration. *Restoration Ecology*, *18*(4), 439–448. DOI: 10.1111/j.1526-100X.2008.00473.x

Siller-Clavel, P., Badano, E. I., Villarreal-Guerrero, F., Prieto-Amparán, J. A., Pinedo-Alvarez, A., Corrales-Lerma, R., Álvarez-Holguín, A., & Hernández-Quiroz, N. S. (2022). Distribution patterns of invasive buffelgrass (Cenchrus ciliaris) in Mexico estimated with climate niche models under the current and future climate. *Plants*, *11*(9), 1160. DOI: 10.3390/plants11091160 PMID: 35567161

Silva-Castro, G. A., Cano, C., Moreno-Morillas, S., Bago, A., & García-Romera, I. (2022). Inoculation of indigenous arbuscular mycorrhizal fungi as a strategy for the recovery of long-term heavy metal-contaminated soils in a mine-spill area. *Journal of Fungi (Basel, Switzerland)*, *9*(1), 56. DOI: 10.3390/jof9010056 PMID: 36675877

Silveira, M. J., Florêncio, F. M., de Carvalho Harthman, V., & Thiébaut, G. (2023). Responses of three invasive alien aquatic plant species to climate warming and plant density. *Journal of Plant Research*, *136*(6), 817–826. DOI: 10.1007/s10265-023-01482-4 PMID: 37505305

Simberloff, D. (2013). *Invasive species: what everyone needs to know*. OUP Us. DOI: 10.1093/wentk/9780199922017.001.0001

Simberloff, D. (2021). Maintenance management and eradication of established aquatic invaders. *Hydrobiologia*, *848*(1), 1–12. DOI: 10.1007/s10750-020-04352-5 PMID: 32836349

Simberloff, D., Martin, J. L., Genovesi, P., Maris, V., Wardle, D. A., Aronson, J., Courchamp, F., Galil, B., García-Berthou, E., Pascal, M., Pyšek, P., Sousa, R., Tabacchi, E., & Vila, M. (2013). Impacts of biological invasions: What's what and the way forward. *Trends in Ecology & Evolution*, *28*(1), 58–66. DOI: 10.1016/j.tree.2012.07.013 PMID: 22889499

Simberloff, D., & Stiling, P. (1996). How risky is biological control? *Ecology*, *77*(7), 1965–1974. DOI: 10.2307/2265693

Simberloff, D., & Von Holle, B. (1999). Positive interactions of nonindigenous species: Invasional meltdown? *Biological Invasions*, *1*(1), 21–32. DOI: 10.1023/A:1010086329619

Simonson, W. D., Miller, E., Jones, A., García-Rangel, S., Thornton, H., & McOwen, C. (2021). Enhancing climate change resilience of ecological restoration—A framework for action. *Perspectives in Ecology and Conservation*, *19*(3), 300–310. DOI: 10.1016/j.pecon.2021.05.002

Simpson, E. H. (1949). Measurement of diversity. *Nature*, *163*(4148), 688. DOI: 10.1038/163688a0

Singh, V., Mina, U., Meena, R., & Koul, M. (2025). Understanding the Nuances and Mitigation Measures of Plant Invasion: A Step Towards Sustainable Development. Green Equilibrium: Deciphering Earth's Ecosystems for Sustainable Tomorrow, 197-224.

Singh, A., Vaish, B., & Singh, R. P. (2016). Eco-restoration of degraded lands through microbial biomass: An ecological engineer. *Acta Biomedica Scientia*, *3*(1), 133–135.

Singh, K., Sobuj, N., & Byun, C. (2024). Native plants do not benefit from arriving early, but invasives pay to arrive late. *Ecological Indicators*, *166*, 112453. DOI: 10.1016/j.ecolind.2024.112453

Singh, L. R., & Singh, O. M. (2013). Datura stramonium: An overview of its phytochemistry and pharmacognosy. *Research Journal of Pharmacognosy and Phytochemistry*, *5*(3), 143–148.

Sirbu, C., Miu, I. V., Gavrilidis, A. A., Gradinaru, S. R., Niculae, I. M., Preda, C., Oprea, A., Urziceanu, M., Camen-Comanescu, P., Nagoda, E., Sirbu, I. M., Memedemin, D., & Anastasiu, P. (2022). Distribution and pathways of introduction of invasive alien plant species in Romania. *NeoBiota*, *75*, 1–21. DOI: 10.3897/neobiota.75.84684

SIS, M. (2024). Global plant responses to intensified fire regimes.

Skurski, T. C., Rew, L. J., & Maxwell, B. D. (2014). Mechanisms underlying non-indigenous plant impacts: A review of recent experimental research. *Invasive Plant Science and Management*, *6*(1), 89–105. DOI: 10.1614/IPSM-D-13-00099.1

Smith, J. T., Kleinhesselink, A. R., Maestas, J. D., Morford, S. L., Naugle, D. E., & White, C. D. (2024). Using Satellite Remote Sensing to Assess Shrubland Vegetation Responses to Large-Scale Juniper Removal in the Northern Great Basin. *Rangeland Ecology and Management*, *97*, 123–134. DOI: 10.1016/j.rama.2024.08.010

Smith, L. A. (2024). Wizards of Climate Science: Ancient Magicians in the Court of Big Science. *Zygon*, *59*(1). Advance online publication. DOI: 10.16995/zygon.11699

Smith, M. C., Julian, P., DeAngelis, D., & Zhang, B. (2024). Ecological benefits of integrative weed management of Melaleuca quinquenervia in Big Cypress National Preserve. *BioControl*, *69*(3), 293–301. DOI: 10.1007/s10526-023-10229-y

Smith, P., Calvin, K., Nkem, J., Campbell, D., Cherubini, F., Grassi, G., Korotkov, V., Le Hoang, A., Lwasa, S., McElwee, P., Nkonya, E., Saigusa, N., Soussana, J.-F., Taboada, M. A., Manning, F. C., Nampanzira, D., Arias-Navarro, C., Vizzarri, M., House, J., & Arneth, A. (2020). Which practices co-deliver food security, climate change mitigation and adaptation, and combat land degradation and desertification? [– Wiley Online Library]. *Global Change Biology*, *26*(3), 2020. DOI: 10.1111/gcb.14878 PMID: 31637793

Snorek, J., Freidberg, S., & Smith, G. (2024). Relationships of regeneration in Great Plains commodity agriculture. *Agriculture and Human Values*, *41*(4), 1–16. DOI: 10.1007/s10460-024-10558-3

Sobia, H. N., Khan, S., & Nadeem, F. (2018). Use of malabar nut (Justicia adhatoda L.) from traditional medicine to current pharmacopeia–A review study. *International Journal of Chemical and Biochemical Science*, *13*, 46–51.

Society for Ecological Restoration (SER). (2004). The SER International Primer on Ecological Restoration. Society for Ecological Restoration International. Retrieved from http://c.ymcdn.com/sites/www.ser.org/resource/resmgr/custompages/publications/SER_Primer.pdf

Sokolov, V., Ehrich, D., Yoccoz, N. G., Sokolov, A., & Lecomte, N. (2012). Bird communities of the Arctic shrub tundra of Yamal: Habitat specialists and generalists. *PLoS One*, *7*(12), e50335. DOI: 10.1371/journal.pone.0050335 PMID: 23239978

Solís-Rodríguez, U. R. J., Ramos-Zapata, J. A., Hernández-Cuevas, L., Salinas-Peba, L., & Guadarrama, P. (2020). Arbuscular mycorrhizal fungi diversity and distribution in tropical low flooding forest in Mexico. *Mycological Progress*, *19*(3), 195–204. DOI: 10.1007/s11557-019-01550-x

Solves, J.-P. (2020). *A Preliminary Flora for Las Cienegas National Conservation Area and Studies on the Life History of the Endangered Huachuca Water Umbel.* Arizona State University.

Song, F., Li, J., Fan, X., Zhang, Q., Chang, W., Yang, F., & Geng, G. (2016). Transcriptome analysis of *Glomus mosseae/Medicago sativa* mycorrhiza on atrazine stress. *Scientific Reports*, *6*(1), 20245. DOI: 10.1038/srep20245 PMID: 26833403

Song, X., Yuan, Z.-Q., Fang, C., Li, X., Zhao, Y.-Y., Li, F.-M., Sardans, J., & Peñuelas, J. (2025). How to develop nature-based solutions for revegetation on abandoned farmland in the Loess Plateau of China? *Journal of Environmental Management*, *373*, 123737. DOI: 10.1016/j.jenvman.2024.123737 PMID: 39706001

Song, Z., Bi, Y., Zhang, J., Gong, Y., & Yang, H. (2020). Arbuscular mycorrhizal fungi promote the growth of plants in the mining associated clay. *Scientific Reports*, *10*(1), 2663. DOI: 10.1038/s41598-020-59447-9 PMID: 32060306

Soni, M., Sheetal, K., Renjith, P., & Subbulakshmi, V. Birbal, Nathawat, N., Panwar, N., & Dagar, J. C. (2024). Domestication of wild halophytes for profitable biosaline agriculture. In halophytes vis-à-vis saline agriculture: perspectives and opportunities for food security, (pp. 479-505). Springer.

Soti, P. G., Purcell, M., & Jayachandran, K. (2020). Soil biotic and abiotic conditions negate invasive species performance in native habitat. *Ecological Processes*, *9*(1), 1–9. DOI: 10.1186/s13717-020-00220-1

Souza, C. S., Marins, G., Camargo, I. F. L., Lima, L. B., Gomes, A. S., & Borghetti, F. (2025). Allelopathic hypotheses revisited: The interactions between native and exotic species in the Brazilian savanna. *Anais da Academia Brasileira de Ciências*, *97*(3), e20240252. DOI: 10.1590/0001-3765202520240252 PMID: 40561269

Spagnoletti, F. N., Carmona, M., Balestrasse, K., Chiocchio, V., Giacometti, R., & Lavado, R. S. (2021). The arbuscular mycorrhizal fungus Rhizophagus intraradices reduces the root rot caused by Fusarium pseudograminearum in wheat. *Rhizosphere*, *19*, 100369. DOI: 10.1016/j.rhisph.2021.100369

Spear, D., Foxcroft, L. C., Bezuidenhout, H., & McGeoch, M. A. (2013). Human population density explains alien species richness in protected areas. *Biological Conservation, 159*, 137–147. DOI: 10.1016/j.biocon.2012.11.022

Staeck, L. (2022). *Fascination Amazon River: Its People, Its Animals, Its Plants*. Springer Nature. DOI: 10.1007/978-3-662-64452-2

Stager, M. (2023). Predicting the Great Lakes wetlands' resilience to climate change in response to Phragmites australis subsp. australis removal. *Surgery Journal (New York, N.Y.)*.

Stapleton, J. J. (2000). Soil solarization in various agricultural production systems. *Crop Protection (Guildford, Surrey)*, *19*(8-10), 837–841. DOI: 10.1016/S0261-2194(00)00111-3

Stehn, I. J. (2020). An ecological study of Tarchonanthus camphoratus L. in the Northern Cape Province of South Africa.

Stout, J. C., & Tiedeken, E. J. (2017). Direct interactions between invasive plants and native pollinators: Evidence, impacts and approaches. *Functional Ecology*, *31*(1), 38–46. DOI: 10.1111/1365-2435.12751

Strömberg, C. A. (2011). Evolution of grasses and grassland ecosystems. *Annual Review of Earth and Planetary Sciences, 39*(1), 517–544. DOI: 10.1146/annurev-earth-040809-152402

Stromberg, J. C., Beauchamp, V. B., Dixon, M. D., Lite, S. J., & Paradzick, C. (2007). Importance of low-flow and high-flow characteristics to restoration of riparian vegetation along rivers in arid southwestern United States. *Freshwater Biology, 52*(4), 651–679. DOI: 10.1111/j.1365-2427.2006.01713.x

Suárez-Ronay, S. H., Medina-Villar, S., & Corona, M. E. P. (2024). The role of fire in the germination of invasive plants in Mediterranean environments: A meta-analysis. *Forest Ecology and Management, 569*, 122168. DOI: 10.1016/j.foreco.2024.122168

Subramanian, M., Portia, R., Mishra, B. R., & Selvakumar, P. P. V., R., & T. C., M. (2025). Assessment and Evaluation in Personalized STEAM Education. In N. Son (Ed.), *Integrating Personalized Learning Methods Into STEAM Education* (pp. 103-122). IGI Global Scientific Publishing. https://doi.org/DOI: 10.4018/979-8-3693-7718-5.ch005

Suding, K. N. (2011). Toward an era of restoration in ecology: Successes, failures, and opportunities ahead. *Annual Review of Ecology, Evolution, and Systematics, 42*(1), 465–487. DOI: 10.1146/annurev-ecolsys-102710-145115

Suding, K. N., Gross, K. L., & Houseman, G. R. (2004). Alternative states and positive feedbacks in restoration ecology. *Trends in Ecology & Evolution, 19*(1), 46–53. DOI: 10.1016/j.tree.2003.10.005 PMID: 16701225

Suding, K., Higgs, E., Palmer, M., Callicott, J. B., Anderson, C. B., Baker, M., Gutrich, J. J., Hondula, K. L., LaFevor, M. C., Larson, B. M. H., Randall, A., Ruhl, J. B., & Schwartz, K. Z. (2015). Committing to ecological restoration. *Science, 348*(6235), 638–640. DOI: 10.1126/science.aaa4216 PMID: 25953995

Sueltenfuss, J. P., Ocheltree, T. W., & Cooper, D. J. (2020). Evaluating the realized niche and plant–water relations of wetland species using experimental transplants. *Plant Ecology, 221*(1), 53–65. DOI: 10.1007/s11258-020-01015-2

Sun, J., Li, W., Li, C., Chang, W., Zhang, S., Zeng, Y., Zeng, C., & Peng, M. (2020). Effect of different rates of nitrogen fertilization on crop yield, soil properties and leaf physiological attributes in banana under subtropical regions of China. *Frontiers in Plant Science, 11*, 613760. DOI: 10.3389/fpls.2020.613760 PMID: 33408734

Sun, K., Pan, Y.-T., Jiang, H.-J., Xu, J.-Y., Ma, C.-Y., Zhou, J., Liu, Y., Shabala, S., Zhang, W., & Dai, C.-C. (2024). Root endophyte-mediated alteration in plant H2O2 homeostasis regulates symbiosis outcome and reshapes the rhizosphere microbiota. *Journal of Experimental Botany*, *75*(10), 3153–3170. DOI: 10.1093/jxb/erae069 PMID: 38394357

Sutton, G. F. (2021). Prioritising biological control agents for release against Sporobolus pyramidalis and.

Su, Y., Zhang, C., Cescatti, A., Yu, K., Ciais, P., Smith, T., Shang, J., Carnicer, J., Liu, J., Chen, J. M., Green, J. K., Wu, J., Ponce-Campos, G. E., Zhang, Y., Zuo, Z., Liao, J., Wu, J., Lafortezza, R., Yan, K., & Zhou, W. (2025). Pervasive but biome-dependent relationship between fragmentation and resilience in forests. *Nature Ecology & Evolution*, •••, 1–15. DOI: 10.1038/s41559-025-02776-7 PMID: 40629088

Świerszcz, S., Czarniecka-Wiera, M., Szymura, T. H., & Szymura, M. (2024). From invasive species stand to species-rich grassland: Long-term changes in plant species composition during Solidago invaded site restoration. *Journal of Environmental Management*, *353*, 120216. DOI: 10.1016/j.jenvman.2024.120216 PMID: 38290260

Sych, M. (2024). Ecological Colonialism and the Impact of the 'Englishman's Foot': A Literary Ecology of an Invasive Species. DOI: 10.7939/r3-6zp1-3v39

Szymura, T. H., Szymura, M., Zając, M., & Zając, A. (2018). Effect of anthropogenic factors, landscape structure, land relief, soil and climate on risk of alien plant invasion at regional scale. *The Science of the Total Environment*, *626*, 1373–1381. DOI: 10.1016/j.scitotenv.2018.01.131 PMID: 29898544

Taillardat, P., Thompson, B. S., Garneau, M., Trottier, K., & Friess, D. A. (2020). Climate change mitigation potential of wetlands and the cost-effectiveness of their restoration. *Interface Focus*, *10*(5), 2020. DOI: 10.1098/rsfs.2019.0129 PMID: 32832065

Tang, J. Q., Guo, X. C., Lu, X. Y., Liu, M. C., Zhang, H. Y., Feng, Y. L., & Kong, D. L. (2020). A review on the effects of invasive plants on mycorrhizal fungi of native plants and their underlying mechanisms. *Acta Phytoecologica Sinica*, *44*(11), 1095–1112. DOI: 10.17521/cjpe.2020.0166

Tanner, E. (2023). *How can sustainable intensification practices affect the environmental and socioeconomic performance of smallholder farms in southern Togo?* ETH Zurich.

Tanner, R. A., & Gange, A. C. (2013). The impact of two non-native plant species on native flora performance: Potential implications for habitat restoration. *Plant Ecology*, *214*(3), 423–432. DOI: 10.1007/s11258-013-0179-9

Tariq, A., Graciano, C., Sardans, J., Zeng, F., Hughes, A. C., Ahmed, Z., Ullah, A., Ali, S., Gao, Y., & Peñuelas, J. (2024). Plant root mechanisms and their effects on carbon and nutrient accumulation in desert ecosystems under changes in land use and climate. *The New Phytologist*, *242*(3), 916–934. DOI: 10.1111/nph.19676 PMID: 38482544

Tartaglia, E. S., & Aronson, M. F. (2024). Plant native: Comparing biodiversity benefits, ecosystem services provisioning, and plant performance of native and non-native plants in urban horticulture. *Urban Ecosystems*, *27*(6), 2587–2611. DOI: 10.1007/s11252-024-01610-5

Tataridas, A., Jabran, K., Kanatas, P., Oliveira, R. S., Freitas, H., & Travlos, I. (2022). Early detection, herbicide resistance screening, and integrated management of invasive plant species: A review. *Pest Management Science*, *78*(1), 13–24. DOI: 10.1002/ps.6963 PMID: 35510308

Taylor, R. M., Stavert, J. R., Pou, D. L., & Wilmshurst, J. M. (2025). Ecology, history, threats, and management of gumland ecosystems in Aotearoal New Zealand. New Zealand Journal of Ecology, 49(1), 3589.

Tedersoo, L., Bahram, M., & Zobel, M. (2020). How mycorrhizal associations drive plant population and community biology. *Science*, *367*(6480), eaba1223. DOI: 10.1126/science.aba1223 PMID: 32079744

TEEB (The Economics of Ecosystems and Biodiversity). (2010). The Economics of Ecosystems and Biodiversity Ecological and Economic Foundations. Pushpam Kumar (Ed.). Earthscan.

Teixeira, L. H., Yannelli, F. A., Ganade, G., & Kollmann, J. (2020). Functional diversity and invasive species influence soil fertility in experimental grasslands. *Plants*, *9*(11), 1511. DOI: 10.3390/plants9010053 PMID: 31906387

Tercero, M. C., Álvarez-Rogel, J., Conesa, H., Ferrer, M., Calderón, A., López-Orenes, A., & González-Alcaraz, M. (2015). Response of biogeochemical processes of the water-soil-plant system to experimental flooding-drying conditions in a eutrophic wetland: The role of Phragmites australis. *Plant and Soil*, *396*(1-2), 109–125. DOI: 10.1007/s11104-015-2589-z

Tezel, A., Papadonikolaki, E., Yitmen, I., & Hilletofth, P. (2020). Preparing construction supply chains for blockchain technology: An investigation of its potential and future directions. [– Springer]. *Frontiers of Engineering Management*, *7*(4), 2020. DOI: 10.1007/s42524-020-0110-8

Thanigaivel, A., Senthil-Nathan, S., Vasantha-Srinivasan, P., Edwin, E. S., Ponsankar, A., Selin-Rani, S., Pradeepa, V., Chellappandian, M., Kalaivani, K., Abdel-Megeed, A., Narayanan, R., & Murugan, K. (2017). Chemicals isolated from Justicia adhatoda Linn reduce fitness of the mosquito, Aedes aegypti L. *Archives of Insect Biochemistry and Physiology*, *94*(4), e21384. DOI: 10.1002/arch.21384 PMID: 28266058

Thapa, L. B., Kaewchumnong, K., Sinkkonen, A., & Sridith, K. (2020). "Soaked in rainwater" effect of *Ageratina adenophora* on seedling growth and development of native tree species in Nepal. *Flora (Jena)*, *263*, 151554. DOI: 10.1016/j.flora.2020.151554

Thompson, B. K., Olden, J. D., & Converse, S. J. (2021). Mechanistic invasive species management models and their application in conservation. *Conservation Science and Practice*, *3*(11), e533. DOI: 10.1111/csp2.533

Tiawoun, M. A., Malan, P. W., Moshobane, M. C., Ramarumo, L. J., Comole, A. A., & Naidoo, K. K. (2024). Ecological traits and socio-economic impacts of the alien invader weed Parthenium hysterophorus L. in South Africa's rangeland ecosystems: A review. *Diversity*, *16*(4), 205. DOI: 10.3390/d16040205

Timmis, K., & Ramos, J. L. (2021). *The soil crisis: the need to treat as a global health problem and the pivotal role of microbes in prophylaxis and therapy* (Vol. 14). Wiley Online Library.

Tittensor, D. P., Walpole, M., Hill, S. L. L., Boyce, D. G., Britten, G. L., Burgess, N. D., Butchart, S. H. M., Leadley, P. W., Regan, E. C., Alkemade, R., Baumung, R., Bellard, C., Bouwman, L., Bowles-Newark, N. J., Chenery, A. M., Cheung, W. W. L., Christensen, V., Cooper, H. D., Crowther, A. R., & Ye, Y. (2014). A mid-term analysis of progress toward international biodiversity targets. *Science*, *346*(6206), 241–244. DOI: 10.1126/science.1257484 PMID: 25278504

Tobbala, D. E., Khesha, G. M., Salama, R. S., & Badawy, A. A. (2024). Reinventing Concrete: Sustainable Nano-Alumina from Metallic Waste-Cans and Nano-Ferrite for Next-Generation Green Magnetite Heavyweight Concrete. *Journal of Building Engineering*, *93*, 109829. DOI: 10.1016/j.jobe.2024.109829

Tomat-Kelly, G., & Flory, S. L. (2023). Research gaps limit understanding of invasion-fire cycles. *Biological Invasions*, *25*(3), 693–711. DOI: 10.1007/s10530-022-02951-y

Topalović, O., Koorem, K., & Wilschut, R. A. (2023). Drivers of host-microbiome interactions in the rhizosphere. *Frontiers in Plant Science*, *14*, 1294388. DOI: 10.3389/fpls.2023.1294388 PMID: 37828922

Török, P., Brudvig, L. A., Kollmann, J. N., Price, J., & Tóthmérész, B. (2021). The present and future of grassland restoration. *Restoration Ecology*, *29*, e13378. DOI: 10.1111/rec.13378

Torres, N., Herrera, I., Fajardo, L., & Bustamante, R. O. (2021). Meta-analysis of the impact of plant invasions on soil microbial communities. *BMC Ecology and Evolution*, *21*(1), 1–13. DOI: 10.1186/s12862-021-01899-2 PMID: 34496752

Tripathi, R. S., Khan, M. L., & Yadav, A. S. (2012). Biology of *Mikania micrantha* HBK: A review. In *Invasive alien plants: An ecological appraisal for the Indian subcontinent* (pp. 99–107). CABI: Cambridge, MA, USA.

Trotta, A. (2024). Environmental impact bonds: Review, challenges, and perspectives. *Current Opinion in Environmental Sustainability*, *66*, 101396. DOI: 10.1016/j.cosust.2023.101396

Trowbridge, C. C., Stanley, A., Kaye, T. N., Dunwiddie, P. W., & Williams, J. L. (2017). Long-term effects of prairie restoration on plant community structure and native population dynamics. *Restoration Ecology*, *25*(4), 559–568. DOI: 10.1111/rec.12468

Tsynkevych, S. (2018). Modeling the risks and damages from a "potential" invasive plant species: yellow starthistle in British Columbia.

Tu, M., Hurd, C., & Randall, J. M. (2001). Weed control methods handbook: Tools and techniques for use in natural areas. The Nature Conservancy. Retrieved from https://tncweeds.ucdavis.edu/handbook.html

Turner, W., Rondinini, C., Pettorelli, N., Mora, B., Leidner, A. K., Szantoi, Z., Buchanan, G., Dech, S., Dwyer, J., Herold, M., Koh, L. P., Leimgruber, P., Taubenboeck, H., Wegmann, M., Wikelski, M., & Woodcock, C. (2015). Free and open-access satellite data are key to biodiversity conservation. *Biological Conservation*, *182*, 173–176. DOI: 10.1016/j.biocon.2014.11.048

Uddin, M. N., Robinson, R. W., & Asaeda, T. (2020). Nitrogen immobilization may reduce invasibility of nutrient enriched plant community invaded by Phragmites australis. *Scientific Reports*, *10*(1), 1601. DOI: 10.1038/s41598-020-58523-4 PMID: 32005878

Uddin, M. S., & Rashid, M. H. U. (2023). Effects of auxins on the stem cuttings of Justicia Adhatoda L. *Bangladesh Journal of Botany*, *52*(3), 861–865. DOI: 10.3329/bjb.v52i3.68947

Uddin, M., & Robinson, R. (2017). Changes associated with Phragmites australis invasion in plant community and soil properties: A study on three invaded communities in a wetland, Victoria, Australia. *Limnologica*, *66*, 24–30. DOI: 10.1016/j.limno.2017.07.006

Ullah, A., Gao, D., & Wu, F. (2024). Common mycorrhizal network: The predominant socialist and capitalist responses of possible plant-plant and plant-microbe interactions for sustainable agriculture. *Frontiers in Microbiology*, *15*, 1183024. DOI: 10.3389/fmicb.2024.1183024 PMID: 38628862

Ullah, F., Baloch, M., Ullah, S., Kabir, M., & Khan, N. H. (2022). Eucalyptus plantation in Pakistan: Holistic view of environmental socioeconomic and medicinal perspective. *Biometrical Letters*, *8*(1), 41–49.

Ullah, M. S., Farooque, A. A., Javed, Q., Ullah, I., Bo, Y., Jabran, K., & Sun, J. (2024). Under Biological Invasion: Impacts of Litter Decomposition Mediated by Invasive Plant Species on Soil Nutrients and Functional Growth Traits of both Invasive and Native Plant Species. *Russian Journal of Ecology*, *55*(2), 89–100. DOI: 10.1134/S1067413624020061

Ullah, R., Khan, N., Hewitt, N., Ali, K., Jones, D. A., & Khan, M. E. H. (2022). Invasive Species as Rivals: Invasive Potential and Distribution Pattern of Xanthium strumarium L. *Sustainability (Basel)*, *14*(12), 7141. DOI: 10.3390/su14127141

Umair, M., Altaf, M., Ahsan, T., Bussmann, R. W., Abbasi, A. M., Gatasheh, M. K., & Elrobh, M. (2024). Study of medicinal plants used in ethnoveterinary medical system in riverine areas of Punjab, Pakistan. *Journal of Ethnobiology and Ethnomedicine*, *20*(1), 48. DOI: 10.1186/s13002-024-00686-9 PMID: 38715115

Underwood, E. C., Klinger, R. C., & Brooks, M. L. (2019). Effects of invasive plants on fire regimes and postfire vegetation diversity in an arid ecosystem. *Ecology and Evolution*, *9*(22), 12421–12435. DOI: 10.1002/ece3.5650 PMID: 31788187

Urcelay, C., Longo, S., Geml, J., Tecco, P. A., & Nouhra, E. (2017). Co-invasive exotic pines and their ectomycorrhizal symbionts show capabilities for wide distance and altitudinal range expansion. *Fungal Ecology*, *25*, 50–58. DOI: 10.1016/j.funeco.2016.11.002

Usman, M., Ho-Plágaro, T., Frank, H. E., Calvo-Polanco, M., Gaillard, I., Garcia, K., & Zimmermann, S. D. (2021). Mycorrhizal symbiosis for better adaptation of trees to abiotic stress caused by climate change in temperate and boreal forests. *Frontiers in Forests and Global Change*, *4*, 742392. DOI: 10.3389/ffgc.2021.742392

Utkhamthiang, K., Chaimanee, V., Hermhuk, S., & Kamyo, T. (2025). Assessment of Invasive Species Severity along the Nature Trail at the Doi Chiang Dao Biosphere Reserve, Chiang Mai Province. *Thai Forest Ecological Research Journal*, *9*(1), 55–72. DOI: 10.34044/tferj.2025.9.1.6271

Van der Heijden, M. G., Bardgett, R. D., & van Straalen, N. M. (2008). The unseen majority: Soil microbes as drivers of plant diversity and productivity in terrestrial ecosystems. *Ecology Letters*, *11*(3), 296–310. DOI: 10.1111/j.1461-0248.2007.01139.x PMID: 18047587

Van der Putten, W. H. (2010). Impacts of soil microbial communities on exotic plant invasions. *Trends in Ecology & Evolution*, *25*(9), 512–519. DOI: 10.1016/j.tree.2010.06.006 PMID: 20638747

Van Driesche R, Center T (2013) Biological control of invasive plants in protected areas. Plant invasions in protected areas: patterns, problems and challenges: 561-597

Van Driesche, R., & Hoddle, M. (2009). *Control of pests and weeds by natural enemies: an introduction to biological control.* John Wiley & Sons.

Van Niekerk, G., Young, E., & Fourie, H. (2024). *Veld.* Penguin random house South Africa. (Pg, 320). ISBN: 1779890222, 9781779890221

Van Noordwijk, M., Ekadinata, A., Leimona, B., Catacutan, D., Martini, E., Tata, H. L., Öborn, I., Hairiah, K., Wangpakapattanawong, P., Mulia, R., Dewi, S., Rahayu, S., & Zulkarnain, T. (2020). Agroforestry options for degraded landscapes in Southeast Asia. *Agroforestry for Degraded Landscapes: Recent Advances and Emerging Challenges-Vol.*, *1*, 307–347. DOI: 10.1007/978-981-15-4136-0_11

Van Wilgen, N. J., Faulkner, K. T., Robinson, T. B., South, J., Beckett, H., Janion-Scheepers, C., . . . Richardson, D. M. (2022). Climate change and biological invasions in South Africa. In *Invasive species and global climate change* (pp. 158-187). GB: CABI.

van Wilgen, B. W., Cole, N. S., Baard, J., Cheney, C., Engelbrecht, K., Stafford, L., Turner, A. A., van Wilgen, N. J., & Wannenburgh, A. M. (2025). Progress towards the control of invasive alien species in the Cape Floristic Region's protected areas. *Biological Invasions*, *27*(1), 8. DOI: 10.1007/s10530-024-03459-3 van Wilgen, B. W., & Richardson, D. M. (2014). Challenges and trade-offs in the management of invasive alien trees. *Biological Invasions, 16*(3), 721–734. DOI: 10.1007/s10530-013-0615-8

Varghese, M., Kamalakannan, R., & Suraj, P. (2024). Eucalyptus: Taxonomy, Geographic Distribution, Domestication, Breeding, Ecology and Economic Importance. In *Economically Important Trees: Origin, Evolution, Genetic Diversity and Ecology* (pp. 81-128). Springer.

Velez-Gavilan, J. (2024). *Phragmites australis* (common reed). DOI: 10.1079/cabicompendium.40514

Velez-Gavilan, J. (2024). *Sorghum Halepense (Johnson Grass)*. CABI Compendium. DOI: 10.1079/cabicompendium.50624

Venette, R. C., Gordon, D. R., Juzwik, J., Koch, F. H., Liebhold, A. M., Peterson, R. K., Sing, S. E., & Yemshanov, D. (2021). Early intervention strategies for invasive species management: connections between risk assessment, prevention efforts, eradication, and other rapid responses. *Invasive species in forests and rangelands of the United States: A comprehensive science synthesis for the United States forest sector*, 111-131.

Vilà, M., Trillo, A., Castro-Díez, P., Gallardo, B., & Bacher, S. (2024). Field studies of the ecological impacts of invasive plants in Europe. DOI: 10.3897/neobiota.90.112368

Vilà, M., Espinar, J. L., Hejda, M., Hulme, P. E., Jarošík, V., Maron, J. L., Pergl, J., Schaffner, U., Sun, Y., & Pyšek, P. (2011). Ecological impacts of invasive alien plants: A meta-analysis of their effects on species, communities and ecosystems. *Ecology Letters, 14*(7), 702–708. DOI: 10.1111/j.1461-0248.2011.01628.x PMID: 21592274

Vilà, M., & Hulme, P. E. (2017). Non-native species, ecosystem services, and human well-being. In *Impact of biological invasions on ecosystem services* (pp. 1–14). Springer. DOI: 10.1007/978-3-319-45121-3_1

Vila, M., & Weiner, J. (2004). Are invasive plant species better competitors than native plant species?–evidence from pair-wise experiments. *Oikos, 105*(2), 229–238. DOI: 10.1111/j.0030-1299.2004.12682.x

Vitti, S., Pellegrini, E., Casolo, V., Trotta, G., & Boscutti, F. (2020). Contrasting responses of native and alien plant species to soil properties shed new light on the invasion of dune systems. *Journal of Plant Ecology, 13*(6), 667–675. DOI: 10.1093/jpe/rtaa052

Volesky, L. A., Iqbal, S., Kelly, J., & Geddes, P. (2018). Relationships of native and exotic strains of Phragmites australis to wetland ecosystem properties. *Wetlands*, *38*(3), 577–589. DOI: 10.1007/s13157-018-1001-1

Wade, D. T. (2020). What is rehabilitation? An empirical investigation leading to an evidence-based description. *Clinical Rehabilitation*, *34*(1), 3–11. DOI: 10.1177/0269215515601176 PMID: 32037876

Wafer, A. (2025). Into the weeds: integrating evolutionary ecology into weed science The Ohio State University].

Waheed, M., Haq, S. M., Arshad, F., Bussmann, R. W., Ali, H. M., & Siddiqui, M. H. (2023). Phyto-ecological distribution patterns and identification of alien invasive indicator species in relation to edaphic factors from semi-arid region. *Ecological Indicators*, *148*, 110053. DOI: 10.1016/j.ecolind.2023.110053

Waheed, M., Haq, S. M., Arshad, F., Jameel, M. A., Siddiqui, M. H., Bussmann, R. W., Manshoor, N., & Alamri, S. (2023). Where will threatened Aegle marmelos L., a tree of the semi-arid region, go under climate change? implications for the reintroduction of the species. *Land (Basel)*, *12*(7), 1433. DOI: 10.3390/land12071433

Waheed, M., Haq, S. M., Arshad, F., Vitasović-Kosić, I., Bussmann, R. W., Hashem, A., & Abd-Allah, E. F. (2024). Xanthium strumarium L., an invasive species in the subtropics: Prediction of potential distribution areas and climate adaptability in Pakistan. *BMC Ecology and Evolution*, *24*(1), 124. DOI: 10.1186/s12862-024-02310-6 PMID: 39390368

Wails, C. N., Baker, K., Blackburn, R., Del Vallé, A., Heise, J., Herakovich, H., Holthuijzen, W. A., Nissenbaum, M. P., Rankin, L., Savage, K., Vanek, J. P., & Jones, H. P. (2021). Assessing changes to ecosystem structure and function following invasion by Spartina alterniflora and Phragmites australis: A meta-analysis. *Biological Invasions*, *23*(9), 2695–2709. DOI: 10.1007/s10530-021-02540-5

Walia, K., & Kumari, Y. Garima, & Mehta, A. (2025). Ecosystem Recovery and Resilience After Forest Fires. In Forest Fire and Climate Change: Insights into Science (pp. 119-145). Springer.

Walker, B., Holling, C. S., Carpenter, S. R., & Kinzig, A. (2004). Resilience, adaptability, and transformability in social–ecological systems. *Ecology and Society*, *9*(2), 5. https://www.ecologyandsociety.org/vol9/iss2/art5/. DOI: 10.5751/ES-00650-090205

Waller, J. M. (2024). *Big Intel: How the CIA and FBI Went from Cold War Heroes to Deep State Villains*. Simon and Schuster.

Walter, I., & Calvo, R. (2009). Biomass production and development of native vegetation following biowaste amendment of a degraded, semi-arid soil. *Arid Land Research and Management*, *23*(4), 297–310. DOI: 10.1080/15324980903231827

Walters, C. J., & Holling, C. S. (1990). Large-scale management experiments and learning by doing. *Ecology*, *71*(6), 2060–2068. DOI: 10.2307/1938620

Wan, F.-H., & Yang, N.-W. (2016). Invasion and management of agricultural alien insects in China. *Annual Review of Entomology*, *61*(1), 77–98. DOI: 10.1146/annurev-ento-010715-023916 PMID: 26527302

Wang, Y., Lin, J., Yang, F., Tao, S., Yan, X., Zhou, Z., & Zhang, Y. (2022). Arbuscular mycorrhizal fungi improve the growth and performance in the seedlings of Leymus chinensis under alkali and drought stresses. *PeerJ*, *10*, e12890.2-1114.

Wang, C. L. (2021). New frontiers and future directions in interactive marketing: Inaugural Editorial. *Journal of Research in Interactive Marketing*, *15*(1), 2021. DOI: 10.1108/JRIM-03-2021-270

Wang, C., Wang, W., Sardans, J., Ouyang, L., Tong, C., Asensio, D., Gargallo-Garriga, A., Wiesmeier, M., & Peñuelas, J. (2020). Higher fluxes of C, N and P in plant/soil cycles associated with plant invasion in a subtropical estuarine wetland in China. *The Science of the Total Environment*, *703*, 135027. DOI: 10.1016/j.scitotenv.2020.139124 PMID: 32388112

Wang, C., Xiao, H., Liu, J., Wang, L., & Du, D. (2015). Insights into ecological effects of invasive plants on soil nitrogen cycles. *American Journal of Plant Sciences*, *6*(1), 34–46. DOI: 10.4236/ajps.2015.61005

Wangchu, L., Tasung, A., Kumar, A., Angami, T., Suryawanshi, M. A., Jini, D., Singh, R., Alone, R. A., Bhagawati, S. K., & Khan, R. TRAINING MANUAL ON.

Wang, F. (2017). Occurrence of arbuscular mycorrhizal fungi in mining-impacted sites and their contribution to ecological restoration: Mechanisms and applications. *Critical Reviews in Environmental Science and Technology*, *47*(20), 1901–1957. Advance online publication. DOI: 10.1080/10643389.2017.1400853

Wang, J., Raes, B., Debrabandere, C., van Aken, V., Jaramillo-Toro, S., Waldherr, S., Horemans, B., & Springael, D. (2025). Niche creation improves bioaugmentation of an organic micropollutant degrader in oligotrophic waters. *The ISME Journal*, *•••*, wraf140. DOI: 10.1093/ismejo/wraf140 PMID: 40613637

Wang, L., Rinklebe, J., Tack, F. M. G., & Hou, D. (2021). A review of green remediation strategies for heavy metal contaminated soil. *Soil Use and Management*, *37*(1), 1–15. DOI: 10.1111/sum.12717

Wang, M., Feng, Y., & Li, X. (2006). Effects of soil phosphorus level on morphological and photosynthetic characteristics of *Ageratina adenophora* and *Chromolaena odorata*. *Ying Yong Sheng Tai Xue Bao. Journal of Applied Ecology*, *17*, 602–606. PMID: 16836086

Wang, M., Lamit, L. J., Lilleskov, E. A., Basiliko, N., Moore, T. R., Bubier, J. L., Guo, G., Juutinen, S., & Larmola, T. (2024). Peatland fungal community responses to nutrient enrichment: A story beyond nitrogen. *Global Change Biology*, *30*(11), e17562. DOI: 10.1111/gcb.17562 PMID: 39492595

Wang, R., Zou, R., Liu, J., Liu, L., & Hu, Y. (2021). Spatial distribution of soil nutrients in farmland in a hilly region of the Pearl River Delta in China based on geostatistics and the inverse distance weighting method. *Agriculture*, *11*(1), 50. DOI: 10.3390/agriculture11010050

Wang, S. (2025). Multi-Level Governance of Managing Invasive Alien Species. In *SUSTAINABLE MANAGEMENT OF INVASIVE SPECIES* (pp. 17–44). World Scientific. DOI: 10.1142/9781800615847_0002

Wang, S., Augé, R. M., & Toler, H. D. (2017). Arbuscular mycorrhiza formation and its function under elevated atmospheric O3: A meta-analysis. *Environmental Pollution*, *226*, 104–117. DOI: 10.1016/j.envpol.2017.03.074 PMID: 28411495

Wang, S., Li, L., & Zhou, D. W. (2021). Root morphological responses to population density vary with soil conditions and growth stages: The complexity of density effects. *Ecology and Evolution*, *11*(15), 10590–10599. DOI: 10.1002/ece3.7868 PMID: 34367599

Wang, X., Fang, L., Beiyuan, J., Cui, Y., Peng, Q., Zhu, S., & Zhang, X. (2021). Improvement of alfalfa resistance against Cd stress through rhizobia and arbuscular mycorrhiza fungi co-inoculation in Cd-contaminated soil. *Environmental Pollution*, *277*, 116758. DOI: 10.1016/j.envpol.2021.116758 PMID: 33652182

Wang, Y., Derakhshan, A., & Zhang, L. J. (2021). Researching and practicing positive psychology in second/foreign language learning and teaching: The past, current status and future directions. *Frontiers in Psychology*, *12*, 2021. DOI: 10.3389/fpsyg.2021.731721 PMID: 34489835

Wang, Y., Wang, M., Li, Y., Wu, A., & Huang, J. (2018). Effects of arbuscular mycorrhizal fungi on growth and nitrogen uptake of Chrysanthemum morifolium under salt stress. *PLoS One*, *13*(4), e0196408. DOI: 10.1371/journal.pone.0196408 PMID: 29698448

Wan, J.-Z., Wang, C.-J., & Yu, F.-H. (2017). Wind effects on habitat distributions of wind-dispersed invasive plants across different biomes on a global scale: Assessment using six species. *Ecological Informatics*, *42*, 38–45. DOI: 10.1016/j.ecoinf.2017.09.002

Wan, X., & Zhang, L. (2021). Global new species of vascular plants published in 2020. *Shengwu Duoyangxing*, *29*(8), 1003–1010. DOI: 10.17520/biods.2021156

Waseem, M., Yao, G.-Q., Hasan, M. M., Li, Y.-R., Qi, S.-H., Cai, L., Yang, D., & Fang, X.-W. (2024). Divergent hydraulic and gas-exchange strategies in two closely related Salix species. *Journal of Plant Ecology*, *17*(6), rtae075. DOI: 10.1093/jpe/rtae075

Watson G, (2024). "Phosphorus Dynamics in the Sediment of a Lake Erie Coastal Wetland."

Wedding, L. M., Moritsch, M., Verutes, G., & Arkema, K. (2021). Incorporating blue carbon sequestration benefits into sub-national climate policies. [– Elsevier]. *Environmental Science & Policy*, •••, 2021.

Wehi, P. M., Kamelamela, K. L., Whyte, K., Watene, K., & Reo, N. (2023). Contribution of Indigenous Peoples' understandings and relational frameworks to invasive alien species management. *People and Nature*, *5*(5), 1403–1414. DOI: 10.1002/pan3.10508

Wehr, N. H., Boone, H. M., van den Bosch, M., Zubiria Perez, A., Wehr, K. L., Wehr, S. R., & Belant, J. L. (2025). Interspecific carnivore competition and ungulate predation correlate with predator species richness. *Ecology*, *106*(6), e70136. DOI: 10.1002/ecy.70136 PMID: 40528493

Weidlich, E. W., Flórido, F. G., Sorrini, T. B., & Brancalion, P. H. (2020). Controlling invasive plant species in ecological restoration: A global review. *Journal of Applied Ecology*, *57*(9), 1806–1817. DOI: 10.1111/1365-2664.13656

Wei, W., Zhu, P., Chen, P., Huang, Q., Bai, X., Ni, G., & Hou, Y. (2020). Mixed evidence for plant–soil feedbacks in forest invasions. *Oecologia*, *192*(3), 651–663. DOI: 10.1007/s00442-020-04703-y PMID: 32642792

Wei, X. P., Yu, J., Gao, T. G., Zhang, B. G., & Zhang, Z. (2020). First report and eradication of an invasive weed, *Solanum carolinense* L. *Zhiwu Jianyi*, *34*(3), 61–63.

Were, D., Kansiime, F., Fetahi, T., Cooper, A., & Jjuuko, C. (2019). Carbon sequestration by wetlands: A critical review of enhancement measures for climate change mitigation. *Earth Systems and Environment*, *3*(2), 327–340. DOI: 10.1007/s41748-019-00094-0

Westbrook, A. S., & DiTommaso, A. (2023). Hybridization in agricultural weeds: A review from ecological, evolutionary, and management perspectives. *American Journal of Botany*, *110*(12), e16258. DOI: 10.1002/ajb2.16258 PMID: 38031455

Westbrook, A. S., Milbrath, L. R., Weinberg, J., & DiTommaso, A. (2023). Biology of Invasive Plants 3. *Vincetoxicum nigrum* (L.) Moench and *Vincetoxicum rossicum* (Kleopow) Barbarich. *Invasive Plant Science and Management*, *16*(1), 3–26. DOI: 10.1017/inp.2023.7

Wied, J. P., Perotto-Baldivieso, H. L., Conkey, A. A., Brennan, L. A., & Mata, J. M. (2020). Invasive grasses in South Texas rangelands: Historical perspectives and future directions. *Invasive Plant Science and Management*, *13*(2), 41–58. DOI: 10.1017/inp.2020.11

Williams, B. K., Szaro, R. C., & Shapiro, C. D. (2009). *Adaptive management: The U.S. Department of the Interior technical guide*. U.S. Department of the Interior.

Wilson, J. R. U., Holst, N., & Rees, M. (2005). Determinants and patterns of population growth in water hyacinth. *Weed Research*, *47*(5), 391–400. DOI: 10.1111/j.1365-3180.2007.00582.x

Wilson, J. R., Foxcroft, L. C., Geerts, S., Hoffman, M. T., MacFadyen, S., Measey, J., Mills, A., Richardson, D. M., Robertson, M. P., & van Wilgen, B. W. (2020). The role of environmental factors in promoting and limiting biological invasions in South Africa. In *Biological invasions in South Africa* (pp. 355–385). Springer International Publishing. DOI: 10.1007/978-3-030-32394-3_13

Wilson, S. (2024). *Biology, Ecology, and Management of Prosapia bicincta (Hemiptera: Cercopidae) in Hawai'i Rangelands*. University of Hawai'i at Manoa.

Wolkovich, E. M., & Cleland, E. E. (2014). Phenological niches and the future of invaded ecosystems with climate change. *AoB Plants*, *6*, plu013. DOI: 10.1093/aobpla/plu013 PMID: 24876295

Wolsiffer, S. K. (2025). *Wildland fuel loading estimates along a gradient of forest cover types and landscape factors in Harpers Ferry National Historic Park, West Virginia and Rock Creek Park*. USA Virginia Tech.

Wong, S., Kaur, J., Kumar, P., Karremans, A. P., & Sharma, J. (2024). Distinct orchid mycorrhizal fungal communities among co-occurring *Vanilla* species in Costa Rica: Root substrate and population-based segregation. *Mycorrhiza*, *34*(3), 1–22. DOI: 10.1007/s00572-024-01147-7 PMID: 38664239

Wortley, L., Hero, J.-M., & Howes, M. (2013). Evaluating ecological restoration success: A review of the literature. *Restoration Ecology, 21*(5), 537–543. DOI: 10.1111/rec.12028

Wright, B. R., Latz, P. K., Albrecht, D. E., & Fensham, R. J. (2021). Buffel grass (Cenchrus ciliaris) eradication in arid central Australia enhances native plant diversity and increases seed resources for granivores. *Applied Vegetation Science, 24*(1), e12533. DOI: 10.1111/avsc.12533

Wubs, E. J., van der Putten, W. H., Bosch, M., & Bezemer, T. M. (2016). Soil inoculation steers restoration of terrestrial ecosystems. *Nature Plants, 2*(8), 16107. DOI: 10.1038/nplants.2016.107 PMID: 27398907

Xia, T., Wang, Y., He, Y., Wu, C., Shen, K., Tan, Q., Kang, L., Guo, Y., Wu, B., & Han, X. (2020). Invasive plant experiences greater benefits of root morphology from enhancing nutrient competition associated with arbuscular mycorrhizae in karst soil than a native …. *PLoS One, 15*(10), e0239778. DOI: 10.1371/journal.pone.0234410 PMID: 32516341

Xie, T., Wang, Q., Ning, Z., Chen, C., Cui, B., Bai, J., Shi, W., & Pang, B. (2021). Artificial modification on lateral hydrological connectivity promotes range expansion of invasive Spartina alterniflora in salt marshes of the Yellow River delta, China. *The Science of the Total Environment, 751*, 141673. DOI: 10.1016/j.scitotenv.2020.144476 PMID: 33460837

Xue, L., Zhang, Y., Wei, F., Shi, G., Tian, B., Yuan, Y., Jiang, W., Zhao, M., Hu, L., Xie, Z., & Gu, H. (2024). Recent Progress on Plant Apomixis for Genetic Improvement. *International Journal of Molecular Sciences, 25*(21), 11378. DOI: 10.3390/ijms252111378 PMID: 39518931

Xu, H.. (2022). A global meta-analysis of the impacts of exotic plant invasions on soil: Increased nutrient content and microbial activity but decreased plant diversity. *The Science of the Total Environment*.

Xu, H., Ding, H., Li, M., Qiang, S., Guo, J., Han, Z., Huang, Z., Sun, H., He, S., Wu, H., & Wan, F. (2006). The distribution and economic losses of alien species invasion to China. *Biological Invasions, 8*(7), 1495–1500. DOI: 10.1007/s10530-005-5841-2

Xu, H., Liu, Q., Wang, S., Yang, G., & Xue, S. (2022). A global meta-analysis of the impacts of exotic plant species invasion on plant diversity and soil properties. *The Science of the Total Environment, 807*, 150786. DOI: 10.1016/j.scitotenv.2021.152286

Xu, H., Qiang, S., Genovesi, P., Ding, H., Wu, J., Meng, L., Han, Z., Miao, J., Hu, B., Guo, J., Sun, H., Huang, C., Lei, J., Le, Z., Zhang, X., He, S., Wu, Y. U., Zheng, Z., Chen, L., & Pysek, P. (2012). An inventory of invasive alien species in China. *NeoBiota*, *15*, 1–26. DOI: 10.3897/neobiota.15.3575

Xu, H., Xu, F., Lin, T., Xu, Q., Yu, P., Wang, C., Aili, A., & Zhao, X. (2023). A systematic review and comprehensive analysis on ecological restoration of mining areas in the arid region of China: Challenge, capability and reconsideration. [– Elsevier]. *Ecological Engineering*, •••, 2023.

Xu, Z., Guo, X., Caplan, J. S., Li, M., & Guo, W. (2021). Novel plant-soil feedbacks drive adaption of invasive plants to soil legacies of native plants under nitrogen deposition. *Plant and Soil*, *467*(1), 347–362. DOI: 10.1007/s11104-021-05057-x

Yadav, S. P. S., Mehata, D. K., Pokhrel, S., Ghimire, N. P., Gyawali, P., Katel, S., & Timilsina, U. (2024). Invasive alien plant species (Banmara): Investigating its invasive potential, ecological consequences on biodiversity, and management strategies. *Journal of Agriculture and Food Research*, *15*, 101031. DOI: 10.1016/j.jafr.2024.101031

Yang, F., Liu, M., Wang, X., Hong, Y., Yao, Q., Chang, X., Shi, G., Chen, W., Tian, B., & Hegazy, A. (2024). Differences in the Microbial Composition and Function of the Arundo donax Rhizosphere Under Different Cultivation Conditions. *Microorganisms*, *12*(12), 2642. DOI: 10.3390/microorganisms12122642 PMID: 39770844

Yang, X.-Q., Kushwaha, S., Saran, S., Xu, J., & Roy, P. (2013). Maxent modeling for predicting the potential distribution of medicinal plant, Justicia adhatoda L. in Lesser Himalayan foothills. *Ecological Engineering*, *51*, 83–87. DOI: 10.1016/j.ecoleng.2012.12.004

Yeasmin, R., Bonser, S. P., Motoki, S., & Nishihara, E. (2019). Arbuscular mycorrhiza influences growth and nutrient uptake of asparagus (Asparagus officinalis L.) under heat stress. *HortScience*, *54*(5), 846–850. DOI: 10.21273/HORTSCI13587-18

Yelenik, S. G., Stock, W. D., & Richardson, D. M. (2004). Ecosystem level impacts of invasive Acacia saligna in the South African fynbos. *Restoration Ecology*, *12*(1), 44–51. DOI: 10.1111/j.1061-2971.2004.00289.x

Ye, X. Q., Yan, Y. N., Wu, M., & Yu, F. H. (2019). High capacity of nutrient accumulation by invasive *Solidago canadensis* in a coastal grassland. *Frontiers in Plant Science*, *10*, 575. DOI: 10.3389/fpls.2019.00575 PMID: 31134115

Yin, Z., Zhang, Y., Hu, N., Shi, Y., Li, T., & Zhao, Z. (2021). Differential responses of 23 maize cultivar seedlings to an arbuscular mycorrhizal fungus when grown in a metal-polluted soil. *The Science of the Total Environment*, *789*, 148015. DOI: 10.1016/j.scitotenv.2021.148015 PMID: 34051499

Yong, J., Wen-Jie, W., Yan-Hong, L., & Shong-Lian, B. (2012). Influence of light quality and quantity on germination and seedling status of invasive species Eupatorium adenophorum. *Bulletin of Botanical Research*, *32*(4), 415.

Yooyongwech, S., Samphumphuang, T., Tisarum, R., Theerawitaya, C., & Cha-Um, S. (2017). Water-deficit tolerance in sweet potato [Ipomoea batatas (L.) Lam.] by foliar application of paclobutrazol: Role of soluble sugar and free proline. *Frontiers in Plant Science*, *8*, 1400. DOI: 10.3389/fpls.2017.01400 PMID: 28848596

Young, T. P., Petersen, D. A., & Clary, J. J. (2005). The ecology of restoration: Historical links, emerging issues, and unexplored realms. *Ecology Letters*, *8*(6), 662–673. DOI: 10.1111/j.1461-0248.2005.00764.x

Yu, H., Le Roux, J. J., Jiang, Z., Sun, F., Peng, C., & Li, W. (2021). Soil nitrogen dynamics and competition during plant invasion: Insights from Mikania micrantha invasions in China. *The New Phytologist*, *229*(2), 841–854. DOI: 10.1111/nph.17125 PMID: 33259063

Zachman, A. N. (2024). *Fire Mediated Forest Structure Influences the Dispersal Dynamics of a Bunchgrass*. The University of North Carolina at Chapel Hill.

Zahoor, M. A. A., Haidar, S. H., Rafique, S., Ahmed, U., Umber Rauf, M. A., Hanif, U., & Shehryar, M. (2025). Biotechnological Advances in Wildlife Conservation: Genetic Engineering, Cloning, Ecosystem Restoration, and Nanoparticle Applications. *Scholars Academic Journal of Biosciences*, *4*(4), 386–416. DOI: 10.36347/sajb.2025.v13i04.005

Zambelli, A., Nocito, F. F., & Araniti, F. (2025). Unveiling the multifaceted roles of root exudates: Chemical interactions, allelopathy, and agricultural applications. *Agronomy (Basel)*, *15*(4), 1–30. DOI: 10.3390/agronomy15040845

Zargar, S. A., Malik, R. A., Khuroo, A. A., Ganie, A. H., & Reshi, Z. A. (2025). Alien flora of the Himalayan highlands: Naturalised and invasive plants in the Trans-Himalayan region of Ladakh, India. *Biological Invasions*, *27*(4), 1–19. DOI: 10.1007/s10530-025-03562-z

Zedler, J. B., & Callaway, J. C. (1999). Tracking wetland restoration: Do mitigation sites follow desired trajectories? *Restoration Ecology*, *7*(1), 69–73. DOI: 10.1046/j.1526-100X.1999.07108.x

Zedler, J. B., & Kercher, S. (2005). Wetland resources: Status, trends, ecosystem services, and restorability. *Annual Review of Environment and Resources*, *30*(1), 39–74. DOI: 10.1146/annurev.energy.30.050504.144248

Zeng, Z., Liang, Z., Chen, Y., Shu, Q., Li, J., Tso, N., Chen, M., Zhang, S., Tan, X., Qiong, L., & Wang, J. (2025). Adaptation analysis of two Asteraceae invasive plants in Lhasa, Tibet. *PeerJ*, *13*, e19667. DOI: 10.7717/peerj.19667 PMID: 40642327

Zenni, R. D., Ziller, S. R., da Rosa, C. A., Sühs, R. B., Puechagut, P. B., Marterer, B. T., Heringer, G., de Matos, A. C. L., Bueno, M. L., Almeida, J. W., Silva, C. H. T., Legracie-Júnior, J. R., & Chapla, T. E. (2024). Invasive non-native species in Brazil: An updated overview. *Biological Invasions*, *26*(8), 1–9. DOI: 10.1007/s10530-024-03302-9

Zhan, F., Li, B., Jiang, M., Li, T., He, Y., Li, Y., & Wang, Y. (2019). Effects of arbuscular mycorrhizal fungi on the growth and heavy metal accumulation of bermudagrass [Cynodon dactylon (L.) Pers.] grown in a lead–zinc mine wasteland. *International Journal of Phytoremediation*, *21*(9), 849–856. DOI: 10.1080/15226514.2019.1577353 PMID: 30994000

Zhang, C., van der Heijden, M. G., Dodds, B. K., Nguyen, T. B., Spooren, J., Valzano-Held, A., & Berendsen, R. L. (2024b). A tripartite bacterial-fungal-plant symbiosis in the mycorrhiza-shaped microbiome drives plant growth and mycorrhization. *Microbiome*, *12*(1), 13. DOI: 10.1186/s40168-023-01726-4 PMID: 38243337

Zhang, D., Ge, W., & Zhang, Y. (2020). Evaluating the vegetation restoration sustainability of ecological projects: A case study of Wuqi County in China. [– Elsevier]. *Journal of Cleaner Production*, *264*, 2020. DOI: 10.1016/j.jclepro.2020.121751

Zhang, D., Lyu, Y., Li, H., Tang, X., Hu, R., Rengel, Z., Zhang, F., Whalley, W. R., Davis, W. J., & Cahill, J. F.Jr. (2020). Neighbouring plants modify maize root foraging for phosphorus: Coupling nutrients and neighbours for improved nutrient–use efficiency. *The New Phytologist*, *226*(1), 244–253. DOI: 10.1111/nph.16206 PMID: 31536638

Zhang, F., Li, Q., Yerger, E. H., Chen, X., Shi, Q., & Wan, F. (2018). AM fungi facilitate the competitive growth of two invasive plant species, *Ambrosia artemisiifolia* and *Bidens pilosa*. *Mycorrhiza*, *28*(8), 703–715. DOI: 10.1007/s00572-018-0866-4 PMID: 30220052

Zhang, H. Y., Goncalves, P., Copeland, E., Qi, S. S., Dai, Z. C., Li, G. L., & Thomas, T. (2020b). Invasion by the weed *Conyza canadensis* alters soil nutrient supply and shifts microbiota structure. *Soil Biology & Biochemistry*, *143*, 107739. DOI: 10.1016/j.soilbio.2020.107739

Zhang, P., Li, B., Wu, J., & Hu, S. (2019). Invasive plants differentially affect soil biota through litter and rhizosphere pathways: A meta-analysis. *Ecology Letters*, *22*(1), 200–210. DOI: 10.1111/ele.13181 PMID: 30460738

Zhang, Q., He, J., & Pan, Q. (2024a). The role of arbuscular mycorrhizal fungi in exotic plant invasion. [JSCET]. *Journal of Sustainable Civil Engineering & Technology*, *3*(2), 43–51. DOI: 10.24191/jscet.v3i2.43-51

Zhang, X., Fang, F., Sha, L., & Nizamani, M. M. (2025). Ecological dynamics and microbial community composition of Heliotropium arboreum in the coastal ecosystems of Hainan province. *Frontiers in Microbiology*, *16*, 1611262. DOI: 10.3389/fmicb.2025.1611262 PMID: 40678047

Zhang, X., Liu, K., Wang, S., Wu, T., Li, X., Wang, J., Wang, D., Zhu, H., Tan, C., & Ji, Y. (2022). Spatiotemporal evolution of ecological vulnerability in the Yellow River Basin under ecological restoration initiatives. [– Elsevier]. *Ecological Indicators*, *135*, 2022. DOI: 10.1016/j.ecolind.2022.108586

Zhang, X., Ward, B. B., & Sigman, D. M. (2020). Global nitrogen cycle: Critical enzymes, organisms, and processes for nitrogen budgets and dynamics. *Chemical Reviews*, *120*(2), 987–1016. DOI: 10.1021/acs.chemrev.9b00613 PMID: 31549814

Zhang, Y., Wang, X., Hu, R., & Pan, Y. (2017). Stemflow volume per unit rainfall as a good variable to determine the relationship between stemflow amount and morphological metrics of shrubs. *Journal of Arid Environments*, *141*, 1–6. DOI: 10.1016/j.jaridenv.2017.02.002

Zhang, Z., Liu, Y., Brunel, C., & van Kleunen, M. (2020c). Soil-microorganism-mediated invasional meltdown in plants. *Nature Ecology & Evolution*, *4*(12), 1612–1621. DOI: 10.1038/s41559-020-01311-0 PMID: 33020599

Zhang, Z., Pan, X., Blumenthal, D., Van Kleunen, M., Liu, M., & Li, B. (2018). Contrasting effects of specialist and generalist herbivores on resistance evolution in invasive plants. *Ecology*, *99*(4), 866–875. DOI: 10.1002/ecy.2155 PMID: 29352479

Zhang, Z., Pan, X., Zhang, Z., He, K. S., & Li, B. (2015). Specialist insect herbivore and light availability do not interact in the evolution of an invasive plant. *PLoS One*, *10*(9), e0139234. DOI: 10.1371/journal.pone.0139234 PMID: 26407176

Zhang, Z., Zou, Y., & Xiao, W. (2023). Exploration of a virtual restoration practice route for architectural heritage based on evidence-based design: A case study of the Bagong House. *Heritage Science*, *11*(1), 1–12. DOI: 10.1186/s40494-023-00878-8 PMID: 36844367

Zhao, S., & Riaz, M. (2024). Plant–soil interactions and nutrient cycling dynamics in tropical rainforests. In *Environment, Climate, Plant and Vegetation Growth* (pp. 229-264). Springer DOI:DOI: 10.1007/978-3-031-69417-2_8

Zhao, Y. Z., Liu, M. C., Feng, Y. L., Wang, D., Feng, W. W., Clay, K., & Kong, D. L. (2020). Release from below-and aboveground natural enemies contributes to invasion success of a temperate invader. *Plant and Soil*, *452*(1-2), 19–28. DOI: 10.1007/s11104-020-04520-5

Zheng, J., & Li, C. (2025). Plant functional diversity as an important mediator of invasive effects on productivity in riparian grasslands. *Biological Invasions*, *27*(5), 128. DOI: 10.1007/s10530-025-03586-5

Zheng, Y. L., Feng, Y. L., Liu, W. X., & Liao, Z. Y. (2009). Growth, biomass allocation, morphology, and photosynthesis of invasive *Eupatorium adenophorum* and its native congeners grown at four irradiances. *Plant Ecology*, *203*(2), 263–271. DOI: 10.1007/s11258-008-9544-5

Zheng, Y., Hu, H. W., Guo, L. D., Anderson, I. C., & Powell, J. R. (2017). Dryland forest management alters fungal community composition and decouples assembly of root-and soil-associated fungal communities. *Soil Biology & Biochemistry*, *109*, 14–22. DOI: 10.1016/j.soilbio.2017.01.024

Zhou, Q., Wang, Y., Li, X., Liu, Z., Wu, J., Musa, A., Ma, Q., Yu, H., Cui, X., & Wang, L. (2020). Geographical distribution and determining factors of different invasive ranks of alien species across China. *The Science of the Total Environment*, *722*, 137929. DOI: 10.1016/j.scitotenv.2020.137929 PMID: 32208272

Zhu, T. R., Litton, C. M., Giardina, C. P., & Trauernicht, C. (2021). Moisture availability and ecological restoration limit fine fuels and modelled wildfire intensity following non-native ungulate removal in Hawaii. *Journal of Applied Ecology*, *58*(10), 2207–2219. DOI: 10.1111/1365-2664.13952

Zhyvko, M. (2024). The cost of fragmentation: ecology and the impact on the global economy.

Zilio, M. I., Restrepo, A. F. S., Gervazoni, P., Minuti, G., Muzon, J., Djeddour, D., Mc Kay, F., & Sosa, A. (2025). Economic loss of ecosystem services caused by yellow flag iris and white ginger lily, two invasive alien plants in natural ecosystems of Argentina. *Journal for Nature Conservation*, *84*, 126789. DOI: 10.1016/j.jnc.2024.126789

Ziska, L. (2022). *Invasive species and global climate change*. CABI. DOI: 10.1079/9781800621459.0000

Zolper, T. J., Smith, D. L., Jackson, P. R., & Cupp, A. R. (2022). Performance of a carbon dioxide injection system at a navigation lock to control the spread of aquatic invasive species. *Journal of Environmental Engineering*, *148*(4), 04022011. DOI: 10.1061/(ASCE)EE.1943-7870.0001987

Zou, J., Ziegler, A. D., Chen, D., McNicol, G., Ciais, P., Jiang, X., Zheng, C., Wu, J., Wu, J., Lin, Z., He, X., Brown, L. E., Holden, J., Zhang, Z., Ramchunder, S. J., Chen, A., & Zeng, Z. (2022). Rewetting global wetlands effectively reduces major greenhouse gas emissions. *Nature Geoscience*, *15*(8), 627–632. DOI: 10.1038/s41561-022-00989-0

Zou, Y. N., Wu, Q. S., & Kuča, K. (2021). Unravelling the role of arbuscular mycorrhizal fungi in mitigating the oxidative burst of plants under drought stress. *Plant Biology*, *23*(S1), 50–57. DOI: 10.1111/plb.13161 PMID: 32745347

About The Contributors

Sana Basharat is working as a postdoctoral researcher at college of tropical agriculture and forestry at hainan university, haikou, hainan, China. She got her doctoral degree from Department of botany, university of agriculture faislabad, Pakistan. Her research direction involved structural and functional Strategies in Cenchrus Species to Combat Environmental Extremities Imposed by Multiple Abiotic Stresses. She had published more than 15 research articles in renowned journals including Plants, Pakistan Journal botany, Acta Physiologiae Plantarum, so on.

***

Usharani Bhimavarapu is an Associate Professor at Koneru Lakshmaiah Education Foundation in Andhra Pradesh, India, where she teaches Computer Science and Engineering. With a career spanning 17 years, she has become a respected authority in Data Mining, Machine Learning, and Data Structure. Dr. Bhimavarapus research prowess is evidenced by her publication of over 100 papers in SCI, SCIE, and Scopus indexed journals. Furthermore, her dedication to education is evident through her authorship of 25 books on programming languages like CPP, Java, Python, HTML, and CSS.

A.Rohini Working as Senior Assistant Professor in Environmental Science, department of Humanities and Sciences CVR College of Engineering from 2007 to till date Got Ph.D in 2023 from Osmania University Title of the thesis Ecological Studies on Safilguda Lake with Reference to Water Quality I have 18 years of teaching experience in Environmental science for B.Tech students My areas of research are Environmental pollution Limnolgy Climate change Algal Studies Biodiversity

Farooq Ahmad, Associate Professor, Department of Botany, University of Agriculture Faisalabad.

Muhammad Sajid Aqeel Ahmad is a distinguished botanist with a strong academic and professional background. He holds a Ph.D. Botany from University of Agriculture, Faisalabad and has 25 years of experience in Plant Ecology and Environmental Biology. Dr. Ahmad has made significant contributions to species distribution modelling and has published numerous papers in reputable journals. He is currently Associate Professor of Botany at University of Agriculture, Faisalabad], where he continues to excel in his field.

Muaz Ameen is a Research Assistant/Lab Technician at the Central Lab of the Post Graduate Research Station (PARS), University of Agriculture Faisalabad, Pakistan. He completed a Master of Philosophy in Botany at the same university, focusing on the effects of exogenous selenium applications on plant morpho-physiological and biochemical responses during drought stress. His research interests include plant stress, plant-environment interactions, biochemistry, molecular biology, and agriculture sciences.

Ansa Asghar is done her PhD. in Botany from University of Agriculture, Faisalabad, Pakistan. and now work as visiting lecturer in University of Education, Faisalabad. the mailing address is "ansaasghar1122@gmail.com".

Jagruti K. Barot is an Assistant Professor in the PG Department of Biosciences at Veer Narmad South Gujarat University, Surat, India, since 2008. She was awarded a Ph.D. in Zoology in December 2011 and secured the 54th rank in the UGC-CSIR NET examination for Lectureship (LS) in June 2011. Dr. Barot has published 15 research papers in reputed national and international journals and has presented over 10 papers at various national and international conferences. She has successfully guided two M.Phil. students, one Ph.D. student and currently supervises three Ph.D. candidates. Additionally, she has supervised more than 40 Master's Dissertation projects. Her academic research has been supported by a UGC-funded Minor Research Project, and she is currently working on another ongoing project. Dr. Barot is committed to advancing research in her field and fostering academic growth through her guidance of graduate and doctoral students.

Laxmi Biban is working as Assistant Professor in Botany . She has done her 10th and 12th education from Jawahar Navodaya Vidyalaya Haryana, after which she has obtained B.Sc. Medical, M.Sc. Botany, CSIR-NET-JRF Botany (Life Science) and Ph.D. in Botany degree from Kurukshetra University Kurukshetra, Dr. Biban has published 35 research papers in national and international journals, most of which have been published by foreign publishers like Springer, Elsevier, Wiley, IEEE etc. Most of the research papers are Scopus, Web of Science, SCI indexed and UGC-Care listed. Dr. Biban has authored two books in her subject. She has also registered 4 patents till now. She has also completed many research oriented training/FDP/orientation/ refresher programmes related to her subject. Dr. Biban has presented her research work in more than 35 national/ international seminars/ Conferences also.

P. P. Dodia is an Associate Professor in the Department of Zoology at Sir P. P. Institute of Science, M. K. Bhavnagar University. With over 25 years of teaching and research experience, Dr. Dodia has significantly contributed to the field of Biological science through numerous research publications and book chapters. He has guided and continues to mentor Ph.D. and postgraduate students, fostering academic excellence and research innovation. Beyond academia, Dr. Dodia is actively involved in biodiversity awareness and conservation outreach, working to bridge the gap between science and society.

Amina Farooq are currently doing Ph. D. Botany from University of Agriculture, Faisalabad.

Sadia Faryad is a PhD. Scholar at University of Agriculture, Faisalabad

Anum Javaid is doing PhD. in Botany from University of Agriculture, Faisalabad, Pakistan. 38000

Shaista Md. Mumtaz Khan is an Assistant Professor in the MBA Department at Datta Meghe Institute of Management Studies, Nagpur, Maharashtra (Postal Code: 440022). With over 15 years of teaching experience, she brings a wealth of knowledge and expertise to the field of management education. Currently pursuing her Ph.D., Ms. Khan is deeply committed to academic excellence, research, and the holistic development of her students.

M. R. Arun completed his BE, ME and PhD from Anna University, Chennai. He completed PG Diploma (PGDEEM) from Annamalai University, M.Sc. Counselling Psychology from TNOU University, M.A. Public Administration from Madras University and B.Lib.Sc. from Alagappa University. He is in the teaching profession for the past 17 years, published around 79 papers in Journals, Patents and Conferences, Active participant crossing more than 130 events and internationally networked through various professional bodies. He is skilled and technically proficient in various simulation software associated to Research in Engineering. He organized and coordinated around 15 events in working institution.

Authored 9 Books, 3 Book Chapters and he has been technically integrated in various international journals as an Editorial Board Member and Reviewer. His area of interest includes Video Processing, Remote Automation, Wireless Communication, Quantum Computing Machine Learning & Artificial Intelligence.

Vishal M. Makwana is currently working at the Department of Biosciences, Veer Narmad South Gujarat University. He has done his PhD with specializing in the urban ecology of birds. He has made significant scholarly contributions, publishing six research articles and six book chapters in renowned national and international journals and books. His research interests span diverse and critical fields, including ornithology, entomology, and environmental toxicology. He is a sought-after speaker, having delivered 7 invited talks and presented his research at 7 prominent national and international conferences and workshops. He has reviewed for many International journals of repute.

T.C. Manjunath was born on 6th February 1967. He did his BE from RV College of Engg. in 1989, ME in Robotics & Automation from LD College of Engg. in 1992 with First Rank and PhD in Systems & Control Engg. from IIT Bombay in the year 2007. He is currently working as the Dean Research R D and as Professor in the Dept. of Computer Science & Engineering with specialization in IoT, Cyber Security & Blockchain Technology in Rajarajeswari College of Engineering, Bengaluru, Karnataka. He has completed B.E. Degree from R.V. College of Engg., M.E. Govt.'s LD College of Engg., Gujarat Univ., Ahmedabad in First Rank and Ph.D. in Systems and Control Engineering from the Indian Institute of Technology Bombay IIT Bombay He has got a teaching academic, research and administrative experience of more than 35+ years 10 years as Principal, 8 years as HoD rest as faculty and has published more than 1200+ research papers in various national & international journals and conferences, with 175+ patents, 25 book-chapters, 25 awards and also published 25 text books in various field of science, engineering and technology, completed projects worth Rs. 9 Crores till date apart from guiding 10 PhD scholars. He is a member of 30 Professional Society Chapters and is a fellow of IE, IETE, AMIEE and a chartered engineer. He got the best research scholar award in 2006 in IIT Bombay amongst all the research scholars of the institute. He has got 10 awards in academics and research activities.

Garima Mishra is a dedicated MBA student at Datta Meghe Institute of Management Studies, located in Nagpur, Maharashtra (Postal Code: 440025). She is focused on building a strong foundation in business and management principles, with the aim of contributing effectively to the corporate sector. Garima is committed to academic growth and aspires to develop the skills necessary for a successful professional career.

P. Selvakumar is currently an Associate Professor in the Department of Science and Humanities at Nehru Institute of Technology, Coimbatore, Tamil Nadu, India. Since 2023, he has served as the Department IQAC Coordinator at the institute. Prior to joining Nehru Institute of Technology, he held positions as an Assistant Professor in the Department of Science and Humanities various engineering colleges in Coimbatore, Tamil Nadu, India, spanning from 2011 to 2023. Dr.P. Selvakumar's academic and research interests lie primarily in the field of phytochemistry. He has made significant contributions to his field, with an impressive record that includes 30 journal publications, 12 patents, and 20 conference papers. Additionally, he has organized 5 seminars and conferences, further demonstrating his commitment to academic and scientific exchange. Throughout his career, Dr. P.Selvakumar has been recognized for his dedication and excellence in teaching, having been honored with several awards, including the Best Teacher Awards. He holds a prominent position in the academic community, contributing actively to research.

Zahida Perveen is doing PhD. in Botany from University of Agriculture, Faisalabad, Pakistan. 38000

Anjum Parvez is an Associate Professor at Law College Dehradun, Uttaranchal University, located in Dehradun, Uttarakhand. With a strong academic background and expertise in legal education, Dr. Parvez is committed to shaping future legal professionals through quality teaching and research. He contributes actively to academic development, curriculum enhancement, and legal discourse within the institution.

Kailash P. Patel, Professor and Head, Department of Biosciences, Veer Narmad South Gujarat University

Rajeshree Patel holds a Ph.D. in Botany and has eight years of experience in teaching and research. Her main research interests include mangrove ecosystems and their restoration, ethnobotany, plant physiology, and organic farming. She has worked on studies related to mangrove biodiversity and ecological restoration, as well as traditional plant use and sustainable agricultural practices. Dr. Patel is involved in both academic teaching and field-based research, and she has presented her work at various academic forums. She continues to explore plant-based solutions to environmental challenges and aims to contribute to sustainable development through her research and teaching efforts.

Ahmed Raza has completed his M.phil (Masters) in Botany from University of Agriculture, Faisalabad, Pakistan. 38000

Swathi is an accomplished professor with over 13 years of teaching and 4 years of research experience in Commerce and Management. Specialized in Accounting, Finance, Taxation, and digital education. Passionate about academic excellence, research, and innovative teaching practices. Actively involved in curriculum development, academic leadership, and national/international conferences and publications.

Muhammad Sufyan is doing PhD. in Botany from UCollege of Biological sciences, China Agricultural University, Beijing, China

Usman Tufail currently doing M. phil in University of Agriculture, Faisalabad.

Index

A

B

C

E

G

H

I

K

M

N

P

R

S

U

W

www.ingramcontent.com/pod-product-compliance
Lightning Source LLC
LaVergne TN
LVHW080613300926
849162LV00005B/1147